The Multinuclear Approach
to NMR Spectroscopy

NATO ASI Series
Advanced Science Institutes Series

A series presenting the results of activities sponsored by the NATO Science Committee, which aims at the dissemation of advanced scientific and technological knowledge, with a view to strengthening links between scientific communities.

The series is published by an international board of publishers in conjunction with the NATO Scientific Affairs Division

A	Life Sciences	Plenum Publishing Corporation
B	Physics	London and New York
C	Mathematical and Physical Sciences	D. Reidel Publishing Company Dordrecht, Boston and Lancaster
D	Behavioural and Social Sciences	Martinus Nijhoff Publishers
E	Engineering and Materials Sciences	The Hague, Boston and Lancaster
F	Computer and Systems Sciences	Springer Verlag
G	Ecological Sciences	Heidelberg

Series C: Mathematical and Physical Sciences No. 103

[NATO Advanced Study Institute on the Multinuclear Approach to NMR Spectroscopy (1982: Stirling, Central Region, Scotland)]

The Multinuclear Approach to NMR Spectroscopy

edited by

Joseph B. Lambert

Department of Chemistry, Northwestern University,
Evanston, Illinois, U.S.A.

and

Frank G. Riddell

Department of Chemistry, University of Stirling,
Stirling, Scotland

QC762
N36
1982

D. Reidel Publishing Company

Dordrecht / Boston / Lancaster

Published in cooperation with NATO Scientific Affairs Division

Proceedings of the NATO Advanced Study Institute on
The Multinuclear Approach to NMR Spectroscopy
Stirling, Scotland
August 23-September 3, 1982

Library of Congress Cataloging in Publication Data

NATO Advanced Study Institute (1982 : Stirling, Central Region,
    Scotland)
    The multinuclear approach to NMR spectroscopy.

    (NATO ASI series. Series C, Mathematical and physical sciences ; v. 103)
    Includes index.
    1. Nuclear magnetic resonance spectroscopy—Congresses.
I. Lambert, Joseph B.   II.   Riddell, Frank G.   III.   Title.
IV. Series.
QC762.N36   1982   543′.0877   83-3224
ISBN 90-277-1582-3

Published by D. Reidel Publishing Company
P.O. Box 17, 3300 AA Dordrecht, Holland

Sold and distributed in the U.S.A. and Canada
by Kluwer Boston Inc.,
190 Old Derby Street, Hingham, MA 02043, U.S.A.

In all other countries, sold and distributed
by Kluwer Academic Publishers Group,
P.O. Box 322, 3300 AH Dordrecht, Holland

D. Reidel Publishing Company is a member of the Kluwer Academic Publishers Group

---

All Rights Reserved
Copyright © 1983 by D. Reidel Publishing Company, Dordrecht, Holland
and copyrightholders as specified on appropriate pages within
No part of the material protected by this copyright notice may be reproduced or utilized
in any form or by any means, electronic or mechanical, including photocopying, recording
or by any informational storage and retrieval system, without written permission from the
copyright owner.

Printed in The Netherlands

CONTRIBUTORS

Dr. C. Brevard, Bruker Spectrospin SA, 34, Rue de l'Industrie,
67160 Wissembourg, France

Dr. Torbjörn Drakenberg, Physical Chemistry 2, Chemical Center,
P.O.B. 740, S-220 07 Lund 7, Sweden

Professor Paul D. Ellis, Department of Chemistry,
University of South Carolina, Columbia, South Carolina 29208,
U.S.A.

Professor J. A. Elvidge, Department of Chemistry,
University of Surrey, Guildford, Surrey GU2 5XH, England

Professor S. Forsén, Physical Chemistry 2, Chemical Center,
P.O.B. 740, S-220 07 Lund 7, Sweden

Professor R. K. Harris, School of Chemical Sciences,
University of East Anglia, Norwich NR4 7TJ, England

Dr. Harold C. Jarrell, Division of Biological Sciences,
National Research Council, Ottawa K1A 0R6, Canada

Professor R. G. Kidd, Department of Chemistry,
University of Western Ontario, London, Ontario N6A 5B7, Canada

Professor Walter G. Klemperer, Department of Chemistry,
University of Illinois, Urbana, Illinois 61801, U.S.A.

Professor Joseph B. Lambert, Department of Chemistry,
Northwestern University, Evanston, Illinois 60201, U.S.A.

Professor Pierre Laszlo, Institut de Chimie,
Universite de Liege, Sart-Tilman par 4000, Liège 1, Belgium

Professor Robert L. Lichter, Department of Chemistry,
Hunter College, 695 Park Avenue, New York, New York 10021, U.S.A.

Professor Otto Lutz, Universität Tübingen,
Physikalisches Institut, D-7400 Tübingen 1, West Germany

Dr. K. J. Packer, School of Chemical Sciences,
University of East Anglia, Norwich NR4 7TJ, England

Professor J. Reisse, Université Libre de Bruxelles,
Faculté des Sciences Appliquées, Service de Chimie Organique,
Avenue F.-D. Roosevelt 50, 1050 Bruxelles, Belgium

CONTRIBUTORS

Dr. Ian C. P. Smith, Division of Biological Sciences,
National Research Council, Ottawa K1A 0R6, Canada

Dr. Graham A. Webb, Department of Chemistry,
University of Surrey, Guildford, Surrey GU2 5XH, ENGLAND

PREFACE

The field of nuclear magnetic resonance has experienced a number of spectacular developments during the last decade. Fourier transform methodology revolutionized signal acquisition capabilities. Superconducting magnets enhanced sensitivity and produced considerable improvement in spectral dispersion. In areas of new applications, the life sciences particularly benefited from these developments and probably saw the largest increase in usage. NMR imaging promises to offer a noninvasive alternative to X rays. High resolution is now achievable with solids, through magic angle spinning and cross polarization, so that the powers of NMR are applicable to previously intractable materials such as polymers, coal, and other geochemicals. The ease of obtaining relaxation times brought an important fourth variable, after the chemical shift, the coupling constant, and the rate constant, to the examination of structural and kinetic problems in all fields. Software development, particularly in the area of pulse sequences, created a host of useful techniques, including difference decoupling and difference nuclear Overhauser effect spectra, multidimensional displays, signal enhancement (INEPT), coupling constant analysis for connectivity (INADEQUATE), and observation of specific structural classes such as only quaternary carbons. Finally, hardware development gave us access to the entire Periodic Table, to the particular advantage of the inorganic and organometallic chemist.

At the NATO Advanced Study Institute at Stirling, Scotland, the participants endeavored to examine all these advances, except imaging, from a multidisciplinary point of view. Thus oxygen-17 and the Group IV elements were presented as subjects of use to, though not limited to, both the inorganic and the organic chemist; nitrogen, deuterium, and tritium to the organic and biochemist; the transition, alkali, and alkaline earth metals and the halogens to the inorganic, organic, and biochemist. Theory was not given short shrift, with broad presentations on the chemical shift, the coupling constant, and solid state methodology. Areas such as relaxation and dynamic processes also cut across all fields. These lectures served as the inspiration for the chapters presented in this volume. We hope that they show the vibrancy of the field of NMR today.

The editors are indebted to the NATO Scientific Affairs division for a grant in support of this Institute and to Janet H. Goranson for production of the entire camera-ready manuscript. Financial support also was provided by British Petroleum, IBM Instruments, Nicolet Magnetics Corp., Varian Asso., Wilmad Glass Co. Inc. (Sponsors), Bruker Instruments Inc., Gallaher Ltd.,

Glaxo PLC, ICI, JEOL, Monsanto Co., Norell Inc., Stohler Isotope Chemical Inc., and Wiley Books (Contributors).

Joseph B. Lambert
Evanston, Illinois
October, 1982

CONTENTS

Contributors v

Preface vii

Properties of Magnetically Active Nuclides xi

1. High Resolution Multinuclear Magnetic Resonance: Instrumentation Requirements and Detection Procedures (C. Brevard) 1
2. The Calculation and Some Applications of Nuclear Shielding (G. A. Webb) 29
3. Calculations of Spin-Spin Couplings (G. A. Webb) 49
4. Relaxation Processes in Nuclear Magnetic Resonance (J. Reisse) 63
5. Dynamic NMR Processes (J. B. Lambert) 91
6. Nuclear Magnetic Resonance in Solids (K. J. Packer) 111
7. Applications of High Resolution Deuterium Magnetic Resonance (H. C. Jarrell and I. C. P. Smith) 133
8. Deuterium NMR of Anisotropic Systems (H. C. Jarrell and I. C. P. Smith) 151
9. Tritium Nuclear Magnetic Resonance Spectroscopy (J. A. Elvidge) 169
10. Nitrogen Nuclear Magnetic Resonance Spectroscopy (R. L. Lichter) 207
11. Application of ^{17}O NMR Spectroscopy to Structural Problems (W. G. Klemperer) 245
12. The Alkali Metals (P. Laszlo) 261
13. Alkaline Earth Metals (O. Lutz) 297
14. The Alkaline Earth Metals--Biological Applications (T. Drakenberg and S. Forsén) 309
15. Group III Atom NMR Spectroscopy (R. G. Kidd) 329
16. Solution-State NMR Studies of Group IV Elements (Other than Carbon) (R. K. Harris) 343
17. High Resolution Solid-State NMR Studies of Group IV Elements (R. K. Harris) 361
18. Group V Atom NMR Spectroscopy Other than Nitrogen (R. G. Kidd) 379
19. Group VI Elements Other than Oxygen (O. Lutz) 389
20. The Halogens--Chlorine, Bromine, and Iodine (T. Drakenberg and S. Forsén) 405
21. Transition Metal NMR Spectroscopy (R. G. Kidd) 445
22. Cadmium-113 Nuclear Magnetic Resonance Spectroscopy in Bioinorganic Chemistry. A Representative Spin 1/2 Metal Nuclide (P. D. Ellis) 457

Participants 525

Index 531

Properties of Magnetically Active Nuclides[a,b]

Element	Atomic Weight	Spin	Natural Abundance (%)	Receptivity (vs. ^{13}C)	Quadrupole Moment (10^{-28} m^2)	Gyromagnetic Ratio (10^7 rad T^{-1}s^{-1})	Resonance Frequency (^{1}H TMS 100 MHz)
Hydrogen	1	1/2	99.985	5.68×10^3	—	26.7510	100.0000
Hydrogen	2	1	0.015	8.2×10^{-3}	2.73×10^{-3}	4.1064	15.351
Hydrogen	3	1/2	—	—	—	28.5335	106.663
Helium	3	1/2	0.00014	3.26×10^{-3}	—	−20.378	76.178
Lithium	6	1	7.42	3.58	$-8. \times 10^{-4}$	3.9366	14.716
Lithium	7	3/2	92.58	1.54×10^3	-4.5×10^{-2}	10.3964	38.864
Beryllium	9	3/2	100.	78.8	5.2×10^{-2}	−3.759	14.052
Boron	10	3	19.58	22.1	7.4×10^{-2}	2.8740	10.744
Boron	11	3/2	80.42	7.54×10^2	3.55×10^{-2}	8.5794	32.072
Carbon	13	1/2	1.108	1.00	—	6.7283	25.145
Nitrogen	14	1	99.63	5.69	1.6×10^{-2}	1.9331	7.226
Nitrogen	15	1/2	0.37	2.19×10^{-2}	—	−2.7116	10.137
Oxygen	17	5/2	0.037	6.11×10^{-2}	-2.6×10^{-2}	−3.6264	13.556
Fluorine	19	1/2	100.	4.73×10^3	—	25.181	94.094
Neon	21	3/2	0.257	3.59×10^{-2}	$9. \times 10^{-2}$	−2.1118	7.894
Sodium	23	3/2	100.	5.25×10^2	0.12	7.0761	26.452
Magnesium	25	5/2	10.13	1.54	0.22	−1.6375	6.122
Aluminum	27	5/2	100.	1.17×10^3	0.149	6.9704	26.057
Silicon	29	1/2	4.70	2.09	—	−5.3146	19.867
Phosphorus	31	1/2	100	3.77×10^2	—	10.8289	40.481
Sulfur	33	3/2	0.76	9.73×10^{-2}	-5.5×10^{-2}	2.0534	7.676
Chlorine	35	3/2	75.53	20.2	-8.0×10^{-2}	2.6210	9.798

Element	Atomic Weight	Spin	Natural Abundance (%)	Receptivity (vs. ^{13}C)	Quadrupole Moment (10^{-28} m^2)	Gyromagnetic Ratio (10^7 rad T^{-1}s^{-1})	Resonance Frequency (^{1}H TMS 100 MHz)
Chlorine	37	3/2	24.47	3.8	-6.32×10^{-2}	2.1817	8.156
Potassium	39	3/2	93.1	2.69	5.5×10^{-2}	1.2483	4.666
Potassium	40	4	0.012	3.52×10^{-3}	$(-)^c$	-1.552	5.801
Potassium	41	3/2	6.88	3.28×10^{-2}	6.7×10^{-2}	0.6851	2.561
Calcium	43	7/2	0.145	5.27×10^{-2}	-0.05	-1.8001	6.729
Scandium	45	7/2	100.	1.71×10^3	-0.22	6.4982	24.292
Titanium	47	5/2	7.28	0.864	0.29	±1.5084	5.639
Titanium	49	7/2	5.51	1.18	0.24	±1.5080	5.638
Vanadium	50	6	0.24	0.755	±0.21	2.6491	9.970
Vanadium	51	7/2	99.76	2.15×10^3	-5.2×10^{-2}	7.0362	26.303
Chromium	53	3/2	9.55	0.49	$±3 \times 10^{-2}$	-1.5120	5.652
Manganese	55	5/2	100.	9.94×10^2	0.55	6.6195	24.745
Iron	57	1/2	2.19	4.2×10^{-3}	—	0.8661	3.238
Cobalt	59	7/2	100.	1.57×10^3	0.40	6.3472	23.727
Nickel	61	3/2	1.19	0.24	0.16	-2.3904	8.936
Copper	63	3/2	69.09	3.65×10^2	-0.211	7.0965	26.528
Copper	65	3/2	30.91	2.01×10^2	-0.195	7.6018	28.417
Zinc	67	5/2	4.11	0.665	0.15	1.6737	6.257
Gallium	69	3/2	60.4	2.37×10^2	0.178	6.420	24.001
Gallium	71	3/2	39.6	3.19×10^2	0.112	8.158	30.497
Germanium	73	9/2	7.76	0.617	-0.2	-9.331	3.488
Arsenic	75	3/2	100.	1.43×10^2	0.3	4.5804	17.123
Selenium	77	1/2	7.58	2.98	—	5.1018	19.072
Bromine	79	3/2	50.54	2.26×10^2	0.33	6.7023	25.054

Element	Atomic Weight	Spin	Natural Abundance (%)	Receptivity (vs. ^{13}C)	Quadrupole Moment (10^{-28} m^2)	Gyromagnetic Ratio (10^7 rad $T^{-1}s^{-1}$)	Resonance Frequency (^1H TMS 100 MHz)
Bromine	81	3/2	49.46	2.77×10^2	0.28	7.2246	27.007
Krypton	83	9/2	11.55	1.23	0.15	-1.029	3.848
Rubidium	85	5/2	72.15	43	0.25	2.5828	9.655
Rubidium	87	3/2	27.85	2.77×10^2	0.12	8.7532	32.721
Strontium	87	9/2	7.02	1.07	0.36	-1.1593	4.334
Yttrium	89	1/2	100.	0.668	—	-1.3108	4.900
Zirconium	91	5/2	11.23	6.04	-0.21	-2.4868	9.296
Niobium	93	9/2	100.	2.740×10^3	-0.2	6.5476	24.476
Molybdenum	95	5/2	15.72	2.88	±0.12	1.7433	6.517
Molybdenum	97	5/2	9.46	1.84	±1.1	-1.7799	6.654
Technetium	99	9/2	100.	1.562×10^{3d}	-0.19^d	6.0211	22.508
Ruthenium	99	5/2	12.72	0.83	7.6×10^{-2}	-1.2343	4.614
Ruthenium	101	5/2	17.07	1.56	0.44	-1.3834	5.171
Rhodium	103	1/2	100.	0.177	—	-0.8520	3.185
Palladium	105	5/2	22.23	1.41	0.8	-0.756	4.576
Silver	107	1/2	51.82	0.195	—	-1.0828	4.048
Silver	109	1/2	48.18	0.276	—	-1.2448	4.654
Cadmium	111	1/2	12.75	6.93	—	-5.6714	21.201
Cadmium	113	1/2	12.26	7.6	—	-5.9328	22.178
Indium	113	9/2	4.28	83.8	1.14	5.8493	21.866
Indium	115	9/2	95.72	1.89×10^3	0.83	5.8618	21.913
Tin	115	1/2	0.35	0.695	—	-8.792	32.86
Tin	117	1/2	7.61	19.54	—	-9.5319	35.632

Element	Atomic Weight	Spin	Natural Abundance (%)	Receptivity (vs. ^{13}C)	Quadrupole Moment (10^{-28} m^2)	Gyromagnetic Ratio (10^7 rad T^{-1}s^{-1})	Resonance Frequency (^{1}H TMS 100 MHz)
Tin	119	1/2	8.58	25.2	—	-9.9756	37.291
Antimony	121	5/2	57.25	5.20 × 10^2	-0.53	6.4016	23.931
Antimony	123	7/2	42.75	1.11 × 10^2	-0.68	3.4668	12.959
Tellurium	123	1/2	0.89	0.89	—	-7.0006	26.170
Tellurium	125	1/2	7.0	12.5	—	-8.4398	31.550
Iodine	127	5/2	100.	5.3 × 10^2	-0.79	5.3525	20.009
Xenon	129	1/2	26.44	31.8	—	-7.4003	27.658
Xenon	131	3/2	21.18	3.31	-0.12	2.1939	8.200
Cesium	133	7/2	100.	2.69 × 10^2	-3. × 10^{-3}	3.5087	13.116
Barium	135	3/2	6.59	1.83	0.18	2.6575	9.934
Barium	137	3/2	11.32	4.41	0.28	2.9728	11.113
Lanthanum	138	5	0.09	0.43	-0.47	3.5295	13.194
Lanthanum	139	7/2	99.91	3.36 × 10^2	0.21	3.7787	14.126
Praseody-mium	141	5/2	100.	1.66 × 10^3	-5.9 × 10^{-2}	7.836	29.291
Neodymium	143	7/2	12.17	2.31	-0.48	1.455	5.438
Neodymium	145	7/2	8.3	0.37	-0.25	0.895	3.346
Samarium	147	7/2	14.97	1.26	-0.21	1.104	4.128
Samarium	149	7/2	13.83	0.59	6. × 10^{-2}	0.880	3.289
Europium	151	5/2	47.82	4.83 × 10^2	1.16	6.634	24.801
Europium	153	5/2	52.18	45.3	2.9	2.930	10.952
Gadolinium	155	3/2	14.73	0.23	1.6	1.022	3.820
Gadolinium	157	3/2	15.68	0.48	2.	1.277	4.775

Element	Atomic Weight	Spin	Natural Abundance (%)	Receptivity (vs. ^{13}C)	Quadrupole Moment (10^{-28} m^2)	Gyromagnetic Ratio (10^7 rad T^{-1} s^{-1})	Resonance Frequency (^{1}H TMS 100 MHz)
Terbium	159	3/2	100.	3.31 × 10^2	1.3	6.067	22.679
Dysprosium	161	5/2	18.88	0.45	1.4	0.881	3.295
Dysprosium	163	5/2	24.97	1.59	1.6	1.226	4.584
Holmium	165	7/2	100.	1.03 × 10^3	2.82	5.487	20.513
Erbium	167	7/2	22.94	0.66	2.83	0.773	2.890
Thullium	169	1/2	100.	3.21	–	-2.21	8.272
Ytterbium	171	1/2	14.27	4.05	–	4.72	17.612
Ytterbium	173	5/2	16.08	1.14	(-)^b	1.31	4.852
Lutetium	175	7/2	97.41	1.56 × 10^2	5.68	3.05	11.407
Lutetium	176	7	2.59	5.14	8.1	2.10	7.872
Hafnium	177	7/2	18.50	0.88	4.5	0.95	4.008
Hafnium	179	9/2	13.75	0.27	5.1	-0.609	2.518
Tantalum	181	7/2	99.988	2.04 × 10^2	3.	3.2073	11.990
Tungsten	183	1/2	14.28	5.89 × 10^{-2}	–	1.1145	4.166
Rhenium	185	5/2	37.07	2.8 × 10^2	2.8	6.0255	22.525
Rhenium	187	5/2	62.93	4.90 × 10^2	2.6	6.0862	22.752
Osmium	187	1/2	1.64	1.14 × 10^{-3}	–	0.6105	2.282
Osmium	189	3/2	16.1	2.13	0.8	2.0773	7.765
Iridium	191	3/2	37.3	2.3 × 10^{-2}	1.5	0.539	1.718
Iridium	193	3/2	62.7	5.0 × 10^{-2}	1.4	0.391	1.871
Platinum	195	1/2	33.8	19.1	–	5.7412	21.462
Gold	197	3/2	100.	6.0 × 10^{-2}	0.58	0.357	1.729
Mercury	199	1/2	16.84	5.42	–	4.7912	17.911
Mercury	201	3/2	13.22	1.08	0.5	-1.7686	6.612

Element	Atomic Weight	Spin	Natural Abundance (%)	Receptivity (vs. ^{13}C)	Quadrupole Moment (10^{-28} m^2)	Gyromagnetic Ratio (10^7 rad T^{-1}s^{-1})	Resonance Frequency (^{1}H TMS 100 MHz)
Thallium	203	1/2	29.50	2.89×10^2	–	15.3078	57.224
Thallium	205	1/2	70.50	7.69×10^2	–	15.4584	57.787
Lead	207	1/2	22.6	11.8	–	5.5797	20.858
Bismuth	209	9/2	100.	7.77×10^2	-0.4	4.2986	16.069
Polonium	209	1/2	–	–	–	–	–
Uranium	235	7/2	0.72	4.9×10^{-3}	4.1	0.479	1.791

[a]Most values taken from C. Brevard and P. Granger, "Handbook of High Resolution Multinuclear NMR," Wiley-Interscience: New York, 1981, pp. 80-211.
[b]Some values taken from the Bruker NMR-NQR Periodic Table; R. K. Harris and B. E. Mann, "NMR and the Periodic Table," Academic Press: London, 1978, pp. 5-7; J. A. Pople, W. G. Schneider, and H. J. Bernstein, "High-resolution Nuclear Magnetic Resonance," McGraw-Hill: New York, 1959, pp. 480-485; R. K. Harris, private communication.
[c]Poorly known or unknown.
[d]K. J. Franklin, C. J. L. Lock, B. G. Sayer, and G. J. Schrobilgen, J. Am. Chem. Soc., 104, pp. 5303-5306 (1982).

CHAPTER 1

HIGH RESOLUTION MULTINUCLEAR MAGNETIC RESONANCE: INSTRUMENTATION REQUIREMENTS AND DETECTION PROCEDURES

C. Brevard

Bruker Spectrospin, 34, rue de l'Industrie
67160 Wissembourg, France

ABSTRACT

The entire Periodic Table is now available to NMR spectroscopists and chemists. This availability, however, hides some traps that need to be appreciated. A clear knowledge of both spectrometer capabilities and frontiers of high resolution detection will lessen the experimentalist's burden. An examination of multinuclear instrumentation will include discussion of computer and spectrometer hardware, probeheads, traps, and hints (frequency range, receptivity, detectability, and magic pulse sequences). By way of example, applications will be discussed of exotic INEPT experiments and ruthenium NMR spectroscopy.

NUCLEAR PARAMETERS

With each magnetically active isotope X, one can associate a set of constants which are of prime importance when starting multinuclear NMR.

(i) A nuclear spin quantum number I_X, which determines the maximum observable component of the nuclear angular momentum P_X^{max} (eq. 1, in which μ_{Imax} is the maximum observable magnetic

$$P_X^{max} = \frac{\mu_{Imax}}{\gamma_X} = \hbar I_X \tag{1}$$

moment of the isotope and γ_X is the magnetogyric ratio of the isotope). The ratio γ_X may be either positive (^{13}C, ^{1}H, ...) or negative (^{15}N, ^{29}Si, ^{109}Ag, ...), and the I_X values range from 1/2 to 9/2.

Practically, the NMR active isotopes can be divided into two classes:

$I_X = 1/2$: dipolar isotope

$I_X > 1/2$: quadrupolar isotope with an associated electric quadrupole moment Q_X, which reflects the nonspherical symmetry of the nuclear charge distribution. This differentiation ($I \geqslant 1/2$) will be essential in multinuclear observation, as 87 of the 116 magnetically active nuclei possess a spin value $I_X > 1/2$ and hence a quadrupole moment Q_X, which will mainly determine, a priori, the detectability of the resonance line through its linewidth.

A complete compilation of γ_X, I_X, and Q_X values will be found in references 1, 2, 3, and 4 and in the table that precedes this chapter.

(ii) A spin-lattice relaxation T_{1X} and a spin-spin relaxation time T_{2X} with, generally, $T_{1X} = T_{2X}$. The different mechanisms contributing to the spin-lattice relaxation process will be detailed in Chapter 4. Nevertheless, the magnetic properties of the isotope under study can be roughly sketched as:

<u>dipolar</u>: "inter"nuclei relaxation only, via surrounding species, long T_{1X} (Ag$^+$: 900 sec!) and sharp lines, or

<u>quadrupolar</u>: "intra"nucleus relaxation via the interaction of Q_X with fluctuating local electric field gradients, short T_{1X} (covalent chlorine: 10^{-6} sec!) and broad lines.

Of course, reverse situations (chemical shielding anisotropy for spin 1/2 or low Q_X values for spin $> 1/2$) can be encountered and a critical evaluation of the expected linewidth is a must before starting any experiment.

(iii) A natural abundance $a_X\%$ expressed as:

$$a_X\% = \frac{\text{number of nuclei (isotope)}}{\text{total number of nuclei (element)}}$$

(iv) A resonance frequency ν_X expressed for a given static magnetic field B_0 (eq. 2, in which σ_X is the shielding constant

$$\nu_X = \frac{\gamma_X}{2\pi} B_0 (1-\sigma_X) \qquad (2)$$

of isotope X in a given compound, which depends upon the electronic environment of X in this compound). The σ_X values can be quite large (^{103}Rh: 10,000 ppm; ^{59}Co: 18,000 ppm) and care must be taken to avoid folded lines when searching unknown resonances for such isotopes (vide infra). On the other hand, in multinuclear applications, the resonance frequency for a given reference compound is often expressed as Ξ(MHz) (3,4), which represents the reference frequency value in a magnetic field for which the protons of TMS resonate at exactly 100 MHz. This quantity allows for a fast calculation of the resonance frequency value corresponding to the available spectrometer.

(v) Homonuclear (J_{X-X}) or heteronuclear (J_{X-Y}) coupling constants can be present, provided that the resonance lines are sharp enough to allow the detection of such a coupling mechanism. Care must also be exercised in multinuclear applications if such a coupling exists; as an example $^2J_{Sn-Sn}$ can be as large as 30,000 Hz (5)! A medium-sized natural abundance value $a_X\%$ can give rise to a treacherous spectrum appearance like the ^{195}Pt spectrum of $[Pt_9(CO)_{18}]^{2-}$ (6).

(vi) Finally, the spectrometer has to be taken as a real parameter in itself with its own hardware and software capabilities and its available magnetic field strength which will govern such observation parameters as sensitivity enhancement or acoustic ringing phenomena.

SENSITIVITY, RECEPTIVITY, DETECTABILITY

Sensitivity: S_X

At constant field B_0 the resonance strength of a given isotope X is proportional to the quantities in eq. 3, in which S_X

$$S_X = \gamma_X^3 \, I_X \, (I_X + 1) \qquad (3)$$

represents the intrinsic sensitivity of the isotope. Of course, a relative sensitivity S_X^R can be defined, depending upon the strength of the static field B_0. It is generally admitted that S_X^R is given by eq. 4 (7).

$$S_X^R \sim (B_0)^{3/2} \qquad (4)$$

Receptivity: R_X

This is the parameter one has to consider in multinuclear observation, eq. 5. If enriched samples are used, a_X is then

$$R_X = S_X \cdot a_X \qquad (5)$$

replaced by the enrichment value.

A relative receptivity $R_X^Y = a_X S_X / a_Y S_Y$ can be defined, the Y species being the isotope of reference. ($R_X^{13C} = 3.9\ \gamma_X^3 a_X I_X (I_X + 1)\ 10^{-28}\ \text{rad}^{-3}\text{T}^{-3}\text{s}^{-3}\%$).

Reference 4 gives the I_X, Q_X, R_X, and Ξ_X values for all magnetically active isotopes, and most values also are found in the table preceding this chapter.

Detectability: D_X

In fact, most multinuclear studies are directed towards a better understanding of the structural or dynamical parameters of the compounds under study. Then, the signal over noise of the experiment can be redefined in terms of spectrometer time compared to the information obtained during the allotted period of time. This concept introduces a new parameter, namely the detectability D_X of a given isotope for a given compound (eq. 6,

$$D_X = f[N,\ \Delta\nu^{-1},\ T_{1X}^{-1}] R_X \qquad (6)$$

in which N is the number of resonating spins in the detection coil and $\Delta\nu$ is the width of the resonance line.

A later section will envisage how to increase this D_X factor, but anyone can foresee the prime importance of the $\Delta\nu$ factor: as the resonance line becomes sharper, it will be more easily detected, whatever the detection method. As an example, if a signal with a 1 Hz linewidth (a) gives a signal to noise ratio of 10 after 10 scans (n) in Fourier mode, the same signal with a 10 Hz halfwidth (b) will give the same signal to noise ratio after 1,000 scans (n'), eq. 7 for Lorentzian line shapes.

$$n' = \left(\frac{b}{a}\right)^2 \cdot n \qquad (7)$$

This is very important when observing quadrupolar isotopes for which $\Delta\nu$ can vary drastically from compound to compound upon the electronic symmetry around the isotope under study. Figure 1 exemplifies this situation.

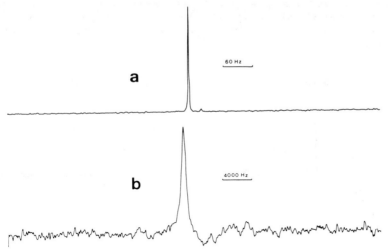

Figure 1. The ^{99}Ru spectra of (a) RuO_4--1 M in CCl_4, 40 s accumulation, (b) $Ru(Cp)_2$--saturated solution in C_6H_6, 13 h accumulation.

THE MULTINUCLEAR OBSERVATION GAME

When starting an NMR experiment on an isotope different from proton, fluorine, phosphorus, or carbon, one will be faced with a lancinating question: how to enhance the quantitative (D_X) response. The obvious answer will be to choose the isotope and the recording scheme which will give the stronger and sharper line.

In most cases, the NMR spectrum will be acquired via the Fourier transform mode, and the following paragraph summarizes the basic concepts which will be used throughout.

The Fourier Transform Method

In the rotating frame (8) a strong radiofrequency field B_1 of duration τ (sec) will force the magnetic moments to precess around it. As a result, the macroscopic magnetization is tilted from its Boltzmann position along B_0 by a so-called pulse angle α (Figure 2, eq. 8), and a NMR signal is picked up by the

$$\alpha = \frac{\gamma_X B_1 \tau}{2\pi} \qquad (8)$$

detection coil during a time t_A (Acquisition time). When $\alpha = 90°$ (maximum signal amplitude), the corresponding pulse width is labeled as τ_{90}. Correspondingly, τ_{180} will invert the macroscopic equilibrium magnetization, namely a population inversion (Figure 3).

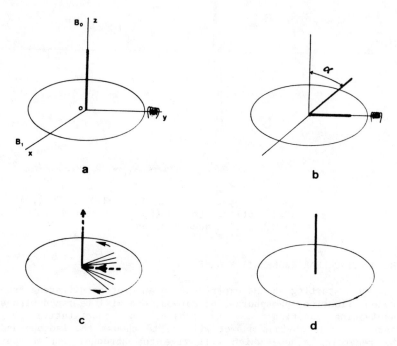

Figure 2. Pulse angle in the rotating frame. (a) equilibrium, (b) pulse angle α, (c) return to initial state via T_1 (z axis) and T_2 (xy plane), (d) after $5T_1$, the spin system is again at equilibrium.

After the pulse, the tilted magnetization will (i) recover exponentially back to its equilibrium value along the z axis via a spin-lattice relaxation mechanism (time constant T_1) and (ii) disappear in the xy plane due to a spin-spin relaxation mechanism (time constant T_2) with $T_2 \leqslant T_1$.

If a second pulse is applied before the spin system has reached its initial equilibrium value along B_0, the coadded signal will be weaker compared to the first one. A multiscan acquired signal will then depend upon the rate of recovery of the magnetization along the z (B_0) axis and hence upon the T_{1X} value. A practical guideline is given by the Ernst equation (9), eq. 9,

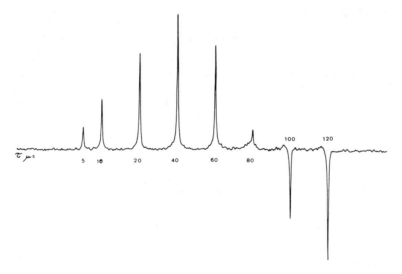

Figure 3. ^{95}Mo NMR. Pulse angle variation and signal intensity. Sample MoO$_4$, 1 M in D$_2$O. Maximum signal amplitude gives τ_{90} (40 µs); τ_{180} = 80 µs and τ_{270} = 120 µs.

$$\cos \alpha_{opt} = e^{-t_w/T_{1x}} \qquad (9)$$

in which t_w stands for the overall time between two successive pulses (acquisition time + waiting time).

Once the proper pulse angle has been determined, the accumulation can be started allowing the signal to noise to increase as the square root of the number of accumulations, or, in a practical way, allowing the spectrometer time to increase by a factor of four each time one wants to double the signal over noise of a given experiment.

On the other hand multinuclear observation can lead to extensive chemical shift scales to deal with (for example ^{59}Co: 18,000 ppm; ^{195}Pt: 13,000 ppm; ^{99}Ru: 9,000 ppm) and then the folding phenomenon (Ref. 4, p. 26) has to be carefully discriminated, especially for high sensitivity isotopes such as cobalt, platinum, or tin. A useful procedure, once a peak has been detected, is to reduce the spectral width to a minimum value around the peak position and to check that it does not change.

The Spectrometer

The Computer. Modern computer-controlled NMR spectrometers allow the implementation of practically any sophisticated pulse sequences such as INEPT (10) or INADEQUATE (11). Very precise clock timing together with the ability to pulse simultaneously the emitter and the decoupler with their own relative rf phases are then a must for the computer. This task is appertaining to the computer pulse generator which is now a standard item on commercial systems.

On the other hand, the computer should allow a large spectral region to be observed when searching for unknown resonances or broad lines; as an example, a spectral width of 100 kHz is not uncommon in multinuclear NMR.

The Probeheads. This accessory is in fact one of the most important parts of the spectrometer for multinuclear observation. The entire frequency range for magnetically active isotopes is covered by, generally, one or two "multinuclear probes," and a parameter of prime importance is the nominal 90° pulse the probe is giving throughout its entire frequency range. In fact, it has been shown (Ref. 8, p. 67) that the usable frequency range Δ(Hz) over which the applied pulse has the expected 90° values (uniform power distribution) is defined by eq. 10. The situation can

$$\Delta \ll \frac{1}{4 \cdot \tau_{90}} \tag{10}$$

become very critical when the spectral widths are increased up to 100 kHz and when the 90° pulse with values τ are above 60 µs.

These long 90° pulsewidth values can lead to a big drop in sensitivity when the carrier frequency is far from the resonance line (Figure 4). Moreover, long 90° and 180° pulses can introduce important rf phase problems when starting multi-pulse, multi-phase experiments. A careful probe design giving the shortest nominal τ_{90} can avoid this problem, as exemplified in Figure 5.

The probeheads and accompanying spectrometer hardware should also be as flexible as possible to allow $\{^1H,X\}$ decoupling, Y observe, or $\{X\},X$ homonuclear decoupling experiments.

A nice example of such a flexibility is given in Figure 6.

The Magnet. A major breakthrough in multinuclear NMR has been the introduction of superconducting magnets. In fact, they not only increase the intrinsic sensitivity but also allow the

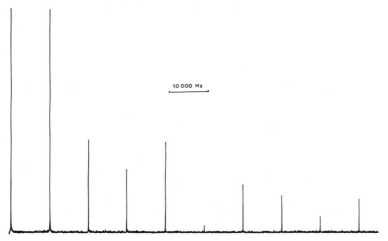

Figure 4. Influence of the pulse angle value and carrier frequency position on the signal to noise. The figure represents the ^{95}Mo power spectrum of a 2 M solution of MoO_4. The carrier frequency has been sequentially shifted by a 10 kHz step over the entire spectral width (100 kHz) every 100 scans. The transformed FID then provides the power distribution of the emitter across the 100 kHz spectral width with τ_{90} = 40 μs.

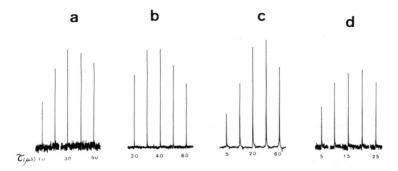

Figure 5. The τ_{90} values for ^{183}W (16.6 MHz), ^{109}Ag (18.6 MHz), ^{15}N (25.3 MHz), and ^{31}P (161.9 MHz). Nominal field value: B_0 = 9.4 T. All the τ_{90} values are in the range 20-40 μs from ^{183}W to ^{31}P (10 mm multinuclear probe).

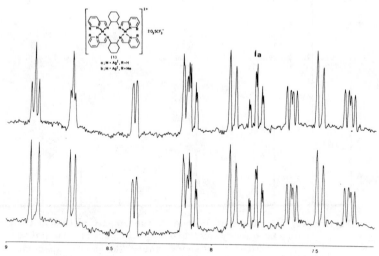

Figure 6. The 250 MHz proton spectrum (aromatic part) of the silver dication compound shown in the insert. (Bottom) normal acquisition, (top) with ^{109}Ag decoupling. Due to the equal (50%) natural abundance of ^{107}Ag, ^{109}Ag isotopes, the imine protons (8.4 and 8.6 ppm) show a characteristic triplet-like structure from the decoupling of half of these protons coupled with ^{109}Ag. The spectrum has been recorded via the decoupling coil of a 15 mm multinuclear probe and decoupled through the observing coil tuned to the ^{109}Ag frequency.

experimentalist to run samples in the unlocked mode, due to the inherent stability of a superconducting magnet. This is very important, for example, for organometallic chemists who are often obliged to prepare their samples under dry-box conditions and with uncommon solvents. Figure 7 shows what one can expect for such unlocked recording conditions.

Choosing the Best Suitable Isotope

We shall often be faced with the choice of the isotope to select (^{95}Mo or ^{97}Mo, ^{107}Ag or ^{109}Ag, ^{99}Ru or ^{101}Ru, etc.). A straightforward indication is furnished by the receptivity factor R_X, but in many cases, and especially when the receptivities of both isotopes do not differ by more than a factor 2 to 4, some considerations have to be paid to the nuclear parameters of the

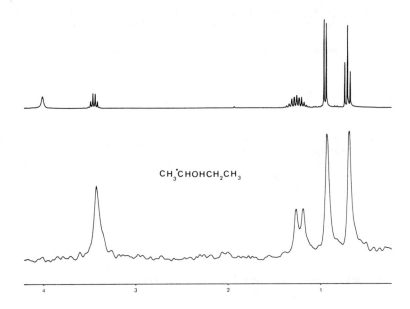

Figure 7. The 250 MHz ^{1}H (top) and 38.39 MHz (bottom) ^{2}H spectra of 2-butanol. The deuterium spectrum (^{1}H broadband decoupled) has been accumulated via the lock coil of the proton probe during 40 h in the unlocked mode. The magnet stability allows an easy separation of the two nonequivalent deuterons α to the asymmetric carbon (0.08 ppm). Note the coalescence of the hydroxyl resonance in the deuterium spectrum due to the exchange.

potential candidates in order to fulfill the optimum detectability conditions.

Dipolar isotopes. Owing to their sharp lines, spin 1/2 isotopes have to be considered superior to quadrupolar isotopes (e.g., ^{14}N and ^{15}N). Moreover, they can furnish very useful structural information via homo or heteronuclear couplings (12,13). However, their long T_1 values can be a serious drawback and the main target, when observing dipolar isotopes, will be to find a way to free the accumulation process from the long T_{1X} value.

Quadrupolar isotopes. A frequent situation in multinuclear observation compels the NMR spectroscopist to decide between two quadrupolar isotopes X and X'. Apart from the intrinsic receptivity (R_X, $R_{X'}$), the detectability (D_X, $D_{X'}$) will mainly depend on the relative resonance linewidths of isotopes X and X'.

For such quadrupolar isotopes, the linewidth $\Delta\nu_{1/2}$ is given by eq. 11, in which χ represents the nuclear quadrupole coupling

$$\Delta\nu_{1/2} = \frac{3\pi}{10} \cdot \frac{(2I+3)}{I^2(2I-1)} \chi^2 (1 + \frac{\eta^2}{3})\tau_c \qquad (11)$$

constant defined by $\chi = e^2 q_{zz} Q/h$ (q_{zz}, q_{xx}, q_{yy} are the principal components of the field gradient tensor), η is the asymmetry parameter for q_{ii} ($0 < \eta < 1$) and is generally close to 0: $\eta = (q_{xx} - q_{yy})/q_{zz}$, and τ_c is the isotropic tumbling correlation time.

In a given molecule, for two different quadrupolar isotopes, the linewidths will be then proportional to $Q^2(2I+3)/I^2(2I-1)$.

Hence, the following cases may be considered. (i) Same I values, different Q values (e.g., ^{95}Mo, ^{97}Mo, ^{99}Ru, ^{101}Ru...). The isotope with the smaller Q value will give the sharper line, the $\Delta\nu_{1/2X}$, $\Delta\nu_{1/2X'}$ linewidths are in the ratio $(Q_X/Q_{X'})^2$ (Figure 8a).

(ii) different I values, different Q values. This is the general case and the $Q^2(2I+3)/I^2(2I-1)$ factor indicates which of the two isotopes will have the sharper linewidth. A good example is furnished by the two isotopes of antimony, ^{121}Sb, ^{123}Sb (Figure 8b). If the spin value of ^{123}Sb were identical to ^{121}Sb (5/2), one should expect a linewidth value for the ^{123}Sb resonance of $220 \times (0.68/0.53)^2 = 361$ Hz. In fact, the spin value of ^{123}Sb (7/2) reduces this figure to an experimental value of 160 Hz, in close agreement to the expected theoretical value of:

$$220 \cdot (\frac{Q_{123}}{Q_{121}})^2 \cdot \frac{2I_{123} + 3}{I_{123}^2(2I_{123}-1)} \cdot \frac{I_{121}^2(2I_{121}-1)}{2I_{121} + 3} = 153 \text{ Hz}.$$

Detectability Enhancement

If we recall the detectability formula (eq. 6), the experimentalist has on hand a palette of parameters to play with.

(i) The number of spins N which resonate in the detection coil can be increased by increasing the diameter of the sample tube. The D_X factor will then be increased as the square of the ratio of the tube's diameter. Thus, going from a 10 to a 15 mm

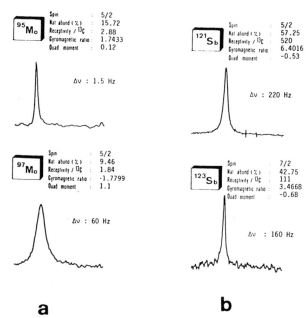

Figure 8. Influence of spin value I_X and quadrupole moment Q_X on linewidth. (a) ^{95}Mo, ^{97}Mo (sample MoO_4^-). (b) ^{121}Sb, ^{123}Sb (sample $SbCl_6^-$). Reprinted from Ref. 4 with the permission of Wiley-Interscience.

tube increases D_X by a factor 2.25. This factor may be slightly less as 15 mm probeheads are more prone to pulse imperfections or longer nominal 90° pulses due to a less favourable probe Q factor (Figure 9.)

(ii) For quadrupolar isotopes, a reduction of $\Delta\nu$ will bring a corresponding increase in D_X. As $\Delta\nu = f(Q^2, I) \cdot \tau_c$, the only way to decrease $\Delta\nu$ for a given compound, once the best detectable isotope has been chosen, is to decrease τ_c by either reducing the viscosity of the solution or by heating the sample; this last procedure is the one of choice when dealing with stable compounds dissolved in a high-boiling point solvent (Figure 10).

(iii) For dipolar isotopes (long T_{1X}), one can quench the dipole-dipole relaxation pathway by adding to the solution a small amount of a relaxation reagent which brings an extra and very effective relaxation scheme via an electron-nucleus contact interaction. However, care is to be taken in choosing the relaxation reagent, especially for solutions of organometallic

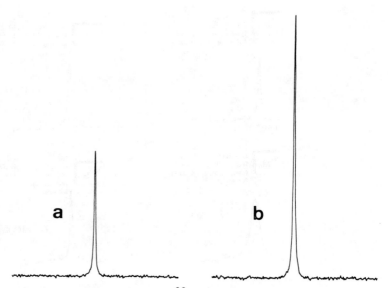

Figure 9. The 11.67 MHz ^{39}K spectrum of a 2 M solution of KBr. (a) 10 mm probe; (b) 15 mm probe. Same acquisition parameters for both spectra (90° pulse, 20 scans).

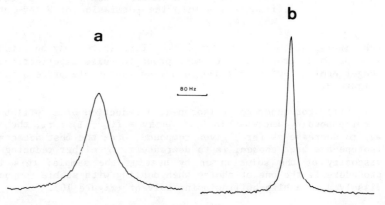

Figure 10. The 54.24 MHz ^{17}O spectrum of H_2O. (a) 297 K, (b) 358 K. (500 scans, 90° pulse).

complexes, to avoid any exchange phenomenon between the transition metal (Cr, Ni, or Fe) of the relaxation reagent and the metal site of the solute.

Another way to shorten the long T_{1X} of dipolar isotopes is to help develop a potential <u>intermolecular</u> dipole-dipole interaction in the solution. A handy way to establish this interaction is to dissolve the compound in a <u>protio</u> solvent instead of the deutero analogue.

(iv) All the above described procedures are based on physical manipulations of the sample; a more subtle way to increase the detectability factor D_X is to increase the receptivity factor R_X itself, by using intra- or intermolecular interactions involving the X isotope and another abundant nucleus (generally ^1H).

A very well known procedure is to broadband-decouple the protons to generate a potential nuclear Overhauser effect on isotope X. This NOE effect can reach a maximum theoretical value equal to $\eta_{NOE} = \gamma_H/2\gamma_X$. Such a NOE effect can result from intra- or intermolecular dipole-dipole interactions. For low γ, spin 1/2 nuclei dissolved in protio solvents (see above), irradiation of the <u>proton</u> solvent resonance sometimes develops an intermolecular NOE effect with a corresponding sizable increase in signal intensity (Figure 11).

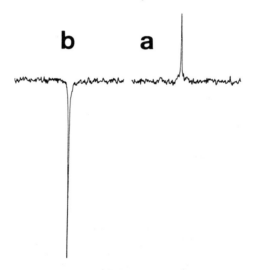

Figure 11. The 12.5 MHz yttrium spectrum of $Y(NO_3)_3$, 1 M in H_2O. (a) normal acquisition; (b) with H_2O protons irradiated. Intermolecular $\{^1H\}$-Y NOE is equal to -3.6 (theoretical maximum value: -10.2).

Another technique which has already found numerous applications was proposed by Freeman (10). It uses a scalar $^nJ_{X-H}$ coupling to force the X isotope energy levels to follow the inversion of the corresponding coupled proton energy levels. This so-called INEPT pulse sequence increases the receptivity of the isotope X by a factor equal to $\eta_{IN} = \gamma_H/\gamma_X$ for an AX system. As an example, a rhodium resonance will be enhanced by a factor equal to 31.4, which means a decrease in spectrometer time of nearly 1000!

In fact, many of the newly proposed pulse sequences are based on the spin echo phenomenon (14). Figure 12a-e visualizes the echo formation in the rotating frame which follows the classical $(90_x° - \tau - 180_x° - \tau - AQT)$ pulse sequence where the subscripts x and y represent the rotating frame axis (x is the B_1 axis, y the detection axis). At time 2τ following the first $90_x°$ pulse, we shall expect again an in-phase spin system. Thus we shall detect a coherent, negative NMR signal on the -y axis. This refocusing process is independent of the τ value used in the sequence, and the echo will be detected as long as the spin-

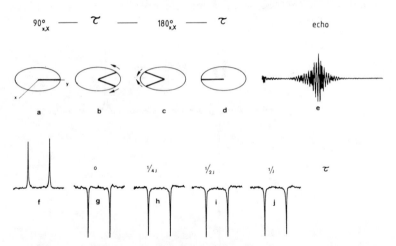

Figure 12. Echo sequence and echo formation (sample $CHCl_3$, ^{13}C observation). The echo (e) has been taken immediately following the 180° pulse to show the magnetization build-up with its maximum at τ second. Spectrum f represents a 90° pulse. Spectra g to j show the lack of influence of τ value on the relative phase of the echo spectrum (compare with Figure 14) and the 180° phase difference between the first echo ("negative echo") and the normally acquired spectrum (f).

spin relaxation process will leave some magnetization in the xy plane (Figure 12g-j).

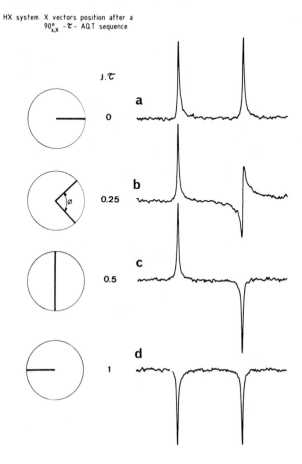

Figure 13. Influence of the delay time τ in a ($90_X^\circ - \tau -$ AQT) sequence for a coupled AX system (CHCl$_3$). Observed isotope: ^{13}C. For spectra b and c, one transition has been deliberately phased in absorption-mode to visualize the ϕ value clearly ($\phi = 2\pi \cdot J \cdot \tau$ radians).

If we choose now an AX spin system presenting a scalar $^n J_{X-H}$ coupling, apart from the defocusing process arising from the spin-spin interaction and magnet inhomogeneities, the J_{X-H} coupling will also defocus the two X_A, X_B transitions with an angular velocity $\omega = \pm 2\pi \cdot J/2$ rad/sec. Then, after a time B following an initial $90_X°$ pulse, the pair of $X_A X_B$ vectors will be at $\pm \pi \cdot J \cdot \tau$ radians from its initial position along the +y axis, and thus the two X_A, X_B lines will dephase one from the other by $\phi = 2\pi \cdot J \cdot \tau$ radians. This situation is exemplified in Figure 13. Of course, a normal spin echo sequence will bring the X_A, X_B lines in phase again, irrespective of the τ value. On the other hand, a 180° proton pulse applied in synchronism with the $180°_{x,X}$ pulse will invert the H_A, H_B states and consequently exchange the X_A, X_B states. There will be no echo formation at time 2τ after the initial 90° pulse but the relative phase of the two vectors will then depend on the chosen τ value; thus after a $90°_{x,X} - \tau - 180°_{x,X} 180°_{x,H} - \tau -$ AQT sequence the relative phase of the X_A, X_B transitions will depend on $\tau(J)$ as $\phi = 4 \cdot \pi \cdot J \cdot \tau$ radians (Figure 14). The INEPT sequence is written as $90°_{x,H} - \tau - (180°_{x,H} 180°_{x,X}) - \tau - (90°_{y,H} - 90°_{x,X}) -$ AQT and will depend on the key value $\tau = 1/4 J_{X-A}$ (AX system, $I_A = 1/2$, $I_X = 1/2$), which will allow the desired proton population inversion by applying the $90°_{y,H}$ pulse.

The corresponding pulse train together with the motion of the H_A, H_B vectors in the rotating frame and correlated INEPT spectrum for X isotope is shown in Figure 15.

If the X isotope has a spin value $I > 1/2$ or if it has a spin value $I = 1/2$ but has been enriched to bring its a% value close to 100% (e.g., ^{15}N) and gives an AX_2-like spectrum, the τ value is no longer equal to $1/4J$. As an example, the driving τ value in a AX_2 system, where $A = {}^1H$ or ^{31}P and $X = {}^{15}N$, 100% enrichment, will be found by setting $\omega = 2\pi \cdot J \cdot \tau = \pm \pi/4$. Under these conditions, $\tau = 1/8J$. In that case, $\tau = 1/4J$ will give no ^{15}N signal enhancement at all in an INEPT experiment. Table 1 gives the τ values when $I_X \geq 1/2$.

Table I

I_X	1/2	1	3/2	2	5/2
$J \cdot \tau$	0.25	0.125	0.088	0.069	0.056

The population inversion of the X isotope energy levels following an INEPT sequence implies the "up-down" appearance of the coupled INEPT spectrum. A decoupled spectrum can be obtained after a suitable delay T which will bring again the X_A, X_B components in phase. Table II gives the T values when X is coupled to $n = 1, 2, 3 \cdots 12$ equivalent spin 1/2 isotopes. The refocused

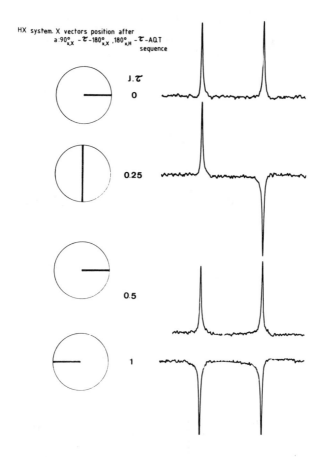

Figure 14. Influence of the delay time τ in a $90_X°-\tau-180_X°$, $180_H°-\tau$-AQT. The reversal of the two vector motions due to the hard 180° proton pulse is transposed into a phase difference in the transformed spectra with $\phi = 4\pi \cdot J \cdot \tau$ radians (compare with Figure 13). Sample $CHCl_3$, ^{13}C observed. $\tau_{90,X} = 20$ μs, $\tau_{90,H} = 40$ μs.

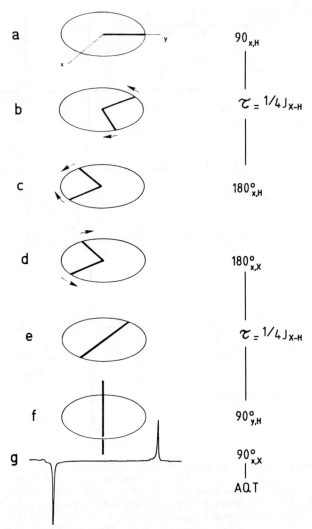

Figure 15. INEPT sequence. (a to f) motion of the proton vectors (AX system). (g) polarized X spectrum (sample $CHCl_3$, $\tau_{90,X}$ = 20 μs, $\tau_{90,H}$ = 40 μs, J_{13C-H} = 210.4 Hz).

decoupled INEPT sequence is then written as $(90°_{x,H}$ - τ - $(180°_{x,H} 180°_{x,X})$ - τ - $90°_{y,H} 90°_{x,X}$ - T - (BB) - AQT).

Table II

n	1	2	3	4	6	9	12
J.T.	0.5	0.25	0.196	0.166	0.133	0.125	0.096

Another useful technique to increase X receptivity is to use a Selective Population Inversion (SPI) pulse sequence (15) on one proton transition. A long (150 to 250 ms), soft pulse is applied via the decoupler to excite and invert one transition of the coupled proton spectrum selectively. The pulse sequence is visualized in Figure 16 together with a $\{^1H\}-^{14}N$ SPI experiment run on a $NH_4NO_3/H_2O/H^+$ solution. The gain in sensitivity is quite appreciable and equals $(\gamma_X \pm \gamma_H)/\gamma_X$ for the $X_{A(B)}$ transitions of an H-X type spectrum.

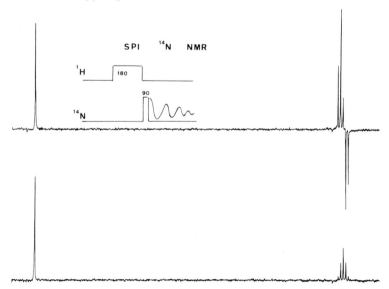

Figure 16. The 18.07 MHz ^{14}N spectrum of a $NH_4NO_3/H_2O/H^+$ solution. (Bottom) normal acquisition, (top) with selective population inversion of the outer transition of the proton NH_4^+ triplet. The NO_3^- line quantitates the sensitivity enhancement (180° proton pulse: 60 ms; 90° ^{14}N pulse: 50 μs).

If one compares the two methods, the SPI technique can be set up on any spectrometer but is rather demanding as far as experimental conditions are concerned (precise calibration of the decoupler, exact knowledge of the proton transition frequency, solvent effects). On the other hand, the INEPT technique is a straightforward, nonselective method but it requires a computer able to phase shift both the decoupler and the emitter frequency and to pulse simultaneously the two channels. On the other hand, the experimental waiting time between two trains of pulses for both sequences only depends on T_{1H} instead of T_{1X} for normal acquisition. As T_{1H} is much shorter than T_{1X}, a fast pulsing rate is possible.

(v) Acoustic ringing is a very annoying problem encountered for low observation frequencies at any magnetic field. In fact, this spurious ringing comes from electromagnetic generation of ultrasonic standing waves in the probehead. This acoustic energy, in the presence of the static field, is converted to an oscillating magnetic field which is picked up by the detection coil as a spurious signal. A critical survey of this ringing problem has been published recently (16). It has been shown that the ringing pulse is proportional to the <u>transmitter pulse amplitude and to the square of the static field intensity</u> (B_0) for observing frequencies in the range 5 to 20 MHz.

This spurious signal can be viewed as a probe response which consists of a broad-frequency envelope, centered around the carrier frequency, <u>in phase with the emitter pulse and the NMR signal</u>. If this ringing response is of the order of the T_2^* of the observed isotope, the FID will be completely masked. Otherwise, a very serious perturbation of the baseline will occur after transforming the accumulated FID's (rolling baseline).

Several pulse sequences have been proposed to attenuate this effect. As an example one takes into account the proportionality of the ringing duration with the pulse length. Then a pulse sequence $[(90° - AQT)_3 - (270° - AQT)]_n$ will theoretically substract the ringing signals (same phase) and add the NMR signal (opposite phase). One can also use an echo sequence by choosing the τ value greater than the ringing value (Ref. 4, p. 56).

In general, all the proposed pulse sequences will only attenuate the ringing and will not suppress it completely. On the other hand, a critical evaluation of probe design and probe material can contribute to getting rid of important ringing problems. In any case, a ringing duration of ca. 200 to 50 μs must be expected with standard, multinuclear high resolution probeheads in the 5-20 MHz range, and the best method to eliminate most of the problems is to introduce a delay d between the end of the pulse and the start of the acquisition sufficient enough to allow the ringing pulse amplitude to decay towards a zero value. Of course, linewidths greater than $1/\pi d$ (1,600 Hz for d = 200 µs) will be hardly detected. For multiline spectra, this delay introduces ineluctably a phase twist across the entire spectral width, which cannot be corrected by the computer phase correction routines.

EXAMPLES OF APPLICATIONS

In general, heavy elements such as ^{103}Rh, $^{107,109}Ag$, ^{207}Pb, ^{199}Hg, etc. will show important $\Delta\delta/K$ factors, between 1 and 3 ppm/K. Then whatever the spin value of the isotope and the recording technique, a precise determination of the chemical shift will require a careful control of the sample temperature: otherwise a serious line broadening will occur. Figure 17 shows such an effect for Te NMR.

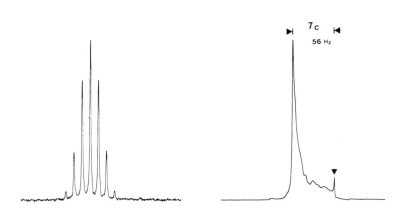

Figure 17. The 78.91 MHz ^{125}Te spectrum of dimethyl tellurium. (Left) ^1H coupled, (right) ^1H broadband decoupled. The broadening of the resonance due to the $\Delta\delta/K$ factor is clearly visible after switching the decoupler on (vertical arrow).

Spin 1/2 Isotopes and INEPT Sequence

Organometallic chemistry provides a vast area where transition metals play a very important role as key centers for catalytic or rearrangement processes. Then it is interesting to get a deeper insight of the metal site itself with respect to its electronic and structural properties. NMR spectroscopy can help to collect such information.

If the metal is in a favorable environment, $^nJ_{M-H}$ couplings can be detected and then an INEPT experiment will be the method of choice to observe low γ, spin 1/2 transition metal nuclei.

The gain in sensitivity can be quite large as expected from the γ_H/γ_M ratio. Table III exemplifies such a gain and Figure 18 shows a $\{^1H\}-^{109}Ag$ INEPT experiment compared to normal acquisition.

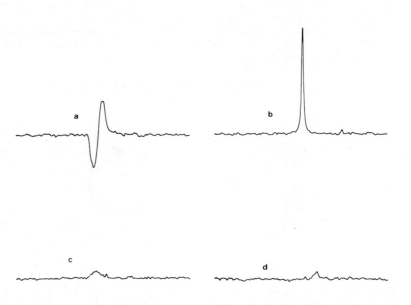

Figure 18. The 18.62 MHz ^{109}Ag spectrum of silver dication 1b (see Figure 6). (a) INEPT ^1H coupled spectrum, doublet separation: $2 \cdot ^3J_{Ag-Himine}$, (b) INEPT refocused, ^1H decoupled spectrum, (c) normal ^1H coupled spectrum, (d) normal ^1H decoupled spectrum. Recording conditions: 100 scans. $\tau_{90.Ag}$: 40 μs, $\tau_{90.^1H}$: 47 μs, τ delay: 0.030 s, T delay: 0.030 s, recycle delay: 3 s. The gain in sensitivity when using the INEPT sequence is self-evident.

Table III. Sensitivity Factor (η_{IN}) and Gain in Time G^{Acc} Expected in $\{^1H\}$ or $\{^{31}P\}$ INEPT Experiment for Representative Transition Metals M

M	$R_M^{13}C$	η^1H_{IN}	$\eta^{31}P_{IN}$	G^{Acc}_{1H}	G^{Acc}_{31P}
^{57}Fe	4.2×10^{-3}	30.95	12.53	961	157
^{103}Rh	0.177	31.52	12.76	992	163
^{109}Ag	0.276	21.4	8.70	462	76
^{183}W	5.9×10^{-2}	24.0	9.72	576	94
^{187}Os	10^{-3}	43.7	17.7	1918	314

Up to now, very interesting results have been obtained via INEPT silver NMR on silver dications complexes (17). In fact, ^{31}P can also be used to polarize the low γ nucleus as very often phosphine ligands stabilize transition metal complexes with well characterized $^nJ_{X-31P}$ couplings. Of course, the gain in sensitivity will be smaller compared to proton INEPT (lower γ_{31P}/γ_M ratio), but still high enough (Table III) to try the experiment as soon as a $^nJ_{X-31P}$ coupling exists, since the experimental set up is not very critical. A recent paper (18) provides all the necessary experimental details together with examples of $\{^{31}P\}^{183}W$, $\{^{31}P\}^{57}Fe$, and $\{^{31}P\}^{103}Rh$ INEPT experiments. Figure 19 shows the net gain obtained with $\{^{31}P\}^{15}N$ polarization transfer on a HMPT solution.

Spin > 1/2

A number of useful quadrupolar isotopes can be envisaged to gain electronic or structural information. Very recently, ruthenium NMR was detected (19,20), and this technique has rapidly proved to be an excellent analytical or structural tool in ruthenium chemistry (21). Figure 20 gives an idea of the actual ruthenium chemical shift scale. A word of caution is necessary to warn the spectroscopist who wants to start ruthenium NMR, as K^+ resonates at 11,340 ppm from the ruthenium reference RuO_4. Any ruthenium NMR study will then have to avoid potassium salts of ruthenium compounds to free the ^{99}Ru results from any K^+ data!

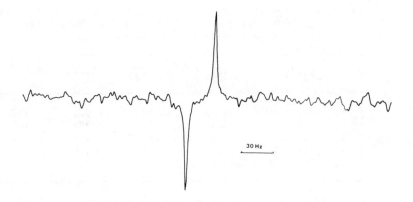

Figure 19. The 25.35 MHz ^{15}N spectrum of a 40/60 (v/v) solution of HMPT in C_6D_6 with 10^{-2} M $(Cr(acac)_3)$ added. (Top): $\{^{31}P\}$-^{15}N INEPT spectrum, (bottom) normal acquisition. 500 scans, 20 s delay for each spectrum. $J_{15N-31P}$ = 27.3 Hz. τ delay = 9 ms, $\tau_{90,31P}$ = 40 µs, $\tau_{90,15N}$ = 50 µs.

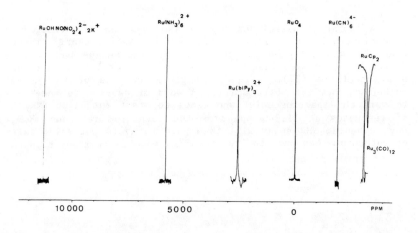

Figure 20. Ruthenium chemical shift scale (^{99}Ru spectra). The peak at 11,340 ppm is the K$^+$ resonance of the ruthenium salt.

REFERENCES

(1) K. Lee, W. A. Anderson, "Handbook of Chemistry and Physics," p. E 69, 55th edition, C.R.C. Press, Cleveland.
(2) G. H. Fuller, J. Phys. Chem. Ref. Data (National Bureau of Standards, Washington), 5, 835 (1976).
(3) "NMR and the Period Table," R. K. Harris and B. E. Mann, Eds., Academic Press, 1978.
(4) C. Brevard and P. Granger, "Handbook of High Resolution Multinuclear NMR," Wiley-Interscience, 1981.
(5) P. Pregosin and H. Rüegger, in press.
(6) C. Brown, B. T. Heaton, P. Chini, A. Fuornazalli, and G. Longoni, J. Chem. Soc. Chem. Commun., 309 (1977).
(7) A. Abragam, "Les Principes du Magnétisme Nucléaire," Presses Universitaires de France, 1961, p. 88.
(8) T. C. Farrar and E. D. Becker, "Pulse and Fourier Transform NMR," Academic, New York, 1971.
(9) R. R. Ernst and W. A. Anderson, Rev. Sci. Inst., 37, 93 (1966).
(10) G. A. Morris and R. Freeman, J. Am. Chem. Soc., 101, 760 (1979).
(11) A. Bax, R. Freeman, and S. P. Kempsell, J. Am. Chem. Soc., 102, 4849 (1980).
(12) J. Lefebvre, F. Chauveau, P. Doppelt, and C. Brevard, J. Am. Chem. Soc., 103, 4589 (1981).
(13) P. R. Sethruaman, M. A. Leparulo, M. T. Pope, F. Zonnevijlle, C. Brevard, and J. Lemerle, J. Am. Chem. Soc., 103, 7665 (1981).
(14) E. L. Hahn, Phys. Rev., 80, 580 (1950).
(15) K. G. R. Pachler and P. L. Wessels, J. Magn. Reson., 12, 337 (1973).
(16) E. Fukushima and S. B. W. Roder, J. Magn. Reson., 33, 199 (1979).
(17) C. Brevard, G. C. Van Stein, and G. Van Koten, J. Am. Chem. Soc., 103, 6746 (1981); G. C. Van Stein, G. Van Koten, and C. Brevard, J. Organometal. Chem., 226, C.27 (1982); G. Van Koten, H. Van der Poel, and C. Brevard, Inorg. Chem., 21, 2878 (1982).
(18) R. Schimpf and C. Brevard, J. Magn. Reson., 47, 108 (1982).
(19) R. W. Dykstra and A. M. Harrison, J. Magn. Reson., 45, 108 (1981).
(20) C. Brevard and P. Granger, J. Chem. Phys., 75, 417 (1981).
(21) P. Granger and C. Brevard, Inorg. Chem., in press.

CHAPTER 2

THE CALCULATION AND SOME APPLICATIONS OF NUCLEAR SHIELDING

G. A. Webb
Department of Chemistry
University of Surrey, Guildford, Surrey, England

ABSTRACT

Ramsey provided the first attempt at formulating the nuclear shielding tensor by quantum mechanical procedures. Although some knowledge of this tensor is important for an understanding of chemical shifts, Ramsey's model has not proved to be very useful for most chemically interesting molecules. Developments in this field stem from the introduction of gauge dependent atomic orbitals in the MO description by Pople. In his model Pople provides an account of nuclear shielding in which local and nonlocal atomic contributions to the shielding tensor may be interpreted in a chemically meaningful manner. In recent years various perturbation models using MO descriptions of numerous degrees of sophistication have been applied to chemical shift calculations. Most of the calculations have been concerned with nuclei from the first row of the Periodic Table, e.g., ^{13}C, ^{14,15}N, ^{17}O. Recently, attention has been turned to some second row nuclei, e.g., ^{29}Si and ^{31}P. The various MO procedures currently available for chemical shift calculations are addressed, and some of the advantages and disadvantages of the commonly employed techniques are indicated.

INTRODUCTION

The development of sensitive experimental techniques such as Fourier transform NMR, during the past ten years, has made possible a wide interest in nuclei other than protons. Experimental chemical shifts are commonly available for the ^{13}C, ^{14}N, ^{15}N, ^{17}O, ^{19}F, ^{29}Si, and ^{31}P nuclei. The halogens and most metals are also receiving NMR attention these days (1).

Nuclear shielding provides a basis for an understanding of the electronic and molecular environments of a nucleus. By studying the NMR spectra of several different nuclei in a given molecule it is possible to obtain an intimate knowledge of the electronic distribution in, and structure of, that molecule. A reliable theory of nuclear shielding will have a number of chemical applications, such as the identification of conformation or structure of the species present in a given sample by comparison of the observed and calculated shieldings.

There are a number of problems to be overcome in producing a satisfactory theoretical account of nuclear shielding. In one category are those produced by shielding contributions arising from the environment of the molecule under investigation. Theoretical calculations are normally based on the concept of a molecule in a vacuum. Thus in comparing the results of such calculations with experimental NMR data some allowance should be made for specific medium effects, such as hydrogen bonding, non-specific medium effects, and possible electric field effects.

A second group of problems arises in the choice of molecular orbital model on which the calculations are to be based. Included in this category are considerations of which terms to include in the shielding expression and whether <u>ab initio</u> or semi-empirical procedures should be used to evaluate them.

Some possible solutions to these various difficulties are discussed in this report.

SOME BASIC CONSIDERATIONS OF NUCLEAR SHIELDING

Nuclear shielding is often discussed in the terminology developed by Ramsey (2). This approach leads to the definition of a nuclear shielding parameter σ, as the sum of contributions from a "diamagnetic" term σ^d and a "paramagnetic" term σ^p, eq. 1, in which σ^d corresponds to the free rotation of electrons

$$\sigma = \sigma^d + \sigma^p \qquad (1)$$

about the nucleus in question and σ^p describes the hindrance of this rotation caused by other nuclei and electrons in the molecule.

In practice it is not easy to employ eq. 1 directly due to three main problems associated with it. (i) For molecules of a reasonable size σ^d and σ^p become large and of opposite sign, such that σ becomes the relatively small difference between two large terms and is subject to considerable error. (ii) In order to evaluate σ^p it is necessary to have some knowledge of the electronic ground state of the molecule and all of its excited states up to, and including, the continuum. From the few detailed calculations available it appears that contributions from the continuum may be at least as important as those arising from the discrete excited states. In general little is known about the high energy excited states of a molecule or the continuum. (iii) The relative values of σ^d and σ^p calculated within Ramsey's framework depend upon the origin chosen for the calculation. This can obviously present difficulties when comparing results from different authors. In addition, due to the practical choice of atomic orbital functions, the value of σ is usually found to depend upon the choice of origin also. Only the use of a complete basis set of atomic orbitals can remove this difficulty.

On account of these problems the Ramsey formulation is not frequently employed in NMR discussions of nuclear shielding. However, it has been useful in connection with molecular beam studies to provide absolute nuclear shielding scales for various nuclei (3). These investigatons depend upon the measurement of molecular spin-rotation interactions, which allow σ^p to be estimated for small, fairly symmetric molecules. This value of σ^p may be combined with a theoretically calculated value of σ^d to provide an estimate of σ from eq. 1.

The use of self-consistent field molecular orbital theory (SCF-MO) implies a number of approximations which are necessary, since the Schrödinger equation is generally unsolvable for three or more bodies. Thus SCF MO's represent an optimum approximation to the many-electron wave function.

Further approximations enter when the effect of the applied magnetic field on the molecular electrons is considered. At this stage either of two quantum mechanical perturbation theories is usually introduced to the calculation of nuclear shielding. These are necessary to describe the perturbations on the approximate many-electron wave function of the molecule, which are produced by the applied field and the nuclear magnetic moments.

The perturbation theories in question are the Raleigh-Schrödinger sum-over-states perturbation theory (SOS) (4) and the self-consistent-field finite perturbation theory (FPT) (5).

CALCULATIONS GIVING GAUGE INDEPENDENT RESULTS

The effect of the applied magnetic field on the atomic orbitals, ϕ_μ, was described by London (6) in terms of eq. 2, in which

$$\chi_\mu = \phi_\mu \exp\left[\frac{-ie}{\hbar}(\underline{A}_j \cdot \underline{r})\right] \qquad (2)$$

χ_μ is the corresponding atomic orbital in the presence of the applied field, $\underline{A}_j$ refers to the vector potential of the field at nucleus j, and $\underline{r}$ is the position vector of an electron in χ_μ.

The orbitals described by eq. 2 are usually referred to, anomalously, as Gauge Invariant Atomic Orbitals (GIAO's). Such orbitals are incorporated into SOS and FPT calculations to give nuclear shielding data which are gauge independent (7). Ditchfield (8) has successfully used GIAO's in ab initio FPT calculations of nuclear shielding in some small molecules as shown in Table 1. This work represents a substantial advance on earlier FPT calculations (9) using normal atomic orbitals, which produced very gauge dependent results.

It has recently been demonstrated (10) that ab initio SOS calculations including configuration interaction can reproduce the FPT results for HF, H_2, NH_3 and CH_4 to within 10%.

The use of GIAO's gives rise to expressions for the components of the nuclear shielding tensor which may be discussed in terms of localised electronic currents. Although these terms carry the names diamagnetic and paramagnetic as used in Ramsey's theory they must not be confused with Ramsey's formulation.

Thus ab initio calculations with GIAO's can provide a very satisfactory account of nuclear shielding for small molecules containing atoms from the first two rows of the Periodic Table.

For larger molecules, and for those containing heavier nuclei, it seems that semi-empirical calculations will be the most feasible and cost effective method for some time yet.

SEMI-EMPIRICAL CALCULATIONS OF NUCLEAR SHIELDING

For a nucleus A the shielding tensor components in the FPT approach may be given by eq. 3 (11, 12), in which α, β refer to the cartesians, x, y, or z. The first two terms on the right-hand side of eq. 3 are one-centre diamagnetic and paramagnetic contributions, respectively, the next four terms refer to

Table 1. Ab Initio FPT Calculations of Nuclear Shielding

Molecule	Nucleus of Interest	STO-5G opt. scale	LEMAO-5G opt. scale	4-31G	Experiment
CH_4	C	0.0	0.0	0.0	0.0
	H	0.00	0.00	0.00	0.00
C_2H_6	C	-11.8	2.2	-7.4	-8.0
	H	0.05	0.03	-0.35	-0.75
CH_3F	C	-50.5	-75.0	-65.4	-77.5
	F	49.2	128.7	74.0	76
	H	-1.18	-3.29	-4.04	-4.00
C_2H_4	C	-111.6	-152.5	-130.8	-125.6
	H	-5.21	-6.20	-5.61	-5.18
H_2CO	C	-147.9	-250.4	-199.6	-197
	O	-757.6	-860.5	-858.8	-580 to -600
	H	-9.34	-10.02	-10.39	-9 to -10
C_2H_2	C	-56.9	-90.0	-75.2	-76
	H	-1.41	-2.27	-1.25	-1.35
HCN	C	-97.5	-147.5	-123.5	-120±10
	N	-266.5	-360.5	-308.4	-297±20
	H	-3.10	-3.79	-2.95	-2.83
NH_3	N	0.0	0.0	0.0	0.0
	H	1.01	0.13	0.86	0.05
OH_2	O	0.0	0.0	0.0	0.0
	H	1.37	-1.45	0.06	-0.60
FH	F	0.0	0.0	0.0	0.0
	H	2.20	-4.16	-2.12	-2.50

$$\sigma_{A\alpha\beta} = \sigma_{A\alpha\beta}^{dAA} + \sigma_{A\alpha\beta}^{pAA} + \sum_{C \neq A} \left[\sigma_{A\alpha\beta}^{dAC} + \sigma_{A\alpha\beta}^{pAC} + \sigma_{A\alpha\beta}^{dCC} + \sigma_{A\alpha\beta}^{pCC} \right]$$

$$+ \sum_{D \neq A \neq C} \left[\left(\sigma_{A\alpha\beta}^{dDC} + \sigma_{A\alpha\beta}^{pDC} \right) \right] \quad (3)$$

shielding currents on atoms A and C and are thus two-centre terms. The final two terms refer to the effect of the induced current density between atoms D and C on the shielding of atom A. Hence these are three-centre terms. These last terms are most likely to make a significant shielding contribution when atoms A, C, and D are bonded together, probably in a linear arrangement with multiple bonding between C and D.

INDO-parameterised calculations based on eq. 3 give very good results for some electron-deficient boron hydrides (Table 2) (13). Significant contributions to the shielding are obtained from the two- and three-centre terms. Hence current densities in relatively remote parts of the molecule play a significant role in determining the boron chemical shifts.

Table 2. Calculated and Experimental Chemical Shifts for Some Representative Boron-Containing Compounds

Compound	σ_A^{AA}	σ_A^{CC}	σ_A^{AC}	σ_A^{CD}	σ_A	σ_A(calc)	σ_A(exp)
$B_5H_{11}(B_1)$	30.55	4.86	25.81	29.17	90.39	−81.03	−72.8
*$B_5H_9(B_1)$	17.70	4.44	17.66	43.68	74.49	−65.13	−70.2
*$B_4H_{10}(B_1)$	9.10	3.74	26.63	22.57	62.04	−52.68	−59.3
$B_{10}H_{14}(B_2)$	19.54	6.19	16.85	35.31	77.89	−68.53	−53.3
*$B_5H_9(B_2)$	−2.70	4.22	23.85	23.36	48.45	−39.09	−30.6
*$B_4H_{10}(B_2)$	−4.42	3.38	28.18	14.16	42.31	−32.95	−24.4
$B_5H_{11}(B_3)$	−10.32	3.43	26.68	13.13	32.92	−23.56	−17.0
$B_{10}H_{14}(B_5)$	−4.96	4.73	18.05	25.36	43.19	−33.83	−16.8
$B_5H_{11}(B_2)$	−11.61	4.73	20.33	20.45	33.90	−24.54	−10.0
$B_{10}H_{14}(B_6)$	−15.57	4.24	19.15	21.52	29.34	−19.98	−7.8
*$B_{10}H_{14}(B_1)$	−10.23	5.99	9.19	33.15	38.10	−28.74	−6.2
B_2H_6	−25.40	2.82	26.43	5.51	9.36	0.00	0.0
$B(C_2H_3)_3$	−36.66	−2.96	5.24	1.97	−32.40	41.76	37.7
$(CH_3)_2$-BC_3H_3	−57.84	1.99	0.69	6.06	−49.10	58.46	53.5
$B(CH_3)_3$	−69.25	−0.46	−1.44	10.88	−60.26	69.62	68.5

In these calculations the INDO β parameter was optimised for some of the molecules considered in order to obtain the most satisfactory results. This represents an improvement over earlier INDO-parameterised FPT calculations in which it was found to be necessary to optimise all of the parameters to obtain satisfactory shielding data (14-17). In this earlier work not all of the terms given in eq. 3 were included in the calculations. The two-centre terms $\sigma_{A\alpha\beta}^{dCC}$ and $\sigma_{A\alpha\beta}^{pCC}$ were estimated by the dipole approximation (18) and the other two- and three-centre terms were ignored.

The results presented in Table 2 augur well for the future of semi-empirical calculations of nuclear shielding. However, at present the most frequently encountered semi-empirical MO calculations of nuclear shielding are based upon Pople's independent electron model (19,20). On this basis local, nonlocal, and interatomic contributions to nuclear shielding arise in the manner proposed by Saika and Slichter, eq. 4. (21). The diffi-

$$\sigma = \sigma^d(\text{loc}) + \sigma^d(\text{nonloc}) + \sigma^d(\text{inter}) + \sigma^P(\text{loc}) + \sigma^P(\text{nonloc}) + \sigma^P(\text{inter}) \quad (4)$$

culties noted with eq. 1 are not applicable to eq. 4. Thus eq. 4 is more suitable for discussions of chemical shifts.

The local terms in eq. 4 arise from electronic currents localised on the atom containing the nucleus of interest. Similarly $\sigma^d(\text{nonloc})$ and $\sigma^P(\text{nonloc})$ are contributions from the currents on neighbouring atoms. Finally $\sigma^d(\text{inter})$ and $\sigma^P(\text{inter})$ are due to shielding currents not localised on any of the atoms in the molecule, e.g., ring currents. These latter two terms usually only produce a shielding contribution of a few ppm at most, which is important for protons due to their small range of chemical shifts. Other nuclei have chemical shift ranges of several hundred ppm, and thus interatomic contributions are negligible by comparison.

Although the nuclear shielding parameter, σ, is a tensor quantity, in an NMR experiment on a nonviscous sample only its average value is measured. Thus we need only to be concerned with rotationally averaged values of the terms in eq. 4 when considering high resolution NMR spectra. Nevertheless it should be mentioned that in experiments where molecules are aligned, such as those on solids and nematic phases, it is possible to measure the tensor components of σ and thus to estimate its anisotropy.

The rotationally averaged values of the local diamagnetic and paramagnetic terms for nucleus A are given by eq. 5 and eq. 6, in which $p_{\mu\mu}$ is the charge density in the atomic orbital

$$\sigma_A^d(\text{loc}) = \frac{\mu_0 e^2}{12\pi m} \sum_\mu^A p_{\mu\mu} \langle \mu | r_{\mu A}^{-1} | \mu \rangle \quad (5)$$

μ at an average distance of $r_{\mu A}$ from nucleus A and C_{X,A_j} is the LCAO coefficient of the P_X orbital on atom A in the molecular orbital j, etc. The summation over B includes A. It is apparent that the summation is zero unless both atoms A and B possess valence p electrons. It also follows from the ordering of the coefficients in eq. 6 that $\pi \rightarrow \pi^*$ transitions do not contribute to the summation.

$$\sigma_A^P(\text{loc}) = \frac{\mu_0 e^2 \hbar^2}{6\pi m^2} \langle r^{-3} \rangle_{np} \sum_j^{\text{occ}} \sum_k^{\text{unocc}} (E_k - E_j)^{-1}$$

$$[(C_{Y,A_j}C_{Z,A_k} - C_{Z,A_j}C_{Y,A_k})\sum_B(C_{Y,B_j}C_{Z,B_k} - C_{Z,B_j}C_{Y,B_k})$$

$$+ (C_{Z,A_j}C_{X,A_k} - C_{X,A_j}C_{Z,A_k})\sum_B(C_{Z,B_j}C_{Y,B_k} - C_{X,B_j}C_{Z,B_k})$$

$$+ (C_{X,A_j}C_{Y,A_k} - C_{Y,A_j}C_{X,A_k})\sum_B(C_{X,B_j}C_{Y,B_k} - C_{Y,B_j}C_{X,B_k})] \quad (6)$$

E_k and E_j refer to the energies of the molecular orbitals k and j which are unoccupied and occupied respectively. The excitation energies, (E_k-E_j), are those for excited singlet states which are mixed with the ground state by an external magnetic field. Hence,

$$E_k - E_j = \varepsilon_k - \varepsilon_j - J_{jk} + 2K_{jk}, \quad (7)$$

in which ε_k and ε_j are eigenvalues of the unperturbed molecule and J_{jk} and K_{jk} are respectively the Coulomb and exchange integrals, eq. 8 and eq. 9.

$$K_{jk} = \iint \psi_j^*(1)\psi_k^*(2)\frac{1}{r_{12}} \psi_k(1)\psi_j(2)d\tau_1 d\tau_2 \quad (8)$$

$$J_{jk} = \iint \psi_j^*(1)\psi_k^*(2)\frac{1}{r_{12}} \psi_j(1)\psi_k(2)d\tau_1 d\tau_2 \quad (9)$$

In eq. 6 $\langle r^{-3} \rangle_{np}$ is the mean inverse cube radius for the p orbital on atom A with primary quantum number n. This quantity is usually evaluated from the expression given by eq. 10.

$$\langle r^{-3} \rangle_{np} = \frac{1}{3} \left(\frac{Z_{np}}{na_0} \right)^3 \quad (10)$$

The matrix elements in eq. (5) are similarly obtained from eq. 11, in which Z_μ is the effective nuclear charge for the

$$\langle \mu | r_{\mu A}^{-1} | \mu \rangle = \frac{Z_\mu}{n^2 a_0} \quad (11)$$

atomic orbital μ on A and a_0 is the Bohr radius.

The value of Z_μ may be obtained from Slater's rules (22) according to which the 2s and 2p orbitals of first row atoms experience the same effective nuclear charge, given by eq. 12,

$$Z_{2s} = Z_{2p} = Z^o + 0.35\, q_A^{net} \tag{12}$$

in which Z^o is the effective nuclear charge for the free atom and q_A^{net} is the net charge on atom A. Eq. (12) may also be written as eq. (13). Alternatively Z_μ can be evaluated from Burns's rules (23), eqs. 14 and 15, in which the values of the constants

$$Z_{2p} = A - 0.35 P_{2s2s} - 0.35(P_{2p_x2p_x} + P_{2p_y2p_y} + P_{2p_z2p_z}) \tag{13}$$

$$Z_{2s} = B - 0.4 P_{2s2s} - 0.35(P_{2p_x2p_x} + P_{2p_y2p_y} + P_{2p_z2p_z}) \tag{14}$$

$$Z_{2p} = C - 0.5 P_{2s2s} - 0.35(P_{2p_x2p_x} + P_{2p_y2p_y} + P_{2p_z2p_z}) \tag{15}$$

A, B, and C in eqs. 13 to 15 are given for some first row atoms by:

	A	B	C
Boron	3.65	3.40	3.35
Carbon	4.65	4.40	4.35
Nitrogen	5.65	5.40	5.35
Oxygen	6.65	6.40	6.35
Fluorine	7.65	7.40	7.35

The nonlocal contributions to σ^d and σ^p, given in eq. 4, are usually estimated by assuming that the induced moments in the electrons on atom B can be replaced by a point dipole. The effect of this dipole on the shielding of nucleus A is found to be negligible for most first row atoms, although it can be of some importance in cases of multiple bonding as shown in Table 3.

Consequently eqs. 5 and 6 are frequently employed to calculate nuclear shielding by semi-empirical methods usually in conjunction with eqs. 7 to 15. These may be evaluated by any semi-empirical molecular orbital method such as INDO, CNDO, MINDO, etc. Good agreement with experimental ^{13}C, ^{14}N, ^{17}O, and ^{19}F chemical shifts have been reported using the CNDO/S and INDO/S parameterizations (7). In these calculations it is found that the chemical shift differences experienced by a given type of nucleus in a range of related molecules arise from changes in $\sigma_A^p(loc)$ except for protons. Consequently $\sigma_A^d(loc)$ remains approximately constant for a given nucleus, e.g., ^{13}C, in a variety of molecules.

Table 3. CNDO/S Calculations of Contributions to Pople's Shielding Equation for Some ^{13}C Nuclei

Molecule	σ^d(loc)	σ^d(nonloc)	σ^p(loc)	σ^p(nonloc)	σ_{total}
CH_4	260.50	0.0	-127.08	0.0	133.42
C_2H_6	260.33	0.0	-123.45	0.05	136.83
C_2H_2	260.62	-0.04	-144.18	-4.36	112.04
CH_3F	259.08	-0.13	-129.96	-1.09	127.89
HCN	259.59	0.16	-172.87	-7.10	79.78
C_2H_4	260.40	0.01	-211.46	-0.21	48.73
C_6H_6	260.17	-0.01	-190.32	1.92	71.76
H_2CO	258.39	-0.09	-223.27	5.79	40.82

AVERAGE EXCITATION ENERGY (AEE) CALCULATIONS

The AEE approximation is often invoked to simplify eq. 6. This involves replacing $(E_k - E_j)$ by a mean value ΔE. Since overlap is neglected in the semi-empirical calculations used, it follows that

$$\sum_{j}^{occ} c_{\mu j} c_{\lambda j} + \sum_{k}^{unocc} c_{\mu k} c_{\lambda k} = \delta_{\mu \lambda}. \qquad (16)$$

By incorporating eq. 16 and the AEE approximation, eq. 6 becomes eq. 17, in which the summation over B includes A and all the

$$\sigma_A^p(loc) = - \frac{\mu_0 e^2 \hbar^2}{8\pi m^2 \Delta E} \langle r^{-3} \rangle_{np} \sum_{B} Q_{AB} \qquad (17)$$

other atoms in the molecule. The bond-order charge-density terms Q_{AB} are given by eq. 18. Most attempts at evaluating $\sigma_A^p(loc)$

$$Q_{AB} = \frac{4\delta_{AB}}{3} (P_{xAxB} + P_{yAyB} + P_{zAzB})$$

$$- \frac{2}{3} (P_{xAxB} P_{yAyB} + P_{xAxB} P_{zAzB} + P_{yAyB} P_{zAzB})$$

$$+ \frac{2}{3} (P_{xAyB} P_{xByA} + P_{xAzB} P_{xBzA} + P_{yAzB} P_{yBzA}) \qquad (18)$$

have used the AEE approximation. Various approaches have been adopted to evaluate eq. 18 including π electron calculations together with σ bond polarizations and all valence electron methods such as extended Hückel, CNDO, INDO, and MINDO.

In general these methods are reasonably successful in accounting for gross chemical shift trends in series of closely related molecules. However, their value is limited by the necessity of choosing a value for ΔE, for which there is no <u>a priori</u> method.

For first row atoms a combination of eqs. 5 and 11 gives eq. 19, for which Z can be found from eqs. 12 to 15. Similarly eqs. 10 and 17 produce eq. 20. Eqs. 19 and 20 are suitable for use

$$\sigma_A^d(\text{loc})(\text{ppm}) = 17.7501 \sum_\mu^A P_{\mu\mu} Z_\mu n^{-2} \qquad (19)$$

$$\sigma_A^p(\text{loc})\Delta E(\text{ppm}\cdot\text{eV}) = 30.1885 Z_{2p}^3 \sum_B Q_{AB} \qquad (20)$$

with semi-empirical molecular orbital data.

Calculations on molecules containing elements with d orbitals are more complicated and often require further approximations. The chemical shifts of ^{29}Si and ^{31}P are often discussed in conjunction with AEE calculations in which only those terms involving p orbitals on the atom of interest are considered. However, recent calculations involving d orbitals in an expression comparable to eq. 6 have revealed the importance of d electron contributions from excited states to ^{29}Si and ^{31}P screening (24).

EXPERIMENTAL CHEMICAL SHIFT RANGES

In discussing chemical shifts it is usual to refer to a high frequency, or low field, shift from a reference signal as a positive one. Thus deshielding of a nucleus gives rise to a positive chemical shift.

The range of chemical shifts experienced by a given nucleus depends to a large extent upon its position in the periodic table and thus is related to its electronic structure, which in turn controls its chemistry.

Some approximate chemical shift ranges, in ppm, are as follows:

H, 20
Li, 10   B, 200   C, 650   N, 1000   O, 1500   F, 800
Na, 20   Al, 500 Si, 400 P, 700   S, 650   Cl, 1000

Some of the heavier nuclei have much larger ranges, e.g., Sn, 2300; Pb, 10,000; Co, 18,000; and Tl, 7000 ppm.

The very small ranges found for H, Li, and Na are principally due to the fact that these atoms do not normally have valence p electrons. Consequently as eq. 20 shows, $\sigma^P(loc)$ does not contribute to the shielding of these nuclei. Thus the relatively small changes occurring in $\sigma^d(loc)$ determine the chemical shift range for these nuclei.

It is noteworthy that N, O, and F have larger ranges than B and C and that a parallel situation occurs amongst the third row elements. These ranges reflect the presence of nonbonding electrons on N, O, and F. In general these electrons give rise to low energy $n \rightarrow \pi^*$ transitions and thus to large values of $\sigma^P(loc)$ which cause nuclear deshielding. If the lone pairs are involved in bonding then this contribution to $\sigma^P(loc)$ is removed and the nuclear shielding increases. Thus the lone pair of electrons provide an extended range of nuclear shielding values.

This effect is well demonstrated in nitrogen NMR where the N atom of the pyridinium ion is more shielded by about 115 ppm than that of pyridine, and the pyrrole nitrogen resonance is about 170 ppm to low frequency of that from pyridine. Similarly the high frequency shift of 250 ppm observed for NO_2^- compared with NO_3^- is largely due to a low energy excitation involving a non-bonding orbital in NO_2^-.

The very large chemical shift ranges observed for the heavy metals undoubtedly reflect the greater polarizability of the p and d valence orbitals and their varied commitments to chemical bonding.

SOME APPLICATIONS OF CALCULATED NUCLEAR SHIELDINGS

Calculations of $\sigma^d(loc)$ and $\sigma^P(loc)$ based upon eqs. (5) and (6) account satisfactorily for the gross ^{13}C chemical shift trends in several series of molecules (7). Anisotropy calculations using the CNDO/S parameterization show that, for planar molecules, the most shielded component of the tensor is perpendicular to the molecular plane. This result can be reconciled with the fact that $\pi \rightarrow \pi^*$ transitions do not contribute to $\sigma^P(\text{loc})$. Thus the out-of-plane component of the shielding tensor is determined by high energy transitions involving σ electrons, which produce small values for $\sigma^P(\text{loc})$ and thus a high shielding contribution for this component. The observed trend of ^{13}C shielding anisotropies, $(CH_3)_2CO > CH_3CHO > CH_3COOH > HCOOCH_3 > (CH_3CO)_2O$, is reproduced by these calculations (7).

The importance of one-centre exchange integrals in calculating the molecular excited electronic states is demonstrated by a comparison of some CNDO/S and INDO/S results for ^{13}C nuclear screening (25). In general the agreement between the experimental and calculated results is best when the INDO/S parameters are employed (26) (Table 4).

The use of MINDO/3 parameterized calculations for nuclear screening is not recommended (27) because of the poor predictability of the molecular excited states by the MINDO/3 procedure.

Many calculations of ^{13}C chemical shifts have employed the AEE approximation. Often a value of 10 eV has been chosen for ΔE. This value may be arbitrarily adjusted for molecules with low-lying excited states. In a series of closely related molecules the ^{13}C chemical shifts may depend in a linear fashion on the carbon charge densities, as in eq. 21, in which $\Delta \delta$ is the

$$\Delta \delta = K \delta q \tag{21}$$

difference in shielding and δq is the charge density difference --both referred to a standard molecule. For conjugated molecules the values of the proportionality factor K range from 160 to 200 ppm per unit π electron charge for various series of molecules.

Eq. 21 will only be applicable for series of molecules in which ΔE and the bond-order terms given in eq. 18 are reasonably constant. It relies upon the dependence of $\sigma^P(\text{loc})$ upon $\langle r^{-3} \rangle_{2p}$, which is closely related to electron denity. The 2p orbitals expand as electrons are added to them. The resulting decrease in $\langle r^3 \rangle_{2p}$ diminishes $\sigma^P(\text{loc})$ and thereby increases the total nuclear shielding.

Various additivity relationships have been reported for ^{13}C chemical shifts of closely related molecules (28). Reactivity parameters, electronegativities, electric field effects, atom-atom polarizabilities, and other semi-empirical parameters have also been used to discuss ^{13}C chemical shift trends (7).

Table 4. Diamagnetic and Paramagnetic Contributions to the ^{13}C Shielding Tensors of Some Unsaturated Molecules

Molecule	σ(loc) (ppm)	INDO/S data σ(loc) (ppm)	σ(nonloc) (ppm)	Total σ (ppm)	Expt.[a] σ(obs) (ppm)
$H_2C=CH_2$	260.41	-199.26	-0.26	60.89	74.0
$CH_3\overset{*}{C}H=CH_2$	260.62	-190.61	-0.13	69.88	81.4
$(CH_3)_2C=\overset{*}{C}H_2$	260.81	-190.87	1.43	71.39	87.0
$CH_2=C=\overset{*}{C}H_2$	260.83	-177.78	1.33	84.37	122.8
$CH_2=CH-CH=\overset{*}{C}H_2$	260.48	-179.05	1.56	82.98	80.2
$CH_3\overset{*}{C}H=CH_2$	260.12	-195.22	-0.13	64.77	61.1, 63.7
cis-$CH_3CH=\overset{*}{C}HCH_3$	260.32	-194.24	-0.22	65.87	73.5, 72.5
trans-$CH_3CH=\overset{*}{C}HCH_3$	260.32	-192.79	-0.20	67.30	72.3, 71.0
$CH_2=CH-\overset{*}{C}H=CH_2$	260.07	-187.94	2.08	74.21	59.6
C_6H_6	260.12	-187.42	1.19	73.88	68.1
$(CH_3)_2\overset{*}{C}=CH_2$	259.84	-199.36	1.39	61.86	55.6
$(CH_3)_2C=\overset{*}{C}(CH_3)_2$	260.23	-197.05	-0.28	62.89	73.6
$CH_2=\overset{*}{C}=CH_2$	259.63	-274.44	-0.59	-15.41	-16.2
$CH_3-N\equiv\overset{*}{C}$	260.39	-188.09	-4.47	67.83	38.1
$CH_3-\overset{*}{C}\equiv N$	258.85	-158.55	-5.70	94.60	79.6
$H-\overset{*}{C}\equiv N$	259.07	-155.35	-5.88	97.85	85.9
$HC\equiv CH$	260.67	-134.45	-4.15	122.08	120.0

[a] The experimental shielding data are obtained from σ_{obs} = $-(\delta_x-\delta_{ethylene})$ + 74.0, in which δ_x values are expressed with respect to TMS. The figure of 74 ppm is the <u>ab initio</u> value for the ^{13}C shielding of ethylene.

Calculations based upon eqs. 5 and 6 in conjunction with CNDO/S and INDO/S parameters produce reasonable agreement with various series of nitrogen chemical shifts (29-31).

For linear species, N_2, HCN, CH_3CN, CH_3NC, and NO_2^+, the major contribution to σ^p arises from the lowest energy $\sigma \to \pi^*$ transition. However for CH_3NC and NO_2^+ there are also contributions from higher energy $\pi \to \sigma^*$ transitions. In the case of the bent ion, NO_2^-, the in-plane components are dominated by $\sigma \to \pi^*$ transitions, whereas several $\sigma \to \pi^*$ transitions contribute to the out-of-plane component of $\sigma^P(loc)$.

For heterocyclic molecules the major contributions to $\sigma^P(loc)$, for "pyridine-type" nitrogen atoms, arises from $n \to \pi^*$ transitions with significant amounts coming from the $\pi \to \sigma^*$ and $\sigma \to \pi^*$ transitions. The effective removal of the lone-pair on "pyrrole-type" nitrogen atoms leaves the $\pi \to \sigma^*$ and $\sigma \to \pi^*$ transitions as the dominant ones. The higher energy $\sigma \to \sigma^*$ and $n \to \sigma^*$ transitions provide only minor contributions (32).

Although the CNDO/S calculations provide reasonable agreement with nitrogen chemical shifts, the agreement is improved by means of INDO/S calculations (33). Again, the use of MINDO/3 calculations is not recommended (27).

The deshielding of the pyridine nitrogen, with respect to that of the pyridinium ion, arises mainly from $n \to \pi^*$ transitions with substantial contributions from $n \to \sigma^*$ transitions. The $\sigma \to \sigma^*$, $\sigma \to \pi^*$, and $\pi \to \sigma^*$ transitions all produce shielding contributions for pyridine relative to its ion. It is noteworthy that AEE calculations predict a substantial deshielding contribution for pyridine from a $\sigma \to \pi^*$ transition, which serves to illustrate the shortcomings of AEE calculations. However, in general the AEE approach gives a reasonable account of nitrogen chemical shift trends along series of closely related molecules.

SOME CALCULATIONS OF MEDIUM EFFECTS ON NUCLEAR SHIELDING

The effects of solvents on the nuclear shielding of solute molecules may be categorised into specific and nonspecific classes. The specific interactions include hydrogen-bonding, protonation, etc. From the theoretical standpoint these may be considered by the supermolecule approach.

Nonspecific influences can arise from electronic interactions between the dipole moments of the solute and solvent molecules. The effect of the corresponding reaction field on nuclear shielding may be discussed in terms of the solvaton model (34).

Concerning hydrogen-bonding effects on nuclear shielding a good example is provided by formamide. <u>Ab initio</u> calculations using GIAO's within the FPT framework have been reported for formamide in the presence of its first hydration shell (35). With the exception of the formyl proton, which is not involved in hydrogen-bonding, significant variations in the shielding of the formamide nuclei are produced by hydration. A large increase is observed in the ^{17}O shielding and smaller decreases in that of the ^{13}C and nitrogen nuclei as hydration occurs. These predictions are in qualitative agreement with experiment.

Table 5. Calculated Nuclear Shielding of Hydrated Formamide

Nucleus	Formamide	I	Formamide (H_2O) II · III		IV	Formamide $(H_2O)_4$
N	235.99	234.51	232.45	233.20	235.24	225.03
C	110.42	107.03	109.25	109.08	108.87	103.05
O	-216.50	-169.88	-169.27	-195.70	-201.71	-88.47
H(c)	23.27	23.16	23.31	23.40	23.21	23.22
$H(N)_c$	27.32	27.24	27.02	25.24	27.61	24.99
$H(N)_t$	27.20	27.01	26.99	27.29	25.01	24.60

For nonspecific solvent-solute interactions the solvaton model provides a satisfactory account of nuclear shielding changes (36). As in the reaction field model, the strength of the solvent-solute interaction depends upon the dielectric constant, ε, of the medium.

An example of the satisfactory nature of the agreement between the observed and calculated nitrogen shielding changes of some nitroalkanes is provided in Table 6. The calculated results are obtained from INDO/S-parameterised SOS calculations involving

TABLE 6. Solvent Effects on Nitrogen Shieldings of Some Nitroalkanes

Compound	Solvent (0.3M Solutions)	ε (30°C)	Observed N Shielding w.r.t. CH_3NO_2	Calc.
CH_3NO_2	DMSO	45.8	-2.01 ± 0.12	-0.6
	DMF	37.5	-0.69 ± 0.13	-0.5
	None	35.9	0.0000	-0.4
	CH_3CN	36.6	$+0.20 \pm 0.13$	-0.4
	Acetone	20.4	$+0.77 \pm 0.10$	0.0
	CH_2Cl_2	9.50	$+3.21 \pm 0.12$	$+1.3$
	CH_2Br_2	6.78	$+3.41 \pm 0.12$	$+2.3$
	$CHCl_3$	5.07	$+3.79 \pm 0.13$	$+4.3$
	ether	4.79	$+3.91 \pm 0.13$	$+4.8$
	CCl_4	2.71	$+7.10 \pm 0.11$	$+8.8$
$CH_3CH_2NO_2$	DMSO	45.8	-11.37 ± 0.16	-10.6
	None	28.1	-10.25 ± 0.10	-10.4
	Acetone	20.4	-9.37 ± 0.11	-10.2
	CCl_4	2.71	-4.09 ± 0.12	-4.4
$CH_3(CH_2)_2NO_2$	DMSO	45.8	-10.09 ± 0.21	-9.8
	None	23.2	-7.73 ± 0.10	-9.3
	Acetone	20.4	-8.31 ± 0.14	-9.2
	CCl_4	2.71	-3.77 ± 0.16	-1.5
$CH_3(CH_2)_3NO_2$	None		-7.90 ± 0.11	
	Acetone	20.4	-7.91 ± 0.16	
	CCl_4	2.71	-3.87 ± 0.16	
$CH_3(CH_2)_5NO_2$	Acetone	20.4	-8.10 ± 0.15	
	None		-6.44 ± 0.12	
	CCl_4	2.71	-3.97 ± 0.19	
$(CH_3)_2CHNO_2$	None	25.5	-19.45 ± 0.10	-19.3
	Acetone	20.4	-19.40 ± 0.15	-19.2
	CCl_4	2.71	-14.73 ± 0.14	-15.1
nitrocyclo-hexane	Acetone	20.4	-18.36 ± 0.22	
	None		-16.27 ± 0.11	
	CCl_4	2.71	-13.63 ± 0.25	
$(CH_3)_3CNO_2$	DMSO	45.8	-28.20 ± 0.17	-27.9
	Acetone	20.4	-25.95 ± 0.11	-27.4
	None		-25.51 ± 0.11	
	CCl_4	2.71	-21.57 ± 0.12	-21.4

the solvaton model. The experimental results were obtained from concentric spherical sample and reference containers in order to remove bulk susceptibility effects from observed shielding changes (37).

The nitrogen shielding of the nitroalkanes decreases as ε increases. This observation is consistent with the induced increase in the polarity of the nitroalkanes. As this occurs the positive charge on the nitrogen atom is enhanced; thus $\langle r^{-3}\rangle_p$ becomes larger and $\sigma^P(loc)$ increases in magnitude, giving rise to a decrease in the nitrogen shielding. Results of solvaton calculations are available for some ^{13}C (38), ^{17}O (39), and ^{19}F (39) shieldings.

The substantial shielding variations predicted by these calculations imply that caution should be excercised in comparing uncorrected chemical shifts from solutions in different solvents.

REFERENCES

1. "NMR and the Periodic Table," R. K. Harris and B. E. Mann, Eds., Academic Press, London, 1978.
2. N. F. Ramsey, Phys. Rev., 78, p. 689 (1950).
3. W. H. Flygare, Chem. Rev., 74, p. 653 (1974).
4. J. O. Hirschfelder, W. Byers-Brown, and S. T. Epstein, Adv. Quant. Chem., 1, p. 255 (1964).
5. J. A. Pople, J. W. McIver, and N. S. Ostlund, J. Chem. Phys., 49, p. 2960 (1958).
6. F. London, J. Phys. Radium, Paris, 8, p. 397 (1937).
7. K. A. K. Ebraheem and G. A. Webb, Progr. NMR Spectrosc. 11, p. 149 (1977).
8. R. Ditchfield, Mol. Phys. 27, p. 789 (1974).
9. R. Ditchfield, D. P. Miller, and J. A. Pople, J. Chem. Phys., 53, p. 613 (1970).
10. H. Fukui, H. Yoshida, and K. Muira, J. Chem. Phys., 74, p. 6988 (1981).
11. R. Ditchfield and P. D. Ellis, Top. ^{13}C NMR Spectrosc., 1, p. 1 (1974).
12. A. R. Garber, P. D. Ellis, K. Seidman, and K. Schade, J. Magn. Reson., 34, p. 1 (1979).
13. P. D. Ellis, Y. C. Chou, and P. A. Dobson, J. Magn. Reson., 39, p. 529 (1980).
14. K. Seidman and G. E. Maciel, J. Am. Chem. Soc., 99, p. 659 (1977).
15. G. E. Maciel, J. L. Dallas, R. L. Elliott, and H. C. Dorn, J. Am. Chem. Soc., 95, p. 5857 (1973).
16. G. E. Maciel and H. C. Dorn, J. Magn. Reson., 24, p. 251 (1976).

17. G. E. Maciel, J. L. Dallas, and D. P. Miller, J. Am. Chem.Soc., 98, p. 5074 (1976).
18. J. A. Pople, Proc. Roy. Soc., Ser. A, 239, p. 541 (1957).
19. J. A. Pople, J. Chem. Phys., 37, p. 53, p. 60 (1962).
20. J. A. Pople, Mol. Phys., 7, p. 301 (1964).
21. A. Saika and C. P. Slichter, J. Chem. Phys., 22, p. 26 (1954).
22. J. C. Slater, Phys. Rev., 36, p. 57 (1930).
23. G. Burns, J. Chem. Phys., 41, p. 1521 (1964).
24. D. J. Reynolds and G. A. Webb, unpublished results.
25. M. Jallali-Heravi and G. A. Webb, Org. Magn. Reson., 11, p. 34 (1978).
26. M. Jallali-Heravi and G. A. Webb, Org. Magn. Reson., 11, p. 524 (1978).
27. M. Jallali-Heravi and G. A. Webb, Org. Magn. Reson., 12, p. 174 (1979).
28. F. W. Wehrli and T. Wirthlin, "Interpretation of ^{13}C NMR Spectra," Heyden, London, 1976.
29. M. Witanowski, L. Stefaniak, and G. A. Webb, Ann. Rep. NMR Spectrosc., 7, p. 117 (1977).
30. M. Witanowski, L. Stefaniak, and G. A. Webb, Ann. Rep. NMR Spectrosc., 11B, p. 1 (1981).
31. K. A. K. Ebraheem and G. A. Webb, Org. Magn. Reson., 9, p. 248 (1977).
32. E. A. K. Ebraheem, G. A. Webb, and M. Witanowski, Org. Magn. Reson., 11, p. 27 (1978).
33. M. Jallali-Heravi and G. A. Webb, J. Magn. Res. 32, p. 429 (1978).
34. H. A. Germer, Theoret. Chim. Acta., 34, p. 145 (1974).
35. F. R. Prado, C. Giessner-Prettre, A. Pullman, J. F. Hinton, D. Horspool, and K. R. Metz, Theoret. Chim. Acta, 59, p. 55 (1981).
36. I. Ando and G. A. Webb, Org. Magn. Reson., 15, p. 111 (1981).
37. M. Witanowski, L. Stefaniak, B. Na Lamphun, and G. A. Webb, Org. Magn. Reson., 16, p. 57 (1981).
38. M. Jallali-Heravi and G. A. Webb, Org. Magn. Reson., 13, p. 116 (1980).
39. I. Ando, M. Jallali-Heravi, M. Kondo, B. Na Lamphun, S. Watanabe, and G. A. Webb, Org. Magn. Reson., 14, p. 92 (1980).

CHAPTER 3

CALCULATIONS OF SPIN-SPIN COUPLINGS

G. A. Webb
Department of Chemistry
University of Surrey, Guildford, UK

ABSTRACT

The quantum mechanical background to the theory of spin-spin couplings was provided by Ramsey. The total coupling may be expressed as a sum of contributions from contact, orbital, and dipolar interactions. As in the case of nuclear shielding, spin-spin coupling is a second-order electronic property of a molecule. Thus perturbation approximations are necessary for a detailed description of spin-spin coupling. Satisfactory MO descriptions of the various spin-spin coupling terms have been developed by Pople and co-workers and applied to nuclei from the first and second rows of the Periodic Table. An account of frequently encountered techniques will be presented with a view to providing an interpretation of spin-spin couplings in a chemically interesting way.

INTRODUCTION

Nuclear spin-spin coupling is produced by an indirect interaction between the spins of neighbouring nuclei. The first nucleus perturbs the electrons in the bonds joining the spin coupled nuclei, and the electrons in turn produce a magnetic field at the second nucleus. The energy of the interaction between nuclei A and B, E_{AB}, is given by eq. 1. Unlike the

$$E_{AB} = hJ_{AB}\underline{I}_A \cdot \underline{I}_B \tag{1}$$

nuclear shielding tensor, $\underline{\sigma}$, J_{AB} is independent of the applied magnetic field; consequently the gauge problem does not enter into discussion of spin-spin interactions. Unlike $\underline{\sigma}$, values of J can be obtainable directly from NMR spectra.

A satisfactory theory of spin-spin couplings will have a number of structural applications in chemistry. The problems encountered in the calculation of spin-spin couplings and the comparison of calculated and observed data are similar to those noted for nuclear shieldings (1). The SOS and FPT procedures have been commonly employed for the calculation of spin-spin couplings. Most calculations have been performed at the semi-empirical level where INDO parameters appear to be the most satisfactory ones to employ. Recently, <u>ab initio</u> calculations have received some impetus, and for small molecules, at least, the future looks rather hopeful (2).

Concentration on the electronic factors involved in nuclear spin-spin couplings is achieved by means of the reduced coupling constant, K_{AB}, eq. 2, in which γ_A and γ_B are the magnetogyric

$$k_{AB} = \frac{2\pi J_{AB}}{\hbar \gamma_A \gamma_B} \quad (2)$$

ratios of the coupled nuclei.

BASIC COUPLING INTERACTIONS

Calculations of spin-spin couplings are usually based upon Ramsey's formulation (3). The appropriate interactions are described by the following Hamiltonian, eq. 3. The first two

$$\hat{H} = \hat{H}_1^{(a)} + \hat{H}_1^{(b)} + \hat{H}_2 + \hat{H}_3 \quad (3)$$

terms on the right-hand side of eq. 3 account for the interaction between the nuclear magnetic moments and the field produced by the orbital motion of the electrons. Thus these are referred to as the orbital terms, $\hat{H}_1^{(a)}$ and $\hat{H}_1^{(b)}$, which respectively are bi- and mononuclear, eqs. 4 and 5, in which μ_B is the Bohr

$$\hat{H}_1^{(a)} = \frac{\mu_0^2 e \hbar \mu_B}{16\pi^2} \sum_{ABk} \gamma_A \gamma_B r_{kA}^{-3} r_{kB}^{-3} [(\hat{\underline{I}}_A \cdot \hat{\underline{I}}_B)(\underline{r}_{kA} \cdot \underline{r}_{kB})$$

$$- (\hat{\underline{I}}_A \cdot \underline{r}_{kB})(\hat{\underline{I}}_B \cdot \underline{r}_{kA})] \quad (4)$$

$$\hat{H}_1^{(b)} = \frac{\mu_0 \mu_B \hbar}{2\pi i} \sum_{Ak}\sum \gamma_A r_{kA}^{-3} \hat{I}_A \cdot (\underline{r}_{kA} \times \hat{\nabla}_k) \tag{5}$$

magneton, $\hat{I}$ is the nuclear spin operator, r_{kA} refers to the separation between electron k and nucleus A, and ∇_k is del, the vector operator for the electrons.

The contribution $\hat{H}_2$ accounts for the dipole-dipole interaction between the nuclear and electron spins. This term applies only to extra nuclear electrons and is referred to as the dipolar term, eq. 6, in which $\hat{S}$ is the electron spin operator.

$$\hat{H}_2 = \frac{\mu_0 \mu_B \hbar}{2\pi} \sum_{Ak}\sum \gamma_A \left[\frac{3(\hat{S}_k \cdot \underline{r}_{kA})(\hat{I}_A \cdot \underline{r}_{kA})}{r_{kA}^5} - \frac{\hat{S}_k \cdot \hat{I}_A}{r_{kA}^3} \right] \tag{6}$$

For electrons present at the nuclear site the interaction between the nuclear and electron spins is given by $\hat{H}_3$. This operator encompasses the Dirac delta function, $\delta(\underline{r}_{kA})$, which picks out the value of the electronic function at the nucleus in an integration over the coordinates of electron k. Since a contribution from $\hat{H}_3$ implies that the nucleus and electron are in direct contact with each other, $\hat{H}_3$ is referred to as the contact term, eq. 7.

$$\hat{H}_3 = \frac{4}{3} \mu_0 \mu_B \hbar \sum_{Ak}\sum \gamma_A \delta(\underline{r}_{kA}) \hat{\underline{S}}_k \cdot \hat{\underline{I}}_A \tag{7}$$

The total nuclear spin-spin coupling energy, E_{AB}, may be obtained from second-order perturbation theory (3) by eq. 8, in

$$E_{AB} = \sum_B \frac{\langle o | \hat{H} | n \rangle \langle n | \hat{H} | o \rangle}{E_o - E_n} \tag{8}$$

which $\langle o |$ and $\langle n |$ refer to the ground and excited states with energies E_o and E_n respectively. E_{AB} is related to J_{AB} and K_{AB} by means of eqs. 1 and 2.

As in the case of calculations of σ, one difficulty associated with the use of eq. 8 is the necessity to consider infinite summations over excited electronic states including those of the continuum. In SOS calculations this problem may be dealt with by using a truncated number of excited states or by means of an Average Excitation Energy approach as noted for calculations of σ.

The alternative procedure of FPT calculations avoids this difficulty by not requiring the explicit calculation of excited state functions (4,5).

SOS CALCULATIONS

By means of an independent electron model and a minimum basis set of valence shell atomic orbitals, Pople and Santry (6) developed the MO expressions appropriate to SOS calculations of spin-spin coupling. A truncated set of excited states is produced by promoting a single electron at a time from occupied to unoccupied MO's. In keeping with the nature of the independent electron model, only one-centre integrals are included in the calculations.

On this basis $K_{AB}^{1(a)}$ depends only upon the electronic ground state of the molecule and it becomes negligible in magnitude due to the presence of the factor, r_B^{-3}, eq. 9, in

$$K_{AB}^{1}(a) = \frac{\mu_0 e^2}{24\pi^2 m} \sum_{j}^{occ} \sum_{\lambda\mu} C_{j\lambda} C_{j\mu} \langle \lambda \mid \underline{r}_A \cdot \underline{r}_B (r_A^{-3} r_B^{-3}) \mid \nu \rangle \rangle \qquad (9)$$

which λ and μ are p orbitals on atom A at a mean distance of r_A with LCAO coefficients $C_{j\lambda}$ and $C_{j\mu}$ respectively in the j occupied MO's. The corresponding expression for $K_{AB}^{1(b)}$ is given by eq. (10), in which ν and σ refer to p orbitals on atom B and

$$K_{AB}^{1}(b) = -\frac{\mu_0 \mu_B^2}{3\pi^2} \sum_{j}^{occ} \sum_{k}^{unocc} (^1\Delta E_{j \to k})^{-1} \sum_{\lambda\mu\nu\sigma}$$

$$C_{j\lambda} C_{k\mu} C_{k\nu} C_{j\sigma} \langle \lambda \mid r_A^{-3} \hat{\underline{M}}_A \mid \mu \rangle \langle \nu \mid r_B^{-3} \hat{\underline{M}}_B \mid \sigma \rangle \qquad (10)$$

and $\underline{M}_A h$ is the orbital angular momentum operator describing the motions of the p electrons about A, $^1\Delta E_{j \to k}$ refers to the singlet excitation energy between the filled and empty MO's, j and k respectively.

The dipolar contribution is given by eq. 11, in which α and β refer to components along the x, y, or z directions.

$$(K_{AB}^2)_{\alpha B} = - \frac{\mu_0^2 \mu_B^2}{12\pi^2} \sum_{j}^{occ} \sum_{k}^{unocc} (^3\Delta E_{j\to k})^{-1} \sum_{\lambda\mu\nu\sigma} c_{j\lambda} c_{k\mu} c_{k\nu} c_{j\sigma}$$

$$\times \langle \lambda | \underline{r}_{A\alpha}\underline{r}_{A\beta} - \underline{r}_A^2 \delta_{\alpha\beta}) r_A^{-5} | \mu \rangle \langle \nu | 3 r_{B\alpha} r_{B\beta} - \underline{r}_B^2 \delta_{\alpha\beta}) r_B^{-5} | \sigma \rangle \tag{11}$$

The expression for the contact term is given by eq. 12, in which S_A and S_B refer to S orbitals on atoms A and B, respectively.

$$K_{AB}^3 = - \frac{16\mu_0^2 \mu_B^2}{9} \langle S_A | \delta(\underline{r}_A) | S_A \rangle \langle S_B | \delta(\underline{r}_B) | S_B \rangle$$

$$\times \sum_{j}^{occ} \sum_{k}^{unocc} (^3\Delta E_{j\to k})^{-1} c_{jS_A} c_{kS_A} c_{kS_B} c_{jS_B} \tag{12}$$

Eqs. 9, 10, and 11 show that the orbital and dipolar contributions to spin-spin coupling depend upon the coupled atoms having valence p electrons. Thus these contributions are not expected to be significant, at this level of approximation, when the couplings involve hydrogen nuclei.

Since both orbital and dipolar terms depend upon the inverse cube of the radius of the p orbitals on atoms A and B, it is expected that these terms will be of most importance when multiple bonding occurs. Their contributions may also be expected to vary periodically as implied by the data presented in Table 1. Thus heavy nuclei can be expected to have very large couplings with significant contributions from orbital, dipolar, and contact interactions. The latter interactions depend upon $S^2(o)$, as shown by eq. 13, in which $S_A^2(o)$ represents the s electron den-

$$K_{AB}^3 = - \frac{4}{9} \mu_0^2 \mu_B^2 S_A^2(o) S_B^2(o) \Pi_{AB} \tag{13}$$

sity at nucleus A. If the value of $^3\Delta E_{j\to k}$ is taken as the difference in the one-electron energies $(\varepsilon_k - \varepsilon_j)$ then the mutual polarisability, Π_{AB}, is given by eq. (14). If the contact

$$\Pi_{AB} = 4 \sum_{j}^{occ} \sum_{k}^{unocc} (\varepsilon_k - \varepsilon_j)^{-1} c_{jS_A} c_{kS_A} c_{jS_B} c_{kS_B} \tag{14}$$

interaction makes the dominant contribution to the spin-spin

Table 1. Some Values of the One-Center Integrals $S^2(o)$ and $\langle r^{-3}\rangle_p$ in (atomic units)$^{-3}$

Nucleus	$S^2(o)$	$\langle r^{-3}\rangle_p$
^{11}B	1.408	0.775
^{13}C	2.707	1.692
^{15}N	4.770	3.101
^{17}O	7.638	4.974
^{19}F	11.966	7.546
^{29}Si	3.807	2.041
^{31}P	5.625	3.319
^{35}Cl	10.643	6.710
^{129}Xe	26.710	17.825

coupling then eq. (13) demonstrates the existence of a relationship between the s character of the hybridised orbitals on the atoms concerned and their nuclear spin-spin coupling.

FPT CALCULATIONS

FPT calculations of spin-spin couplings were introduced by Pople et al. (4,5). This procedure together with INDO wavefunctions constitutes the most widely used approach to calculations of couplings. An analogous procedure, known as the self consistent perturbation theory (SCPT), has some computational advantages over the original FPT approach and is also commonly encountered (7,8). Details of the quantum mechanical and computational aspects of these calculations have been reviewed elsewhere (9,10).

Within the SCPT procedure the orbital term is presented by eq. 15, in which the P's refer to the first-order elements of the

$$K_{AB}^1 = \frac{\mu_0^2 \mu_B^2}{6\pi^2} \langle r_A^{-3}\rangle_p \langle r_B^{-3}\rangle_p [P_{x_B y_B} + P_{z_B y_B} + P_{x_B z_B}] \quad (15)$$

charge density, bond-order matrix. The dipolar expression for the z component is given by eq. 16, in which the superscripts α and β refer to the spin orbitals with α and β spin respectively

$$(K_{AB}^2)_{zz} = \frac{\mu_0^2 \hbar^2 \mu_B^2}{20\pi^2} \langle r_A^{-3}\rangle_p \langle r_B^{-3}\rangle_p$$

$$[2P_{z_B z_B}^{\alpha\alpha} - P_{x_B x_B}^{\alpha\alpha} - P_{y_B y_B}^{\alpha\alpha} + 3P_{x_B z_B}^{\alpha\beta} + 3Q_{y_B z_B}^{\alpha\beta}] \tag{16}$$

and $Q^{\alpha\beta}$ is the imaginary part of the first-order matrix.

The contact term is given by eq. 17. The integrals $S^2(o)$

$$K_{AB}^3 = \frac{8}{9} \mu_0^2 \mu_B^2 S_A^2(o) S_B^2(o) P_{s_A s_B}^{\alpha} \tag{17}$$

in eqs. 13 and 17 and $\langle r^{-3}\rangle_p$ in eqs. 10, 11, 15, and 16 are often treated as parameters which are evaluated from the best least-squares agreement between the calculated and experimental couplings.

THE RESULTS OF SOME CALCULATED COUPLINGS

The results of some SOS calculations are given in Table 2 (11). The data are derived from INDO-parameterised calculations and show the importance of the noncontact terms for the couplings considered. In all cases the noncontact contributions are not negligible and for couplings involving fluorine they frequently dominate the contact term. Usually, in saturated molecules $^{13}C-^{13}C$ couplings are determined almost entirely by the contact interaction whereas the noncontact terms become important when multiple bonding occurs. In the case of bicyclobutane the C_1-C_3 coupling is dominated by the noncontact interactions to give a negative value for the coupling as confirmed experimentally (12).

The three coupling mechanisms play similar variable roles in the case of $^{15}N-^{13}C$ couplings. However, in this case the lone-pair electrons can make a significant contribution to the contact term (13,14,15). An example of the importance of the lone-pair electrons is afforded by a comparison of the $^1J(^{15}N-^{13}C)$ results for pyridine and the pyridinium ion. For the former the contact contribution is calculated by the SOS-INDO procedure to be 0.04 Hz, whereas it is predicted to be −14.49 Hz for the latter (14). This result is in reasonable agreement with the experimental values of 0.45 Hz and −11.9 Hz for these two couplings. The difference between the two couplings arises from the presence of a low energy $n \to \pi^*$ transition in pyridine, which gives a large positive contribution to the contact term, and thus largely can-

Table 2. SOS-INDO Calculations of Various Contributions, in Hz, to Some Spin-Spin Couplings

Coupling	Molecule	Contact	Orbital	Dipolar	Total
$^1J(^{13}C-^{13}C)$	C_2H_2	69.43	4.13	2.84	76.40
	C_2H_4	29.75	-3.76	0.93	26.92
	C_2H_6	11.99	-0.55	0.30	11.74
$^1J(^{19}F-^{13}C)$	CH_3F	-126.30	-4.15	6.92	-123.50
	CF_4	-85.33	-20.17	2.40	-103.1
$^1J(^{15}N-^{13}C)$	H_2CN_2	8.21	-1.36	0.24	7.09
$^1J(^{15}N-^{15}N)$	H_2CN_2	-6.12	-0.42	0.30	-6.24
$^1J(^{19}F-^{15}N)$	NF_3	-80.68	-3.77	5.96	-78.49
$^2J(^{19}F-^{19}F)$	CH_2F_2	2.90	43.72	23.72	70.75
	CHF_3	-1.27	22.62	19.76	41.14
$^1J(^{19}F-^{19}F)$	F_2	-470.9	1631.8	903.5	2064.4

cels the effects of other, negative contributions. In the pyridinium ion, the absence of the lone pair removes the n → π* transition and a large negative contact contribution remains. In general, singly bonded $^{15}N-^{13}C$ couplings are predicted to be negative in sign, in agreement with experiment. A small, positive coupling can occur, due to the contact contribution, if a lone pair is present in an orbital with s character on the nitrogen atom. Table 3 shows the importance of the noncontact contributions to multiply bonded $^{15}N-^{13}C$ couplings (15). In all cases the $^1J(^{15}N\equiv^{13}C)$ couplings are predicted to be negative due to the dominance of the noncontact terms.

In the case of some $^1J(N-N)$ data negative values are normally predicted and their magnitudes are determined by the contact interaction (16). The presence of lone-pair electrons in an orbital with s character means a negative contact contribution, whereas a lone pair in an atomic p orbital can produce a positive contact interaction.

Due to the often significant contributions from the noncontact terms and the variability of the contribution of the nonbonding electrons to the contact term, a simple empirical relationship, such as eq. 18, between coupling and s character

$$80 \left| ^1J(^{15}N-^{13}C) \right| = \%s_N\%s_C \qquad (18)$$

of the intervening bond is not expected to be generally applicable to couplings including nitrogen and other atoms bearing lone pairs.

Table 3. Some SCPT-INDO Calculations of $^{15}N-^{13}C$ Couplings in Hz

Molecule	Coupling	J_C	J_O	J_D	J_{Total}	J_{Exp}
CH_3CN	$^1J(N\equiv C)$	2.331	-9.291	-14.952	-21.912	-17.5
	$^2J(N-C)$	3.211	0.252	0.088	3.551	+3.0
CH_3CH_2CN	$^1J(N\equiv C)$	2.586	-9.151	-14.862	-21.427	(16.4)
CH_3NC	$^1J(N-C)$	-13.201	0.297	-0.150	-13.054	-10.6
	$^1J(N\equiv C)$	14.852	-7.895	-12.439	-5.481	-8.8
CH_3CH_2NC	$^1J(N\equiv C)$	14.665	-8.326	-12.617	-6.278	(7.4)
$CH_2=C=CHNC$	$^1J(N-C)$	-18.946	0.322	-0.162	-18.786	(20.0)
	$^1J(N\equiv C)$	14.801	-7.694	-12.472	-5.366	(6.6)
Ph-NC	$^1J(N-C)$	-14.707	0.329	-0.139	-14.517	(18.5)
	$^1J(N\equiv C)$	13.971	-6.553	-12.216	-4.798	(7.3)
cyclohexyl-NC	$^1J(N-C)$	-10.849	0.278	-0.129	-10.700	(8.1)
	$^1J(N\equiv C)$	14.258	-7.225	-12.612	-5.578	(6.3)
o-di(Me)-phenyl-CNO	$^1J(N\equiv C)$	-29.813	-22.837	-19.533	-72.184	(77.5)

Another simple empirical relationship which is often invoked in discussions of spin-spin couplings is the Karplus equation (17). For saturated X-C-C-Y fragments, the value of $^3J(X-Y)$ may be related to the dihedral angle, ϕ, by an expression of the type given by eq. 19, in which A, B, and C are empirical constants,

$$^3J(X-Y) = A\cos 2\phi + B\cos\phi + C \qquad (19)$$

generally B is negative, and $|A| > |B|$. Thus values of $^3J(X-Y)$ are at a maximum for $\phi = 180°$ and a minimum when ϕ is close to 90°. In general, relationships similar to eq. 19 are reasonably satisfactory for the dihedral angle dependence of $^3J(H-H)$ values. However, some SOS-INDO calculations on fluoroethanes have shown a much more complicated $^3J(F-F)$ dependence on the F-C-C-F dihedral angle (18) as shown in Figures 1 and 2. Recent calculations of $^1J(P-N)$, $^1J(P-C)$, $^1J(P-F)$, $^1J(Si-C)$, and $^1J(Si-F)$ (22-24) have revealed the importance of including d orbitals for P and Si in a description of these couplings. The role of the d orbitals is especially important for $^1J(P-F)$ and $^1J(Si-F)$, because of $p_\pi-d_\pi$

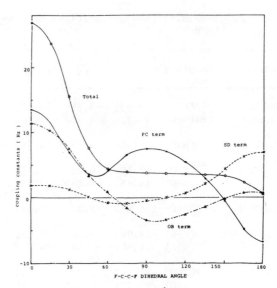

Figure 1. Plots of calculated J_{FF}^{vic} vs. the FCCF dihedral angle, θ, for $F_2CF^*-CH_2F^*$. Reprinted with permission from Ref. 18. Copyright 1973, American Chemical Society.

back-bonding.

The importance of the noncontact interactions is shown for all of these couplings involving second row nuclei. In the case of the fluorine couplings, the dominant contribution is found to arise from the orbital term (4).

In conclusion, it appears that semi-empirical MO calculations can provide reasonably satisfactory accounts of spin-spin couplings in a number of cases, ^2J data being a notorious exception (1,9). In those cases where satisfactory results are obtained they imply the need for caution in the use of empirical relationships between couplings and structural features of the molecules concerned.

MEDIUM EFFECTS ON SPIN-SPIN COUPLING

As in the case of nuclear shielding, the solvaton model has been applied to calculations of spin-spin couplings (19). The reaction field and other similar models have been used, together with INDO-FPT calculations of the contact contribution, to a number of couplings with some success (20).

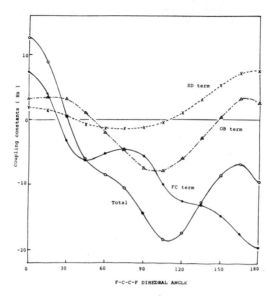

Figure 2. Plots of calculated J_{FF}^{vic} vs. the FCCF dihedral angle, θ, for F_2CF^*-F^*C=CF_2. Reprinted with permission from Ref. 18. Copyright 1973, American Chemical Society.

In general, spin-spin couplings appear to be less sensitive than chemical shifts to changes of solvent. This is borne out by some FPT-INDO calculations of $^1J(N{\equiv}C)$ using the solvaton model (21). The results presented in Table 4 show that $^1J(N{\equiv}C)$ is predicted to become more negative as the dielectric constant (ε) of the medium increases. As ε varies from 1 to 80, $^1J(N{\equiv}C)$ may change by up to 2 Hz. The isocyanide couplings are found to be more sensitive than those of the cyanides to a change in ε. This is consistent with the usually accepted, more polar nature of isocyanides.

Although the calculated variation in $^1J(N{\equiv}C)$, as a function of ε, is not large, it is comparable to the $^1J(N-C)$ data for aniline and formamide which is attributed to specific hydrogen bonding effects. Thus, the nonspecific influences, included in the solvaton model, could be as important as hydrogen bonding, and perhaps conformational effects, on $^1J(N{\equiv}C)$.

Table 4. Results of Some FPT-INDO Calculations, Using the Solvaton Model, of the Effect of the Dielectric (ε) of the Medium on Some Values of $^1J(N\equiv C)$ in Hz

Molecule	ε	J_C	J_O	J_D	J_{Tot}
CH_3NC	1	14.817	−8.429	−12.583	−6.195
	4	14.074	−8.371	−12.628	−6.925
	8	13.935	−8.364	−12.639	−7.068
	20	13.850	−8.359	−12.647	−7.156
	40	13.822	−8.359	−12.649	−7.186
	80	13.807	−8.358	−12.650	−7.201
Ph−NC	1	13.717	−5.632	−11.329	−3.244
	4	12.937	−5.633	−11.383	−4.079
	8	12.789	−5.634	−11.396	−4.241
	20	12.699	−5.635	−11.403	−4.339
	40	12.669	−5.636	−11.406	−4.373
	80	12.654	−5.636	−11.408	−4.390
o−di(Me)−phenyl−NC	1	13.817	−6.517	−12.280	−4.980
	4	13.018	−6.561	−12.372	−5.915
	8	12.878	−6.567	−12.387	−6.076
	20	12.793	−6.572	−12.397	−6.176
	40	12.765	−6.573	−12.401	−6.209
	80	12.750	−6.574	−12.402	−6.226
CH_3CN	1	2.316	−9.334	−15.021	−22.039
	4	1.772	−9.105	−14.784	−22.117
	8	1.678	−9.063	−14.741	−22.126
	20	1.619	−9.037	−14.714	−22.132
	40	1.600	−9.028	−14.705	−22.133
	80	1.590	−9.024	−14.701	−22.135
$CH_2=CHCN$	1	2.079	−9.187	−15.184	−22.292
	4	1.568	−8.978	−14.938	−22.348
	8	1.478	−8.939	−14.893	−22.354
	20	1.423	−8.915	−14.865	−22.357
	40	1.405	−8.907	−14.856	−22.358
	80	1.396	−8.903	−14.852	−22.359

REFERENCES

1. G. A. Webb in "NMR and the Periodic Table," R. K. Harris and B. E. Mann, Eds., Academic, London, p. 49, 1978.
2. J. Kowalewski, Ann. Rep. NMR, 12, p. 81 (1982).
3. N. F. Ramsey, Phys. Rev., 91, p. 303 (1953).
4. J. A. Pople, J. W. McIver, and N. S. Ostlund, J. Chem. Phys., 49, p. 2960 (1968).
5. J. A. Pople, J. W. McIver, and N. S. Ostlund, J. Chem. Phys., 49, p. 2965 (1968).
6. J. A. Pople and D. P. Santry, Mol. Phys., 8, p. 1 (1964).
7. A. C. Blizzard and D. P. Santry, J. C. S. Chem. Commun., p. 87 (1970).
8. A. C. Blizzard and D. P. Santry, J. Chem. Phys., 55, p. 950 (1971).
9. J. Kowalewski, Progr. NMR Spectrosc., 11, p. 1 (1977).
10. P. D. Ellis and R. Ditchfield, Top. ^{13}C NMR Spectrosc., 2, p. 433 (1976).
11. A. D. C. Towl and K. Schaumberg, Mol. Phys., 22, p. 49 (1971).
12. J. M. Schulman and M. D. Newton, J. Am. Chem. Soc., 96, p. 6295 (1974).
13. J. M. Schulman and T. Venanzi, J. Am. Chem. Soc., 98, p. 4701 (1976).
14. T. Khin and G. A. Webb, Org. Magn. Reson., 10, p. 175 (1977).
15. T. Khin and G. A. Webb, Org. Magn. Reson., 11, p. 487 (1978).
16. T. Khin and G. A. Webb, J. Magn. Reson., 33, p. 159 (1979).
17. M. Barfield and M. Karplus, J. Am. Chem. Soc., 91, p. 1 (1969).
18. K. Hirao, H. Nakatsuji, and H. Kato, J. Am. Chem. Soc., 95, p. 31 (1973).
19. I. Ando and G. A. Webb, Org. Magn. Reson., 15, p. 111 (1981).
20. M. Barfield and M. D. Johnston, Chem. Rev., 73, p. 53 (1973).
21. S. N. Shargi and G. A. Webb, Org. Magn. Reson., 19, p. 126 (1982).
22. S. Duangthai and G. A. Webb, Org. Magn. Reson., 20, p. 33, (1982).
23. S. Duangthai and G. A. Webb, Org. Magn. Reson., in press.
24. S. Duangthai and G. A. Webb, Org. Magn. Reson., in press.

CHAPTER 4

RELAXATION PROCESSES IN NUCLEAR MAGNETIC RESONANCE

J. REISSE

Université Libre de Bruxelles, Belgium

> Indeed if it were not for the prevalence of relaxation, physicists might have abandoned the field of magnetic resonance to chemists long ago.
>
> C.P. Poole and H.O. Farrach (1971)

ABSTRACT

This chapter is devoted to a general discussion about relaxation in nuclear magnetic resonance with a special emphasis on the physical signification of concepts, equations, and terms usually used in this field. Systems which are moved away from equilibrium after a change in one of the variables of state will reach the equilibrium by relaxation processes. Therefore, relaxation is of great importance in chemistry and physics. It plays a special role in NMR because this spectroscopy is characterized by a very low-frequency domain. In NMR, relaxation is generally described by the Bloch equations, which lead to a clear distinction between transverse and longitudinal relaxation. Though they work well for a semiquantitative description of relaxation processes, the Bloch equations are far from general: nonexponential decays are frequently observed for relaxing spin systems. Theories of relaxation require the definition of correlation $G(\tau)$ and spectral density $J(\omega)$ functions. They also require the calculation of transition probabilities per second for jumps between the Zeeman levels. The nature of the coupling Hamiltonian determines the relaxation mechanism, while the analytical form of $J(\omega)$ determines the frequency dependence of transition probabilities. The efficiency of the various relaxation mechanisms depends on the nature of the observed nuclei and also on the value of the applied field B_0. The case of nuclei with $I > 1/2$ is particularly interesting because the quadrupolar mechanism is the only one which is due to an electrical coupling between the lattice and spin system.

INTRODUCTION

For many years, relaxation studies in nuclear magnetic resonance were mostly performed by physicists. Even now a survey of the NMR literature shows that among chemists relaxation studies remain less popular than chemical shift or coupling constant studies. We shall not argue about the relative interest of relaxation studies compared with other NMR studies, but we can certainly claim that relaxation time measurements lead to information of great importance for chemists as well as for physicists (and even for biologists and more recently for physicians).

From a theoretical point of view, relaxation studies require methods which are fundamentally different from those used to interpret other NMR parameters like σ or J. The interpretation of σ or J values implies the use of theoretical models essentially based on the "isolated molecule concept." The σ and J are molecular constants, while relaxation times are parameters related to properties of a large ensemble of nuclei in interaction with their surroundings. Relaxation parameters like T_1 or T_2 have no direct meaning at the molecular level: they characterize the behaviour of macroscopic properties. Of course, σ and J also are measured on statistical ensembles of molecules and not on a single molecule. A correct theory of σ or J must certainly take into account the distribution of molecules among the vibrational levels and the influence of intermolecular interactions, but these contributions may be considered as small perturbations.

The interpretation of T_1 and T_2 requires the use of theoretical approaches like thermodynamics of irreversible processes, quantum mechanical description of large ensembles, or semiclassical description of the temporal evolution of expectation values. All these approaches are never used in the interpretation of σ or J.

Many books, review articles, and papers devoted to relaxation in NMR have been published since the "historical paper" of Bloembergen, Purcell, and Pound (1). It is pertinent to question the necessity of writing another review article on this subject, especially if we take into account the fact that many contributors to this volume will illustrate some applications of relaxation time measurements in their own field of interest.

This review article will be more oriented towards the physical meaning and the role of relaxation in NMR than on the extensive derivations of formulae which are given with details in basic treatises or review articles like those cited in the reference list (1-10). Due to the general scope of this volume, some emphasis will be put on the relaxation phenomena for nuclei characterized by $I > 1/2$.

ABOUT THE NATURE, THE IMPORTANCE, AND THE GENERALITY OF RELAXATION PROCESSES

Following Leffler and Grunwald (11), relaxation can be defined as "the return to equilibrium of a system that has been slightly perturbed by the imposition of a change in one of the variables of state." Relaxation corresponds to an irreversible process which is usually studied by following the temporal evolution of the system after a sudden disturbance. The disturbance consists of the change of one physical variable of the system (i.e., temperature, pressure, external electric or magnetic field, etc.)

The sudden mixing of two chemical species (A and B) initially separated by a mechanical barrier is a very simple example of a change of one of the constraints acting on the system. If A is able to react with B, the system will evolve and the study of this temporal evolution is called chemical kinetics. If A is only able to mix with B, the system will also evolve in order to reach a lower free energy state. This new state is generally characterized by a different temperature from the initial state (the mixing, in an adiabatic calorimeter, being endothermic or exothermic, rarely athermic). The temporal evolution of temperature T will inform us about the temporal evolution of the system towards its equilibrium state. The (T, t) curve can be described as a relaxation function of the system (associated with energy relaxation).

Relaxation studies could be named physical kinetics but it is clear that the distinction between chemical and physical kinetics is arbitrary. Chemical kinetics and physical kinetics (relaxation studies) are strongly interconnected domains. <u>Without relaxation, no system would be at equilibrium in a nonstationary world like ours.</u> This remark leads to the conclusion that relaxation phenomena are of fundamental importance and of great generality.

RELAXATION IN NUCLEAR MAGNETIC RESONANCE: A QUALITATIVE DESCRIPTION

An NMR experiment always starts by the immersion of the sample into the static magnetic field B_0. Under the influence of B_0, the energy degeneracy of the Zeeman nuclear levels is suddenly suppressed. Just after the immersion of the spin system in B_0 (at time $t = 0$), all the Zeeman levels are equally populated and this situation corresponds to an out-of-equilibrium state. The spin system is abnormally "hot." To reach the equilibrium state which corresponds to the well known Boltzmann distribution of the nuclei among the Zeeman levels, the spin system has to cool down and give up its excess energy to its surroundings. The surroundings are described as the lattice. This exchange of energy requires an interaction (also called coupling) between the two systems, i.e., the spin system and the lattice. Following Van Vleck, for relaxation to occur "the spins must be on speaking terms with the lattice" (4).

The cooling down of the spin system implies that the lattice is able to warm up. Obviously, the total energy of the larger system (the spin system + lattice) remains constant during the experiment. The energy transfer corresponds to the transformation of potential energy into kinetic energy and the lattice becomes warmer in the true sense of the word. The temperature increase is deceptively small but, in principle at least, this relaxation process could be studied by plotting the temperature of the lattice as a function of time, just as in the mixing process described above. The measurement of T as a function of t is generally replaced by the measurement of the z component of the macroscopic magnetization of the spin system as a function of t. This procedure gives quantitative information about the relaxation process via a measurement not on the lattice but on the spin system. Due to the relative "size" of the two systems, the "smaller" (spin) is much more sensitive to the energy exchange than the "larger" (lattice).

What we have described here is a relaxation process which corresponds to an energy relaxation. In NMR this energy relaxation is described as longitudinal (because the z direction, identical to the B_0 direction, is known as the longitudinal direction). The term spin-lattice relaxation is also frequently used.

Without a careful examination of the system under study it is impossible to predict anything about the shape of the curve $M_z = f(t)$. A priori, the only certainties we have are the following: when the spin system is submitted to B_0, at $t = 0$, M_z is equal to 0 (because all the Zeeman levels are equally populated) but after an infinite time, M_z must reach its equilibrium value M_0. These limiting conditions are compatible with an infinity of $M_z = f(t)$ curves. The choice of a particular analytical function to describe this curve is obviously not arbitrary. This problem will be discussed later.

In NMR spectroscopy, it is necessary to define two different relaxation functions depending on the component of the magnetization vector in which we are interested.

If an ensemble of spins, at equilibrium in the presence of a static B_0 field, is submitted to a B_1 rotating field orthogonal to B_0 and characterized by a well defined rotational frequency (the Larmor frequency, ω_0, rad sec^{-1}), the macroscopic magnetization vector with initial components $(0, 0, M_z = M_0)$ is tilted away from the z direction. The new components of the magnetization vector are $(M_x, M_y, M_z < M_0)$. This new macroscopic state is characterized by higher energy $(M_z < M_0)$ and by lower entropy of the spin system $(M_x \neq 0, M_y \neq 0)$ than the initial state. Qualitatively, and using a semiclassical description of the NMR experiment, it is easy to realize that the new state is effectively characterized by a lower entropy. The nonzero values of the transverse components of the magnetization vector require some phase coherence between the individual magnetic moments in precessional motion around the z direction. The state characterized by $M_x \neq 0$ (and therefore $M_y \neq 0$) has a higher order than the $M_x = M_y = 0$ state.

The new state of the spin system with its higher energy and lower entropy can only be reached if work is done on the system via the rotating B_1 field. The $(M_x, M_y, M_z < M_0)$ state is stationary (in the rotating frame), i.e., $M_x \neq f(t)$, $M_y \neq f(t)$, and $M_z \neq f(t)$, if the larger system (spin + lattice) remains nonisolated (with a continuous flow of energy via the rotating field) and if a coupling exists between the spin and the lattice systems. This coupling provides the only opportunity for the spin system to evacuate a part of the energy received towards the lattice and therefore to prevent the saturation situation.

If we take the $(M_x, M_y, M_z < M_0)$ stationary state in the presence of B_0 and B_1 into consideration, we can now examine the evolution of the system after the sudden cut off of the B_1 rotating field. The system evolves towards the equilibrium situation characterized by $M = (0, 0, M_0)$. The relaxation of the system can be followed by measuring not only the function $M_z = f(t)$ but also $M_x = f(t)$ or $M_y = f(t)$.

For reasons of symmetry, the temporal evolution of M_x and M_y must be identical in an isotropic gaseous or liquid sample which is submitted to B_0 along the z direction. Strong B_1 fields are able to introduce an anisotropy in the x, y plane but this problem will not be considered here. It is discussed in reference 4 (p. 109).

As we have seen above, the temporal evolution of M_z is due to the longitudinal relaxation. The temporal evolution of M_x (or M_y), which is described as <u>transverse relaxation</u>, is fundamentally different. It corresponds to the loss of phase coherence between the individual magnetic moments. Transverse relaxation can be described as an entropy relaxation in the sense that the final state $(0, 0, M_0)$ is characterized by a higher entropy than the initial state (M_x, M_y, M_z). As in the case of the temporal evolution of M_z, the analytical form of the $M_y = f(t)$ function is not predictable without careful analysis. This problem will also be discussed later.

We would like to emphasize one important difference between the two relaxation processes. A priori, it is conceivable that transverse relaxation is achieved at, let us say, 99% while longitudinal relaxation is only achieved at 75%, but the inverse cannot exist. <u>Absence of phase coherence does not require thermal equilibrium, while thermal equilibrium requires absence of phase coherence</u>.

For some solid state samples, transverse relaxation may be very efficient, while longitudinal relaxation does not take place at a measurable rate. This behaviour is easily understandable if we take into account that direct interactions between the individual moments may lead to the loss of phase coherence (initially induced by B_1) without any change in the energy of the spin system. Transverse relaxation does not necessarily require the presence of the lattice and the expression <u>spin-spin relaxation</u> frequently used in place of transverse relaxation illustrates this characteristic. Nevertheless, the term spin-spin relaxation can be criticized and the possibility of a transverse decay in the absence of any lattice has been questioned even in the solid phase (3, p. 29, and 4, p. 108). In the gas or the liquid phase, the loss of phase coherence is always mediated by the lattice and therefore the term spin-spin relaxation is confusing.

WHY RELAXATION PLAYS SO IMPORTANT A ROLE IN NMR COMPARED TO OTHER ABSORPTION SPECTROSCOPIES

Figure 1 gives a qualitative description of the various processes which lead to an energy transfer between two states of a quantum mechanical system whatever the system may be. For example, it could correspond to the two Zeeman states of an $I = 1/2$ spin system or to the S_0 and S_1 electronic states of a molecular system.

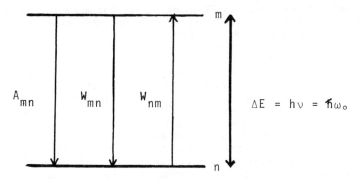

Figure 1. Transition probabilities per second in a two-level system, in the absence of any electromagnetic oscillating field. A_{mn} corresponds to spontaneous emission. W_{mn} and W_{nm} correspond to nonradiative transitions.

The spontaneous emission is characterized by a probability, A_{mn}, which depends not only on the matrix element connecting the two states but also on the third power of the frequency. A_{mn} is important in UV spectroscopy, in which spontaneous emission corresponds to the fluorescence phenomena.

In NMR, A_{mn} is negligible. Following an estimation of Abragam (2), $A_{mn} \simeq 10^{-25}$ sec^{-1} for a proton in a B_0 field of 0.75 T. Therefore, the spontaneous emission cannot play any role in the longitudinal relaxation process. A_{mn} is defined for a system in free space, without any electromagnetic field of thermal origin. Such a situation does not correspond to any real situation and probability of induced absorption and induced emission by <u>incoherent coupling</u> between the quantum mechanical system and the electromagnetic field should also be considered (2). Nevertheless, for a spin system in a B_0 field, this coupling cannot play any role in the longitudinal relaxation mechanism because the distance between the individual magnetic moments is small compared with the wavelength of the electromagnetic field susceptible of inducing the transition (2).

Therefore, A_{mn} being negligible and the process just discussed being inefficient, the only coupling which can lead to a polarization of the spin system placed in a B_0 field (and which can also prevent the saturation during an NMR experiment) is the spin-lattice coupling. From this point of view, the NMR absorption spectroscopy plays a particular role in comparison with all the other absorption spectroscopies and this particularity is a consequence of the very low-frequency domain of the nuclear magnetic resonance. In UV spectroscopy for example, spontaneous emission is the process which prevents saturation even if, in some cases, the so-called nonradiative processes (i.e., relaxation processes) also play a role. Obviously, the "risk" of saturation in high-frequency UV spectroscopy is negligible with respect to the very low-frequency NMR, at least for a two-level system. In NMR spectroscopy, the attainment of the Boltzmann distribution in the B_0 field can be frequently observed for nuclei characterized by a slow longitudinal relaxation process (like ^{129}Xe (44)). Those who are interested in recording the UV, IR, and MO absorption spectra of molecules at normal temperatures are very rarely concerned with the question of knowing if their samples are at equilibrium when they start their experiments. As we said, this question is sometimes relevant in NMR. The reason is double. First, NMR is a Zeeman spectroscopy and the sample is generally introduced into the B_0 field just before the experiment. Secondly, NMR is a magnetic spectroscopy. Within an excellent level of accuracy, UV, IR, and MO spectroscopies can be described as being electrical spectroscopies in the sense that the moments of transition are electrical. Therefore, the coupling between the lattice and electronical, vibrational, and rotational degrees of freedom is also electrical. In NMR, the coupling between the lattice and the nuclear Zeeman degrees of freedom is magnetic in all cases but one, and this magnetic coupling is relatively inefficient. The exception, the only electrical coupling which plays a role in NMR, is the quadrupolar coupling between the nuclear quadrupole (when it exists) and the lattice, via electric field gradients. When it operates, this electrical coupling is generally much more efficient than the magnetic coupling. This problem will be discussed in a later section.

The importance of energy relaxation in NMR has a direct consequence: the lattice plays a fundamental role in this spectroscopy. The lattice acts as an energy donor or acceptor. It is a better acceptor than a donor. This fact sometimes described as a paradox is in no way strange or difficult to understand at least qualitatively (12). It is the direct consequence of the Boltzmann distribution of the various spins among the Zeeman energy levels when the spin system is at equlibrium with the lattice. For a spin system distributed between two energy levels (I = 1/2), the relative population of the lower level a and the higher level b is given by eq. 1, in which ΔE is the energy

$$\frac{N_a}{N_b} = \exp \frac{\Delta E}{kT} > 1 \qquad (1)$$

difference between the two energy levels. At equilibrium, the rate of the a → b process is equal to the rate of the b → a process. Therefore, the ratio of the rate constants (or probabilities per second), W, for the two processes is given by eq. 2.

$$\frac{W_{b \to a}}{W_{a \to b}} = \frac{N_a}{N_b} > 1 \qquad (2)$$

Each b → a transition in the spin system corresponds to the acceptance by the lattice of an amount of energy equal to ΔE, while each a → b transition corresponds to an energy transfer, ΔE, from the lattice to the spin system. Therefore, if the spin system is itself at equilibrium and also at equilibrium with the lattice, the lattice must be a better energy acceptor than a donor.

The quantum mechanical calculation of $W_{b \to a}$ and $W_{a \to b}$ is difficult (2, p. 284). Many approximate theories lead to $W_{a \to b} = W_{b \to a}$ and this fact is at the origin of the so-called "paradox." As we have already said, in the liquid or the gas phase, the lattice also plays a role relative to transverse relaxation but in this case its role is different. It only acts as a "shaker" contributing to the loss of phase coherence between the individual nuclear magnetic moments in precession around B_0. This loss of phase coherence also implies a → b and b → a transitions, at least if B_0 is perfectly homogeneous.

It could be argued that, due to the fact that transitions between the energy levels are necessary for both transverse and longitudinal relaxation, the two processes are much more similar than previously stated. Considering the kind of interactions and coupling between the lattice and the spin system which are responsible for longitudinal and transverse relaxation, it is true that from a mechanistical point of view the two are effectively very similar. However and as we showed, the physical signification of the two relaxation processes is completely different.

The lattice with its dynamic structure also plays a role with respect to the transverse relaxation when B_0 is not perfectly homogeneous (as is always the case with real NMR experiments). Due to magnetic field inhomogeneity, the various individual nuclear magnetic moments are not subjected to the same B_0. Their precessional motions are therefore different. If we call isochromat an ensemble of moments submitted to the same B_0 field and therefore precessing at the same angular frequency, it is obvious that translational motions of magnetic moments from one isochromat to another will contribute to the loss of phase coherence in each isochromat. In this case the lattice once again plays a role because its own motion is the cause of the displacement of the nuclei from one isochromat to another. The diffusion of nuclei in the sample is therefore a factor which influences transverse relaxation (and not longitudinal relaxation). This observation lies behind the well known spin-echo method in the presence of a field gradient for the determination of diffusion coefficients (13).

RELAXATION FUNCTIONS AND RELAXATION TIMES

Longitudinal relaxation and transverse relaxation are generally introduced via the phenomenological Bloch equations. These equations, originally developed to describe the temporal evolution of macroscopic variables are also applicable to expectation values of microscopic variables. Let us call M_x, M_y, and M_z the components of the macroscopic magnetization vector, B_0 the static magnetic field along z, and B_1 the rotating magnetic field in the xy plane, at an angular velocity $-\omega$. Then, the Bloch equations take the very simple form of eq. 3, in which T_1

$$\frac{dM_x}{dt} = \gamma (M_y B_0 + M_x B_1 \sin \omega t) - \frac{M_x}{T_2} \qquad (3a)$$

$$\frac{dM_y}{dt} = \gamma (M_x B_1 \cos \omega t - M_1 B_0) - \frac{M_y}{T_2} \qquad (3b)$$

$$\frac{dM_z}{dt} = \gamma (- M_x B_1 \sin \omega t - M_y B_1 \cos \omega t) - \frac{M_z - M_0}{T_1} \qquad (3c)$$

is the longitudinal relaxation time and T_2 is the transverse relaxation time. The relaxation functions as they appear in the Bloch equations are exponential functions, and therefore knowledge of the initial and final conditions and of one rate constant ($R_1 = 1/T_1$ or $R_2 = 1/T_2$) is sufficient to characterize each relaxation function. As pointed out by Deutch and Oppenheim (8), it is surprising that a set of such simple linear differential equations is sufficient to describe a process which depends on a complex ensemble of couplings at the microscopic level between the spin system and the lattice degrees of freedom. First of all it is important to realize that the Bloch equations are not valid in all circumstances. As shown by Senitzky (14, 15), the Bloch equations are a special case of a general class of equations describing the temporal evolution of a two-level spin system, driven by external fields and coupled to a relaxation mechanism. The Bloch equations in their simple form (eq. 3) are not applicable to describe situations commonly encountered, even in the case of $I = 1/2$ spin systems. As it will be shown later, cross-relaxation and cross-correlation are examples of cases where eq. 3 is not applicable.

It is tempting to inquire the necessary conditions which force a spin system to relax exponentially (we shall consider now the longitudinal relaxation only). The answer to this question is simple. A single exponential relaxation function is observed when a spin temperature can be defined. This problem is discussed in detail in reference 3 (p. 116) and we will only give here a brief summary of the arguments developed by Slichter. Let us consider a spin system characterized by a set of Zeeman energies E_a, E_b, E_c, ... at thermal equilibrium at temperature T. The probability of finding the system in the level E_n (with n = a, b, c ...) is denoted p_n (eq. 4). The ratio of two p_n values,

$$\sum_n p_n = 1 \tag{4}$$

let us say p_a and p_b, is given by eq. 5. From eqs. 4 and 5, we

$$\frac{p_a}{p_b} = \frac{e^{-E_a/kT}}{e^{-E_b/kT}} \tag{5}$$

have eq. 6, in which $Z = \Sigma(n)\, e^{-E_n/kT}$ is the __partition function__

$$p_a = \frac{e^{-E_a/kT}}{\sum_n e^{-E_n/kT}} = \frac{e^{-E_a/kT}}{Z} \tag{6}$$

of the spin system. By definition, the spin systems for which eq. 5 is valid for all pairs of energy levels can be characterized by a spin temperature even when the spin system is not at equilibrium with the lattice. In other words, even if the spin temperature, T_S, and the lattice temperature, T_L, are not equal, the spin systems for which eq. 7 is valid (for all pairs of energy levels) are characterized by a spin temperature T_S.

$$\frac{p_a}{p_b} = \frac{e^{-E_a/kT_S}}{e^{-E_b/kT_S}} \tag{7}$$

If $T_S > T_L$, the spin system is out of equilibrium relative to the lattice, and the longitudinal relaxation will correspond to a temperature decrease of the spin system and a temperature increase of the lattice (as we have already said, the temperature increase of the lattice is negligible and the heat capacity of the lattice is generally considered as being infinite for this reason).

The relaxation of a spin system, for which a temperature can be defined and whose Hamiltonian $\hat{H}$ has eigenvalues E_n, can be described as soon as a "master" equation is introduced. This master equation relates the temporal dependence of the probability p_n on the probability per second that the lattice induces a transition in the spin system from n to m (W_{nm}) or from m to n (W_{mn}). This master equation has the form of a linear rate equation (eq. 8). To compute the relaxation function, the

$$\frac{dp_n}{dt} = \sum_m (p_m W_{mn} - p_n W_{nm}). \tag{8}$$

temporal evolution of the average energy $\langle E \rangle$ of the spin system is estimated, taking into account that $\langle E \rangle = \Sigma_{(n)} \, p_n \, E_n$. Eq. 9, in which $\beta = 1/kT_S$, relates $d\langle E\rangle/dt$ to the temperature variation of the spin system during relaxation.

$$\frac{d\langle E\rangle}{dt} = \frac{d\langle E\rangle}{d\beta} \frac{d\beta}{dt} \tag{9}$$

The calculation procedure is very simple and leads to an equation relating T_1^{-1} to W_{mn}, E_n, and E_m (eq. 10). It is inter-

$$\frac{1}{T_1} = \frac{1}{2} \frac{\sum_{m,n} W_{mn}(E_m - E_n)^2}{\sum_n E_n^2} \tag{10}$$

esting to observe that eq. 8 would imply in itself that the longitudinal relaxation must be a sum of exponentials. Eq. 10 shows that the assumption of a spin temperature, which can be defined for all the relaxation path, leads to a single exponential relaxation function.

Many spin systems are not characterized by one spin temperature and the description of such systems may require the definition of an ensemble of spin temperatures (10). We will not consider this problem here.

As it appears in the case we have just discussed, the fundamental point in relaxation problems is the link between the transition probabilities per second (W_{mn}) and relaxation time or times. This fundamental problem was examined 27 years ago by Lurçat is a very concise and important paper which is not frequently cited (30). The only point we would like to insist on now is the following: <u>a nonexponential relaxation process is far from exceptional</u>. Sometimes, a good choice of experimental conditions can transform a complex relaxation function into an exponential or at least into a function which can be approximated by an exponential. Sometimes, motional characteristics of the lattice lead to a monoexponential behaviour while the same spin system immersed in another lattice relaxes according to a nonexponential law. It appears therefore that "special conditions," prescribed or "natural," may simplify the relaxation law. However, it is not always easy to predetermine what are these special conditions and if it is reasonable or not to postulate a monoexponential behaviour. The experimental test remains very important. It is perhaps useful to point out here that a computer programme written with the object of determining the "best monoexponential function" corresponding to an ensemble of points obviously always generates monoexponential functions even if the point distribution corresponds to a biexponential or to an elephant!

Before discussing some multiexponential relaxation functions, we will consider some specific aspects of theoretical methods describing the relaxation phenomena in nuclear magnetic resonance. Various methods have been proposed in the literature since the work of Bloembergen, Purcell, and Pound (1). Many of them have some common features like the use of density matrix theory, correlation functions, and their associated spectral density functions. People interested in the foundations of these theories will find some excellent book chapters and review articles in the literature (7-9,21). Slichter's book (Chapter 5) contains a very good introduction to the density matrix which is of crucial importance in all these theories (3). On the other hand, Abragam's book (2) remains a "Bible" in this field. The more recent book of Lenk (5) contains very concise definitions of many terms and concepts widely used in all the relaxation theories. Moreover, this author presents an excellent overview of many recent theories, especially those based on irreversible thermodynamics.

As we previously said, many of these methods present similarities, but their fields of application may slightly differ. All of them can be classified in the very broad domain of nonequilbrium statistical mechanics (including classical and quantum statistical mechanics) and irreversible thermodynamics.

Due to the fundamental importance of correlation functions and spectral densities in all these theories of relaxation, the next section will be devoted to this problem.

CORRELATION FUNCTIONS AND SPECTRAL DENSITIES

At the beginning of this section it is important to point out that the theoretical treatment we will briefly discuss now is sometimes called semiclassical just because the correlation functions are classical. A quantum mechanical treatment has been proposed (2, p. 284) which presents many formal similarities to the semiclassical treatment but which fundamentally differs from it by the way the correlation functions are defined (in terms of time-dependent operators and not of time functions). This distinction is rarely done in review articles about relaxation where the semiclassical approach is generally presented as the only way to handle the problem. To introduce the correlation functions and spectral densities, we will consider a spin system characterized by eigenstates a, b, c .... and corresponding energies E_a, E_b, E_c. A perturbation, time-dependent Hamiltonian $\hat{H}(t)$ acts on this system. $\hat{H}(t)$ corresponds to the coupling of the spin system with the lattice. We shall make the assumption that $\hat{H}(t)$ can be written as a product $A \cdot f(t)$.

Time-dependent perturbation theory leads to eq. 11, which

$$W_{ab} = \frac{1}{\hbar^2} \left| (a|A|b) \right|^2 J(\omega_0) \tag{11}$$

relates the average transition probability per second, W_{ab}, to the off-diagonal matrix element $(a|A|b)$ and $J(\omega_0)$, the spectral density value for $\omega = \omega_0 = |E_a - E_b|/\hbar$. The spectral density, $J(\omega)$, is defined by eq. 12. It is seen that $J(\omega)$ is the Fourier

$$J(\omega) = \int_{-\infty}^{+\infty} G(\tau) e^{-i\omega\tau} d\tau \tag{12}$$

transform of the function $G(\tau)$, the auto-correlation function. The auto-correlation function itself is defined by eq. 13, in

$$G(\tau) = \langle f(t) \cdot f(t+\tau) \rangle_t \tag{13}$$

which the brackets imply an ensemble average taken over time t. The f(t) function is associated with a random process, which is considered as stationary, gaussian, and ergodic. It is very important to realize what these terms mean. The definitions are clearly given in the chapter of the Lenk's book (5) devoted to probabilistic and statistical considerations. Very briefly we can say that a random process is stationary if it is independent of the origin of time. A random process is gaussian if the random variables associated with it can be described according to gaussian probability laws. A random process is ergodic if the time average of a random variable is equal to the ensemble average. For a stationary, random process, eq. 13 can be written in a simpler form (eq. 14).

$$G(\tau) = \langle f(0) \cdot f(\tau) \rangle_0 \tag{14}$$

In his review article, Gordon (9) gave the following qualitative description of a correlation function: "a correlation function describes how long some given property of a system persists until it is averaged out by the microscopic motion of the molecules in the system." The auto-correlation function is sometimes described as a memory function because it can be considered as a measure of how the system loses the memory of its initial state. The expression $\langle f(0) \cdot f(0) \rangle_0$ gives the maximum value of $G(\tau)$ (taken as unity in the case of a normalized $G(\tau)$ function), while $\langle f(0) \cdot f(\infty) \rangle_0$ is equal to 0.

The function $G(\tau)$ depends on the system under study but also on the characteristics of the fluctuating physical quantity which is considered. The choice of a correct analytical form for the $G(\tau)$ function is still a very difficult task. Redfield's comment on this problem (7, p. 28) remains correct, more than 15 years later, when he said that the calculation of spectral densities or correlation functions is "the most difficult problem in any relaxation theory."

Several models lead to an exponential correlation function (eq. 15), in which τ_c is a correlation time. Exponential

$$G(\tau) = G(0) \exp(-\tau/\tau_c) \tag{15}$$

correlation functions are very popular in NMR even if more complex situations have been known for many years (22,23). Sometimes the choice of exponential functions is supported by theoretical arguments; sometimes it is only a choice based on arguments of simplicity.

As far as $G(\tau)$ is defined by eq. 15 (with $G(0) = 1$), $J(\omega)$ is easily calculated by eq. 12. The spectral density is a

Lorentzian function (eq. 16), with a maximum value (equal to $2\tau_c$)

$$J(\omega) = \frac{2\tau_c}{1 + \omega^2\tau_c^2} \tag{16}$$

at $\omega = 0$. If $\omega\tau_c \ll 1$, $J(\omega)$ is constant and equal to $2\tau_c$. <u>The constant $J(\omega)$ domain corresponds to the so-called extreme narrowing conditions.</u> These conditions play a crucial role in relaxation. When they are realized, the relaxation function takes a simpler form (Figure 2). The spectral density of the fluctuating function is "white" in this region. As shown in Figure 2, the $J(\omega)$ curve may be characterized by a "cutting frequency" ω^* with $\omega^* \simeq 1/\tau_c$. For higher frequencies, $J(\omega)$ is negligibly small, while for lower frequencies $J(\omega)$ is approximately constant. The extension of the plateau is therefore a function of τ_c. A small τ_c corresponds to a high ω^* value. Nevertheless, the area of the $J(\omega)$ curve is independent of τ_c and therefore a small τ_c corresponds to a small $J(\omega)$ for the whole of the ω domain, including the plateau. What is important in relaxation is the $J(\omega)$ value of $\omega_0 = |E_a - E_b|/\hbar$ (see eq. 11). This result is easily understandable if we consider the spin system as quantum mechanical. The energy transfer between the spin system and the lattice corresponds to transfer of energy "grains." We shall see later that spectral densities depending on frequency values other than ω_0 (such as $2\omega_0$, $(\omega_0^I + \omega_0^S)$, or $(\omega_0^I - \omega_0^S)$) must also be taken into account.

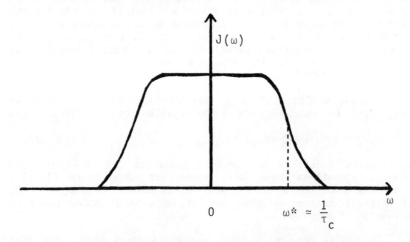

Figure 2. Spectral density function (ω^* corresponds to the "cutting frequency").

RELAXATION MECHANISMS: A GENERAL SURVEY

Relaxation in the liquid phase implies transitions between the various eigenstates a, b, c,... of the spin system. This is true for longitudinal as well as for transverse relaxation. People involved in relaxation theoretical studies are therefore primarily concerned with the coupling of these states under the influence of the perturbation Hamiltonian $\hat{H}(t)$. The first problem consists in the calculation of transition probabilities per second between the various states of the spin system. As we have seen in the previous paragraph, W_{ab} is dependent on two complementary aspects of the $\hat{H}(t)$ function: a good time dependence (in terms of compatibility between ω_0 and τ_c values) and a coupling efficiency with respect to the a, b, c, ... states (nonvanishing off-diagonal matrix elements of the general form $\int \psi_m^* \hat{H}(t) \psi_n d\tau$). The nature of $\hat{H}(t)$ determines what is called the relaxation mechanism.

As a first example, we shall consider the dipole-dipole mechanism. If two spins I and S in the same molecule interact by a dipolar interaction, the perturbation Hamiltonian is time dependent because the $\underline{r}$ vector joining the two spins is characterized by a random motion (Euler's angles θ and ϕ are time dependent). When I and S are in different molecules, $\underline{r}$ becomes time dependent. If θ and ϕ alone are time dependent, the relaxation mechanism is purely intramolecular. If $\underline{r}$ is time dependent, the mechanism is intermolecular. In the first case relaxation is due to molecular rotation, in the second case it corresponds to translation. The theoretical treatment gives different results if the two spins I and S are identical or not (2, p. 291). For two identical spins, at a fixed r distance (intramolecular interaction), R_1^D and R_2^D are given by eqs. 17 and 18. Whereas R_1^D

$$R_1^D = \frac{1}{T_1^D} = \frac{2}{5} \frac{\gamma^4 \hbar^2 I(I+1)}{r^6} \left(\frac{\tau_c}{1+\omega_0^2 \tau_c^2} + \frac{4\tau_c}{1+4\omega_0^2 \tau_c^2} \right) \quad (17)$$

$$R_2^D = \frac{1}{T_2^D} = \frac{1}{5} \frac{\gamma^4 \hbar^2 I(I+1)}{r^6} \left(3\tau_c + \frac{5\tau_c}{1+\omega_0^2 \tau_c^2} + \frac{2\tau_c}{1+4\omega_0^2 \tau_c^2} \right) \quad (18)$$

depends on $J(\omega_0)$ and $J(2\omega_0)$, R_2^D also depends on $J(\omega=0)$.

A qualitative explanation of this peculiar difference between R_1^D and R_2^D is given in reference 6 (p. 48). If $\omega_0 \tau_c \ll 1$ (extreme narrowing conditions), R_1^D becomes equal to R_2^D (eq. 19). Figure 3 illustrates the τ_c dependence of the

$$R_1^D = R_2^D = \frac{2\gamma^4 \hbar^2 I(I+1)}{r^6} \tau_c \qquad (19)$$

functions between brackets in eqs. 17 and 18. We see a behaviour frequently emphasized in review articles and already discussed in the paper of Bloembergen, Purcell, and Pound (1).

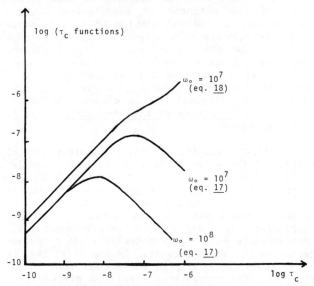

Figure 3. τ_c functions, i.e., functions given between brackets in eqs. 17 and 18. The functions are calculated for two ω_0 values in the case of eq. 17, for only one in the case of eq. 18.

We have previously pointed out that longitudinal relaxation is not always described by a monoexponential function (and therefore by one relaxation time). This case can be illustrated by dipole-dipole relaxation between different spins (I ≠ S) (2, p. 295; 16,17). In this case, the description of the relaxation of I by S requires the definition of two longitudinal relaxation times. The first relaxation time T_1^I depends on the quantum number, S, of spin S (the relaxing spin) and on three spectral densities, respectively $J(\omega_0^I - \omega_0^S)$, $J(\omega_0^I)$, and $J(\omega_0^I + \omega_0^S)$. The second relaxation time, T_1^{IS}, depends on the quantum number, I, of spin I and on two spectral densities, $J(\omega_0^I - \omega_0^S)$ and $J(\omega_0^I + \omega_0^S)$. In the extreme narrowing conditions (2, p. 297), the ratio of the relaxation rates is given by eq. 20. This

$$\frac{R_1^I}{R_1^{IS}} = \frac{2S(S+1)}{I(I+1)} \tag{20}$$

cross-relaxation phenomenon plays an important role in NMR (16,17) and numerous examples have been published in the recent literature. A special choice of experimental conditions can lead to the apparent suppression of cross-relaxation. For example, in the two spin system just discussed, relaxation of I is a monoexponential function if the S spins are irradiated by a strong B_1 field (16, p. 15).

Another behaviour which leads to the definition of multiple relaxation times in the case of dipole-dipole relaxation of identical spins is the so-called cross-correlation phenomenon. Eq. 17 is applicable for two identical interacting nuclei. What happens in the case of, let us say, four identical nuclei placed at the apices of a tetrahedron (CH_4, for example)? It is tempting to replace eq. 17 by a sum over three dipole-dipole interactions (each spin interacts with three partners). This treatment would ignore that reorientations of each $\underline{r}$ vector connecting spin pairs are dependent. The motions of these vectors are correlated. It is necessary to use cross-correlation functions in order to define the spectral density functions correctly. The result in the case of four nuclei at the apices of a tetrahedron is given by Abragam (2, p. 295). The temporal evolution of M_z is described by eq. 21, in which $M_z(t)$ is the M_z component at time

$$M_z(t) - M_0 = (M_z(0) - M_0)(ae^{-\alpha t/T_1} + be^{-\beta t/T_1}) \tag{21}$$

t, $M_z(0)$ is the M_z component at time t = 0, M_0 is the equilibrium M_z value, T_1 is the longitudinal relaxation time in the absence of cross-correlation, and a, b, α, and β are numerical coefficients (a = 0.035; b = 0.965; α = 1.35; β = 0.99). Clearly, the unique exponential obtained by making a = 0 and β = 1 is an excellent approximation, but, strictly speaking, the relaxation behaviour is biexponential in this case.

The dipole-dipole mechanism when $\underline{r}$ is time dependent is discussed in reference 2, p. 301. This mechanism is very efficient when paramagnetic species are present in the solution. These paramagnetic species are sometimes impurities like O_2. They may be voluntarily added and then called relaxation reagents (43, p. 214).

The dipole-dipole mechanism is probably the most popular relaxation mechanism for I = 1/2 nuclei in liquid samples. It is a magnetic relaxation mechanism because the dipolar Hamiltonian corresponds to a magnetic interaction.

Another mechanism, also of magnetic type, is spin-rotation. Its importance is great in the gas phase. At normal temperature it generally plays a less important role in the liquid phase, except in the cases of small molecules, of some ionic species forming transient complexes with other molecules, and of polyatomic groups (like CH_3) in rapid internal motion relative to the rest of the molecule. This mechanism is related to the temporal fluctuation of the molecular rotational moment (spherical top case). It is interesting to note that the associated correlation time, τ_{j}, is different from τ_c. Expressions for R_1^{SR} and sometimes R_2^{SR} are given in references 2, 6, 24, 25, and 44.

The so-called scalar coupling mechanism is also of a magnetic type. It is observed when a spin I is coupled to a spin S and when this coupling is time dependent. The origin of this time dependence may be a fast relaxation of S, when S is a quadrupolar nucleus. In this case the scalar relaxation is said to be of the second kind. When the origin of the time dependence of J is a chemical exchange or an internal motion in the molecule, the scalar coupling is said to be of the first kind.

This classification was introduced by Abragam (2), and the corresponding equations for the relaxation times are derived in his book on p. 307. A short discussion of this mechanism is also given in references 6, 24, 25, and 43. The recent paper of Briguet, Duplan, and Delmau gives an excellent discussion of scalar relaxation of the second kind in $SiCl_4$ and $SiHCl_3$ including cross-relaxation terms (31).

The fourth mechanism of magnetic type is the one depending on the anisotropy of the screening tensor $\underline{\sigma}$ (improperly called chemical shift anisotropy and denoted CSA). The anisotropy of $\underline{\sigma}$ has direct consequences on NMR spectra when they are recorded for liquid crystals and solid state samples. In isotropic phases, the chemical shift only depends on the average value of $\underline{\sigma}$ (i.e., $\text{Tr}\{\sigma\}$). However, because of σ anisotropy, the rapid tumbling motion of the molecule in the gas or the liquid phase generates a fluctuating, induced field at the level of the nucleus. Therefore, the interaction of the spin with this fluctuating field can induce transitions between nuclear Zeeman levels. This leads to a relaxation mechanism that is B_0-dependent because the fluctuating field is an induced field (R_1^{CSA} is itself proportional to B_0^2). The σ tensor anisotropy can be measured by the $\Delta\sigma$ quantity in eq. 22, in which σ_{xx}, σ_{yy}, and

$$\Delta\sigma = \sigma_{zz} - \frac{1}{2}(\sigma_{xx} + \sigma_{yy}) \tag{22}$$

σ_{zz} are the diagonal components of the $\underline{\sigma}$ tensor. For an axial molecule with the symmetry axis in the z direction, eq. 22 becomes eq. 23. For example, $\Delta\sigma$ is equal to 461 ppm and 320 ppm

$$\Delta\sigma = \sigma_\| - \sigma_\perp \qquad (23)$$

for the ^{31}P nucleus in S=PF$_3$ and O=PBr$_3$, respectively (32). The excellent review article of Robert and Wiesenfeld (32) gives a general discussion of the anisotropy of $\underline{\sigma}$ and $\underline{J}$ tensors from the theoretical and experimental point of view. A clear-cut example of a very efficient relaxation process due to the anisotropy of the $\underline{\sigma}$ tensor is given by Gillies et al. (33). For diphenylmercury, at B_0 = 2.35 T, this mechanism is strongly dominant (the $\Delta\sigma$ value for ^{199}Hg is equal to 6800 ± 680 ppm). A nice example of competition between the spin-rotation mechanism and the CSA mechanism is given (34) in the case of ^{205}Tl (cf. Figure 4). The temperature dependence of the two mechanisms SR and CSA is opposite because they are functions of two different correlation times, τ_j and τ_c, which are characterized by opposite temperature dependences. Whereas τ_c becomes shorter when the temperature increases, τ_j becomes longer when the temperature increases (6, p. 64; 43, p. 310). It should be noted the CSA and dipole-dipole relaxation respond to the same τ_c and consequently exhibit the same temperature dependence. At high B_0 fields like those attained nowadays, we can predict that relaxation due to the screening constant anisochrony will play an increasing role.

As in the case of the other mechanisms quickly reviewed in this section, the equations giving R_1^{CSA} and R_2^{CSA} are easily accessible in the literature (6,24,25,33,34, and 43).

The anisotropy of the spin-spin coupling J tensor, well documented in the paper or Robert and Wiesenfeld (32), could lead to a relaxation mechanism but, as far as the author knows, no clear example of the efficiency of such mechanism has been demonstrated. All the mechanisms discussed until now depend on magnetic interactions. The only mechanism which depends on an electric interaction is the so-called quadrupolar mechanism, which is observed in the case of nuclei with $I > 1/2$. The importance of the quadrupolar mechanism for uncommon nuclei leads us to devote the next section to this problem.

QUADRUPOLES AND QUADRUPOLAR RELAXATION

Before discussing quadrupolar relaxation, it could be useful to say a few words about quadrupoles and give some illustrative examples taken from molecular physics. A quadrupole can be visualized as an ensemble of four charges or, what is better, as two opposite dipoles. A pair of two dipoles oriented as in Figure 5 is a good example of a quadrupole. 1,4-dioxane and trans-1,2-dichloroethylene are quadrupolar molecules but also benzene and all the homonuclear diatomic molecules like H_2 or N_2.

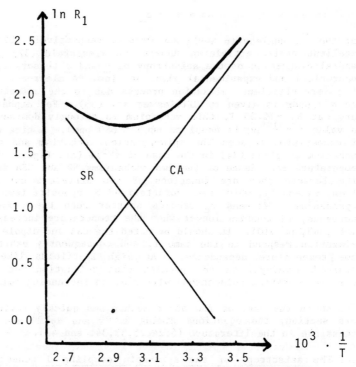

Figure 4. Longitudinal relaxation of ^{205}Tl for $(CH_3)_2TlNO_3$ at 34.7 MHz in D_2O. Example of competition between spin-rotation (SR) and chemical shielding anisotropy (CSA) mechanisms (from Ref. 34).

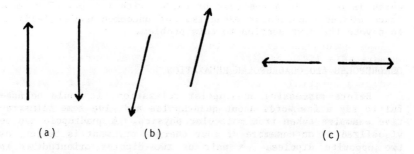

Figure 5. Three configurations of a two dipole system for which the first nonzero electrical moment is a quadrupole.

It is not always well understood that the presence of a charge or the existence of a dipole in a molecule does not exclude the existence of a quadrupole. Any charge distribution can be expanded as a series of l-poles (monopole, dipole, quadrupole, octupole, hexadecapole, and so on), and it is a serious error to limit the series to the first nonzero moment.

The value of the l-pole moment of a charge distribution is independent of the choice of the origin only if this l-pole moment is the first nonzero moment. In other words, the dipole moment of HCl and the quadrupole moment of H_2 are independent of the choice of the origin, but the quadrupole moment of HCl and the hexadecapole moment of hydrogen depend on the choice of the origin.

In a recent work (26,27) we have clearly shown the importance of higher moments than the first nonzero moments in molecular chemical physics. The situation is similar in nuclear physics. The presence of a charge (monopole) does not preclude the presence of higher multipole moments (28).

For nuclei with $I > 1/2$, the charge distribution inside the nucleus is not spherical, and the description of this charge distribution requires the definition of a quadrupolar nuclear moment and also, at least in some cases, of higher moments. In all cases, the dipole moment vanishes (28). The moments are always calculated relative to the center of mass, identical to the center of charge.

The deformed charge distribution is generally axially symmetrical and this fact has an important consequence. It permits to characterize the charge distribution asymmetry by means of only one quantity, Q (called the quadrupolar moment), even if the quadrupolar operator, $Q^{\alpha\beta}$, is a 3x3 matrix (in classical physics, a quadrupole is a second rank tensor). The definition of Q and the explicit form of $Q^{\alpha\beta}$ are given in references 2 and 3.

A quadrupole is able to interact with an electric field gradient (a field gradient is the second derivative of a potential). A fluctuating electric field gradient is thus able to induce transitions between the Zeeman levels of a quadrupolar nucleus oriented by a B_0 magnetic field. This is the physical basis of the so-called quadrupolar relaxation mechanism which corresponds to an electric interaction. The field gradient is generated by the electronic environment of the nucleus. It is defined by the three second derivatives of the potential along the three axes of an internal molecular reference frame. By definition, $(\partial^2 V/\partial z^2) \cdot e^{-1}$ is called the field gradient and denoted as q, while the asymmetry factor, η, is given by eq. 24. For extreme narrowing conditions $R_1^Q = R_2^Q$ is given (2,6,24) by eq. 25, in which $e^2 qQ/\hbar$ is called the quadrupole coupling constant. This constant is a function of a nuclear property, Q,

$$\eta = \frac{\partial^2 V/\partial x^2 - \partial^2 V/\partial y^2}{\partial^2 V/\partial z^2} \tag{24}$$

$$\frac{1}{T_{1Q}} = \frac{1}{T_{2Q}} = \frac{3}{40} \frac{(2I + 3)}{I^2(2I - 1)} \left(1 + \frac{\eta^2}{3}\right) \left(\frac{e^2 qQ}{\hbar}\right)^2 \tau_c \tag{25}$$

but also a function of a molecular property, q, the field gradient along the z direction.

The potential V must satisfy Laplace's equation and therefore eq. 26 holds. Now, if these three terms are equal for sym-

$$\partial^2 V/\partial x^2 + \partial^2 V/\partial y^2 + \partial^2 V/\partial z^2 = 0 \tag{26}$$

metry reason, they must also be equal to zero. Such a situation is observed at the center of a cubic arrangement of identical ligands. Therefore, the value of the quadrupolar relaxation rate is strongly dependent on the environment of the nucleus. Due to the efficiency of the quadrupolar mechanism, linewidths of NMR absorption peaks are generally very large for quadrupolar nuclei. As a first approximation, the full width of a Lorenzian line at half height is given by eq. 27, at least when the transverse

$$W_{1/2} = (\pi T_2)^{-1} \tag{27}$$

relaxation time is so short that the peak broadening due to inhomogeneity in B_0 can be neglected. Linewidths of thousands of Hz are not exceptional for quadrupolar nuclei. The independent determination of Q and q are difficult problems. Chemists are obviously more interested by q and η values, because these are related to the electronic distribution around the nucleus. The theoretical calculation of q is a difficult task especially because the closed shell electrons, in proximity to the nucleus, are distorted from spherical symmetry and also contribute to q. The discussion of this problem is outside the scope of this chapter but examples will be given in other chapters about the importance of the so-called Sternheimer antishielding factor (29). The last comment we would like to make in this section concerns the cases where the quadrupolar relaxation function is a multiexponential function. As previously pointed out, Lurcat (30) was the first to enumerate the maximum number of relaxation times necessary to describe completely the time evolution of

M_z towards equilibrium. This number is equal to $I + 1/2$ for $I = 1/2, 3/2, 5/2, \ldots$ Therefore, the relaxation function is monoexponential for $I = 1/2$ (if we exclude cross-relaxation and cross-correlation) but biexponential for $I = 3/2$, triexponential for $I = 5/2$, and so on.

The $I = 5/2$ and $I = 7/2$ cases have been extensively discussed (35-38). In extreme narrowing conditions a monoexponential relaxation curve is observed but a multiexponential behaviour is predicted as far as $\omega_0 \tau_0 > 1$ at least for $I = 7/2$ nuclei. In the $I = 5/2$ case, the monoexponential behaviour is an excellent approximation in all the ω domain. The theoretical treatment corresponding to the $I = 3/2$ case has been extensively discussed (39, 40) and presents some formal similarities to the three-identical-spin-1/2 system, in the presence of strong cross-correlations (41, 42). It appears that in some circumstances this system is much more complex than what is predicted on the basis of the numbering procedure of Lurçat (30).

The calculated spectrum for $I = 3/2$ including the effect of second-order dynamic frequency shift leads to the prediction that the two spectrum components (degenerated only for extreme narrowing conditions) are characterized by different decay curves after a simple resonant B_1 pulse. The so-called broad component $(3/2-1/2, -3/2-(-1/2))$ decays exponentially while the narrow component $(1/2-(-1/2))$ decays nonexponentially.

This last example certainly corresponds to an extreme case: higher order analysis of relaxation phenomena cannot be applied to all the real and complex systems chemists are interested in.

To conclude not only this section but also this chapter, it seems necessary to emphasize the complexity of relaxation phenomena. The simple Bloch exponential equations are not general at all. It is true that in many circumstances the exponential decay is a good approximation, but it is of fundamental importance to analyse each case carefully both from the theoretical and experimental points of view.

The very recent and very nice work of Chenon, Bernassau, Mayne, and Grant (35) on the reorientation dynamics of liquid CH_2Cl_2 from longitudinal relaxation studies is a perfect example which substantiates this conclusion.

ACKNOWLEDGMENTS

The author wishes to express thanks to the Fonds National de la Recherche Scientifique for the financial support to his laboratory.

REFERENCES

(1) N. Bloembergen, E. M. Purcell, and R. V. Pound, Phys. Rev., 73, pp. 679-712 (1948).
(2) A. Abragam, "Les Principes du Magnétisme Nucléaire," Institut National des Sciences et Techniques Nucleaires et Presses Universitaires de France, Paris, 1961.
(3) C. P. Slichter, "Principles of Magnetic Resonance," Harper, Row and Weatherhill, New York, 1964.
(4) C. P. Poole, Jr., and H. A. Farrach, "Relaxation in Magnetic Resonance," Acadamic Press, New York, 1971.
(5) R. Lenk, "Brownian Motion and Spin Relaxation," Elsevier Scientific Publishing Company, Amsterdam, 1977.
(6) T. C. Farrar and E. Becker, "Pulse and Fourier Transform NMR," Academic Press, New York, 1971.
(7) A. G. Redfield, Advan. Magn. Reson., 1, pp. 1-32 (1965).
(8) J. M. Deutsch and I. Oppenheim, Advan. Magn. Reson., 3, pp. 43-78 (1968).
(9) R. G. Gordon, Advan. Magn. Reson., 3, pp. 1-42 (1968).
(10) J. Jeener, Advan. Magn. Reson., 3, pp. 205-310 (1968).
(11) J. E. Leffler and E. Grunwald, "Rates and Equilbria of Organic Reactions," J. Wiley, New York, 1963.
(12) H. G. Hecht, "Magnetic Resonance Spectroscopy", J. Wiley, New York, 1967.
(13) H. G. Hertz, "Translational Motion as Studied by Nuclear Magnetic Resonance," in "Molecular Motions in Liquids," J. Lascombe, Eds., D. Reidel Pub. Co., Dordrecht, 1974, pp. 337-357.
(14) I. R. Senitzky, Phys. Rev., 134, pp. A816-823 (1964).
(15) I. R. Senitzky, Phys. Rev., 135, pp. A1498-1505 (1964).
(16) J. H. Noggle and R. E. Schirmer, "The Nuclear Overhauser Effect," Academic Press, New York, 1971.
(17) P. S. Hubbard, Proc. Roy. Soc., London, A291, pp. 537-555 (1966).
(18) F. Bloch, Phys. Rev., 102, pp. 104-135 (1956).
(19) R. K. Wangsness and F. Bloch, Phys. Rev., 89, pp. 728-739 (1953).
(20) R. Kubo and K. Tomita, J. Phys. Soc. Japan, 9, pp. 888-919 (1954).
(21) D. Kivelson and K. Ogan, Advan. Magn. Reson., 7, pp. 71-155 (1974).
(22) H. C. Torrey, Phys. Rev., 92, pp. 962-969 (1953).
(23) H. A. Resing and H. C. Torrey, Phys. Rev., 131, pp. 1102-1104 (1963).
(24) J. R. Lyerla and D. M. Grant, Intern. Rev. Sci., 4, pp. 155-197 (1972).
(25) G. A. Webb, in "NMR and The Periodic Table," R. K. Harris and B. E. Mann, Eds., Academic Press, London, 1978.
(26) M.L. Stien, M. Claessens, A. Lopez, and J. Reisse, J. Am. Chem. Soc., to be published (1982).

(27) M. Claessens, L. Palombini, M. L. Stien, and J. Reisse, Nouv. J. Chim., to be published (1982).
(28) A. Bohr and B. R. Mottelson, "Nuclear Structure," Vol. 2. W. A. Benjamin, London, 1975.
(29) R. M. Sternheimer, Phys. Rev., 84, pp. 244-253 (1951); 86, pp. 316-324 (1951); 95, pp. 736-750 (1954).
(30) F. Lurçat, C. R. Acad. Sci., Paris, 240, pp. 2402-2403 (1955).
(31) A. Briguet, J. C. Duplan, and J. Delmau, J. Magn. Reson., 42, pp. 141-146 (1981).
(32) J. B. Robert and L. Wiesenfeld, Phys. Report, to be published.
(33) D. G. Gillies, L. P. Blaauw, G. H. Hays, R. Huis, and A. D. H. Clague, J. Magn. Reson., 42, pp. 420-428 (1981).
(34) F. Brady, R. W. Matthews, M. J. Forster, and D. G. Gillies, Inorg. Nucl. Chem. Letters, 17, pp. 155-159 (1981).
(35) M. T. Chenon, J. M. Bernassau, C. L. Mayne, and D. M. Grant, J. Chem. Phys., to be published (1982).
(36) T. E. Bull, S. Forsén, and D. L. Turner, J. Chem. Phys., 70, pp. 3106-3111 (1979).
(37) B. Halle and H. Wennerström, J. Magn. Reson., 44, pp. 89-100 (1981).
(38) T. Andersson, T. Drakenberg, S. Forsén, E. Thulin, and M. Swärd, J. Am. Chem. Soc., 104, pp. 576-580 (1982).
(39) P. S. Hubbard, J. Chem. Phys., 53, pp. 985-987 (1970).
(40) L. G. Werbelow and A. G. Marshall, J. Magn. Reson., 43, p. 443-448 (1981).
(41) L. G. Werbelow, A. Thevand, and G. Pouzard, J. Chem. Soc. Faraday Trans. II, 75, pp. 971-974 (1979).
(42) L. G. Werbelow, J. Magn. Res., 34, pp. 439-442 (1979).
(43) D. Shaw, "Fourier Transform NMR Spectroscopy," Elsevier, Amsterdam, 1976.
(44) M. Claessens, D. Zimmermann, and J. Reisse, unpublished results.

CHAPTER 5

DYNAMIC NMR PROCESSES

Joseph B. Lambert
Northwestern University
Evanston, Illinois 60201 USA

ABSTRACT

Five trends are examined in the field of dynamic NMR processes. (i) Improved computer programs have been developed for handling complex spin systems. These programs permit direct analysis without synthetic deuteration and often allow differentiation of possible mechanisms of exchange. (ii) Two dimensional analysis of dynamic processes provides an alternative method for the assignment of exchanging nuclei and of the unraveling of complex exchange mechanisms. (iii) The availability of an increased variety of magnetically active nuclei, such as ^{15}N, ^{17}O, and metals, permits analysis of new problems. (iv) The relaxation time has been accepted as an additional observable from which kinetic information can be extracted. Relaxation times offer another method for the identification of exchanging spins, through saturation transfer, and an expanded dynamic range of kinetics. Whereas lineshape analysis has a range of about 10^0 to 10^4 sec^{-1}, the family of relaxation methods has a range of about 10^{-3} to 10^{12} sec^{-1}. (v) Analysis of exchanging spins in the solid, with cross polarization and magic angle spinning, offers both an expanded lower temperature range because of removal of constraints imposed by solution samples, and the opportunity to study the effects of lattice forces on dynamic processes.

The state of the art described by a recent text on the subject of dynamic nuclear magnetic resonance spectroscopy (1) typically included two or three spins in one spectral dimension, examined ^{1}H, ^{19}F, and occasionally ^{13}C or ^{31}P, employed the chemical shift or coupling constant as the dynamic variable, and studied materials in solution. Developments in the field over the past decade have overcome many of these limitations. We will consider in this review the expansion of the DNMR field to include more complex spin systems, two and three dimensionality, new sensor nuclei, the relaxation time as the dynamic variable, and processes in the solid phase.

SOFTWARE DEVELOPMENT FOR COMPLEX SPIN SYSTEMS

This area has been reviewed recently by Binsch and Kessler (2), and many of the examples in this section come from their review. Binsch and co-workers have developed a series of computer programs for the treatment of dynamic NMR spectral data for systems with considerable complexity. Recent versions, DNMR5 and DAVINS (3,4), use raw spectral data for input, obtain rate constants by an iterative procedure, and carry out a complete error analysis. One of the simplest cases is that of n exchanging, noncoupled singlets, as might be found in a ^{13}C spectrum. The calculation requires knowledge of the statistical matrix Q, which specifies which spins interconvert during the dynamic process. Bullvalene (1), for example, has four types of carbons and hence

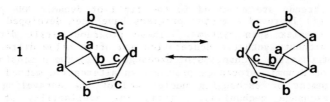

a 4 x 4 statistical matrix. The off-diagonal elements q_{ij} represent the probability that the nucleus j becomes nucleus i as the result of the Cope rearrangement. For example, the single d carbon is always converted to an a carbon, so element ad of the matrix is 1. Of the three a carbons, two become c and one becomes d, so elements ca and da respectively are 2/3 and 1/3. The full matrix is given on the next page. The diagonal elements are obtained from the sum of the off-diagonal columns, with a negative sign. These figures represent the overall probability that a given site will undergo any change. Thus, all of the a nuclei but only a third of the b nuclei are transformed. Because two-thirds of the b nuclei are not involved in any given act of exchange, the b resonance broadens much less than those of the other nuclei, which have diagonal elements of unity (Figure 1) (5). Selective broadening of this type can be used to make unambiguous peak assignments, or, conversely, selective broadening of peaks with known assignments can be interpreted in terms of a

	a	b	c	d
a	-1	0	2/3	1
b	0	-1/3	1/3	0
c	2/3	1/3	-1	0
d	1/3	0	0	-1

specific mechanism (with a known statistical matrix) in cases with competing mechanisms.

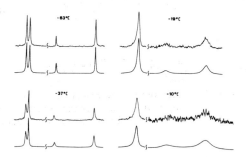

Figure 1. Observed and calculated ^{13}C spectra of bullvalene as a function of temperature. Reproduced from Ref. 5 with the permission of Pergamon Press.

A more complex set of exchanging singlets is provided by tropylium azide (6). The carbons can be interchanged randomly by ionization, or selectively by 1,2, 1,3, or 1,4 shifts of the azide group (Figure 2). The exchange matrices derived from these mechanisms can be used to calculate the expected spectra (Figure 3). In $SO_2/CDCl_3$, both covalent and ionic forms can be observed, and their exchange is adequately accounted for by the random mechanism. In the less polar solvent CD_3CN, the ionic form requires a mixture of mechanisms. The 1,3 shift is preferred to the extent of about 60%, but the 1,2 (30%) and 1,4 (10%) mechanisms are also represented.

Binsch's group has studied a number of particularly complex systems, including N-N rotation in syn- and anti-1,4-dinitrosopiperazine (7) and aryl rotation in arsines such as 2 (8). The C_3 propeller arrangement can undergo exchange by the rate-determining, simultaneous rotation of one, two, or three rings. In an

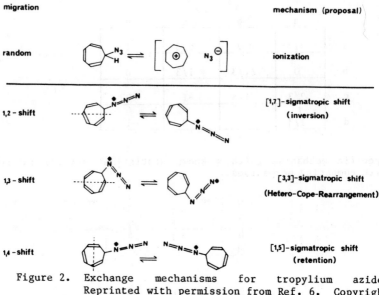

Figure 2. Exchange mechanisms for tropylium azide. Reprinted with permission from Ref. 6. Copyright 1979, American Chemical Society.

edge interchange e (9), one or more rings undergo the equivalent of a 180° rotation. Such a process interchanges the top pair of fluorine atoms with the bottom pair on a given phenyl ring (top and bottom refer to the arrangement with respect to the lone pair on arsenic), but does not interchange the two fluorines within a geminal pair. Thus, the variants of this mechanism (e^1, e^2, and e^3, depending on the number of rings that initially rotate) all give an AB spectrum at fast exchange. A second process of helicity reversal h is the equivalent of an approximately 90° rotation of each ring, in which the transition state has all three rings parallel, with the result of an inversion of chirality. This process interchanges fluorine atoms within a geminal pair

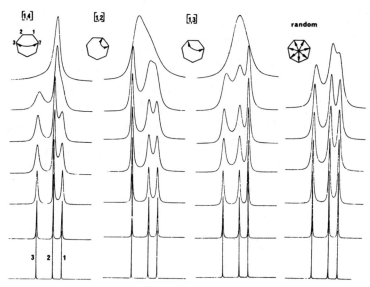

Figure 3. Theoretical ^{13}C spectra for different exchange mechanisms of tropylium azide. Reprinted with permission from Ref. 6. Copyright 1979, American Chemical Society.

but does not interchange the top and bottom pairs with each other. Thus helicity reversal by itself results in two singlets at fast exchange. Helicity reversal can be combined with edge interchange to give three composite mechanisms (eh, e^2h, and e^3h). Only simultaneous operation of e and h can produce the fast exchange singlet, which is the actual observed result. Because the statistical matrices and expected lineshape dynamics are different for the various composite mechanisms, they may be differentiated. Spectral analysis (8) showed that the mechanism of interchange is e^2h, in which two rings flip simultaneously with helicity reversal. On account of coupling between ^{1}H and ^{19}F and the presence of three rings with four fluorine nuclei per ring, 2 had to be analyzed as an $[ABXCDY]_3$ spin system.

By the use of large spin systems, extremely complex exchange mechanisms can be elucidated with a single molecule and with a minimum of deuterium labeling, which was necessary previously to removing J coupling. Furthermore, the possible presence of multiple coalescence processes in complex spectra can improve the accuracy of the resulting activation parameters.

TWO AND THREE DIMENSIONAL DYNAMIC NMR SPECTRA

Although the analysis of complex dynamic behavior can often distinguish possible mechanisms, sometimes the lineshape changes are too subtle for an unambiguous conclusion. Under these circumstances, exchanging sites can be identified either by saturation transfer experiments (vide infra) or by multidimensional displays. In the two dimensional approach, developed by Ernst and co-workers (10), a sequence of three pulses is used (Figure 4a). An initial 90° pulse creates xy magnetization. During a wait time t_1, the nuclei are modulated by precession according to their natural resonance frequencies. Another 90° pulse then returns the modulated x components back to the z direction. During a second and longer wait time τ_m or t_m, nuclei are able to undergo exchange between sites. Finally, a third 90° pulse is administered, and the free induction decay is collected over the time t_2. Both t_1 and t_2 are varied, and Fourier transformation over the two time domains produces the two dimensional spectra shown in Figure 4b for the exchange process of the heptamethylbenzenium ion (3) (1,2 methyl shifts). Peaks along the diagonal

3

of the spectrum come from nuclei that did not undergo exchange during the period t_m (auto-peaks). Those off the diagonal (cross-peaks) derive from nuclei that passed through the t_1 period at one site, underwent an exchange process during t_m, and passed through the t_2 period at a new site.

An off-diagonal peak demonstrates exchange between the sites to which it is connected by vertical and horizontal lines. Thus for the heptamethylbenzenium ion, methyls at site 1 exchange only with those at site 2; those at 4 exchange only with those at 3; those at 2 exchange with those at 1 and 3; and those at 3 exchange with those at 2 and 4. This set of exchanges defines a 1,2-shift mechanism and excludes 1,3-, 1,4-, and random-shift mechanisms.

The two-dimensional approach is advantageous not only when the exchange mechanism is unclear but also when molecules are particularly large and complex, as in biomolecules. The dif-

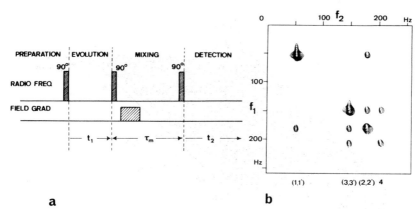

Figure 4. (a) Pulse sequence to obtain information on chemical exchange. (b) Two dimensional contour plot for the exchange of methyl groups in heptamethylbenzenium (3). Reprinted with permission from Ref. 10. Copyright 1979, American Chemical Society.

ficulties of such an experiment are illustrated by the intermolecular exchange of the amide (NH) protons in glutathione (glu-cys-gly) with the solvent water (Figure 5) (11). This experiment used a much more complex pulse sequence than that shown in Figure 4, in order to suppress the large signal from H_2O. The large peaks in the upper left corner are the diagonal peaks from cys-NH and gly-NH. The resonances labeled X and Y are at the appropriate positions to demonstrate exchange of these NH peaks with the solvent peak. The glu-NH peaks were too exchange broadened for observation by this procedure. Peaks Z and W arose from nonexchange phenomena, such as dipolar or J interactions or residual excitation of H_2O by the acquisition pulse.

It is possible to extract quantitative data from a 2D display. For example, in the simplest case of a two-site exchange with equal populations and equal spin-lattice relaxation times, the exchange rate k can be obtained from the ratio of the intensity of an auto-peak to that of a cross-peak (eq. 1) (12).

$$\frac{I(auto)}{I(cross)} = \frac{(1 - kt_m)}{kt_m} \tag{1}$$

Overlap of peaks along the diagonal, as in the spectra of large biomolecules, prevents the use of such an equation, as does the presence of several mechanisms of exchange or cross-relaxation pathways. Alternatively, analysis of the t_m dependence of the

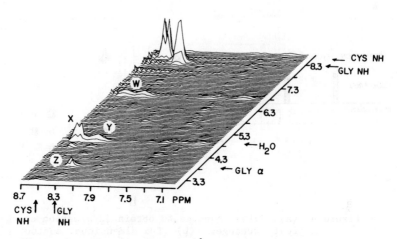

Figure 5. Two dimensional ^{1}H spectrum of glutathione to show proton exchange pathways. Reprinted with permission from Ref. 11. Copyright 1982, American Chemical Society.

cross-peak intensities can yield a quantitative result, but at the cost of a prohibitive amount of spectrometer time to collect 3D data from the three time domains, t_1, t_m, and t_2.

Bodenhausen and Ernst (12) introduced an alternative to the 3D method, which they termed accordion spectroscopy. Reduction in dimension can be obtained by setting $t_m = \kappa t_1$. Incrementation of t_1 then automatically and proportionately increments t_m (Figure 6). In the ordinary 2D experiment shown on the left of the figure, t_m is kept constant as the other time variables are altered. In the accordion experiment shown on the right, t_m is incremented as a multiple of t_1. The appearance of the pulse sequence inspired the name. In this fashion the information from a 3D experiment is reduced to two dimensions. A set of accordion spectra is shown for the ring reversal process in cis-decalin (4). The spectra superficially resemble 2D results, but there is important additional information contained in the lineshapes. Figure 7 shows a 1D cross section at the frequency of C4/C8 (called site B) as a function of temperature (the peak at the left is the auto-peak for B; that at the right is the cross-peak of B with C1/C5 (called site A)). These lineshapes may be analyzed directly to provide rate constants (12). When the diagonal peaks are well resolved, rate constants may be extracted easily from the sum and difference combinations of two resonances (Figure 7). The sum component provides a direct measure of T_1 alone, and the difference component in addition has

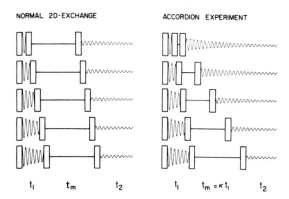

Figure 6. Pulse sequence used for normal (left) and accordion (right) exchange experiments. Reprinted with permission from Ref. 12. Copyright 1982, American Chemical Society.

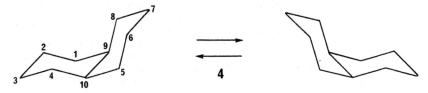

the exchange information (k). The half-heights of these two combinations, $\Delta\nu^\Sigma$ and $\Delta\nu^\Delta$, give the rate constant by eq. 2.

$$k = (\pi/2)(\Delta\nu^\Delta - \Delta\nu^\Sigma) \qquad (2)$$

The complexity and time consumption of 2D experiments will restrict their general application to dynamic problems. In addition to its elegance, however, the 2D experiment can offer the practical result of unraveling complex exchange mechanisms in an unambiguous fashion. Its ability to yield quantitative results in the hands of the rank and file remains to be demonstrated.

CHOICE OF NUCLEUS

Most dynamic studies have used the highly sensitive nuclei ^1H, ^{19}F, and ^{31}P (1). More recently, ^{13}C has become a useful subject, and occasionally ^{11}B has been employed in studies of boranes. The trend for expansion throughout the Periodic Table should continue. A multinuclear capability makes possible the study both of new mechanisms and new molecules.

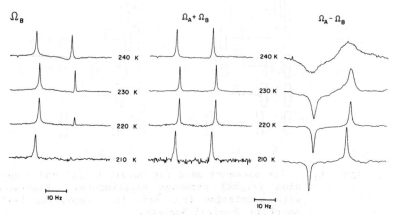

Figure 7. (Left) One dimensional cross section at the C4/C8 ^{13}C frequency (Ω_B) from the accordion spectrum of cis-decalin as a function of temperature. (Middle) The sum of Ω_A (C1/C5) and Ω_B as a function of temperature. (Right) The difference of Ω_A and Ω_B as a function of temperature. Reprinted with permission from Ref. 12. Copyright 1982, American Chemical Society.

We ended the earlier discussion of the rearrangement of tropylium azide with the conclusion that the mechanism was predominantly a 1,3-shift (6). The ^{13}C spectrum, however, did not define how the azido group was transferred from position 1 to position 3. It may undergo a [1,3] sigmatropic shift with retention or inversion, or it may undergo a [3,3] sigmatropic shift with reversal of the point of attachment. Kessler and Feigel (2, 6) proved the [3,3] mechanism by observation of coalescence of the ^{15}N signals for the two relevant nitrogen atoms.

Dynamics in molecules containing metals can rely heavily on the spectral changes of the metallic nucleus. Intermolecular exchange of ligands on mercury, for example, can be followed by examination of the ^{199}Hg spectra. Peringer and Winkler (13) observed ^{199}Hg line broadening of phenylmercuric triflate in the presence of mercuric ion and demonstrated that the dynamic effect arose from the process of eq. 3. A molecule of the type 5 poses

$$C_6H_5HgOSO_2CF_3 + (CF_3SO_2O)_2Hg' \rightleftharpoons$$
$$(CF_3SO_2O)_2Hg + C_6H_5Hg'OSO_2CF_3 \quad (3)$$

a problem if only ^{1}H and ^{13}C are available. Riddell, Gillard, and Wimmer (14) observed the dynamic process of ring reversal by examination of the ^{195}Pt spectrum. Two peaks in the ratio 30/1 ($\Delta G°$ = 1.3 kcal/mol) were observed to coalesce. The major peak was attributed to the C_3 all-chair form, but the minor conformer was not defined. The interconversion barrier ($\Delta G^{\ddagger}$) was measured to be 10.8 or 12.1 kcal/mol, depending on the direction of the reaction. As inorganic chemists continue to expand their use of metallic nuelei, it is expected that further examples of dynamic processes involving such nuclei will be observed.

KINETICS FROM RELAXATION TIMES

One of the major results of the Fourier transform revolution was the acceptance of the relaxation time as a third variable for measurement and analysis, after the chemical shift and the coupling constant. Because the spin-lattice relaxation time T_1 depends on motional activity of a nucleus, it is useful in investigating the dynamic behavior of molecules (15).

Much of the theory in this area has been due to Woessner (16). Although the theory was available two decades ago, routine experimental applications lagged until their development by Grant (17) and others. The relationship between the relaxation time and molecular dynamics results from the dependence of the dipolar or quadrupolar spin-lattice relaxation time on the various modes of internal and overall molecular diffusion. For a rigid molecule that is tumbling in solution isotropically (diffusion coefficients D_1, D_2, and D_3 all equal), the dipolar relaxation time for a carbon atom with one or more attached protons is given by eq. 4, in which n is the number of attached or nearest protons,

$$\frac{1}{T_1(DD)} = n\gamma_C^2\gamma_H^2\hbar^2 r^{-6}/6D \qquad (4)$$

γ is a gyromagnetic ratio, h is Planck's constant, r is the bond distance from carbon to the attached or nearest protons, and D is the overall diffusion coefficient (the effective correlation time $\tau_c = 1/6D$). If an otherwise rigid molecule contains a methyl group that rotates by a series of threefold jumps, the dipolar relaxation time for the methyl carbon is given by eq. 5,

$$\frac{1}{T_1(DD)(CH_3)} = 3\gamma_C^2\gamma_H^2\hbar^2 r^{-6} \left[\frac{A}{6D} + \frac{B + C}{6D + 3/2\ D_i} \right] \quad (5)$$

in which A, B, and C are geometrical constants (15), the overall diffusion constant D comes from relaxation of carbons in the rigid portion of the molecule, and D_i is the methyl jump rate. Alternative expressions are available for stochastic methyl rotation and anisotropic overall diffusion (15). Hence measurement of D for rigid carbons from eq. 4 and of $T_1(DD)(CH_3)$ (=1.988$T_1(CH_3)/\eta$, in which η is the nuclear Overhauser enhancement minus one) gives D_i. Because methyl rotation is a thermally activated process, temperature-dependent measurements of D_i yield activation parameters, eq. 6.

$$D_i = D_0 \exp(-V_0/RT) \quad (6)$$

A wide variety of molecules has now been studied by the Woessner method, such as 6-8 (18-20). The method is restricted

to the methyl group or its symmetry equivalent, and barriers are in the range 0.5-4 kcal/mol (rates faster than 10^8 sec^{-1}). Less symmetrical groups may be studied by application of the Woessner method to quadrupolar relaxation, as of deuterium (15).

Analysis of dipolar or quadrupolar relaxation restricts the dynamic range of rate constants to those on the order of the frequency of the B_0 field (γB_0). Slower rates ($10^2 < k < 10^6$), comparable to those available from lineshape analysis, can sometimes be obtained by analysis of relaxation in the rotating frame ($T_{1\rho}$). Application of a 90° pulse followed by a continuously applied 90° phase-shifted pulse maintains the magnetization vector in the xy plane. The rate of loss of magnetization in the xy plane is then measured (time constant $T_{1\rho}$). Because the magnetic field in the xy plane (B_1) is lower than B_0, rates measured in the rotating frame are slower (on the order of γB_1). For this reason, $T_{1\rho}$ is sensitive to dynamic processes not detected by T_1, in addition to the other components that make up T_1. Both relaxation times ($T_{1\rho}$ and T_1) are measured and subtracted to give $T_{1\rho}$(exchange), the rotating frame relaxation time that is dependent only on the dynamic process (eq. 7). Rate constants can be obtained from eq. 8, in which ω_1 is the spin-lock frequency, and

$$\frac{1}{T_{1\rho}(\text{exch.})} = \frac{1}{T_{1\rho}} - \frac{1}{T_1} \qquad (7)$$

$$\frac{1}{T_{1\rho}(\text{exch.})} = \pi^2(\Delta\nu)^2 \frac{k}{k^2 + \omega_1^2} \qquad (8)$$

$\Delta\nu$ is the slow exchange chemical shift difference between the exchanging nuclei. The advantage of this technique is that rates can be measured at temperatures well above their temperatures of lineshape coalescence. This advantage would prove useful in studies of molecules with especially low coalescence temperatures or in solutions with high freezing points. Measurement of $T_{1\rho}$(exchange) as a function of ω_1 yields both k and $\Delta\nu$, so observation of the slow-exchange limit is not necessary. Amide barriers in ureas (21) and C-C rotational barriers in benzaldehydes (22) were measured well above their coalescence temperatures. The barrier to ring reversal in 1,1-dimethylpiperidinium iodide was measured in aqueous solution, even though the coalescence temperature would have been well below the freezing point of water (23).

Saturation transfer has been used commonly to obtain information about the connectivity of exchanging sites (24). Selective irradiation of one site causes transfer of magnetization to another site with which it is exchanging. The changes in peak intensities may be analyzed to obtain rate constants. For example, measurement of intensities at site A prior to double irradiation at site B ($M_z^A(0)$) and after equilibrium has been attained during irradiation ($M_z^A(\infty)$) can give the rate constant from eq. 9, in which T_{1A} is the relaxation time at site A and

$$M_z^A(\infty)/M_z^A(0) = \tau_{1A}/T_{1A} \qquad (9)$$

τ_{1A} is a total lifetime ($\tau_{1A}^{-1} = \tau_A^{-1} + T_{1A}^{-1}$, and $k_A = \tau_A^{-1}$). This particular approach requires the independent measurement of T_{1A}. If the relaxation times for both exchanging sites are equal ($T_{1A} = T_{1B}$), all necessary quantities are easily accessible. The barriers to bond rotations within cyclophanes (25) and for ring reversal in cis-decalin (26) for example, have been measured in this fashion. Although the method has not been expanded to many cases for which fast exchange data could not also be obtained, there is no reason why it could not be applied quantitatively to noncoalescing cases.

In nonselective double irradiation, exchanging nuclei can exhibit magnetization that is double exponential, eq. 10 (15).

$$M_z^A(t) = M_z^A(0) + C_1\exp(-t/T_I) + C_2\exp(-t/T_{II}) \qquad (10)$$

At slow exchange, T_I and T_{II} correspond respectively to the normal single exponential relaxation times for the two sites, T_{1B} and T_{1A}; the C_1 term drops out for A magnetization and the C_2 term for B magnetization, so that the usual single exponential behavior is observed. At fast exchange, the C_1 term goes to zero and both sites exhibit a relaxation time that is the weighted average of the two relaxations, $T_1^{-1} = 0.5(T_{1A}^{-1} + T_{1B}^{-1})$ for equal populations. This transition from distinct relaxation times to average relaxation times means that the quantities must coalesce at intermediate temperatures. Such behavior has been studied for dimethylformamide (Figure 8) (27). Analysis of the double exponential behavior in the intermediate temperature range can yield the rate of the dynamic process. This method also is potentially applicable to cases in which lineshapes do not coalesce at the highest available temperature, since relaxation-time coalescence occurs well below the temperature of lineshape coalescence.

Relaxation methods can provide rate constants over a range of possibly 15 orders of magnitude (Figure 9), whereas lineshape results are limited to four or five. Thus dynamic range is far superior for the family of relaxation methods. Although the relaxation experiments sometimes are more difficult than the lineshape approach, they can supply kinetics that are well out of the lineshape range (Woessner method), for systems that fail to decoalesce at low temperature (rotating frame), and for systems that fail to coalesce at high temperatures (saturation transfer or relaxation coalescence).

DYNAMIC PROCESSES IN THE SOLID STATE

There are at least two important reasons for expanding high-resolution DNMR studies to the solid state. First, the additional constraints of crystalline lattice forces may have interesting and unanticipated effects on dynamic processes. Second, absence of the need for the solute or solvent to remain in the liquid state means that lower temperatures can be explored

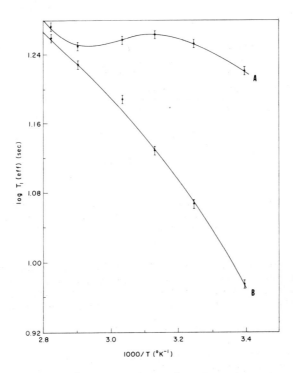

Figure 8. Apparent (single exponent) relaxation times for the methyl carbons of dimethylformamide as a function of temperature. Reproduced from Ref. 27 with the permission of Academic Press.

than were previously possible, so that lower activation energies might be measurable.

The advent of magic angle spinning and cross polarization techniques have brought the solid state close to the liquid state, in terms of resolution and sensitivity. Even in the earliest experiments with these techniques, important differences between solid and liquid spectra were evident for dynamic processes. For example, p-dimethoxybenzene exhibits only one aromatic resonance in solution, even at low temperatures, but two in the solid (28). Rapid C-OCH$_3$ rotation in solution averages the aromatic resonances, but slow rotation in the solid gives rise to distinct resonances from carbons that are syn and anti to OCH$_3$.

The group of Yannoni has been responsible for many recent advances in this field. They observed that the ^{13}C spectrum of

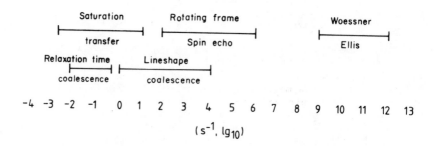

Figure 9. Dynamic range of rate constants (sec^{-1} on a logarithmic scale with base 10) for NMR methods. Reproduced from Ref. 15 with the permission of Verlag Chemie.

the 2-butyl cation in SbF$_5$ exhibits two peaks throughout the observable range in the solid state (-190 to -60°C) (29). Hydride shifts between the central carbons average C1 with C4 and C2 with C3, eq. 11. A second process can bring about interchange

$$CH_3CH_2\overset{+}{C}HCH_3 \rightleftharpoons CH_3\overset{+}{C}HCH_2CH_3 \qquad (11)$$

of the methyl and methylene groups through a methyl shift followed by proton interchange within the protonated cyclopropane intermediate, eq. 12. In solution, this process would have a

$$\overset{*}{C}H_3\text{---}CH_2\overset{+}{C}HCH_3 \rightleftharpoons \begin{array}{c} \overset{*}{C}H_3 \\ \diagup \;\; \diagdown \\ CH_2\overset{+}{\text{---}}CHCH_3 \end{array}$$

$$\updownarrow$$

$$\overset{*}{C}H_3\text{---}CH_2\overset{+}{C}HCH_3 \rightleftharpoons \begin{array}{c} \overset{*}{C}H_2 \\ \diagup \;\; \diagdown \\ CH_3\text{---}\overset{+}{\text{---}}CHCH_3 \end{array} \qquad (12)$$

coalescence temperature of about -60°C, but in the solid the process is still slow at this temperature. Even more dramatic is the suppression of the Cope rearrangement in semibullvalene (9) (30) which has a coalescence temperature of -140°C in solu-

9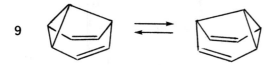

tion (30). Slow-exchange resonances are observed in solid semibullvalene up to its melting point near -85°C. Thus changes that are mainly electronic and involve very little atomic motion can be radically affected by the change in phase. The first observation of a coalescence phenomenon in the solid state was made on the 2-norbornyl cation in SbF_5 (32), but here the behavior closely parallels that in solution. The 6,2 hydride shift is slowed, with T_c about -120°C, but the Wagner-Meerwein shift is still fast at -195°C (Figure 10). The nearly identical activation energies in the liquid and the solid contrast with the observations for the 2-butyl cation, for example, and point up our poor understanding of solid state effects on dynamic processes.

ACKNOWLEDGMENTS

The author is grateful to the National Science Foundation for support of part of the work described herein.

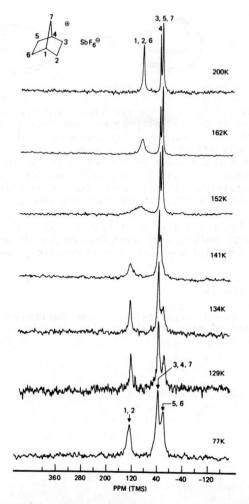

Figure 10. The ^{13}C CPMAS spectrum of the 2-norbornyl cation as a function of temperature. Reprinted with permission from Ref. 32. Copyright 1982, American Chemical Society.

REFERENCES

(1) L. M. Jackman and F. A. Cotton, "Dynamic Nuclear Magnetic Resonance Spectroscopy," Academic Press, New York, 1975.
(2) G. Binsch and H. Kessler, Angew. Chem., Intern. Ed. Engl., 19, pp. 411-428 (1980).
(3) D. S. Stephenson and G. Binsch, J. Magn. Reson., 32, pp. 145-152 (1978); Quantum Chem. Progr. Exch., 10, p. 365 (1978).
(4) D. S. Stephenson and G. Binsch, J. Magn. Reson., 37, pp. 395-407, 409-430 (1980); Quantum. Chem. Progr. Exch., 11, p. 378 (1979).
(5) H. Günther and J. Ulmen, Tetrahedron, 30, pp. 3781-3786 (1974).
(6) M. Feigel, H. Kessler, D. Leibfritz, and A. Walter, J. Am. Chem Soc., 101, pp. 1943-1950 (1979).
(7) D. Höfner, D. S. Stephenson, and G. Binsch, J. Magn. Reson., 32, pp. 131-144 (1978).
(8) E. E. Wille, D. S. Stephenson, P. Capriel, and G. Binsch, J. Am. Chem. Soc., 104, pp. 405-415 (1982).
(9) K. Mislow, Acc. Chem. Res., 9, pp. 26-33 (1976).
(10) B. H. Meier and R. R. Ernst, J. Am. Chem. Soc., 101, pp. 6441-6442 (1979).
(11) J. D. Cutnell, J. Am. Chem. Soc., 104, pp. 362-363 (1982).
(12) G. Bodenhausen and R. R. Ernst, J. Am. Chem. Soc., 104, pp. 1304-1309 (1982).
(13) P. Peringer and P.-P. Winkler, J. Organomet. Chem., 195, pp. 249-252 (1980).
(14) F. G. Riddell, R. D. Gillard, and F. L. Wimmer, J. Chem. Soc. Chem. Commun., pp. 332-333 (1982).
(15) J. B. Lambert, R. J. Nienhuis, and J. W. Keepers, Angew. Chem., Intern. Ed. Engl., 20, pp. 487-500 (1981).
(16) D. E. Woessner, J. Chem. Phys., 37, pp. 647-654 (1962); D. E. Woessner, B. S. Snowden, Jr., and G. H. Meyer, ibid., 50, pp. 719-721 (1969).
(17) K. F. Kuhlmann and D. M. Grant, J. Chem. Phys., 55, pp. 2998 -3007 (1971); J. R. Lyerla, Jr., and D. M. Grant, J. Phys. Chem., 76, pp. 3212-3216 (1972); T. D. Alger, D. M. Grant, and R. K. Harris, ibid., 76, pp. 281-283 (1972).
(18) K. H. Ladner, D. K. Dalling, and D. M. Grant, J. Phys. Chem., 80, pp. 1783-1786 (1976).
(19) N. Platzer, Org. Magn. Reson., 11, pp. 350-356 (1978).
(20) H. Beierbeck, R. Martino, and J. K. Saunders, Can. J. Chem., 58, pp. 102-109 (1980).
(21) P. Stilbs and M. E. Moseley, J. Magn. Reson., 31, pp. 55-61 (1978).
(22) D. M. Doddrell, M. R. Bendall, P. F. Barron, and D. T. Pegg, J. Chem. Soc. Chem. Commun., pp. 77-79 (1979).
(23) D. M. Doddrell, P. F. Barron, and J. Field, Org. Magn. Reson. 13, pp. 119-121 (1980).

(24) S. Forsén and R. A. Hoffman, J. Chem. Phys., 39, pp. 2892-2901 (1963).
(25) S. A. Sherrod and V. Bockelheide, J. Am. Chem. Soc., 94, pp. 5513-5515 (1972).
(26) B. E. Mann, J. Magn. Reson., 21, pp. 17-23 (1976).
(27) J. B. Lambert and J. W. Keepers, J. Magn. Reson., 38, pp. 233-244 (1980).
(28) M. M. Maricq and J. S. Waugh, J. Chem. Phys., 70, pp. 3300-3316 (1979).
(29) P. C. Myhre and C. S. Yannoni, J. Am. Chem. Soc., 103, pp. 230-232 (1981).
(30) R. D. Miller and C. S. Yannoni, J. Am. Chem. Soc., 102, pp. 7396-7397 (1980).
(31) A. K. Cheng, F. A. L. Anet, J. Mioduski, and J. Meinwald, J. Am. Chem. Soc., 96, pp. 2887-2891 (1974).
(32) C. S. Yannoni, V. Macho, and P. C. Myhre, J. Am. Chem. Soc., 104, pp. 907-909 (1982).

CHAPTER 6

NUCLEAR MAGNETIC RESONANCE IN SOLIDS

K. J. Packer

School of Chemical Sciences
University of East Anglia
Norwich, Norfolk, NR4 7TJ, U.K.

ABSTRACT

A qualitative discussion is given of the main anisotropic spin interactions which are important in the NMR of solids. The effects of time dependence, whether thermally produced or imposed by the experimenter, are discussed. The techniques whereby high resolution may be achieved in NMR in solids are outlined as are some features of spin relaxation specific to solids.

INTRODUCTION

From an experimental viewpoint the main distinction between solids and isotropic liquids or solutions in NMR spectroscopy is that the spectra of solids are generally rather broad and featureless. For example, the ^{1}H NMR spectrum of a typical organic solid can be a single line with a half-width of 10^4 Hz or more. This contrasts with the solution-state spectrum of the same material, which may cover only 10^3 Hz or less and comprise many lines with half-widths less than 1 Hz. These general statements have significant exceptions and should be taken only as a guide.

From a theoretical standpoint the significant difference between isotropic liquids or solutions and all other samples, including solids, is that in the latter the nuclei experience one or more orientation-dependent or anisotropic interactions which lead to significant splittings or broadening of the spectra. Clearly, these interactions themselves contain useful information and in this chapter we shall be concerned with the effects of these interactions on spectra and relaxation properties of spin systems and with ways in which we can selectively remove their effects from spectra to reveal other smaller interactions such as the chemical shift, so important in solution-state NMR.

ANISOTROPIC SPIN INTERACTIONS

The three main anisotropic interactions are (i) the magnetic dipole-dipole coupling, (ii) the electric quadrupole coupling ($I > 1/2$), and (iii) the magnetic shielding (chemical shift). These will be dealt with in turn with the emphasis being on qualitative understanding since quantitative expositions abound elsewhere (1-4).

Magnetic Dipole-Dipole Coupling

Figure 1 illustrates the direct magnetic coupling between two nuclear magnetic moments in the presence of the large static field B_0. One nucleus produces a local magnetic field at the site of the other and vice-versa. This local field can be thought of as having two components: (a) static, arising from the z-projection of µ and (b) rotating, arising from the x/y components of µ as it precesses.

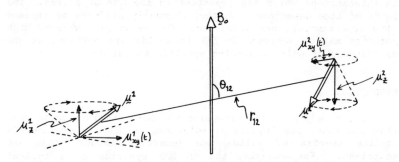

Figure 1. The dipole-dipole interaction between nuclear magnetic moments μ^1 and μ^2. Each moment precesses about B_0 and the static projections, μ_z^1 and μ_z^2, produce static local fields, whilst the rotating xy components produce time-varying local fields. The high-field, secular dipolar coupling energy depends on $(1 - 3\cos^2\theta_{12}/r_{12}^3)$.

In a many-spin system we can readily see how a broad line can be generated by consideration of the z components of the local dipolar field only. Figure 2 illustrates this. Nucleus i experiences a net local field, b_i^ℓ, which is the sum of all local fields (eqs. 1 and 2).

$$b_i^\ell = \sum_{j=1}^{N} b_{ij}^\ell \qquad (1)$$

$$b_{ij}^\ell = \mu_{zj}(1 - 3\cos^2\theta_{ij})/r_{ij}^3 \qquad (2)$$

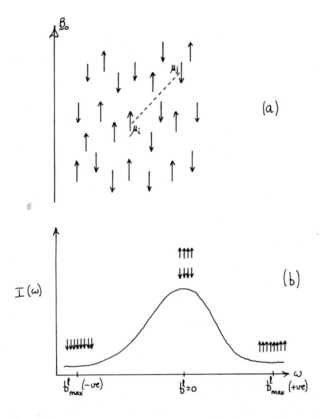

Figure 2. Qualitative illustration of the origin of the broad lines in the NMR spectra of abundant spins in solids. (a) A nuclear magnetic moment, μ_i, sees a net local field arising from all the other spins, typified by μ_j. (b) The resonance absorption is spread out over a range of frequency due to the spread of local field values, largely determined by the statistics of spin-up and spin-down states.

If we consider only the μ_{zj} term, then we can see that the total field, b_1^{ℓ}, will have a range of values depending on the statistics of whether the spins j are "up" or "down."

Apart from the small Boltzmann factor, up and down states are equally probable, thus we may imagine that resonance absorption for spin i will occur at frequency $\omega_i = \gamma[B_0 + b_i^{\ell}]$. Since b_i^{ℓ} has a range of values, so then will ω_i. This is illustrated in Figure 2(b). The width of this resonance absorption thus reflects the root-mean-squared dipolar local field.

The role of the time-varying local fields is somewhat more subtle. If the nuclear spins involved have the same resonance frequency, then the rotating local fields are resonant with the nuclear precession. This is just what is needed to cause the spin to flip from "up" to "down" or vice-versa and this is what they do. They undergo mutually induced, energy-conserving "flip-flop" transitions. In effect they irradiate each other with resonant radiation. These dynamic processes lead to the ability of strongly dipolar-coupled spin systems comprising like nuclei to disperse any local differences in M_z which may occur. This process is known as spin-diffusion and can cause systems with a few potent sources of relaxation to appear to relax at a much faster and often more uniform rate than expected. Examples would be found in the proton relaxation in systems such as solid hydrocarbons with methyl groups acting as relaxation sinks or in solid heterogeneous polymers (5).

For unlike nuclei, e.g., $^{13}C/^1H$, the rotating local fields are nonresonant and cannot bring about the energy-conserving flip-flop transitions. This is why the heteronuclear contribution to a dipolar dominated linewidth is less than for the corresponding homonuclear case. To use the usual terminology, the part of the Hamiltonian corresponding to this process is "nonsecular" for the heteronuclear interaction.

The general problem of calculating the lineshape for dipolar-coupled spins is not capable of analytic solution except for very simple cases, such as isolated pairs. This is because it is a many-body problem (1,2). Useful quantities, in such circumstances, are the moments of the lineshape. In particular the second moment, M_2, which is the mean-squared linewidth, is often used. This, we can see from the above discussion, is a measure of the strength of the dipolar local field, and its utility lies in the fact that its calculation does not require the solution of the many-body problem (1,2). For a single crystal eq. 3 holds,

$$M_2^{SC} \propto \sum_{i<j} (1-3\cos^2\theta_{ij})^2/r_{ij}^6 \qquad (3)$$

and for a powder or polycrystalline sample, eq. 4 holds.

$$M_2^P \propto \sum_{i<j} (r_{ij})^{-6} \qquad (4)$$

The significant facts of the dipolar interaction are that (i) it is a many-body interaction, (ii) it occurs between all magnetic particles in the sample, and (iii) it is a second-rank tensor interaction, i.e., is anisotropic with terms such as $(1-3\cos^2\theta_{ij})$ featuring in it. For spin 1/2 nuclei such as ^1H, ^{13}C, ^{15}N, ^{19}F, ^{29}Si, ^{31}P, etc., it is usually the dominant linewidth-determining interaction in solid samples.

Electric Quadrupole Coupling

Nuclei with I>1/2 have an ellipsoidal charge distribution. The nuclear magnetic moment is parallel to one of the axes of this ellipsoid. In the same way that a charge separation is a dipole and has an energy which is orientation dependent in a uniform electric field, an ellipsoid of charge has a quadrupole moment and has an energy which depends on orientation in an electric field gradient (efg). These facts are illustrated in Figure 3. Nuclei with I>1/2 then are generally subjected to non-zero efg's due to their not being at sites of cubic symmetry, for example, the ^{14}N nucleus in CH_3CN. In solids, these efg's will be static (ignoring any motion of the molecules in the lattice) and the nucleus, placed in an external magnetic field B_0, finds itself suffering a form of nuclear schizophrenia. The magnetic moment wishes to orient along B_0, the quadrupole moment along the efg. Which wins or what compromise is reached depends on the relative magnitudes of the interaction of the magnetic moment with B_0 (Zeeman energy) and of the nuclear electric quadrupole moment with the efg (quadrupole energy). Since quadrupole energies can be often of the order of MHz it can be seen that this problem may not be trivial.

To the extent that the Zeeman energy predominates, then the quadrupole interaction can be treated as a perturbation on the Zeeman levels and it lifts the degeneracy of these (2I + 1) states. Figure 3(b) illustrates this for I = 1 and I = 3/2. Apart from the 1/2 to -1/2 transition for half-integral spins, the quadrupole shifts in the levels and hence the splittings observed in the spectra are orientation dependent. In a powder sample, therefore, broad "powder" spectra are observed, their breadth often making observation difficult by pulse techniques. Indeed, for half-integral spins often only the 1/2 to -1/2 transition is observable.

Magnetic Shielding (Chemical Shift)

Spins are shielded from the applied field B_0 by the surrounding electrons, etc., such that the field experienced by nuclear moment, i, is expressed by eq. 5, in which $\underline{\sigma}_i$ is the

$$\underline{B}_i = (\underline{1} - \underline{\sigma}_i)B_0 \qquad (5)$$

shielding tensor for spin i.

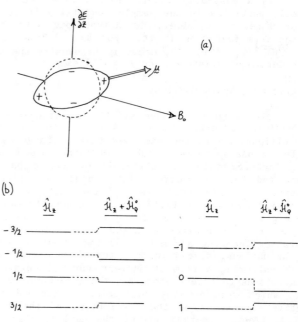

Figure 3. (a) A nuclear electric quadrupole moment showing the equivalent sphere (dotted line) of charge from which it may be considered to have arisen by removal of charge from the poles (say) and adding it to the equator. The magnetic moment is shown as colinear with the major axis of the charge ellipsoid. (b) The effects of the quadrupole interaction on the Zeeman energy levels ($\hat{H}_Z \gg \hat{H}_Q^o$) for I = 1 and 3/2.

The tensor nature of $\underline{\sigma}$ expresses the simple fact that the resonance frequency of a nucleus in a noncubic site of a molecule depends on the orientation of that molecule in B_0. In the most general case it is found that three independent numbers are required to specify the magnitude of the shielding in three molecule-fixed directions. Relative to the laboratory, three more numbers are required to specify the orientation of the molecule.

For suitably large B_0, the resonance frequency can be written as eq. 6 (6), in which $\delta = (\sigma_{11} - \sigma_i)$ and $\eta = (\sigma_{22} - \sigma_{33})/\delta$,

$$\omega = \omega_0 \delta [1/2(3\cos^2\theta - 1) + 1/2(\eta\sin^2\theta\cos^2\phi)] \qquad (6)$$

with $\sigma_i = 1/3(\sigma_{11} + \sigma_{22} + \sigma_{33})$. The convention $|\sigma_{11} - \sigma_i| > |\sigma_{22} - \sigma_i| > |\sigma_{33} - \sigma_i|$ has been used here.

Clearly, if a single crystal sample is studied, then changing its orientation in the field B_0 will yield information on the magnitude and orientation in the crystal lattice of the shielding tensor elements, σ_{jj} (j = 1,2,3). These quantities, σ_{jj} (j = 1,2,3) are known as the principal elements of the shielding tensor. If only a polycrystalline or powder sample is available, then the spectra will be a superposition of lines from all possible orientations, the intensity at any frequency being related to the statistical weight associated with the appropriate orientations giving rise to that frequency. These powder lineshapes are illustrated in Figure 4.

TIME DEPENDENCE AND AVERAGING

As we have indicated, solids generally give very broad, featureless NMR spectra whilst liquids may give very sharp lines, some $\sim 10^5$ times narrower. This would be true, for example, for the proton resonance in ice and liquid water at 273 K. Why is this? The answer can only be in the fact that in the liquid the nuclear spins are undergoing much faster and more developed relative motions, i.e., rotations, translations, etc., and this must clearly lead to an averaging away of the dipolar proton-proton local fields responsible for the broad line in the ice. This indeed is the reason and we should then ask ourselves how fast do the motions have to be and about what directions in space must they occur in order that averaging will take place.

As far as the rate is concerned it is a very simple matter to understand. Everyone is familiar with the idea in the high-resolution NMR of liquids that, if a spin can jump between two resonance frequencies which differ by Δ and does so with an average time between such transfers of τ, then if $\Delta\tau \gg 1$, two lines are seen characteristic of the two frequencies. On the other hand, if $\Delta\tau \ll 1$ then only a single line is observed at the average frequency. For the intermediate situation, $\Delta\tau \cong 1$, considerable broadening of the spectrum results. This well known "chemical exchange" phenomenon in liquid state high resolution NMR is illustrative of a general phenomenon referred to as motional averaging. Whenever a nuclear spin or spins experience an interaction which leads to splittings or broadening in the spectrum and which is made time dependent, in whatever manner,

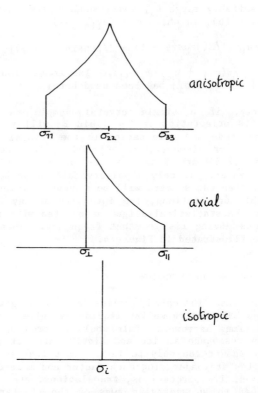

Figure 4. Powder lineshapes arising from the chemical shift interaction. The discontinuities correspond to the principal elements of the shielding tensors. In this example the axial and isotropic shapes are drawn as if they are obtained from the anisotropic case by appropriate averaging, e.g., $\sigma_\perp$ is represented as being the average of σ_{11} and σ_{22} and this could arise if the system had some motion which interchanged σ_{11} and σ_{22}; $\sigma_i = 1/3(\sigma_{11} + \sigma_{22} + \sigma_{33})$.

then when the product $\Omega\tau < 1$, averaging of the effects of the interaction will occur in so far as the spectrum is concerned. In this case τ is, as above, representative of the timescale of the time variations of the interaction involved and Ω is characteristic of the effect it produces in the spectrum when the particular time dependence being considered is too slow to have an effect.

Thus, for the dipolar interaction which produces a broad monotonic lineshape, Ω will be of the order of the width of this line and τ must be of the order of Ω^{-1} before averaging will begin. This can be seen in another way. A broad line, of width Ω, exhibits a free induction decay (FID) which decays with a relaxation time $T_2 \sim \Omega^{-1}$. Suppose we imagine applying a 90° pulse to such a system at time $t = 0$. The FID persists for a time of the order Ω^{-1}. If $\tau \gg \Omega^{-1}$ then, as the FID decays, no jump or motion will occur and, invoking causality, can then not affect the evolution of the FID. On the other hand, if $\tau < \Omega^{-1}$, then the time dependence may affect the FID and, hence, the spectrum.

The extent of the averaging produced by a particular time dependence depends on a number of factors. Consider the case of the proton dipolar couplings in a polycrystalline sample of benzene. It is well known that the least hindered thermally activated motion in this material (neglecting vibrations) is a jumping of the molecules about their sixfold symmetry axes. When $\tau_c(C_6)\omega_D < 1$ then a narrowing of the proton resonance line occurs. This is illustrated qualitatively in Figure 5. Note that well below the temperature for which $\omega_D\tau \sim 1$ the linewidth is temperature independent, as it is also well above that temperature. The reason for the fact that ω_D, the high temperature linewidth, is not zero is because the C_6 motion only makes <u>part</u> of the dipolar interaction time dependent. It would require fast motion about all three dimensions <u>and</u> fast translational diffusion of the molecules finally to reduce the dipolar interactions to their isotropic average, which is zero. Indeed, in materials such as liquid crystals, even though they may approach a normal isotropic fluid in mobility, there is a net, average anisotropy in the system which prevents the complete averaging of the dipolar spin couplings and similar anisotropic interactions. It is important to realise that, although the effects of the dipolar local fields on the spectrum are reduced, finally to zero for isotropic motion, the local fields are still present. Thus the instantaneous local dipolar magnetic field experienced by a proton in liquid benzene is not too different in magnitude from that in the solid--it is just very time dependent.

Before leaving this section we should broaden our view of what effects can produce time dependence and, hence, averaging of anisotropic spin couplings. Clearly, movement of the spins due to thermal motion of the molecules containing them causes the angles and distances, etc., defining the anisotropic spin interactions to be time dependent. There are, however, two approaches available to the NMR spectroscopist to impose time dependence on the spins. The first is just irradiation of spins with radiation at or close to their resonance frequency. In simple, pictorial terms, such irradiation drives transitions of

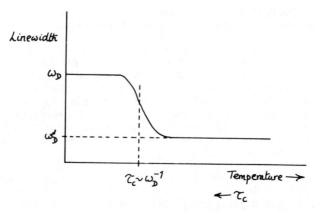

Figure 5. Schmatic illustration of motional narrowing. A system has a low temperature limiting linewidth represented by ω_D. When a motion, modulating the dipolar coupling giving rise to ω_D, occurs at such a rate that $\tau_C \omega_D \sim 1$, a new limiting linewidth, ω_D', is reached with $\omega_D' < \omega_D$. The value of ω_D'/ω_D depends on the spatial extent of the motion.

spins between their energy states, i.e., changes their orientations with respect to B_0 and hence makes any spin couplings time dependent. When the strength of the irradiation field is comparable to the coupling strength, averaging occurs and the effects of the coupling on the spectrum are progressively removed until with a field much larger than the coupling, complete averaging occurs. Such decoupling is of course well known in solution state NMR and in solids the principle is no different. There are, however, differences when it comes to homonuclear decoupling of dipolar interactions, and this will be returned to later on.

The second possibility for the NMR spectroscopist is to impose a macroscopic motion on the sample. This was first done by Andrew (8) and Lowe (9) who demonstrated both experimentally and theoretically that if a sample is rotated at a frequency ω_R about an axis making an angle θ with B_0 then if the spins in the sample experience some second-rank anisotropic interaction which has a strength expressed by Ω, when $\omega_R > \Omega$ the anisotropic parts of the interaction are averaged and scaled by a factor $1/2(3\cos^2\theta - 1)$. Clearly, if θ is chosen to be $\cos^{-1}(\sqrt{3})^{-1}$ (54.7°) then the scaling factor is zero. This angle is known as the "magic angle" and the process as magic angle spinning (MAS) or magic angle rotation (MAR). It is not generally of utility in averaging dipolar interactions since the spinning frequency required is often impractically large. Its present use is most often to remove the effects of chemical shielding anisotropy to yield isotropic average chemical shifts in solid state spectra.

HIGH RESOLUTION NMR IN SOLIDS

Whilst the dipolar interactions are of considerable interest in view of the structural information they contain, the broadening they produce in NMR spectra usually obscures the weaker chemical shift and indirect spin coupling information which is of great utility to the chemist. Over the last fifteen years or so a great deal of effort has been devoted to developing techniques which separately or in appropriate combinations allow high resolution NMR spectra of solids to be observed. The basic principles behind the techniques are just those discussed in the previous section, i.e., the use of appropriate averaging techniques, and we deal with them in turn below.

Multiple Pulse Cycles--Homonuclear Dipolar Decoupling

The use of so-called multiple pulse cycles for the suppression of homonuclear dipolar interactions is described in great detail in a number of places (3,4,10,11,12). All we do here is to consider the ideas behind the approach.

Historically, the first move in the direction of multiple-pulse line narrowing was brought about by the observation and explanation of the formation of a "solid-echo" following a resonant quadrature pulse-pair sequence, i.e., 90_x-τ-90_y, applied to a solid whose FID was dominated by homonuclear dipolar couplings. The significance of this observation was that the rapid decay of the FID, brought about by the dipolar couplings, could be reversed (partially) by the second pulse. This lengthening of the lifetime of the transverse magnetisation corresponds, of course, to a narrowing of the resonance line. Soon afterwards it was demonstrated that a multiple pulse sequence of the form $90_x(-\tau-90_y-\tau-)_n$ prolonged the lifetime of the NMR signal (and hence narrowed the resonance line) to a limit, at small τ, related only to the relaxation of the magnetisation along the B_{1y} field. A substantial averaging of the dipolar interactions was achieved, but everything else of interest was also averaged, including the chemical shift. Out of these investigations, however, grew the concept of pulse sequences consisting of cycles of pulses with the cycles designed so that the average of the dipolar interactions over the cycle tended to zero but that of the chemical shift in particular tended to a scaled but finite value. Such a pulse cycle sequence is illustrated in Figure 6. The crucial parameter is the cycle time, t_c, and in very general terms this cycle time must be made comparable to the timescale associated with the dipolar interactions to be averaged for the averaging to be efficient. Technical limitations of rf power dissipation, system recovery from overload, etc., place limitations on the minimum value of t_c achievable. There have been many such pulse cycle sequences proposed, analysed, and investigated. Of these the so-called MREV-8 and BR-24 sequences probably represent the best compromise between efficiency of averaging and ease of implementation (11).

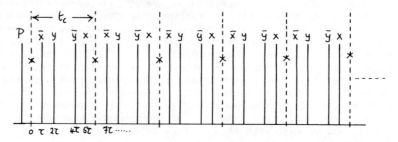

Figure 6. The WAHUHA multiple-pulse-cycle sequence for homonuclear dipolar decoupling. The repeated cycle lies between successive vertical dotted lines. The crosses represent typical signal sampling points. The initial pulse P produces transverse magnetisation and $x, y, \bar{x}, \bar{y}$ correspond to orthogonal directions in the rotating frame for the pulse phases.

Heteronuclear Dipolar Decoupling

An important situation encountered in many materials is one in which the magnetic nuclear isotope to be observed is present in low concentration, e.g., ^{13}C, which has ~1% natural abundance, but is accompanied by a high concentration of another nucleus, e.g., ^{1}H, ^{19}F, etc. In a solid the NMR linewidth for the magnetically dilute spin species is usually large, being dominated by the heteronuclear dipolar coupling to the abundant spin species. This linewidth contribution has exactly the same overall theoretical form as a heteronuclear J coupling, and hence, by irradiating the abundant spins at their resonance frequency with a radiation strength greater that the dipolar couplings, it can be removed. Details of this process and its dependence on a number of parameters can be found elsewhere (4,13). The main experimental difficulty is that the decoupling field strength has to be considerably greater than those employed for solution state J decoupling; 60 kHz would be a fairly common value with up to 100 kHz being necessary for the highest resolution for samples with very strong dipolar couplings. The rf power required to achieve these fields is measured in hundreds rather than tens of watts and it is important to have efficient and robust circuit components and to provide adequate cooling of the sample/coil area, usually using air or nitrogen gas.

Cross Polarisation

A feature of any NMR experiment is its sensitivity. NMR is an inherently insensitive spectroscopic technique due to the small Boltzmann factor at normal temperatures, and any technique for enhancing the signal is welcome. One factor which may limit sensitivity in addition to the small Boltzmann ratio is the value of the spin-lattice relaxation time. This is particularly so for studying systems with intrinsically very weak signals where many repetitions of the experiment must be added to achieve an acceptable signal-to-noise ratio. For dilute spins, such as ^{13}C, in solids spin-lattice relaxation can be very slow and can severely limit the rate at which experiments may be repeated. In addition, being dilute, these spins have an intrinsically weak signal anyway. The abundant spins on the other hand have a very strong signal by comparison. In fact the ratio of the equilibrium magnetisations for two spin species, I and S, is given by eq. 7.

$$\frac{M_0^I}{M_0^S} = \frac{N_I \omega_{0I}^2}{N_S \, \omega_{0S}^2} \tag{7}$$

For a typical organic molecule the ratio $(M_0(^1H)/M_0(^{13}C))$ may be of the order of 2000. In addition the spin-lattice relaxation times of the protons will generally be shorter than the corresponding ^{13}C values. Both of these factors make it sensible to generate the dilute-spin (^{13}C) signal from the abundant spins (^1H). In order to achieve this <u>cross polarisation</u> process, it is necessary simultaneously to irradiate both spin-species with resonant or near-resonant radiation, the amplitudes of these two rf fields satisfying a particular relationship which is dealt with below.

The necessity for this double resonance experiment can be understood by recognising that, in order to exchange polarisation, the different spin species must be able to undergo mutual flip-flop transitions. Clearly, for ^1H and ^{13}C in B_0, this process is not energy conserving and hence will occur only through a highly improbable multispin process or with participation of the lattice. Either makes it a very slow process. However, when irradiated with a large amplitude resonant rf field, a spin system behaves in many respects as if the B_1 field is playing the role of the B_0 field. Thus by simultaneous irradiation of ^1H and ^{13}C, say, with resonant B_1 fields which satisfy the condition of eq. 8, the two spin systems can undergo energy-conserving flip-

$$\gamma(^1H)B_1(^1H) = \gamma(^{13}C)B_1(^{13}C) \tag{8}$$

flop transitions at a rate dictated by the dipolar interaction strengths modified by the B_1 fields. This condition is known as the Hartmann-Hahn matching condition (14). The processes of cross polarisation and heteronuclear dipolar decoupling were first put together, analysed and tested by Waugh and his co-workers. A particularly important paper is reference 15, where all the basic ideas are set out.

There are many ways of carrying out these experiments, but that most commonly used is illustrated in Figure 7. The first stage consists of spin-locking the abundant spins along the rf field $B_1(^1H)$. This is achieved by a $90_x°$ pulse followed immediately by a long $B_{1y}(^1H)$ pulse. In this spin-locked state the proton magnetisation can be thought of as being quantised along $B_1(^1H)$. If we apply the Curie law to both laboratory (B_0) and rotating ($B_1(^1H)$) frames of reference, then eq. 9 follows,

$$M_0 \propto \frac{B_0}{T_L} = \frac{B_1(^1H)}{T_S(^1H)} \tag{9}$$

whence $T_S(^1H) = T_L(B_1(^1H)/B_0)$, in which $T_S(^1H)$ is the effective spin temperature of the system regarded as quantised along $B_1(^1H)$, and T_L is the lattice temperature. Typically we might have $T_S(^1H) = 300 \, (10/10^4) = 0.3K$.

Thus the purpose of the spin-lock preparation is to produce the protons in a state of very low spin temperature (high order) so they are in a favourable situation to transfer order (magnetisation) to the dilute ^{13}C spins.

The next stage in the experiment is the process of polarisation transfer. With the protons still spin-locked, an rf field is applied to the dilute spins (^{13}C) with an amplitude satisfying the Hartmann-Hahn condition. During this so-called "contact" pulse, ^{13}C magnetisation, $M(^{13}C)$, grows out along $B_1(^{13}C)$ until the two spin systems (in thermal contact) are at the same spin temperature. This is so when eq. 10 holds, in which $M_0(^{13}C)$ is

$$M(^{13}C) = (\gamma(^1H)/\gamma(^{13}C)) \cdot M_0(^{13}C) \tag{10}$$

the magnetisation that the ^{13}C spin system would have achieved via spin-lattice relaxation with the lattice at temperature T_L. When the field $B_1(^{13}C)$ is switched on, the effective spin temperature of the ^{13}C spins in their rotating frame, $T_S(^{13}C)$, is infinite (i.e., zero magnetisation along $B_1(^{13}C)$). The growth of $M(^{13}C)$ is the result of $T_S(^{13}C)$ reducing towards $T_S(^1H)$. There

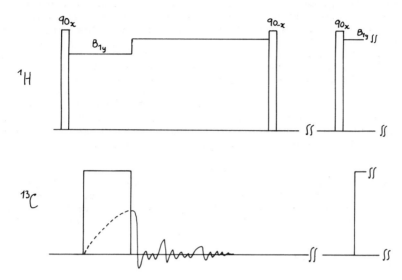

Figure 7. A typical double-resonance pulse sequence used for obtaining high resolution NMR spectra of dilute spins, e.g., ^{13}C, in solids in the presence of abundant spins, e.g., ^{1}H. The abundant spins are spin-locked and a contact pulse is applied to the dilute spins to effect polarisation transfer ($\omega_1(^1H) = \omega_1(^{13}C)$). The resulting FID is recorded in the presence of the dipolar decoupling field. The final 90_{-x} pulse flips any remaining magnetisation back along the z axis and is an alternative to using multiple contacts.

is an increase in the latter, arising from the polarisation transfer, but it is rather small due to the large heat capacity of the abundant spin system [$N(^1H) \gg N(^{13}C)$ and $\gamma(^1H) > \gamma(^{13}C)$].

Following the switching off of the contact pulse, the dilute spin signal may be observed as a FID in the presence of the strong dipolar decoupling due to $B_1(^1H)$.

A number of observations should be made here. Firstly, the typical rate of cross polarisation for carbons in organic materials is such as to require contact times of between 1 and 10 ms for complete equilibrium to be achieved. For nuclei with no directly bonded protons the longer times are appropriate. Secondly, having carried out one contact and acquired an FID (total time required $T_c + t_{aq}$), it is possible in principle to repeat the process of contact plus signal acquistion a number of times without having to wait for spin-lattice relaxation to occur. There are a number of factors limiting this approach in practice. Clearly, the abundant spin reservoir of polarisation is finite. More importantly, the abundant spin spin-locked magnetisation undergoes relaxation which diminishes it towards zero. This process, designated by the time constant $T_{1\rho}(^1H)$, clearly limits the length of time the abundant spins can be held at a low temperature in the rotating frame. For solid organic polymers, for example, it is often the case that the proton $T_{1\rho}$ values prohibit more than a single contact. In some cases cross polarisation may not be possible (16-18).

The final limitation to multiple contact operation is the power-handling capability of the spectrometer, the sample probe, and associated electronics in particular. The acquisition time for a FID is determined by the required or achievable spectral resolution. Even for modest resolution of the order of 10 Hz, t_{aq} would be 0.1 s, so ten contacts would require of the order of 1 s continuous irradiation of ~60 kHz amplitude, say, and a $T_{1\rho}(^1H)$ of the order of 1 s or greater.

Magic Angle Rotation and Powder Samples

When the above techniques are applied to appropriate materials we have still not completed the task of providing the chemist with his familiar high resolution spectrum. This is because the chemical shift is an anisotropic interaction as we have already noted (see Figure 4). This is where magic angle sample spinning is introduced. In the simplest case the sample is spun at a rate comparable to the largest anisotropy in the spectrum, under which conditions the spectra take on the appearance typical of high resolution in liquids with lines at each isotropic average shift value. Figure 8 illustrates the effects of both dipolar decoupling and magic angle sample rotation.

SPIN-RELAXATION IN SOLIDS

As with spins in liquids, relaxation in solids requires the existence of fluctuating spin interactions and the important couplings in this regard are predominantly magnetic dipolar or quadrupolar. There are many discussions of relaxation behaviour elsewhere (1,2,4,19-21) so we shall concentrate here on some questions peculiar to solids.

The first point to be noted is that motion in solids is generally restricted and highly anisotropic. As noted earlier, this means that the anisotropic spin interactions are only partly

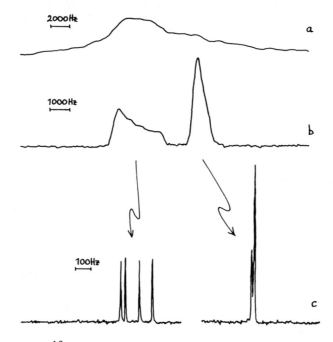

Figure 8. ^{13}C spectra of $Ca(CH_3CO_2)_2 \cdot H_2O$ powder taken under various experimental conditions. (a) Single resonance; (b) cross polarisation with dipolar decoupling; and (c) as for (b) but with magic angle rotation of the sample.

averaged and fluctuating. Generally, the fluctuating part of the interaction determines the spin-lattice relaxation behaviour via its spectral density at the appropriate frequencies whilst the residual, nonaveraged part often dominates the lineshape and linewidth. For example, the C_6 motion in solid benzene causes part of the proton dipolar coupling to be time dependent (eq. 11, in which $\hat{H}_D'(t)$ has a zero average value and is that part of

$$\hat{H}_D(t) = \langle \hat{H}_D \rangle_{C_6} + \hat{H}_D'(t) \tag{11}$$

the interaction modulated by the C_6 motion). The term $\hat{H}_D'(t)$ determines the rate of spin-lattice relaxation via its spectral density $J(\omega)$ (eq. 12), whilst $\langle \hat{H}_D \rangle_{C_6}$, the average over the C_6

$$J(\omega) \propto \int \langle \hat{H}_D'(t) \hat{H}_D'(t+\tau) \rangle e^{i\omega t} d\tau \tag{12}$$

motion is still "secular" or static in character and dominates the linewidth. Thus the definition of a unique T_2 in a solid is usually not possible unless the lineshape is close to a simple function such as a gaussian. The time scale of transverse relaxation is thus usually dominated by the static parts of the interaction.

The presence of a static interaction, $\hat{H}_S^o$, such that the total secular Hamiltonian has the form of eq. 13, gives rise

$$\hat{H}_T^o = \hat{H}_Z + \hat{H}_S^o \tag{13}$$

to the possibility of other forms of relaxation than the conventional T_1 and $T_{1\rho}$ processes. We should recall that these latter processes are the relaxation times of the magnetisation along the B_0 and $B_1(\omega_0)$ fields, being sensitive to fluctuations at frequencies of the order of ω_0 (= γB_0) and ω_1 (= γB_1) respectively.

To measure $T_{1\rho}$ the process of spin-locking is used, as illustrated in Figure 9. In any system, liquid or solid, effective spin-locking requires that the B_1 field strength be larger than the resonance linewidth. If it were not so then the magnetisation components across the line would dephase. Thus for spin-locking of protons in organic solids, B_1 field strengths of up to 100 kHz (=$\gamma_H B_1$) may be required. We might then ask what happens if we reduce B_1 to become comparable to or less than the linewidth. The quantitative details depend on exactly how we carry this out, but qualitatively we find that we transfer some of the spin order represented by the magnetisation M into what is usually referred to as dipolar order. There is no macroscopic magnetisation associated with this state, but the order or entropy is still there. The spins are now ordered locally with respect to their mutual local fields. Raising the B_1 field to a value greater than the local field (~ linewidth) will transfer the order back into Zeeman order along B_1, i.e., a spin-locked state. There are many interesting facets of the states of order in internal secular spin interactions, but all we will mention here are (i) their creation and (ii) relaxation.

There are two main methods of creating a state of dipolar order. The first is so-called Adiabatic Demagnetisation in the Rotating Frame (ADRF) (22), illustrated in Figure 10. After spin-locking, the B_1 field is reduced at an appropriately slow rate. The second technique is a pulse method originally discussed by Jeener and Broekaert (23). The sequence used is 90_x-τ-45_y and it suffices to say that, following this two-pulse preparation applied to a strongly dipolar-coupled set of spins (such as protons in an organic solid) with τ chosen to be around the point of most rapid change of the FID, a substantial transfer

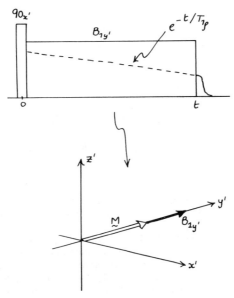

Figure 9. Measurement of $T_{1\rho}$. A $90_{x'}$ pulse is immediately followed by a $B_{1(y')}$ pulse of varying length. The signal obtained after this pulse measures the relaxation of the magnetisation along B_1. The spin-locked state is illustrated in the lower figure.

of Zeeman into dipolar order is achieved. Having created such a state we should note that we have effectively cooled the dipolar energy states of the system; that is, those energy levels associated with $\hat{H}_D^0$ have a population distribution characteristic of a very low temperature, very different from that of the lattice. As a consequence, the dipolar order relaxes to the lattice with a rate often represented by $(T_{1D})^{-1}$, the dipolar-lattice relaxation rate which in general depends on fluctuations at frequencies of the order of ω_L (= γb_{local}) and ω_0.

With the advent of high resolution techniques in solids it became of interest to measure site-specific relaxation times. For the typical double-resonance $^{13}C/^1H$ type of experiments, there are no real problems in measuring ^{13}C T_1's under MAR conditions, except possibly the often very large values. When it comes to $T_{1\rho}$, however, for the dilute spin species, then there can be difficulties in interpretation. Suppose we have created some spin-locked magnetisation say for ^{13}C in a solid, probably via cross polarisation. To measure $T_{1\rho}$ we would switch off the

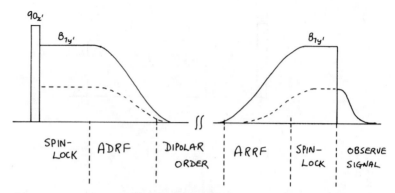

Figure 10. Creation of dipolar order using ADRF. The magnetisation is spin-locked and then the rf field amplitude is reduced to a value less than the dipolar fields at an appropriately slow rate to keep the process adiabatic or isentropic. The dipolar-ordered state persists for times of the order of T_{1D} and, for example, may be examined by a reversal of the same process, ARRF (adiabatic remagnetisation in the rotating frame).

proton rf field for the duration of the relaxation, switching it on again during data acquisition to give resolved chemical shifts. The problem arises from the fact that the strong proton-proton dipolar couplings cause fluctuations in the proton-carbon dipolar interactions. If these fluctuations contain frequencies around $\omega_1(^{13}C)$ then these may be a potent source of relaxation, the spin-locked ^{13}C signal undergoing a cross-relaxation into proton dipolar order and thence to the lattice. This spin-spin process thus may obscure the more informative contribution related to molecular motions. Generally, in highly crystalline organic solids, the spin-spin process dominates, but for glassy polymers the motional contributions may often be obtained (24).

REFERENCES

(1) A. Abragam, "The Principles of Nuclear Magnetism," Clarendon Press, Oxford, 1961.
(2) C. P. Slichter, "Principles of Magnetic Resonance," 2nd ed., Springer-Verlag, 1978.
(3) U. Haberlen, "High-resolution NMR in Solids, Selective Averaging," Suppl. 1 to Advan. Magn. Reson., J. S. Waugh, Ed., Academic Press (1976).

(4) M. Mehring, "NMR Spectroscopy in Solids," NMR Basic Principles and Progress, P. Diehl, E. Fluck, and R. Kosfeld, Eds., 11, Springer-Verlag, 1976.
(5) V. J. McBrierty, Faraday Disc. Chem. Soc., 68, p. 78 (1979).
(6) Reference 3, p. 24.
(7) "Magnetic Resonance in Colloid and Interface Science," H. A. Resing and C. G. Wade, Eds., ACS Symposium Series, No. 34, 1976.
(8) E. R. Andrew, A.. Bradbury, and R. G. Eades, Nature Lond., 182, p. 1659 (1958).
(9) I. J. Lowe, Phys. Rev. Lett. 2, p. 285 (1959).
(10) P. Mansfield, Progr. NMR Spectrosc., 8, p. 41 (1972).
(11) B. C. Gerstein, Phil. Trans. Roy. Soc. Lond., A299, p. 521 (1981).
(12) P. Mansfield, Phil. Trans. Roy. Soc. Lond., A299, p. 479 (1981).
(13) A. N. Garroway, D. L. VanderHart, and W. L. Earl, Phil. Trans. Roy. Soc. Lond., A299, p. 609 (1981).
(14) S. R. Hartmann and E. L. Hahn, Phys. Rev. 128, p. 2042 (1962).
(15) A. Pines, M. G. Gibby, and J. S. Waugh, J. Chem. Phys., 59, p. 569 (1973).
(16) J. Schaefer and E. O. Stejskal, Top. Carbon-13 NMR Spectrosc., 3, p. 283 (1979).
(17) J. R. Lyerla, Contemp. Top. Polym. Sci., 3, p. 143 (1979).
(18) E. O. Stejskal, J. Schaefer, and T. R. Steger, Faraday Symp. Chem. Soc., 13, p. 56 (1978).
(19) H.W. Spiess, NMR Basic Principles and Progress, P. Diehl, E. Fluck, and R. Kosfeld, Eds., Springer-Verlag, 15, p. 55, 1978.
(20) F. Noack, NMR Basic Principles and Progress, P. Diehl, E. Fluck, and R. Kosfeld, Eds., Springer-Verlag, 3, p. 83, 1971.
(21) H. Pfeiffer, NMR Basic Principles and Progress, P. Diehl, E. Fluck, and R. Kosfeld, Eds., Springer-Verlag, 7, p. 53, 1972.
(22) M. Goldman, "Spin Temperature and Nuclear Magnetic Resonance in Solids," Oxford University Press, 1970.
(23) J. Jeener and P. Broekaert, Phys. Rev. , 157, p. 232 (1967).
(24) J. Schaefer, E. O. Stejskal, M. D. Sefeik, and R. A. McKay, Phil. Trans. Roy. Soc. Lond., A299, p. 593 (1981).

CHAPTER 7

APPLICATIONS OF HIGH RESOLUTION DEUTERIUM MAGNETIC RESONANCE

Harold C. Jarrell and Ian C. P. Smith

Division of Biological Sciences
National Research Council of Canada
Ottawa, Canada K1A OR6

ABSTRACT

High resolution ^{2}H NMR has proved to be a valuable source of information on a variety of systems. Examples of the use of ^{2}H NMR to solve stereochemical problems are discussed. In addition, the use of ^{2}H NMR to elucidate chemical reaction mechanisms and biochemical transformations is presented. Relaxation times of ^{2}H have provided a wealth of information on molecular dynamics, some examples of which are discussed.

INTRODUCTION

Much of the interest in deuterium as a nucleus for NMR studies has been associated with systems undergoing relatively slow anisotropic motions. In these systems the quadrupolar interaction between the nuclear quadrupole moment and the electric field gradient at the nucleus is manifested as a shifting of the Zeeman energy levels resulting in broad spectra such as those shown in Fig. 1A-D. However, as the rate of tumbling of the molecule increases, the influence of the quadrupolar interaction is reduced (Figure 1E-F), and in the limit of isotropic motion which is fast compared with the inverse of the quadrupolar coupling constant a relatively narrow high resolution spectrum is observed (Figure 1G). Deuterium has proved to be a useful nucleus for high resolution NMR studies, providing information which would be difficult to obtain by other techniques. Examples of the applications of ^{2}H NMR under high resolution conditions will be discussed in this chapter. Extensive surveys of the numerous applications of ^{2}H NMR may be found in recent reviews (1,2).

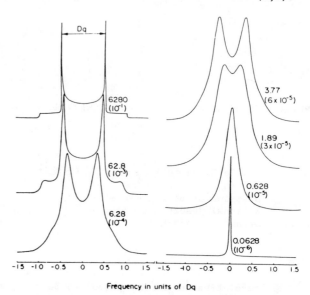

Figure 1. The influence of the correlation time for overall vesicle rotation (τ_r) on the shape of the ^{2}H NMR spectrum calculated for an ordered CD moiety with a residual quadrupolar splitting, D_q, of 10 kHz. The numbers beside the individual spectra represent the quantities $2\pi D_q \tau_r$ and τ_r (in s rad^{-1}) (in brackets). Data taken from Ref. 1.

GENERAL CONSIDERATIONS

The magnetic constants and natural abundance of the hydrogen isotopes are given in Table I. The low magnetogyric ratio and natural abundance of deuterium (a receptivity which is 1.45 x 10^{-6} of that of ^{1}H) would appear to make ^{2}H NMR hopeless from a practical point of view. However, with present day instrumentation observation of ^{2}H at natural abundance is not difficult, as reflected in the natural abundance ^{2}H NMR spectrum of pyridine (Figure 2). In addition because of the relative ease and low cost of labeling with ^{2}H, ^{2}H NMR has proved to be particularly useful. Indeed only slight enrichment leads to significant gains in signal enhancement; an enrichment to 1% leads to an approximate 64-fold signal enhancement and a concomitant ~4100-fold time saving.

Table I. NMR Properties of Hydrogen Isotopes

Isotope	Natural abundance (%)	Magnetic[a] moment μ/μ_N	Magnetogyric ratio $(\gamma/10^7)$ rad $T^{-1}s^{-1}$	Quadrupole moment $Q/10^{-28}$ m^2	Relative[b] receptivity
^{1}H	99.985	4.8371	26.7510	–	1.00
^{2}H	0.015	1.2125	4.1064	2.73×10^{-3}	1.45×10^{-6}
^{3}H	–	5.1594	28.5335	–	–

[a] μ_N, the nuclear magneton, $= 5.05095 \times 10^{-27}$ JT^{-1}.
[b] Relative to ^{1}H.

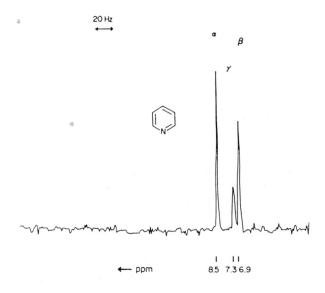

Figure 2. Natural abundance ^{2}H NMR spectrum (15.4 MHz) of neat pyridine. The spectrum was obtained without proton decoupling and required 17,000 accumulations. Note that the ^{1}H-^{2}H spin couplings are not observed. Data from Ref. 1.

Deuterium chemical shifts, in parts per million, are essentially the same as those of the ^{1}H isotope in analogous systems.

However, since the chemical shift is proportional to the magnetogyric ratio (γ), the actual values of the ^{2}H chemical shifts in hertz are only ~15% those of their ^{1}H analogues. Although the frequency range is compressed relative to ^{1}H, at a given field, spectra frequently are well resolved as is observed for pyridine (Figure 2) and 2-propanol (Figure 3). With the high magnetic fields currently available, overlap of resonance lines has become even less of a problem.

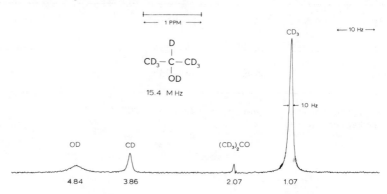

Figure 3. ^{2}H NMR spectrum (15.4 MHz) of neat perdeutero-2-propanol containing a trace of acetone-d_6 as an internal reference. Note that although the spectral width in Hz is small (240 Hz) the individual ^{2}H resonances are well resolved. Data taken from Ref. 1.

Since scalar coupling between identical nuclei is proportional to γ^2, ^{2}H-^{2}H scalar couplings are ~2.4% of the corresponding ^{1}H-^{1}H couplings, usually less than 0.5 Hz, and hence are frequently too small to be detected. Heteronuclear coupling constants, $J(X,^2H)$, are reduced by a factor of 6.51 relative to the corresponding $J(X,^1H)$. The reduced couplings lead to spectral simplification, as is evident in the ^{2}H NMR spectrum of pyridine (Figure 2). Couplings to ^{1}H contribute to the ^{2}H linewidths, which may be narrowed significantly by ^{1}H decoupling.

The ability to obtain chemical shifts from ^{2}H spectra, without resorting to spectral simulation, is a useful first step towards the interpretation of complex ^{1}H NMR spectra (1). In ^{1}H spectra the coupling between magnetically equivalent nuclei is not manifest (3). By substitution with a deuteron, the $J(^1H,^2H)$ can be determined, as for methanol (Figure 4), and the corresponding $J(^1H,^1H)$ calculated according to eq. 1 (1,2).

$$J(^1H, ^1H) = \frac{\gamma_{1_H}}{\gamma_{2_H}} J(^1H, ^2H) \qquad (1)$$

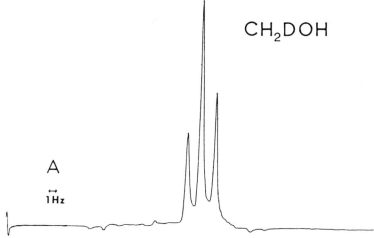

Figure 4. ^2H NMR spectrum (15.4 MHz) of CH_2DOH as a 5% (v/v) solution in carbon tetrachloride. Note the well resolved ^1H-^2H scalar coupling. Data taken from Ref. 1.

APPLICATIONS

Reaction Mechanisms and Kinetics

In studies of reactions involving hydrogen, one has traditionally monitored the disappearance of resonances due to the incorporation of ^2H. In cases where the reaction is incomplete or involves a number of positions, the ^1H NMR spectra may become difficult to interpret. Thus ^2H NMR has provided a particularly convenient means of selectively monitoring the reaction site since separate resonances may be observed. The often negligible scalar couplings are advantageous in such studies.

Stothers and co-workers (3) have examined the stereochemistry and kinetics of the β-enolization of a number of cyclic ketones such as adamantanone in the presence of potassium tert-butoxide in tert-butanol-d$_1$ (Figure 5). The ^1H spectrum exhibits five types of nonequivalent protons. Eight protons are beta to the carbonyl function; the endo-β protons are readily distinguished from the exo-β protons by their larger chemical shifts, induced by shift reagent (Figure 5A). The ^{2}H spectrum of the enolized product (Figure 5B) is very simple, with a single line for each nonequivalent deuteron. Inspection of Figure 5 clearly indicates that enolization, and hence exchange, involves the exo-β protons stereoselectively. By using ^{2}H spectral areas in conjunction with mass spectroscopic results, the rates of exchange at the various positions were determined. More recently ^{2}H NMR has been used to elucidate the details of the reaction mechanism involved in the deamination of α-aminoketones (4).

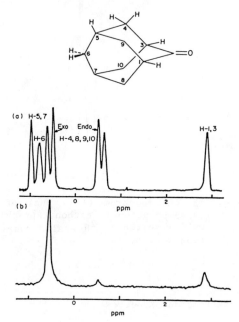

Figure 5. (A) ^{1}H NMR spectrum (100 MHz) of adamantanone; (B) ^{2}H NMR spectrum (15.4 MHz) of deuterated adamantanone resulting from treatment of adamantanone with potassium t-butoxide and t-butanol-d. Both samples contained hexafluorobenzene/chloroform (1:4) as solvent and 0.17 equivalent of Pr(fod)$_3$. Data taken from Ref. 3.

In a biosynthetic study, the stereochemical course of β-lactam formation in Nocardicin A was established (5). By introducing ^{2}H-labeled serine into cultures of Nocardia uniformis the sites of ^{2}H enrichment were easily determined by ^{2}H NMR (Figure 6). The crucial points in the study were the possibilities to distinguish sites A and B (Figure 6) and to follow their evolution from the precursor. The precursor, stereospecifically labeled with ^{2}H, was incorporated into Nocardicin A without configurational change, so that the stereochemical course of β-lactam formation was established.

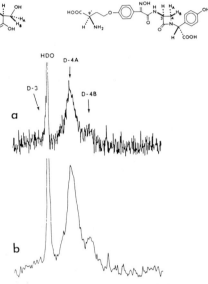

Figure 6. ^{2}H NMR spectrum (46.1 MHz) of (A) Nocardicin A biosynthetically enriched in ^{2}H by growth on a specifically labeled serine; H_A was replaced by ^{2}H; (B) spectrum as in (A) but with 1.5 Hz line broadening to emphasize the resonances D-4A and D-4B. Reprinted with permission from Ref. 5. Copyright 1982, American Chemical Society.

In another mechanistic study the biochemical transformation of dehydrogriseofulvin (Figure 7) was elucidated (6). The precursor was hydrogenated by the organism in a fermentation medium containing D_2O. The ^{2}H spectrum of the product, griseofulvin, shown in Figure 7A, was interpreted as arising from deuterons at the 5'-β and 6' positions; signals were assigned based on ^{1}H studies. The assignment of the 6' deuteron resonance was achieved by selectively removing the labile 5'-β-^{2}H to give Figure 7B. Due to the simplicity of the ^{2}H spectrum and selective monitoring of the reaction sites the course of the biotransformation was unambiguously elucidated.

The applications discussed so far have ranged from a well defined chemical system to more complex biochemical systems. Deuterium NMR has been used in even more complex situations. The products of in vivo mammalian metabolism of the urinary anti-

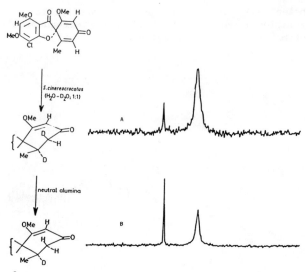

Figure 7. ^{2}H NMR spectrum (15.28 MHz) of the microbial transformation products of dehydrogriseofulvin in chloroform. (A) Product from fermentation showing deuterons at the 5'-β and 6'-β positions; (B) product in (A) after exchange with H_2O under mildly alkaline conditions, showing the loss of the 5'-β deuteron, and hence confirming the assignment of the 6'-β deuteron resonance. Data from Ref. 6.

biotic nalidixic acid have been followed by ^{2}H NMR (1). Specifically labeled drug was administered to female monkeys and the concentrated urine examined by ^{2}H NMR (Figure 8). The results demonstrated that one metabolic pathway involved the hydroxylation of a methyl group. Clearly the approach circumvented the need for complicated separations of metabolic products, since ^{2}H resonances arise only from the drug or its degradation products.

Stereochemistry

Since the ^{2}H spectrum of a compound is relatively simple compared with its ^{1}H analogue, chemical shifts may be measured directly. This can be extremely useful in stereochemical problems. One example, Nocardicin biosynthesis, has already been discussed. The ^{2}H NMR of selectively substituted 2-benzamido-4,5-d_2-norborneols allows easy distinction between the exo and endo deuterons (Figure 9) (1). Such stereochemical information was crucial to mechanistic studies of reactions involving protonated cyclopropanes (4,7).

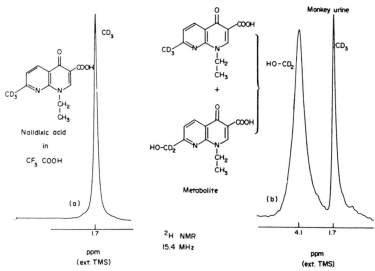

Figure 8. ^{2}H NMR spectra (15.4 MHz) (A) of nalidixic acid deuterated at the aromatic methyl group and (B) of freeze-dried urine from a monkey treated with the deuterated drug. Spectra required 1500 accumulations. Spectrum B indicates the presence of a metabolite of nalidixic acid, and where in the drug the modification occurred. Data taken from Ref. 1.

The presence of an asymmetric centre often induces chemical shift nonequivalence in ^{1}H and ^{2}H spectra, but in the case of ^{1}H, detection of nonequivalence is not always straightforward. The nonequivalence of methylene deuterons, D_A and D_B, of selectively deuterated chiral 1,2-disubstituted propanes (Figure 10) is readily detected, even in the case of rapid interconversion between conformers (1).

The ability to distinguish between deuterons in various environments is particularly useful in the determination of rates of conformational interconversion as is illustrated by the cis-trans isomerization of N-nitroso-2,5,5-d_3 proline (1). The ^{2}H spectrum of 2,5,5-d_3-proline is readily interpreted (Figure 11A), as is that of the Z-form of N-nitrosoproline (Figure 11B). Isomerization occurs by a 180° rotation about the N-N bond. At equilibrium, the presence of both isomers is readily detected, and the Z:E isomer ratio calculated (Figure 9C). The rate of isomerization was conveniently determined from the time course of signal intensities (1).

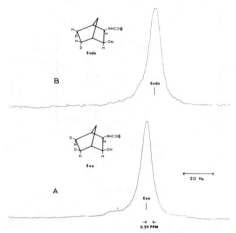

Figure 9. ^{2}H NMR spectra (15.4 MHz) of selectively deuterated exo- (A) and endo- (B) 2-benzamido-4,5-d$_2$-norborneol in chloroform. Data taken from Ref. 1.

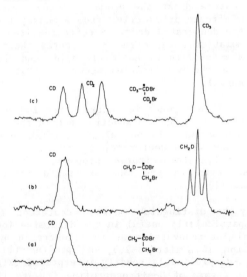

Figure 10. ^{2}H NMR spectra (15.4 MHz) of selectively deuterated 1,2-disubstituted propanes in carbon tetrachloride solution. The nonequivalence of methylene deuterons adjacent to the chiral centre is evident in (C). Data taken from Ref. 1.

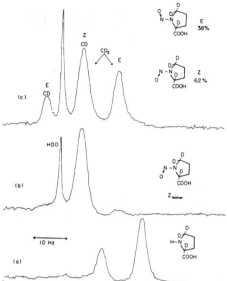

Figure 11. ^{2}H NMR spectra (15.4 MHz) of aqueous solutions of (A) 2,5,5-d_3-proline and (B) N-nitroso-2,5,5-d_3-proline as its Z-isomer; (C) after sample in (B) had equilbrated at room temperature for 24 h. Note that the Z and E isomers are readily detected and may be quantitated. Data taken from Ref. 1.

RELAXATION TIMES

Relaxation of deuterium is almost always dominated by the interaction of the quadrupole moment and the electric field gradient at the nucleus, which is modulated by molecular motions. Since the dipole-dipole interaction depends upon $(\gamma_{2H})^2(\gamma_X)^2$, the ^{2}H-^{1}H and ^{2}H-^{2}H interactions are therefore ~42 and 1778 times weaker, respectively, than the corresponding ^{1}H-^{1}H interaction. The dominance of the quadrupolar relaxation mechanism makes ^{2}H an ideal nucleus for studying molecular motion or exchange processes. Relaxation of ^{2}H has found numerous applications (1,2,14).

In the limit of rapid isotropic motion ($\omega_0^2\tau_c^2 \ll 1$) the spin-lattice (T_1) and spin-spin (T_2) relaxation times are given by eq. 2 (8), in which e^2qQ/h is the quadrupole coupling constant,

$$\frac{1}{T_1} = \frac{1}{T_2} = \frac{3}{8}\left(1 + \frac{\eta^2}{3}\right)\left(\frac{e^2qQ}{h}\right)^2 \tau_c \qquad (2)$$

η the asymmetry parameter of the electric field gradient, and τ_c the correlation time for rapid reorientation. The asymmetry parameter for a C-^{2}H fragment is usually very close to zero, so that only the quadrupole coupling constant is required to calculate τ_c from eq. 2.

The dipolar contribution to the T_1 relaxation time of ^{13}C in a ^{13}C-^{1}H residue is given by eq. 3, in which r is the C-H bond

$$\frac{1}{T_1(^{13}C)} = n\hbar^2 \gamma_C^2 \gamma_H^2 r^{-6} \tau_c \qquad (3)$$

distance and n is the number of protons attached to the ^{13}C nucleus; fast isotropic motion is assumed. Because of the r^{-6} term the ^{13}C relaxation comes from intramolecular interactions so that the ^{2}H and ^{13}C relaxation may be related through eq. 4 (9). Substitution of reasonable values for r and the

$$\frac{T_1(^2H)}{T_1(^{13}C)} = \frac{8}{3} \frac{n\hbar^2(\gamma_C)^2(\gamma_H)^2}{r^6(e^2qQ/h)^2} \qquad (4)$$

quadrupole coupling constants into eq. 4 provides a means of estimating the contribution of interactions other than dipole-dipole to ^{13}C relaxation. Alternatively, a comparison of the corresponding ^{2}H and ^{13}C relaxation times provides a method of calculating the quadrupole coupling constant. Representative examples of the two uses are given in Table II.

Table II. Comparison of Observed ^{13}C T_1 Values with Those Calculated from ^{2}H T_1 Values, and Quadrupole Coupling Constants Calculated from Equation 4[a]

Compound	Observed group	^{13}C T_1 (s)		Quadrupole coupling constant (kHz)	
		Calc.	Obs.	Calc.	Obs.
Acetic acid	CH$_3$	9.7	9.8	176	168
Cyclohexane	CH$_2$	14.6	17.0	183	174
Phenylacetylene	≡C-H	7.8	9.3	254	215
Benzene	CH	33.9	29.3	172	186

[a]Data from Ref. 1.

Deuterium relaxation times have been used to study molecular complexation and binding interactions (1,2). The transverse relaxation time, T_2, has recently been used to study substrates and inhibitors interacting with soybean lipoxygenase (10). The expression for T_2 is given by eq. 5, in which ω_0 is the ^{2}H Larmor

$$\frac{1}{T_2} = \frac{3}{80} \left(\frac{e^2qQ}{h}\right)^2 [3\tau_c + \frac{5\tau_c}{1+\omega_0^2\tau_c^2} + \frac{2\tau_c}{1+4\omega_0^2\tau_c^2}] \quad (5)$$

frequency, and τ_c is the effective correlation time for the motion; the electric field gradient is assumed to have cylindrical symmetry ($\eta=0$). The observed value, $T_{2,ob}$, is given for the binding interaction by eq. 6, in which $T_{2,F}$ and $T_{2,B}$ are the

$$\frac{1}{T_{2,ob}} = \frac{1}{T_{2,F}} + \frac{f}{T_{2,B}+\tau_B} \quad (6)$$

transverse relaxation times of free and bound substrate, respectively, τ_B is the lifetime of the bound molecule, and f is the fraction of the bound moiety. The T_2 values were determined from the linewidths required to fit the ^{2}H spectra to a single Lorentzian function (Figure 12). Since T_2 increased with temperature, τ_B was concluded to be much less than $T_{2,B}$, that is, the interaction was in the fast exchange limit. The variation of $T_{2,ob}$ with f (determined by other studies) permitted $T_{2,B}$ to be calculated for various positions of the fatty acyl chain; typically the linewidth increase was from ~17 Hz (free) to ~2500 Hz (bound). The positional dependence of $T_{2,B}$ indicated the points of the fatty acid which bound strongly to the enzyme (stronger binding leads to decreased mobility and a shorter $T_{2,B}$).

In some cases additional information about the nature of molecular motion can be deduced from relaxation studies. In a recent ^{2}H NMR study of the binding of sugars to wheat germ agglutinin, the association constant, K_d, was calculated from eq. 5 using T_2 values measured from linewidths (11). In addition, details of the molecular dynamics of the bound sugars were delineated. The effective correlation time for the label was calculated from linewidths using eqs. 5 and 6 as described above. The correlation time of the protein was calculated using the Stokes-Einstein relationship, $\tau_R = 4\pi\eta r^3/3kT$, where r is the radius of the protein (assumed to be spherical) and η is the solution viscosity; τ_R was calculated to be 1.6×10^{-8} sec at 35°C. The correlation times for the labels on the bound sugars are given in Table III. It is seen from Table III that the correlation time of ^{2}H attached to the sugar ring agrees to within a factor of two with that of the protein, indicating that the six-membered ring

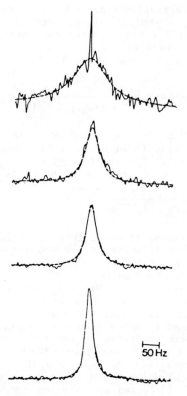

Figure 12. ^{2}H NMR spectra (15.4 MHz) of [9,10,12,13-^{2}H]-linoleic acid in the presence of lipoxygenase from soybean. Spectra are for samples with fatty acid/enzyme mole ratios of (A) ∞; (B) ~272; (C) ~90; and (D) ~45. Dashed lines are single Lorentzian line shapes used by the authors to determine T_2 values. Data from Ref. 10.

does not have any significant degree of motion relative to the protein; there is no rapid motion within the binding site. The effective correlation time of the deuterons of the N-acetyl group is much shorter than that of the protein, indicating internal motion in the binding site. For a methyl group, rotating about its threefold axis at a rate $>10\tau_R$, the effective correlation time τ_c is related to τ_R by eq. 7. The calculated τ_c of 1.7 x

$$\tau_c = \tau_R \left(\frac{3 \cos^2 109° - 1}{2}\right)^2 = 0.11\ \tau_R \qquad (7)$$

10^{-9} s is in excellent agreement with the observed value given in Table III. The close agreement indicates that no additional fast internal motions, which would lead to shorter τ_c values, are associated with the N-acetyl side chain. The results therefore allowed the authors to construct a model for the sugar-lectin interaction.

Table III. T_2 Values and Rotational Correlation Times (τ_c) for Sugars Binding to Wheat Germ Agglutinin[a]

Sugar[b]	$T_{2,F}$[c] (msec)	$T_{2,B}$[c] (msec)	$\tau_{c,F}$[c] (s)	$\tau_{c,B}$[c] (s)
[NAc-^{2}H$_3$] Glc NAc	388	1.6	6×10^{-12}	1.7×10^{-9}
Glc NAc-3-^{2}H	17	0.21	1.3×10^{-10}	3×10^{-8}

[a]Data from Ref. 11.
[b]Glc NAc = N-acetyl-D-glucosamine.
[c]$T_{2,F}$ and $\tau_{c,F}$ correspond to the free sugar; $T_{2,B}$ and $\tau_{c,B}$ refer to the bound sugar.

The deuterium nucleus is an excellent probe of anisotropic motion, since one can readily monitor the stereochemical dependence of ^{2}H relaxation times and then deduce the reorientational parameters (1,2,12). The use of ^{2}H NMR to detect segmental motions was described above. In another study ^{2}H relaxation times (T_1) were measured for a large number of deuterated compounds (12) and analyzed in terms of isotropic or anisotropic motions. Pyridine-d$_5$, for example, exhibited equality in T_1 values for all positions (1.2 ± 0.1 s), which was taken to be indicative of rapid isotropic molecular reorientation. In the case of [2,4,6-^{2}H$_3$]benzaldehyde, the relative T_1 values, para position less than ortho positions, indicated anisotropic motion with rotation about an axis through the para C-^{2}H bond being more rapid than that about the other two orthogonal axes (12).

There have been a large number of ^{2}H NMR studies of water (D$_2$O) interacting with solutes, polymers, and macromolecular structures (1,2). The ^{2}H relaxation times of D$_2$O have been used to detect conformational changes of biopolymers (1,13). For example, Figure 13 shows the T_1 values for water in solutions of L-glutamic acid and poly(L-glutamic acid) as a function of pH. L-Glutamic acid had no effect on the T_1 values of D$_2$O at any pH value. The polymer showed a strong interaction with water at both low and high pH values. At the centre of the helix-coil transition of poly-L-glu no polymer-D$_2$O interaction was detected,

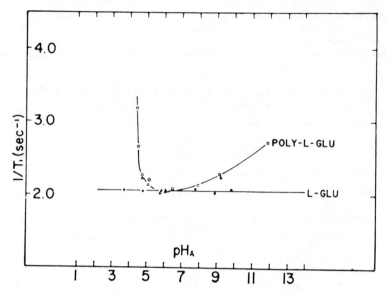

Figure 13. Variation of relaxation rate with apparent pH for solutions of poly-L-glutamic acid and L-glutamic acid in 0.1 N NaCl in D_2O. The decrease in T^{-1} reflects a diminished polymer-D_2O interaction as the pH is varied. The minimum in the T^{-1} vs pH curve occurs at the helix-coil transition of the polymer. Reprinted with permission from Ref. 13. Copyright 1970, American Chemical Society.

as evidenced by a minimum in the T_1^{-1} vs pH curve (Figure 13). A similar manifestation of conformational changes in the solvent's ^2H relaxation times has been observed for other biopolymers (1,13).

SUMMARY

High resolution ^2H NMR has been extremely useful in structural and stereochemical problems in chemistry and biochemistry. It has proved to be particularly powerful for elucidating mechanisms and kinetics of reactions involving hydrogen. The method is clearly of great value for selectively monitoring biotransformations and providing information which would be exceedingly difficult to obtain by other means. One can anticipate a more expanded role for ^2H NMR in such studies. Relaxation times of ^2H yield valuable information on the rates and the nature of molecular reorientation, relaxation pathways of other nuclei, solute-solvent interactions, and factors affecting the electric field gradient at the ^{2}H nucleus (related to bonding in the molecule).

The large number of studies now reported using high resolution ^{2}H NMR obviates a comprehensive coverage here. Further details are available in the earlier (1,2) and more recent (14) reviews.

REFERENCES

(1) H. H. Mantsch, H. Saitô, and I. C. P. Smith, Progr. NMR Spectrosc., 11, pp. 211-272 (1977).
(2) C. Brevard and J. Kintzinger, in "NMR and the Periodic Table," R. K. Harris and B. E. Mann, Eds., Academic Press, New York, 1978, pp. 107-128.
(3) J. B. Stothers and C. T. Tan, J. Chem. Soc. Chem. Commun., p. 738 (1974); A. K. Cheng, J. B. Stothers, and C. T. Tan, Can. J. Chem., 55, pp. 447-453 (1977).
(4) O. E. Edwards, J. Dixon, J. W. Elder, R. J. Kolt, and M. Lesage, Can. J. Chem., 59, pp. 2096-2115 (1981).
(5) C. A. Townsend and A. M. Brown, J. Am. Chem. Soc., 104, pp. 1748-1750 (1982).
(6) Y. Sato, T. Oda, and H. Saitô, J. Chem. Soc. Chem. Commun., pp. 415-417 (1977).
(7) U. Burger, J.-M. Sonney, and P. Vogel, Helv. Chim. Acta, 63, pp. 1006-1015 (1980).
(8) A. Abragam, "The Principles of Nuclear Magnetism," Clarendon Press, Oxford, 1961.
(9) H. Saitô, H. H. Mantsch, and I. C. P. Smith, J. Am. Chem. Soc., 95, pp. 8453-8455 (1973).
(10) T. S. Viswanathan and R. J. Cushley, J. Biol. Chem., 256, pp. 7155-7160 (1981).
(11) K. J. Neurohr, N. Lacelle, H. H. Mantsch, and I. C. P. Smith, Biophys. J., 32, pp. 931-938 (1980).
(12) H. H. Mantsch, H. Saitô, L. C. Leitch, and I. C. P. Smith, J. Am. Chem. Soc., 96, pp. 256-258 (1974).
(13) J. A. Glasel, J. Am. Chem. Soc., 92, pp. 375-381 (1970).
(14) I. C. P. Smith and H. H. Mantsch, in "New Methods and Applications of NMR Spectroscopy," G. C. Levy, Ed., ACS Symposium Series, Washington, 191, 1982, pp. 97-117.

Figures 1, 2, 3, 4, 8, 9, 10 and 11 are reproduced from Ref. 1, with permission from Pergamon Press Ltd.

CHAPTER 8

DEUTERIUM NMR OF ANISOTROPIC SYSTEMS

Harold C. Jarrell and Ian C. P. Smith

Division of Biological Sciences, National Research
Council of Canada, Ottawa, Canada K1A OR6

ABSTRACT

A brief summary of the concepts involved in ^{2}H NMR of solids and semisolid systems is given. Examples of the use of ^{2}H NMR to elucidate motional modes in systems undergoing anisotropic motions are discussed. The application of ^{2}H NMR to characterize partially ordered systems and to detect and quantitate phase transitions is discussed. Finally, two of the newer methods of analyzing ^{2}H powder spectra are demonstrated with examples.

INTRODUCTION

Since about 1975, ^{2}H NMR has proved to be a valuable tool for the study of solids and semisolids. The purpose of this chapter is not to present a review of all previous NMR work in this field, but rather to illustrate the types of information that are obtainable. More exhaustive treatments of the subject can be found in several recent reviews (1-4). Since spin relaxation will be dealt with elsewhere, the subject will be excluded from this chapter. We should mention in advance that the examples presented here reflect our particular bias, but this should not be taken as any indication of limitations in the technique.

PRINCIPLES

The deuterium nucleus, with I=1, has an unsymmetrical charge distribution and therefore has an associated quadrupole moment, Q. The electrostatic interaction of the nuclear quadrupole moment with the electric field gradient (EFG), $V_{\alpha\beta}$, at the nucleus shifts the energies of the spin states. At high magnetic field strengths the Zeeman interaction dominates the quadrupolar interaction so that the latter may be treated as a perturbation giving rise to the energy levels shown in Figure 1 (4) and given by eq. 1, in which m = -I, -I+1,..., I, B_0 is the applied magne-

$$E_m = -g_N \beta_N B_0 m + \frac{eQ}{4I(2I-1)} V^{(2,0)} [3m^2 - I(I+1)] \tag{1}$$

tic field, β_N and g are the nuclear g factor and magnetic moment, and $V^{(2,0)}$ is an element of the EFG tensor. In the principal coordinate system of the EFG (3), eq. 2 holds, in which Θ and

$$V^{(2,0)} = \frac{V_{zz}}{2} [(3\cos^2\Theta - 1) + \eta \sin^2\Theta \cos 2\phi] \tag{2}$$

ϕ are the polar and azimuthal angles, respectively, relating the principal coordinate system to B_0, $|V_{zz}| \geqslant |V_{xx}| \geqslant |V_{yy}|$, V_{zz} = eq and $\eta = (V_{xx} - V_{yy})/V_{zz}$. The allowed transitions ($\Delta m = \pm 1$) for the deuterium nucleus are shown in Figure 1. The two resonance lines of equal intensity are observed with a frequency spacing, $\Delta\nu_Q$, given by eq. 3. Thus, for a homogeneously oriented sample

$$\Delta\nu_Q = \nu_+ - \nu_- = \frac{3}{2} \frac{eQ}{h} V^{(2,0)} \tag{3}$$

of deuterium spins with B_0 parallel to the z axis of the EFG, the line separation is given by eq. 4, in which e^2qQ/h is generally

$$\Delta\nu_Q (B_0 \| z) = \frac{3}{2} \frac{eQ}{h} V_{zz} = \frac{3}{2} \frac{e^2qQ}{h} \tag{4}$$

referred to as the static quadrupole coupling constant. If B_0 is aligned with the x axis ($\Theta=90°$, $\phi=0°$) or the y axis ($\Theta=90°$, $\phi=90°$), $\Delta\nu_Q$ will have a value determined by V_{xx} or V_{yy}. The general expression (3) is given by eq. 5. Frequently the EFG

$$\Delta\nu_Q (\Theta,\phi) = \frac{3}{2} (\frac{e^2qQ}{h}) [(\frac{3\cos^2\Theta - 1}{2}) + \frac{\eta \sin^2\Theta \cos 2\phi}{2}] \tag{5}$$

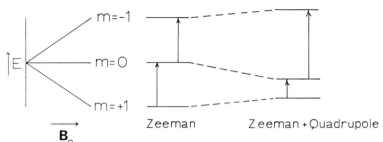

Figure 1. Effect of the quadrupolar interaction on the Zeeman energy levels of an I=1 nucleus.

has axial symmetry, so that the asymmetry parameter η, vanishes ($V_{xx} = V_{yy}$) and eq. 5 reduces to eq. 6. The dependence of $\Delta\nu_Q$

$$\Delta\nu_Q = \frac{3}{2}\frac{e^2qQ}{h}\left(\frac{3\cos^2\theta - 1}{2}\right) \qquad (6)$$

on Θ is shown in Figure 2A. Note that eq. 6 collapses to 0 (centre of Figure 2A) at $\cos^2\theta = 1/3$, which corresponds to an orientation angle of Θ = 54.74°, frequently referred to as the "magic angle."

In general the EFG's are not homogeneously oriented with respect to B_0, but are randomly distributed. For the case of axial symmetry (η=0), the probability distribution of the EFG's as a function of orientation is given by p(Θ) = ½sinΘ. The resulting "powder" spectrum will be given by a superposition of spectra arising from all Θ angles, whose intensities are scaled by p(Θ) (Figure 2B). The least intense resonances correspond to Θ=0, while the most intense resonance corresponds to Θ=90° with a Δν separation given by eq. 7. The static quadrupolar coupling

$$\Delta\nu_{powder} = \frac{3}{4}\frac{e^2qQ}{h} \qquad (7)$$

constant can be determined from polycrystalline samples. In the case where η ≠ 0 the powder pattern is more complicated since both Θ and φ in eq. 5 are involved in the resonance frequencies and in the probability of a given orientation (5).

OBTAINING ^{2}H SPECTRA

In general ^{2}H quadrupolar spectra are orders of magnitude wider than high resolution ^{2}H spectra. The broad distribution of

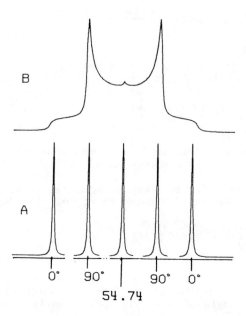

ORIENTATION (θ)

Figure 2. (A) Dependence of the quadrupolar splitting on the angle θ between B_0 and the principal (z) axis of the EFG (η=0) for a homogenously oriented sample of deuterons. (B) Powder spectrum corresponding to a random distribution of deuterons with respect to B_0; axially symmetric EFG.

frequencies results in a fast decaying signal as the Fourier components dephase, so that the loss of signal due to receiver overload and probe ringdown can result in severe spectral distortions. The effect of receiver dead-time on a ^{2}H powder spectrum is represented in Figure 3A. In order to circumvent this problem and the associated spectral distortion, use is made of the quadrupolar echo technique which usually involves a 90° pulse followed after a time τ by a second 90° pulse, phase shifted by π/2 relative to the first. An echo is formed at a time τ after the second pulse; τ≪T_2. Starting the Fourier transform at the echo maximum gives an undistorted spectrum as shown in Figure 3B. Missing the echo maximum will give a distorted spectrum, the degree of which will depend on experimental conditions. A detailed discussion of experimental procedures and potential problems has been presented recently (6).

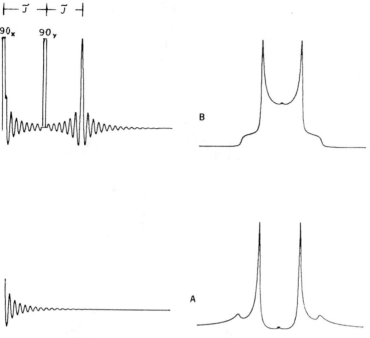

Figure 3. (A) Simulated FID after a 90° pulse and 30 μsec delay corresponding to the receiver dead-time, and the Fourier transform of the FID. (B) Simulated quadrupolar echo and its Fourier transform starting at the echo maximum.

APPLICATIONS

Solids

Many of the systems of major interest are not static but exhibit motions which average to some degree the quadrupolar interaction of ^{2}H. In solids or quasisolids the molecular movements are anisotropic and as a result the quadrupolar interaction is not completely averaged to zero. The type and rate of motion can be deduced from ^{2}H NMR spectra. In a study of hydrated solid B-form DNA, ^{2}H NMR spectra of calf thymus DNA specifically labeled at the C-8 position of the purine bases were examined (7). At all temperatures the spectra yielded a quadrupole splitting of 128 kHz, Figure 4, the value expected for a static deuteron. The authors were able to conclude in conjunction with ^{31}P NMR results that while the backbone of the DNA was reorient-

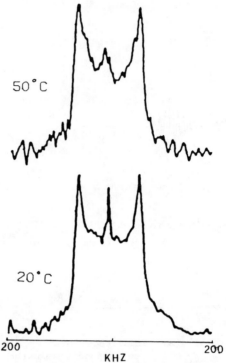

Figure 4. ^{2}H NMR spectra of calf thymus DNA deuterated at C-8 of the purine bases. With permission from Ref. 7. Copyright: Academic Press Inc. (London) Ltd.

ing quickly (10^4 s^{-1}) the bases were not undergoing motions on the order of or greater than 10^6 s^{-1}.

In several studies the detailed nature of motion has been deduced from ^{2}H powder spectra (8-10). For [2,3,5,6-^{2}H$_4$]tyrosine in a bacteriophage coat protein (8), a discrimination between fast continuous diffusion and rapid discrete diffusion was made. Fast continuous diffusion about the C_β-C_γ bond (Figure 5) would give an axially symmetric ^{2}H spectrum with a peak separation given by eq. 8, in which β is the angle between the unique axis

$$\Delta\nu_Q = \frac{3}{4} \frac{e^2qQ}{h} \left(\frac{3\cos^2\beta - 1}{2} \right) \qquad (8)$$

of the static tensor (the C-D bond) and the rotation axis, Figure 5A. If the motion involves discrete 180° jumps, the averaged tensor is calculated by transforming the static tensor in each of the two sites to the flipping frame axis system and then summing the two transformed tensors (8). In this case the expected spectrum is broad and axially asymmetric (effective η = 0.6), as shown in Figure 5C. The experimental spectrum, shown in Figure 5B, is representative of a discrete flipping of the tyrosine ring between two sites.

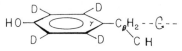

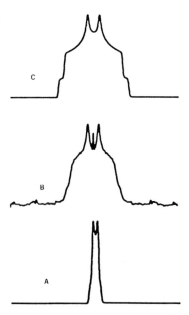

Figure 5. ^{2}H NMR spectra for [2,3,5,6-^{2}H$_4$]tyrosine side chain in bacteriophage coat protein: (A) theoretical spectrum for fast continuous diffusion about the C_β-C_γ bond; (B) experimental spectrum; (C) theoretical spectrum for fast discrete diffusion involving 180° flips about the C_β-C_γ bond. Data from Ref. 8.

Another example of the use of ^{2}H NMR to discriminate between motional modes involves the solid-solid phase transition of nonadecane. The transition involves a change from orthorhombic to hexagonal packing of the alkane. The orthorhombic phase gives rise to a static ^{2}H powder spectrum (Figure 6A). Transition to a hexagonal packing arrangement gives rise to the spectrum shown in Figure 6B. If the alkane were undergoing rapid rotation about its long axis, the ^{2}H spectrum would have a breadth of roughly one half of that of the static pattern, which is not observed. The spectrum instead indicates that the chains are undergoing discrete diffusion about the long molecular axis, resulting in an axially asymmetric ^{2}H powder spectrum. Consideration of the quadrupole splittings from various specifically labeled positions demonstrated that the ends of the molecule are more conformationally disordered than the middle.

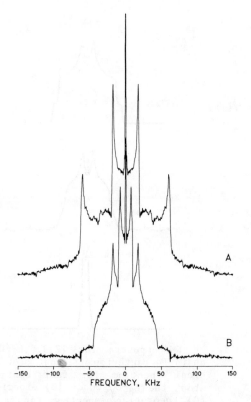

Figure 6. 46.063 MHz ^{2}H NMR spectrum of nonadecane-d_{44} at temperatures below and above that of the solid-solid phase transition: (B) 20°C, orthorhombic packing; (B) 25°C, hexagonal packing. Data from M. G. Taylor, E. C. Kelusky, I. C. P. Smith, H. L. Casal, and D. G. Cameron, J. Chem. Phys., in press.

Liquid Crystals

Deuterium NMR has made considerable contributions to the understanding of the structure and dynamics of liquid crystals and of guest molecules in liquid crystalline matrices (1-3). Liquid crystals (L.C.) are fluid systems which exhibit regular ordering of the constituent molecules. The most complex systems exhibiting this type of organization are biological membranes (2,4). L.C. systems tend to align the constituent rod-like molecules with their long molecular axes parallel. Rotations perpendicular to the long axis are restricted to small angular amplitudes, while those around the long axis are unhindered. The question then is how these motional modes affect the ^{2}H spectrum. Recall that for a single crystal the angular dependence of $\Delta\nu_Q$ was given by eq. 5. The axis of diffusion of the L.C. is called the director, $\underline{n}$. If B_0 is parallel to $\underline{n}$ then the instantaneous position of the EFG tensor is specified by Θ and ϕ. If the molecules fluctuate rapidly about $\underline{n}$ then the static tensor is averaged to a new tensor, axially symmetric with respect to $\underline{n}$; the quadrupolar splitting then is given by eq. 9, in which the

$$\Delta\nu_Q (\Theta,\phi) = \frac{3}{2} \frac{e^2qQ}{h} \left(\frac{\langle 3\cos^2\Theta - 1\rangle}{2} + \eta \frac{\langle \sin^2\Theta\cos 2\phi\rangle}{2} \right) \qquad (9)$$

angular brackets indicate a time average. The fluctuations of the principal axis system are best described by the so-called order parameter S_{ii}, which is defined by eq. 10, in which

$$S_{ii} = \frac{\langle 3\cos^2\Theta_i - 1\rangle}{2} \qquad (i = x,y,z) \qquad (10)$$

$\langle \cos^2\Theta_i \rangle$ denotes the time average angular fluctuations of the i'th coordinate axis with respect to $\underline{n}$. Hence eq. 9 can be rewritten (3) as eq. 11. If $\eta = 0$ then only one order parameter,

$$\Delta\nu_Q (\Theta,\phi) = \frac{3}{2} \frac{e^2qQ}{h} [S_{zz} + \frac{1}{3} \eta(S_{xx} - S_{yy})] \qquad (11)$$

S_{zz}, is required to describe the motion. Again, because many of the systems studied have a random distribution of director axes with respect to B_0, the spectrum is a superposition of resonances associated with the various director orientations. In the case of the C-D bond the dominant axis of the EFG tensor is the bond vector so that S_{zz} may be called S_{CD} and therefore represents the angular fluctuations of the C-D bond vector about $\underline{n}$. The frequency spacing of the peaks of the powder spectrum is given by eq. 12.

$$\Delta\nu_Q = \frac{3}{4} \frac{e^2 qQ}{h} S_{CD} \qquad (12)$$

Lyotropic L.C. systems are of widespread interest because of their similarities to the lipid bilayers in complex biological membranes. Of particular interest is the local order at various positions along the alkyl chains (corresponding to depth). This is demonstrated by the potassium laurate-H_2O system which forms a lamellar lyotropic phase (11). From the quadrupolar splitting for deuterons at various positions of the acyl chain, the S_{CD} can be calculated from eq. 12. The resulting "order-position" profile shown in Figure 7 is indicative of decreasing degrees of local order (fluctuations of increased amplitude) on going from the α to the ω position. The similarity in the S_{CD} values for the 2-7 positions is referred to as the "plateau" and is a characteristic feature for a wide range of L.C. systems, including natural membranes (1,2,4). Order parameters have been used to calculate membrane thickness and the numbers of trans-gauche conformers present in L.C. systems (2,12).

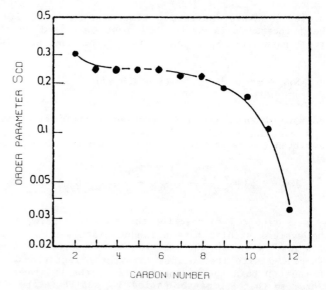

Figure 7. Positional dependence of the bond order parameter, S_{CD}, for potassium laurate-water (79:21 wt %) at 82°C. Data from Ref. 11. (with permission from North Holland Publ. Co.)

In some cases the ^{2}H spectrum is not readily characterized by one S_{CD}, particularly in membrane systems. In order to assist in interpreting and quantitating data from ^{2}H spectra, an analysis based on spectral moments has been developed (13). The n'th moment of a spectrum can be defined by eq. 13, in which ω_0

$$M_n = \frac{\int_0^\infty (\omega-\omega_0)^n \, G(\omega-\omega_0) \, d\omega}{\int_0^\infty G(\omega-\omega_0) \, d\omega} \tag{13}$$

is the Larmor frequency and $G(\omega-\omega_0)$ is the lineshape function. Eq. 13 makes use of the symmetry of ^{2}H spectra about ω_0, to allow odd moments to be calculated. The n'th moment can be related to $\Delta\nu_Q$ (90° orientation) by eq. 14 (13), in which A_n are constants.

$$M_n = A_n \, (2\pi)^n \, \Delta\nu_Q^n \tag{14}$$

For a single powder pattern, measuring $\Delta\nu_Q$ is equivalent to measuring any of the moments. If there is more than one quadrupole splitting in the system, M_n is proportional to the statistical average of $(\Delta\nu_Q)^n$. The average order can be determined; the width of its distribution may be defined by eq. 15 (13). If

$$\Delta_2 = \frac{\langle S_{CD}^2 \rangle - \langle S_{CD} \rangle^2}{\langle S_{CD} \rangle^2} = \frac{M_2}{1.35 \, M_1^2} - 1 \tag{15}$$

there is a unique value of S_{CD}, Δ_2 will be zero. The parameter Δ_2 has been found to be particularly useful in characterizing phase transitions. For example, the gel-liquid crystal phase transition of a biomembrane is manifest by a sharp increase in Δ_2, Figure 8, corresponding to an increase in the distribution of order parameters or quadrupole splittings. The maximum in Δ_2 occurs close to the centre of the phase transition.

Moments can in some cases provide a means of calculating the relative amounts of coexisting phases (14), which is otherwise difficult. An example of the analysis is the bilayer-hexagonal transition of a phosphatidylethanolamine-cholesterol system (14). The phase fractions were calculated according to eq. 16,

$$M_n = fM_n^I + (1-f)M_n^{II} \tag{16}$$

in which M_n are the observed moments, f is the fraction of phase I, and M_n^I and M_n^{II} are the n'th moments of phases I and II, respectively. The calculated f's were used to simulate the experimental spectra as shown in Figure 9.

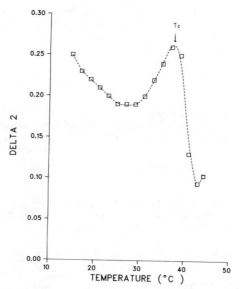

Figure 8. Temperature dependence of the moment parameter Δ_2 for the ^{2}H NMR spectra of A. laidlawii B membranes containing myristic acid-d_{27}, over the range of the gel-liquid crystalline phase transition centered at 37°C.

Guest Molecules in a Liquid Crystalline Matrix

Another application of ^{2}H NMR that has received widespread use is the monitoring of guest molecules in an ordered matrix. Such situations have been used to measure deuterium quadrupolar coupling constants for small molecules (1), and to monitor the conformation or location of the guest molecule (3). In one study a multiply ^{2}H-labeled cholesterol was inserted into a phospholipid matrix (15). The ring system of cholesterol is internally rigid, but reorients rapidly about its long axis. The ^{2}H spectra are governed by this motion as well as the orientation of the carbon-deuterium bond relative to this axis. The variation of the quadrupolar splittings between the deuterium sites for the rigid part of the molecule can be related by eq. 17, in which

$$\Delta\nu_i = \frac{3}{4} \frac{e^2qQ}{h} \frac{(3\cos\alpha_i^2 - 1)}{2} \frac{\langle 3\cos^2\gamma - 1 \rangle}{2} \tag{17}$$

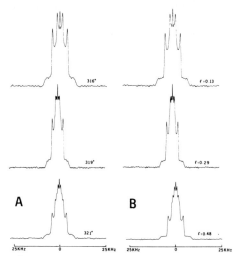

Figure 9. ^{2}H spectra (A) of egg phosphatidylethanolamine-d_4:cholesterol (4:1) (bilayer-hexagonal transition); (B) simulation by adding experimental spectra for each phase (eq. 16). From Ref. 14, reproduced from The Biophysical Journal, 1981, vol. 34, pp. 451-463 by copyright permission of the Biophysical Society.

$\Delta\nu_i$ is the quadrupolar splitting (peak for the 90° orientation of the director with respect to B_0) of the i'th deuteron, α_i is the angle between the i'th C-D bond and the rotation axis, and the last term describes the anisotropic motion of the rotation axis. The last term involves only the motion of the molecule as a whole, and is therefore the molecular order parameter, S_{mol}. It is clear that the quadrupolar splitting of the i'th deuteron relative to the j'th deuteron is determined by a geometrical factor which can be defined by eq. 18. By using the values R_k and

$$R_k = \frac{\Delta\nu_i}{\Delta\nu_j} = \frac{(3\cos^2\alpha_i - 1)}{(3\cos^2\alpha_j - 1)} \qquad (18)$$

the atomic coordinates of the deuterons, the axis of diffusion in the molecular frame was located, as is shown in Figure 10. It had previously been assumed that the C(3)-H(3) bond in cholesterol makes a 90° angle with the axis of rotational diffusion. However, the above analysis indicated that the angle is in fact 79° (15), a result that can significantly affect the value of S_{mol} calculated from eq. 17. The result is useful in that once the axis of diffusion in the molecular frame is known, quadrupole splittings for deuterons at various positions can be calculated from eq. 17 for a given S_{mol}. This is essential for the design of a labeled cholesterol with good spectroscopic characteristics.

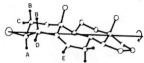

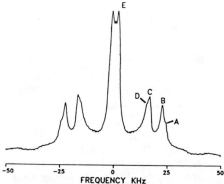

Figure 10. Upper: [2,2,4,4,6-^{2}H$_5$]cholesterol and the determined axis of diffusion; Lower: 46.063 MHz ^{2}H NMR spectrum of cholesterol-d$_5$ in egg phosphatidylcholine (1:1). Data taken from Ref. 15. (with permission from North Holland Publ. Co.)

De-Paking

Our final example involves a recent development in the analysis of ^{2}H powder spectra. Detailed analyses of ^{2}H powder spectra frequently rely upon spectral simulations in which $\Delta\nu_Q$, the linewidth, the lineshape function, and, in the case of overlapping powder spectra, the relative spectral contribution must be fitted. Clearly, this is difficult since only the approximate value of $\Delta\nu_Q$ is easily determined. Recently M. Bloom and co-workers (16) have developed a means of calculating an oriented-sample spectrum from the powder spectrum. Since the orientational frequency distribution of a ^{2}H powder spectrum is scaled by $(3\cos^2\theta-1)/2$ as in eq. 6 (for $\eta=0$), and if one assumes that the linewidth scales in the same way, one half of the powder spectrum can be defined in terms of one component (one orientation) of the spectrum according to eq. 19, in which $F_{90}(x)$ is the

$$G(\omega) = 0.5 \int [3x(\omega-x)]^{-1/2} F_{90}(x)dx; \quad x = \frac{\omega}{1-3\cos^2\theta} \quad (19)$$

lineshape function for the 90° angle (Θ). Details of the calculation are described in the original paper. The technique can, in principle, provide a determination of $\Delta\nu_Q$, the lineshape function, and, in case of multicomponent powder spectra, the relative spectral contributions. Two illustrative examples are presented in Figures 11 and 12. In Figure 11A a theoretical spectrum was generated by adding two powder spectra in a ratio of 2:1, $\Delta\nu_Q^1$ = 25 and $\Delta\nu_Q^2$ = 26 kHz. The powder spectrum was then analyzed by the "de-Paking" procedure, to yield the spectrum of the oriented sample (Figure 11B). The latter yields an accurate determination of the $\Delta\nu_Q$ values, which are not readily determined from the powder spectrum, and also their relative contributions. (Note that in this case the linewidth was intentionally not scaled as $(3\cos^2\Theta-1)/2$, as is assumed in eq. 19, to demonstrate that useful information may still be obtained from the analysis).

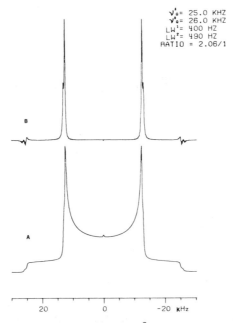

Figure 11. (A) Axially symmetric ^{2}H NMR powder spectrum simulated using residual quadrupolar splittings of 25.0 and 26.0 kHz, and associated angular-independent Lorentzian linewidths of 400 and 500 Hz, respectively. Spectra are in a population ratio of 2:1, respectively; (B) 90°-oriented sample spectrum calculated from (A) according to Ref. 16.

Figure 12A illustrates a real situation. The ^{2}H NMR spectrum is very complex, so that a detailed analysis would be a rather imposing problem. The "de-Paked" spectrum is shown in Figure 12B. The quadrupolar splittings, relative intensities, and approximate linewidths were readily determined (M. Rance, I. C. P. Smith, and H. Jarrell, unpublished results). With this information the source of the spectral complexity was shown to arise from the presence of more than one lipid species in the membrane. The ^{2}H quadrupolar splitting is sensitive to the effects of the lipid head groups. This observation was rather important in as much as similar observations had previously been attributed to other effects.

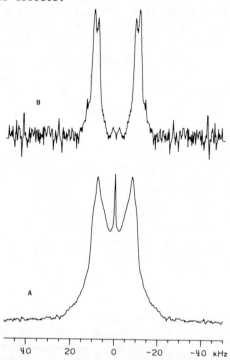

Figure 12. (A) 46.063 MHz ^{2}H NMR spectrum of A. laidlawii B membranes containing [2-^{2}H$_2$]dihydrosterculic acid; (B) 90°-oriented sample spectrum calculated from (A).

SUMMARY

Deuterium NMR has contributed extremely useful information on condensed phases. With the improved instrumentation now available and the associated higher quality of the spectra, increasingly more detailed examinations of solid and semisolid systems are being reported. An area which is beginning to develop is the ^{2}H NMR of macromolecules including nucleic acids (7) and proteins such as collagen (17), bacterial proteins (8-10), and heme-containing proteins (10). In the last case, the proteins have been oriented by the magnetic field so that relatively simple ^{2}H spectra may be obtained.

The use of ^{2}H NMR relaxation times to elucidate the dynamics of anisotropic fluids is just beginning (18,19) and in conjunction with multiple quantum coherence (20) has considerable potential for examining molecular motions.

Recently there have been attempts to use magic angle spinning (6) in combination with multiple quantum transitions (21) and cross polarization from ^{1}H (22) to achieve spectral simplification. Such techniques promise to add new dimensions to ^{2}H NMR of anisotropic systems.

REFERENCES

(1) H. H. Mantsch, H. Saitô, and I. C. P. Smith, Progr. NMR Spectrosc., 11, pp. 211-272 (1977).
(2) J. Seelig, Quart. Rev. Biophys. 10, pp. 353-418 (1977).
(3) J. W. Emsley and J. C. Lindon, "NMR Spectroscopy Using Liquid Crystal Solvents," Pergamon Press, 1975.
(4) I. C. P. Smith, in "Biomembranes," Vol. 12, L. A. Manson and M. Kates, Eds., Plenum Press, New York, in press.
(5) M. H. Cohen and F. Reif, Solid State Phys., 5, pp. 321-438 (1957).
(6) R. G. Griffin, Meth. Enzymol., 72, pp. 108-174 (1981).
(7) J. A. Di Verdi and S. J. Opella, J. Mol. Biol., 149, pp. 307-311 (1981).
(8) C. M. Gall, T. A. Cross, J. A. Di Verdi, and S. J. Opella, Proc. Natl. Acad. Sci. (USA), 79, pp. 101-105 (1982).
(9) D. M. Rice, A. Blume, J. Herzfeld, R. J. Wittebort, T. H. Huang, S. K. Das Gupta, and R. G. Griffin, "Biomolecular Stereodynamcis II," R. H. Sarma, Ed., Adenine Press, New York, 1981, pp. 255-270.
(10) S. Schramm, R. A. Kinsey, A. Kintanar, T. M. Rothgeb, and E. Oldfield, "Biomolecular Stereodynamics II," R. H. Sarma, Ed., Adenine Press, New York, 1981, pp. 271-286.
(11) J. Charvolin, P. Manneville, and B. Deloche, Chem. Phys. Lett. 23, pp. 345-348 (1973).
(12) E. Oldfield, M. Meadows, D. Rice, and R. Jacobs, Biochemistry, 17, pp. 2727-2740 (1978).
(13) J. H. Davis, Biophys. J., 27, pp. 339-358 (1979).
(14) H. C. Jarrell, R. A. Byrd, and I. C. P. Smith, Biophys. J., 34, pp. 451-463 (1981).

(15) M. G. Taylor, T. Akiyama, and I. C. P. Smith, Chem. Phys. Lipids, 29, pp. 327-339 (1981).
(16) M. Bloom, J. H. Davis, and A. L. MacKay, Chem. Phys. Lett., 80, pp. 198-202 (1981).
(17) L. W. Jelinski, C. E. Sullivan, and D. A. Torchia, Biophys. J., 10, pp. 515-529 (1980).
(18) K. R. Jeffrey, Bull. Magn. Reson. 3, pp. 69-82 (1982).
(19) R. Y. Dong, Bull. Magn. Reson. 3, pp. 93-103 (1982).
(20) D. Jaffe, R. R. Vold, and R. L. Vold, J. Magn. Reson. 46, pp. 475-495 (1982); J. Magn. Reson., 46, pp. 496-502 (1982).
(21) R. Eckman, L. Müller, and A. Pines, Chem. Phys. Lett., 74, pp. 376-378 (1980).
(22) L. Müller, R. Eckman, and A. Pines, Chem. Phys. Lett., 76, pp. 149-154 (1980).

CHAPTER 9

TRITIUM NUCLEAR MAGNETIC RESONANCE SPECTROSCOPY

J. A. Elvidge
Chemistry Department, University of Surrey
Guildford, GU2 5XH, U.K.

ABSTRACT

Tritium has a spin quantum number $I = 1/2$ and so is a suitable nuclide for high resolution NMR observation. Its essential nuclear magnetic properties are discussed, as well as its sensitivity to NMR detection. Also discussed are the relaxation characteristics of the ^3H nucleus, spectrometer operating conditions, the chemical shifts of ^3H in compounds, referencing, isotope effects, and coupling constants. The quantitation of ^3H NMR spectra is considered, along with the use of coupled spectra, of resolution enhancement, and of proton spin-decoupled spectra. The significance of nuclear Overhauser effects in these contexts is described. The wide applicability and usefulness of ^3H NMR spectroscopy for the investigation of tritium-labeled compounds, reaction products, and metabolites is illustrated with examples. Tritium has a radioactive nucleus which is a β-emitter, so that precautions are necessary in handling and using tritium-labeled compounds, but these safety measures are not onerous and are easily reduced to a simple standard routine. With this routine, ^3H NMR spectroscopy is a safe, convenient, and direct method for establishing the positions of tritium label in compounds and of giving the relative amounts in each position. The method will also provide stereochemical information. Being rapid and non-destructive, it compares extremely favourably with the tedious methods of stepwise chemical degradation and counting hitherto used.

INTRODUCTION

Tritium (^{3}H or T) finds wide application in tracer studies in chemistry, biochemistry, and medicine (1), as reference to Chemical Abstracts demonstrates (2). This is largely because tritium is an isotope of hydrogen and has a radioactive nucleus with a convenient half-life of 12.35 y. The decay product is ^{3}He.

The magnitude of the half-life of tritium implies a fairly high rate of nuclear disintegration: the maximum specific radioactivity of monotritiated compounds is 29.15 Ci mmol^{-1}, which is 1.078 TBq, where 1 Ci = 3.7 x 10^{10} Bq (1 Bq is defined as 1 disintegration s^{-1}). The only radiation emitted is soft β-radiation (electrons) of mean energy 5.66 keV, which makes tritiated compounds easy and relatively very safe to employ, given proper facilities.

The radioactivity and the virtual absence of natural tritium (<10^{-16}%, from cosmic ray-induced reactions in the upper atmosphere) make it simple to measure the rate of participation of tritiated compounds in reactions by means of scintillation counting and to locate tritiated compounds by autoradiography. Their separation and isolation by t.l.c. is similarly easily followed, and radio gas chromatography is eminently applicable. Even the identification of tritiated compounds in trace amounts can be accomplished, by reverse dilution analysis or by the method of mixed chromatograms. None of these very sensitive and well established techniques, however, can delineate precisely the positions of the label in tritiated compounds, give simultaneously the relative amounts in the various positions, or indicate the stereochemistry of the labeling. Stepwise chemical degradation and counting at each stage has been the only available approach. This is necessarily tedious and often uncertain because of doubts about the integrity of reaction processes employed (1,3-5). Tritium NMR spectroscopy, in contrast, provides a direct, reliable, quick, and nondestructive method of analysis (6-8). Instead of occupying weeks, a full analysis takes only hours (9). The extreme importance of knowing the exact distribution of tritium in tracer compounds is widely recognised (1,5,10,11). This information is now best obtained by ^{3}H NMR spectroscopy, provided sufficient material is available for the sensitivity of the NMR instrumentation.

NUCLEAR MAGNETIC PROPERTIES OF TRITIUM AND ^{3}H NMR SPECTROMETER OPERATION

Like the proton, ^{1}H, the isotopic nucleus ^{3}H--the triton--has a spin I = 1/2. As the triton has an even higher magnetic moment μ_T and therefore a higher magnetogyric constant $\varkappa_T = \gamma_T/2\pi$

$= \mu_T/I \cdot \hbar \cdot 2\pi =$ ca. 4.5414×10^7 Hz T^{-1} than the proton ($\gamma_H =$ ca. 4.2577×10^7 Hz T^{-1}), the triton is an ideal nucleus for high resolution NMR observation. The resonance frequency, $\nu = \gamma B$, for ^3H is higher than that for ^1H at the same field strength B by the factor $\gamma_T/\gamma_H =$ ca. 1.06664, and the spectral dispersion is better by the same factor. (For a spectrometer operating at 90 MHz for protons with a field of 2.11 Tesla, the resonance frequency for ^3H is 96 MHz. With a 9.38 Tesla superconducting magnet (400 MHz for ^1H), the ^3H resonance frequency would be 426.7 MHz.) A further consequence of the high nuclear magnetic moment is the improved sensitivity to NMR detection: this is higher for tritons than for equal numbers of protons at the same field by the factor $(\gamma_T/\gamma_H)^3 = 1.21$ (12). Because tritiated compounds will normally be examined at low isotopic abundance, this sensitivity to NMR detection facilitates analytical use of the technique. However, to obtain adequate spectra it is essential to match the level of labeling to the instrumental capability (see Figure 1). The sensitivity to detection of a particular nucleus improves as the 3/2 power of the spectrometer operating field or frequency (12). Thus, if a ^3H compound bearing 10 mCi per site can be adequately examined by a 60 MHz (1.4 Tesla) Fourier transform NMR spectrometer, operating overnight, with ^1H decoupling and quadrature detection, then if a 400 MHz (9.38 T) FT spectrometer were available, only $10 \times (60/400)^{3/2}$ mCi would be needed, i.e., 10 μCi per site would suffice. Further well known operational points are that signal-to-noise improves as t^2 where t is the spectral acquisition time, and the NMR signal strength is directly proportional to the amount of isotope responsible (13) as we early demonstrated for tritiated water (7). Signal-to-noise (S/N) is commonly defined as S/N = (signal height ÷ extreme noise escursion) x 2.5 and for the 7β-^3H signal in Figure 1 is 47, 4, and 1.6 for the successive dilutions.

For the triton, the (2I + 1) quantum rule indicates two energy states or orientations for the nuclear magnetic vector in an applied magnetic field. The spin quantum number I = 1/2 also implies spherical symmetry, so that the only property of the nucleus dependent upon orientation is the magnetic moment. As a result, the relaxation of tritons from their upper spin state will only be induced by magnetic fields. These need to have components perpendicular to the applied magnetic field and oscillating with the nuclear precessional (Larmor) frequency, ν_T Hz. Such fields will arise in the sample by chance from the collection of nuclear spins, but this mechanism of spin-lattice relaxation will necessarily be slow, as for the proton (with similar requirement and resonance frequency ν_H), because of the stringent conditions. In consequence by the uncertainty principle, the long upper-state lifetime will lead to a very narrow linewidth, similar to that for a corresponding proton. Experiments by us (with Dr. J. M. A. Al-Rawi) have shown that the relaxation times

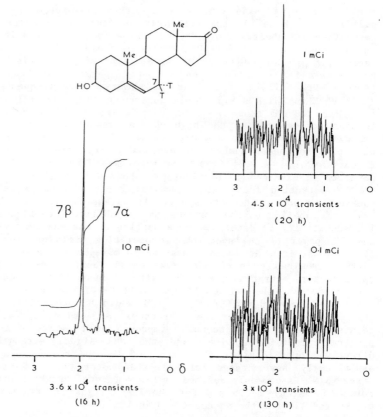

Figure 1. ^{3}H NMR spectra (^{1}H-decoupled) of 3β-hydroxy-[7-^{3}H]androst-5-en-17-one in DMSO-d$_6$ at successive dilutions; acquired with quadrature detection at 96 MHz (2.11 T).

T_1 for tritons in partially monotritiated compounds are indeed similar to, but often slightly longer than, those for corresponding protons (see Table 1). Spectrometer operating conditions, as for acquisition of proton spectra free from differential relaxation effects, will thus apply, but perhaps the pulse length (or flip angle) should be reduced slightly or the delay between pulses slightly increased.

A test sample comprising 20-30 mCi of partially monotritiated dimethyl sulphoxide proves ideal for setting up an FT NMR spectrometer for ^{3}H resonance studies. The sample, easily pre-

Table 1. Experimental Spin-Lattice Relaxation Times for ^{3}H and ^{1}H in the Same Sites, Measured by the Inversion-Recovery Technique on Purified, Partially Monotritiated Compounds

Compound (position tritiated)	T_1 (s) ^{3}H	^{1}H	Correlation coefficients	
Water	0.29(6)	0.21(4)	0.999	0.997
Diethyl malonate (2)	8.63(7)	6.49(6)	0.991	0.998
Dimethyl sulphoxide	0.83(6)	0.66(8)	0.996	0.998
Benzimidazole (2)	1.56(6)	1.79(5)	0.996	0.998

pared by exchange (14), is diluted with distilled perdeuterated dimethyl sulphoxide and sealed (vide infra, Sample Preparation). No noticeable further exchange occurs and the sample, being self-scavenging, is radiolytically stable over very long periods. The field homogeneity and lock are adjusted from the ^{2}H NMR signal as usual. With the pulse point set close to the ^{3}H NMR signal and with adequate ^{1}H noise decoupling, the acquisition parameters can readily be adjusted to give a free induction decay signal which appears on the monitor as a well shaped low frequency sine wave. A few transients will then transform to a properly shaped singlet line. Optimisation of the ^{1}H decoupling power is also facilitated because the test sample will give a proton-coupled ^{3}H NMR signal which is a relatively wide triplet ($^2J_{TH}$ = 13.2 Hz). For a spectrometer with a ^{19}F channel for field-frequency locking, the foregoing test sample would be diluted with an equal volume of trichlorofluoromethane.

To facilitate direct comparison of ^{3}H with ^{1}H NMR spectra, especially where these have not previously been interpreted, identical ppm scales are employed, viz. 21.33 Hz cm^{-1} for ^{3}H and 20.00 Hz cm^{-1} for ^{1}H (ratio, ν_T/ν_H) or appropriate equal fractions thereof.

SAFETY MATTERS: SAMPLE PREPARATION

No new laboratory facilities are required for ^{3}H NMR spectroscopy beyond those necessary for the conduct of work with tritiated compounds. Tritiated compounds recovered from tracer and other experiments are usually in very dilute solution and then will need to be concentrated by freeze-drying and redissolution in a small volume of NMR solvent containing internal ^{1}H reference (i.e., tetramethylsilane or sodium 4,4-dimethyl-4-silapentanesulphonate). This solution, having sufficient total radioactivity (for the number of labeled sites--vide supra) is transferred by syringe to a NMR microcell or, e.g., 3 mm precision tube, and sealed (8). The microcell is mounted in a holder and transferred to a standard NMR tube; alternatively, the tube is mounted in grooved Teflon rings and slid into the NMR tube to the correct, predetermined position (the optimum position with respect to the rf coil). The capped tubes are then placed in Vermiculite in a metal can, which bears identification and hazard marks, and the lid is taped on. The tubes and contents can then be transported safely to the NMR laboratory, examined in the normal way, and returned in the can to the radiochemical laboratory. Handling of the samples at the spectrometer is made exceedingly safe by the double glass containment: the cell in a dropped tube does not break. The spectrometer probe has no bottom holes so that in the unlikely event of a breakage in the probe the sample would be retained and the probe could be removed, placed in a plastic bag, and safely taken to the radiochemical laboratory for decontamination. As a standard precaution, surfaces around the NMR spectrometer are swabbed at intervals and the swabs are subject to scintillation counting. A further precaution is to exhaust the air from the vicinity of the NMR probe to an outside stack or through a large active-charcoal filter. No breakage of tritium samples or contamination has occurred in our NMR laboratory during measurement of very many hundreds of samples over a 12 y period, fully justifying the simple precautions imposed.

Of precision cells for mounting inside NMR tubes (Figure 2), we prefer ones made from 3 mm precision tubing; each length is sealed to a short length of 5 mm tube. The latter is closed with a serum cap and the assembly evacuated through a syringe needle to facilitate subsequent filling (by syringe through the cap) and flame-sealing (15). An inert liquid (CCl_4) or the field-frequency lock solvent (with or without external reference) can be added to the annular space between the cell and NMR tube, as may be required; some liquid in this space is desirable to obviate wobble when the sample is spun in the probe. On special occasions when maximum detection sensitivity is imperative, the bare sealed sample tube may be used in a 3 mm tuned probe, the tube being mounted at its upper end in 5 mm tubing for attachment of a standard spinner (see Figure 2).

CHARACTERISTICS OF ^{3}H NMR SPECTROSCOPY

Triton chemical shifts can be expected to be virtually the same as proton chemical shifts (16). This follows because the shielding of a hydrogen nucleus in a compound in solution is very largely a function of the local environment and the latter is not materially affected by isotopic replacement of the hydrogen. Measurement of the Larmor ratio ν_T/ν_H for a range of partially

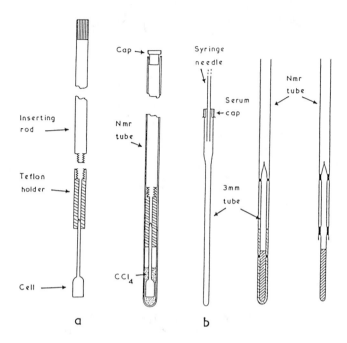

Figure 2. Details of micro cells, (a) commercial cell (100 μl) and insertion tool, (b) 3 mm tube.

monotritiated compounds at constant field gave a value constant to 7 decimal places (16). Later, more precise measurements (17) revealed a small dependence of hydrogen nuclear shielding on bond order whilst confirming that triton and proton chemical shifts were the same to ca. 0.02 ppm. This is the magnitude of the tritium primary isotope effect, which effect is defined as the change in chemical shift resulting from a single isotopic substitution. The effect may also be described in the present context as the upfield shift in a ^{3}H NMR signal resulting from a second tritium substitution at the same site: values of -0.02 to -0.03 ppm were measured for changes such as $RCH_2T \rightarrow RCHT_2$ (17) and $RCHT_2 \rightarrow RCT_3$ (18). The shift in a tritium resonance signal arising from the presence of a second, vicinal triton is ca. -0.01 ppm—a secondary isotope effect. These effects are just resolvable at 96 MHz but would be more readily discerned at higher fields with their resulting, higher, spectral dispersion. In a ^{3}H NMR spectrum acquired with ^{1}H noise decoupling, the signal from a multiply tritiated site such as a methyl group may be resolved as 2 or 3 closely spaced lines, or appear as a stepped envelope or broadened line (see Figure 3), depending on

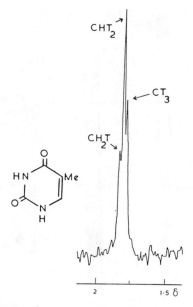

Figure 3. Methyl ^{3}H NMR signal (^{1}H-decoupled) of high specific radioactivity [methyl-^{3}H]thymine in D_2O.

the distribution of label between the possible species RCH_2T, $RCHT_2$, and RCT_3. By subjecting such signals to resolution enhancement (19) and expansion before integration, a detailed analysis of the tritium distribution is possible (18). Often even more helpful for that purpose are the enhanced, proton-coupled, ^{3}H NMR spectra (18).

The virtual identity of triton and proton chemical shifts has important consequences. One is that the enormous range of established correlation information in ^{1}H NMR spectroscopy is immediately applicable to the interpretation of ^{3}H NMR spectra. A second consequence is that no special internal standard is necessary for ^{3}H NMR spectroscopy. Originally we used tritiated water (7) as an expedient, but ordinary TMS (tetramethylsilane) or DSS (sodium 4,4-dimethyl-4-silapentanesulphonate) suffices, provided the spectrometer has a field-frequency lock (^{2}H or ^{19}F) and a ^{1}H channel, besides the ^{3}H channel (as usually will be the case). The frequency of the signal from the ^{1}H internal standard, obtained from the spectrometer output after a few pulses and Fourier transformation, is multiplied by the Larmor ratio, ν_T/ν_H = 1.066639738--the ratio appropriate to TMS, and so to DSS (17)--to provide the corresponding internal ^{3}H reference frequency for the eventual ^{3}H NMR spectrum. The ^{1}H transmitter and matched probe are then changed for the ^{3}H accessories (or other appropriate adjustments are made, depending on the spectrometer). With the same sample (and lock) and hence at exactly the same field, the tritium free induction decay signal is acquired and subsequently transformed. Because the ^{3}H pulse point and its computer address are known, the calculated ^{3}H reference frequency can be assigned to its correct computer address and treated as the internal reference point. Effectively, this ghost referencing procedure is the same as using a partially monotritiated TMS as internal reference, as we have demonstrated (17). Long (20) employed the nonvolatile tritiated hexamethyldisiloxane as reference when using a spectrometer without a ^{1}H channel. Altman (21,22) employed secondary references (which tend to incur slight errors) but has more recently used the ghost referencing technique (23).

The preceding discussion makes clear the fact that the precise values of chemical shift isotope effects will depend on the choice of isotopic reference. Correct choice of the reference is somewhat subtle. In the context of tritium as an isotope at low abundance, pertritiated TMS would clearly be an incorrect choice: partially monotritiated TMS would appear to be best.

As regards coupling of, e.g., protons to tritium nuclei, theory indicates for first row elements of the Periodic Table that all contributions to the coupling constant $J_{NN'}$ are proportional to the product of the magnetogyric constants of the coupled nuclei N and N' (13,24).

The ideal eq. 1 therefore follows. Because the magnetogyric

$$J_{TT} = J_{HT}(\gamma_T/\gamma_H) = J_{HH}(\gamma_T/\gamma_H)^2 \qquad (1)$$

constants refer to isolated, bare nuclei, the values are not directly measurable. However, assuming that the ratio of hydrogen isotopic screenings in compounds must be very close to unity (25), the experimental (average) Larmor ratio $\nu_T/\nu_H = 1.06663975$ is acceptable (16). This value is a good average taken from precision measurements on a variety of partially monotritiated compounds. The eq. 2, which is necessarily approximate, however

$$J_{TT} = J_{TH} \times (\nu_T/\nu_H) = J_{HH} \times (\nu_T/\nu_H)^2 \qquad (2)$$

accurate the Larmor ratio, can then be employed to relate triton and proton coupling constants (14,17). Any departures of observed values from calculated values are, of course, termed coupling constant isotope effects. It is to be noted that the precise values of such isotope effects will depend on choice of Larmor ratio.

Equations analogous to eq. 2 also serve to relate the values of coupling constants between tritium and other elements, e.g., J_{TB} (23), J_{TC} (14), J_{TF} (9), and J_{TP} (18), to the corresponding established proton coupling constants, for structure interpretation and for stereochemical correlation.

Besides its obvious use for determining the stereochemistry of tritium labeling (10), eq. 2 also provides an excellent means of obtaining inaccessible proton coupling constants, e.g., $^2J_{HH}$, where the geminal protons are isochronous (i.e., have the same chemical shift), through partial tritiation and measurement of the splitting in the tritium NMR signal. This procedure (14,17) is superior in accuracy to the conventional approach, using deuterium substitution, measurement of the observed $^2J_{HD}$ coupling constant, and substitution in eq. 3. Here the combination of a

$$J_{HH} = J_{HD} \times (\nu_H/\nu_D) \tag{3}$$

large Larmor frequency ratio, ν_H/ν_D = ca. 6.5144, with the small measured coupling constant magnifies any errors.

With correct spectrometer operation, the intensity of ^1H NMR signals is directly proportional to the amount of nuclide responsible (13). This direct proportionality holds also for ^3H NMR signals (7) because of the close similarity in relaxation mechanisms already discussed. Signals in the ^3H NMR spectrum will of course show the expected couplings to adjacent protons. Because tritium is normally used at low abundance, the resultant distribution of signal intensity amongst lines of multiplets will usually necessitate inconveniently long spectral acquisition times, to build up an adequate signal-to-noise ratio. To concentrate available signal strength into single lines, ^3H NMR spectra are, then, frequently recorded with simultaneous ^1H noise decoupling--compare ^{13}C NMR. As a result, Overhauser effects may arise which alter signal intensities (26,27). For partially tritiated compounds, the nuclear Overhauser effect observed in the ^3H NMR signal intensities is derived from the ^1H induced relaxation of the tritons. Irradiation of the protons increases their population of the upper spin state relative to the lower. To maintain an overall Boltzmann distribution between lower and upper spin states for the whole nuclear spin system, a compensating redistribution of triton spins necessarily occurs to give a greater population in their lower spin state. More rf energy can then be absorbed, with resultant increase in triton signal strength. For the present nuclear system, the maximum Overhauser effect is 47%, given by $\gamma_H/2\gamma_T$ (26). This is not large, when compared with 199% for ^{13}C on ^1H irradiation, and the worst resulting error in comparing the relative intensities of tritium signals would be ±10% (27). Generally the error would be much less, particularly as experiment suggests there is little variation in nuclear Overhauser effect between different tritons in a generally tritiated compound (21-23,27). Indeed, this observation (10,28) led us at first, incautiously, to suppose that there was no Overhauser effect. It is thus evident that differential Overhauser effects are often small or negligible, and indeed this is to be expected (27). Triton relaxation in the average, low isotopic abundance, sample will be mediated by dipole-dipole interaction with the surrounding protons and there will be little or no difference in detailed mechanism between one triton and another. Straightforward integration of the ^{3}H NMR signals in a ^{1}H noise-decoupled spectrum can then be expected to give results of normal accuracy (±2-5%).

Overhauser enhancements can always be suppressed, if required, through use of an inverse gating sequence with delays ⩾10 x T_1 s between acquisitions (29,30). Unfortunately, this process increases the spectrometer operating time ca. 4 times, and more time still is needed for determination of the longest relaxation time, T_1, of the different tritons in the sample, if the acquisition delay is to be set correctly. The suppression procedure is therefore too costly in time to justify routine use (21,27).

Before describing applications of ^{3}H NMR spectroscopy, two further points should perhaps be made. One is that because tritium labeling is normally at low abundance (<0.1%), largely for reasons of the radioactivity, ^{1}H NMR spectra cannot suffice for ^{3}H detection. This is analogous to the fact that ^{13}C at ca. 1% (i.e., in natural abundance) does not intrude usually in ^{1}H NMR spectra of organic compounds. The second point is also obvious, but nevertheless can be overlooked: it is that most ^{3}H NMR spectra will be superpositions of the spectra of differently monolabeled species and not the spectrum of a single molecular species, as is common in ^{1}H NMR. This has significance for the correct interpretation of ^{3}H NMR splitting patterns arising from multiply-labeled compounds or in ^{1}H-coupled ^{3}H NMR spectra (see Figure 4) (17).

APPLICATIONS OF ^{3}H NMR SPECTROSCOPY

To date, the majority of applications of ^{3}H NMR spectroscopy have been analytical. This is simply because the technique is such a direct and convenient method for determining the regio- and stereospecificity of tritium labeling and for measuring the relative amounts of tritium in different positions in labeled molecules. Although NMR methods intrinsically lack high sensitivity, ^{3}H NMR has such considerable advantages to offer over the analysis of [^{3}H]compounds by stepwise chemical degradation

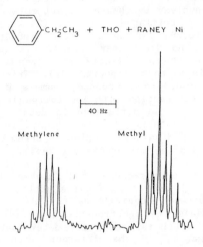

Figure 4. ^{3}H NMR spectrum of $[1',2'-^3H]$ethylbenzene showing the methyl (^{1}H-coupled) and methylene signals, respectively, as a triplet of triplets and a doublet of quartets, which arise from the geminal and the vicinal proton-triton coupling in the species $PhCH_2CH_2T$ and $PhCHTCH_3$ present in the sample.

and counting that the possible necessary use of mCi rather than µCi or lesser amounts of tritiated compounds as tracers may frequently be warranted. Fortunately, tritium is the least toxic and the cheapest radionuclide, and the necessary precautions for its safe use are easily reduced to a simple routine (1,6,8,31).

Tritium NMR spectroscopy has enabled the detailed analysis of many generally tritiated compounds, e.g., [G-^{3}H]polycyclic hydrocarbons (28), to be achieved for the first time. Very close observation of the regio- and stereospecificity of so-called specific tritiation methods is feasible (11): these methods include the use of tritium gas for catalytic reduction of multiple carbon-carbon bonds (11,32), of functional groups such as oxo (reductive amination) and nitrile groups (11), for catalysed tritiodeprotonation (33) and -dehalogenation (11,33) and Wilzbach tritiation (11); hydrogen exchange processes involving tritiated water or acids as source of the isotope (28,34,35,36); borotritide reductions (5,11); and decarboxylation in presence of tritiated water or analogous media (11,37). Tritium NMR is highly suitable for study of the integrity of the [^{3}H]label in substrates which have been modified by chemical or enzymic methods subsequent to their initial tritiation (11). Tritium NMR--of course with the use of tritiated compounds--provides elegant means of elucidating reaction mechanisms, both chemical (16,38) and biochemical (31,39-41). In studies of the mechanisms of catalytic addition (32) and exchange of hydrogen (42-45), including reforming reactions on zeolitic catalysts (46,47), the potentialities of ^{3}H NMR are already well recognised. The use of ^{3}H NMR in the study of the self-radiolysis of [^{3}H]compounds at high specific activity appears promising (7).

A straightforward application of ^{3}H NMR is in the examination of the extent of labeling achieved by simple processes (devised for simple compounds) when applied to more complex structures.

Bromination and tritiodebromination of lysergic acid diethylamide (I) gave a [^{3}H]product which showed a single line at δ 7.22 (see Figure 5) in the ^{3}H NMR spectrum acquired with ^{1}H decoupling (11). This was indicative of completely specific 2-tritiation. Equally selective was the tritiation of a dehydrodiprenorphine with sodium borotritide to give solely [16-^{3}H]-diprenorphine (II), as shown by the single sharp ^{3}H resonance at

(II)

δ 2.37 in the ^{1}H-decoupled ^{3}H NMR spectrum. This reduction must proceed through an imine tautomer (11). Treatment of the labile, light sensitive, antitumour drug vincristine (III) with [O-^{3}H]-trifluoroacetic acid at ambient temperature and back exchange of labile tritium with methanol, provided, as with vinblastine (34), a [^{3}H]labeled, active drug suitable for the setting up of a radioimmunoassay. The ^{3}H NMR spectrum showed that the label was confined to stable aromatic positions in the indole ring alone (see Figure 6). This was in contrast to vinblastine, where the 17 position was also labeled (34).

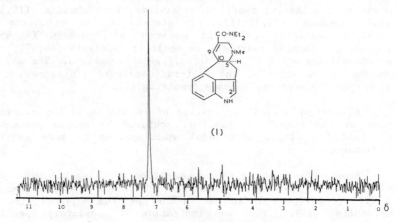

Figure 5. ^{3}H NMR spectrum (^{1}H-decoupled) of [^{3}H]lysergic acid diethylamide (I) in DMSO-d_6 at 96 MHz.

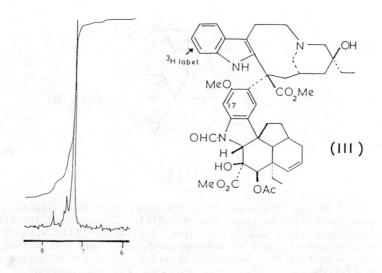

Figure 6. ^{3}H NMR spectrum (^{1}H-decoupled) of [^{3}H]vincristine (III).

The tritiation of ethyl β-carboline-3-carboxylate (IV) by treatment of the 6-bromo derivative with tritium and a catalyst did not proceed regiospecifically to the [6-^{3}H]product: there were two lines in the ^{1}H-decoupled ^{3}H NMR spectrum (see Figure 7). Comparison with the ^{1}H NMR spectrum showed that the extra signal at δ 9.01 came from the 1 position, which was presumably sufficiently acidic to exchange under the hydrogenation conditions employed. Reduction of a carboxyl-protected 4-carboxypyridine with sodium borotritide and removal of labile tritium by exchange provided a tritiated isoguvacine (V) which showed singlets at δ 3.38 and 3.87 in the ^{3}H NMR spectrum acquired with ^{1}H decoupling. The ^{1}H NMR spectrum of V could be assigned unambiguously from the splitting pattern making interpretation of the ^{3}H NMR spectrum immediately obvious (see Figure 8). This result corrects an earlier one (11): it is now recognised that the earlier product was impure.

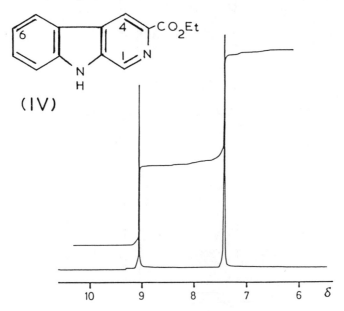

Figure 7. ^{1}H-Decoupled ^{3}H NMR spectrum of ethyl [^{3}H]β-carboline-3-carboxylate (IV).

In all these examples the ^{3}H NMR spectra comprised single lines, indicating the absence of triton-triton coupling, either because there was no double labeling of molecules or the sites were too distant from one another. However, multiple lines can also arise from isotope effects.

By exposure of the antibiotic actinomycin D (VI) to tritium gas (the Wilzbach labeling method) high specific activity [^{3}H]actinomycin D is produced. After back-exchange of any labile

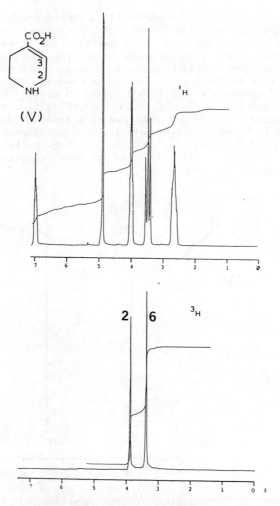

Figure 8. NMR spectra of isoguvacine (V) in D_2O: the ^1H spectrum at 90 MHz with HOD signal at δ 4.9; and the ^3H spectrum of [^3H]isoguvacine at 96 MHz, acquired with ^1H decoupling.

tritium introduced, the product gives signals in the ^1H-decoupled ^3H NMR spectrum (48) which coincide with the singlets at δ 2.16 and 2.57 in the known ^1H NMR spectrum (49), indicating ^3H label only in the 4- and 6-methyl groups, respectively. These signals, plotted at 21.33 Hz cm^{-1} for a spectrum acquired at 96 MHz, appear as closely spaced groups of 3 lines, as shown in Figure 9.

The methyl groups contain sufficient tritium to comprise a mixture of the species CTH_2, CT_2H, and CT_3, the chemical shifts of which are progressively ca. 0.02 ppm to higher field. It may be feasible to quantitate such signals by expansion of the display before integration. Usually, it is best to enhance the resolution, i.e., separate the lines, first, and this is easily done by the procedure of Clin et al. (19), which is applicable to Fourier transform NMR. Briefly, the procedure effects digital filtering by a sinusoidal window. In separating partially overlapped lines, it distorts individual lineshape and can therefore affect intensity, but provided T_2^*, the effective spin-spin relaxation time, is the same for each line (as might be expected for a methyl group), then no net distortion of the relative intensities occurs. By definition, $T_2^* = 1/\pi\Delta_{obs.}$, where $\Delta_{obs.}$ is the measured linewidth (including the field inhomogeneity effect) (26).

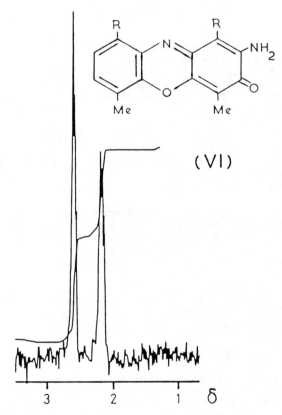

Figure 9. ^3H NMR spectrum of [^3H]actinomycin D (VI) (acquired with ^1H decoupling) in CD_3OH at 96 MHz.

The enhancement procedure is of course equally applicable to ^{1}H-coupled ^{3}H NMR spectra. From the integral trace or printout of such spectra, the complete quantitative analysis of highly tritiated methyl groups can be achieved accurately in a direct way for the first time. The full analysis by these means of [methyl-^{3}H]thymine (VII) provided new information about the

(VII)

mechanism of the labeling process (18). The method also enables the specific radioactivity of compounds to be measured which contain very highly tritiated groups. The conventional approach comprising weighing, and then scintillation counting after serial dilution, can be too inaccurate, or even impractical when the weight of available compound is extremely small. Where it has been feasible to compare the new method with the old, results are in satisfactory agreement (18). Figure 10 shows results for [methyl-^{3}H]methionine (VIII) prepared by methylation of the corresponding thiol with [^{3}H]methyl iodide, in turn obtained from [^{3}H]methanol resulting from reduction of carbon monoxide or dioxide with tritium gas.

In a ^{1}H-decoupled ^{3}H NMR spectrum, true multiplicity will of course arise when there is coupling to magnetic nuclei of other kinds. Thus the doublet signal J = 5.93 Hz, shown in Figure 11, is from the tritium in a product prepared by tritiodehalogenation of 2-bromo-4-fluorobenzoic acid. Evidently the tritium is in the expected 2 position, meta-coupled to fluorine in the 4 position. The corresponding proton-fluorine coupling constant is 5.5 Hz (50) compared with a value of 5.56 Hz calculated from the foregoing tritium-fluorine coupling constant by means of the approximate eq. 4.

$$J_{TF} = J_{HF} \times (\nu_T/\nu_H) \qquad (4)$$

From methyltriphenylphosphonium iodide, prepared by methylation of triphenylphosphine with highly tritiated methyl iodide, the enhanced, coupled ^{3}H NMR spectrum shows fully resolved and interleaved, doublet, triplet, and quartet signals from the PCT$_3$,

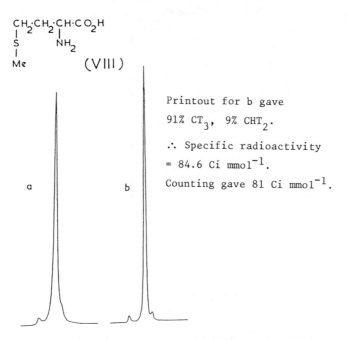

Figure 10. (a) ^{3}H NMR spectrum of [methyl-^{3}H]methionine (VIII) in D_2O; (b) with resolution enhancement, showing CT_3 singlet and CT_2H doublet with $^2J_{TH}$ = 12.8 Hz.

PCT_2H, and $PCTH_2$ groups present (Figure 12), the coupling constants $^2J_{TP}$ and $^2J_{TH}$ happening to be equal (18). The relative intensities of the three signals measured from the integral were 7:41:52%. Taking the specific radioactivity of a tritium atom in a compound as 29.1 Ci mmol^{-1}, the NMR results indicate a specific radioactivity for this methylphosphonium iodide of (3 x 29.1 x 7/100) + (2 x 29.1 x 41/100) + (29.1 x 52/100) = 45.1 Ci mmol^{-1}. By counting the [^{3}H]methyl iodide employed, a value of 45 Ci mmol^{-1} had been estimated for the product.

Usually, multiplicity in a ^{1}H-decoupled ^{3}H NMR spectrum results from there being two tritium atoms on adjacent positions. The steroid product (IX) from reduction of the 6,7-dehydro precursor with tritium, using a β-specific catalyst, gave the spectrum shown in Figure 13. Comparisons with data for related steroids (10) facilitate the interpretation. The pair of intense doublets with $^3J_{TT}$ = 5.6 Hz arise from the major species present, the [6β,7β-^{3}H$_2$]compound. The minor doublets (J = ca. 4 Hz) arise from the product of α-addition of tritium, viz., the [6α,7α-^{3}H$_2$]-

Figure 11. ^{1}H-Decoupled ^{3}H NMR spectrum of 4-fluoro[2-^{3}H]-benzoic acid.

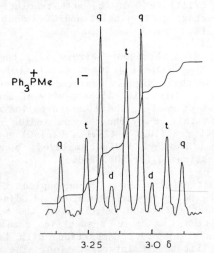

Figure 12. Resolution-enhanced ^{3}H NMR spectrum of [methyl-^{3}H]methyltriphenylphosphonium iodide in DMSO-d_6 at 96 MHz.

compound. The ^{3}H NMR spectrum thus shows that the catalyst was completely regio- but only partially stereospecific.

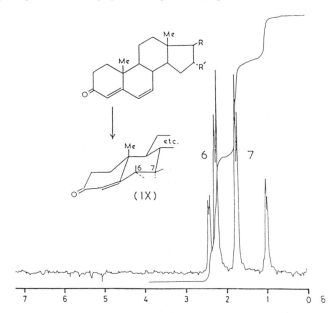

Figure 13. ^{1}H-Decoupled ^{3}H NMR spectrum of the [6,7-^{3}H$_2$]-steroid (IX), showing AB quartet signals from the major [6β,7β-^{3}H$_2$] and minor [6α,7α-^{3}H$_2$]-products.

Saturation of a double bond with a mixture of tritium and hydrogen results in some addition of TH besides T$_2$ (and H$_2$). The ^{3}H NMR spectrum will then show two groups of three lines because the observed spectrum is the superposition of the spectra from the doubly and the two singly tritiated species. The doubly labeled species gives rise to the usual pair of doublet signals. Inside each doublet, but generally off-centre because of the changed isotope effect, is a singlet from the respective singly labeled species (48).

A related example is shown in Figure 14. This is the ^{1}H-decoupled ^{3}H NMR spectrum of a uridine triphosphate (Xa), in which the uracil residue was labeled by exchange and tritiodebromination. The major doublets centred at δ 7.962 and 5.978 (J = 9.2 Hz) arise from [5,6-^{3}H$_2$]uridine triphosphate. Inside the doublets are singlets, at δ 7.959 and 5.982, which derive respectively from [6-^{3}H]- and [5-^{3}H]uridine triphosphate. The minor signals in the spectrum are from the corresponding [^{3}H]uracil species (Xb), there evidently having been some hydrolytic decomposition during preparation or examination of the sample.

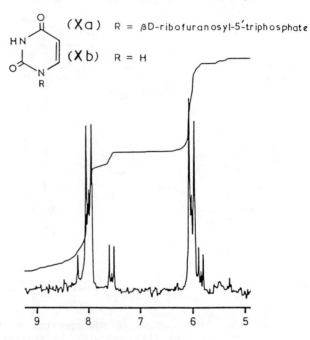

Figure 14. ^{1}H-Decoupled ^{3}H NMR spectrum of [5,6-^{3}H$_2$]uridine triphosphate showing the presence of [5,6-^{3}H$_2$]-uracil impurity, and monolabeled species.

Such examples of superimposed spectra are commonplace in ^{3}H NMR spectroscopy because of the use of the isotope at very low abundance.

DL-[2,3-^{3}H]Ornithine (XI) was prepared from ethyl 2-acetamido-4-cyanobutyrate by tritiation and then reduction with tritium, deprotection, and back-exchange of labile tritium. The ^{1}H-decoupled ^{3}H NMR spectrum shows four main signals, two of which each comprise a pair of lines (see Figure 15). This is not an indication of double labeling because the spacings are unequal and the intensities are incorrect for coupled doublets. The line at δ 4.04 is from the 2-position and that at δ 3.03 is from the 5 position. The lines at δ 2.02 and 1.96 evidently arise from the diastereotopic tritons in the 3-position (the two 3-sites are nonequivalent by virtue of the 2-chirality: that the amino acid is racemic is irrelevant). The pair of lines at δ 1.85 and 1.77 analogously arise from the 4-sites. There is a weak line at δ 3.23 attributable to impurity (2% of total ^{3}H). The whole spectrum (Figure 15) is very obviously the superposition of the individual spectra of the seven species present, namely [2-^{3}H], [3-^{3}H], [3'-^{3}H], [4-^{3}H], [4'-^{3}H], and [5-^{3}H]ornithine and a trace [^{3}H]impurity.

$$H_2N \cdot CH_2CH_2CH_2CH \cdot CO_2H$$
$$\underset{NH_2}{|}$$

(XI)

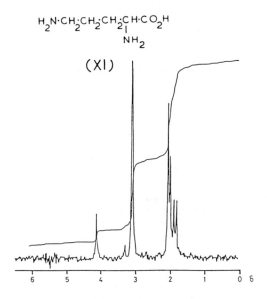

Figure 15. ^{1}H-Decoupled ^{3}H NMR spectrum of DL-[^{3}H]ornithine (XI) in D_2O at 96 MHz.

A further example, which also illustrates a purely NMR advantage of the use of tritium labeling, is that provided by the ^{3}H NMR spectrum (48) of 2-deoxy[2,6-^{3}H]glucose (XII), which

(XII)

afforded the ^{3}H chemical shifts listed in Table 2. The aqueous solution of XII comprises an equilibrium mixture of the 1α and 1β anomers. Hence there are two lines in the ^{1}H-decoupled ^{3}H NMR spectrum from ^{3}H in the 2-axial site, and two lines from ^{3}H in the 2-equatorial site. The 6,6' positions in the side chain methylene group are in principle nonequivalent because of the adjacent chirality; therefore, taking account of there being two 1 anomers, four lines are expected from the 6-CH$_2$ group. All the chemical shifts (Table 2) are of course available by inspection: they point to a slight error in interpreting the complex ^{1}H NMR spectrum where it was considered that the 6 and 6' chemical shifts in the 1α anomer were identical (51). The ^{3}H NMR results show that in dilute solution in D$_2$O at 25°C, these two shifts are different. This is not to denigrate the ^{1}H NMR work but merely to point to a singular usefulness of ^{3}H substitution and ^{3}H NMR. Equally easily checked are the computer analyses of very complex ^{1}H NMR spectra, as from polycyclic aromatic hydrocarbons. Figure 16 shows the ^{1}H NMR spectrum of benz[a]anthracene (XIII) alongside the ^{1}H-decoupled ^{3}H NMR spectrum of [G-^{3}H]benz[a]anthracene, which compound had been obtained by simple catalysed exchange (28). Whilst the ^{1}H-decoupled ^{3}H NMR spectrum gives the chemical shifts of the tritiated positions direct (and the relative degree of labeling, from the line intensities), the assignments can only be made by reference to the computer analysis of the complex ^{1}H spectrum (52), taking account of the effects of solvent, concentration, and temperature differences if any. The readily obtained ^{3}H NMR chemical shifts thus provide an immediate and potentially useful check on the calculated ^{1}H shifts. In the present case the agreement is highly satisfactory.

Table 2. Chemical Shifts for the 2 and 6 Positions in 2-Deoxyglucose in D$_2$O (DSS)

Chemical Shift

δ_T[a]	δ_H[b]	Assignment	Rel. %
1.53	1.51	2ax in 1β anomer	6.1
1.72	1.71	2ax in 1α anomer	6.1
2.15	2.13	2eq in 1α anomer	19.8
2.29	2.27	2eq in 1β anomer	19.8
3.75	3.73	6 in 1β anomer	12.7
3.80	3.81	6 in 1α anomer	14.7
3.86	3.81	6' in 1α anomer	11.2
3.92	3.90	6' in 1β anomer	9.6

[a]At 96 MHz. [b]At 300 MHz (51).

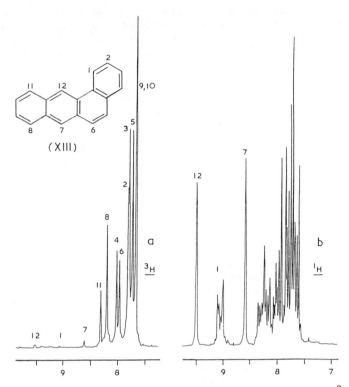

Figure 16. NMR spectra of benz[a]anthracene (XIII); (a) ^{3}H spectrum (^{1}H-decoupled) of the [G-^{3}H]compound, (b) ^{1}H spectrum of XIII.

Occasionally, the chemical shifts obtained directly from the ^{1}H-decoupled ^{3}H NMR spectrum of a [G-^{3}H]labeled compound can be assigned without recourse to analysis of the complex ^{1}H spectrum. This was possible (53) for phenanthridine (XIV).

(XIV)

In appropriate situations, the ^{1}H-coupled ^{3}H NMR spectrum will provide solutions to assignment problems. Examples concerning [^{3}H]atropine (8,11), 5α-dihydro[4,5-^{3}H]testosterone (8), and 5-hydroxy[G-^{3}H]tryptamine (8,11) have already been given. The ^{3}H assignments for [^{3}H]strychnine (XV) were made in part by

(XV)

comparison of the ^{1}H NMR spectrum (at 250 MHz) with the ^{1}H-decoupled ^{3}H NMR lines (11). Useful confirmation of assignments came partly from the ^{1}H-coupled ^{3}H NMR spectrum, using the signal multiplicities, and partly from a specific ^{1}H irradiation experiment. Because a signal at δ 7.21 in the ^{1}H-coupled ^{3}H NMR spectrum was a triplet, it was necessarily assignable to the 2 or the 3 position in XV. Irradiation of 4-^{1}H at δ 8.1 in the proton region (near 90 MHz) (the lowest field signal) failed to decouple the ^{3}H triplet in the ^{3}H region, indicating that the triton responsible could not be in the 3 position, adjacent to 4-H, so that the 2 assignment was made (11).

When the [^{3}H]serotonin derivative (XVI) was prepared by tritiodebromination of the 4,6-dibromo compound, the specific radioactivity was unexpectedly high. Indeed, the ^{1}H-decoupled ^{3}H NMR spectrum showed 3 lines (Figure 17a), at δ 6.63, 6.86, and 7.07. The ^{1}H-coupled ^{3}H NMR spectrum (Figure 17b) showed that the signal at δ 6.63 was a double doublet (J_{TH} 9, 2.2 Hz) and so from the 6 position. By adding D$_2$O to exchange the 1-NH proton and remove the coupling to the 2 position, the signal at δ 7.07 sharpened to a singlet (Figure 17c): hence this third line arose from a 2 triton. The signal at δ 6.86 was then assignable to the 4 position. Thus the labeling problem was solved.

In early studies of the potentialities of tritium labeling and ^{3}H NMR for the solution of biochemical problems, the 1,2-dehydrogenation of testosterone by <u>Cylindrocarpon</u> <u>radicicola</u> was easily and unambiguously shown to proceed by a stereospecific 1α,2β-trans-diaxial elimination (39,48). The substrate fed to the mould comprised 5 parts of [1α,2α-^{3}H$_2$]testosterone (XVII) with 1 part of the [1β,2β-^{3}H$_2$]compound, and the product consisted

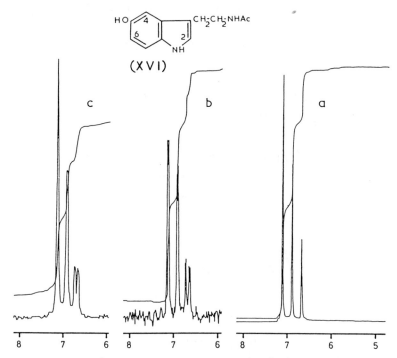

Figure 17. ^{3}H NMR spectra of N-acetyl[^{3}H]serotonin (XVI) in DMSO-d_6; (a) ^{1}H-decoupled, (b) ^{1}H-coupled, and (c) as in b after addition of a trace of D_2O.

of 5 parts of the [2-^{3}H]dehydro compound (XVIII) and 1 part of the corresponding [1-^{3}H]product, as was at once apparent from the ^{3}H NMR spectra (decoupled and coupled) (48).

The biosynthesis of penicillic acid (XIX) from acetate by the mould Penicillium cyclopium had been studied using [^{14}C]-acetate (55,56) and [^{14}C]malonate (57), and also by the use of [^{13}C]labeling and ^{13}C NMR (58). We were therefore particularly interested to discover whether the use of [^{3}H]acetate and ^{3}H NMR might have any advantage. Not only did the new approach point to the correct biochemical pathway in a direct and unambiguous manner, in contrast to the first ^{14}C experiments (55,56), but provided novel information (31). This came from the ^{3}H NMR spectrum (48,59) of the penicillic acid metabolite. The spectrum showed that in the 5-methylene group, the hydrogen trans to C-methyl was largely specifically labeled. The same held when [3,5-^{3}H]orsellinic acid (XX) was fed to the mould. Not only did

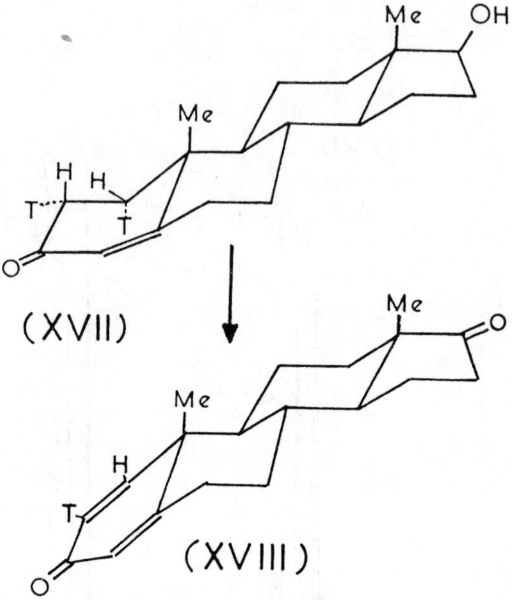

(XVII)

(XVIII)

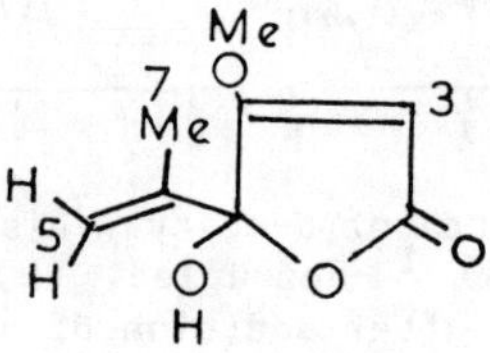

(XIX)

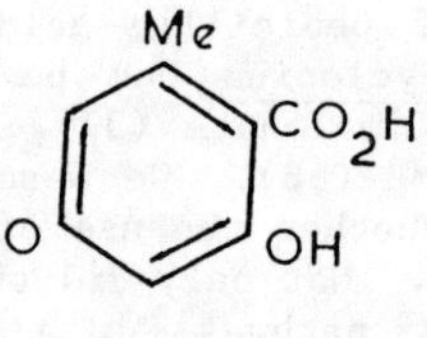

(XX)

these observations powerfully support the suspected role of XX as a true intermediate on the pathway from acetate to XIX, but they had important stereochemical implications for any analysis of the detailed enzymic processes which occur. A further finding from the ^{3}H NMR spectrum was the order of assembly of the C_2 units, as indicated (31) by the diminishing intensity of the ^{3}H NMR signals from the 7, 3, and 5 positions in the [^{3}H]penicillic acid produced. Feeding of [^{3}H]malonate to the mould labeled the 3 position and to a lesser extent the 5 position trans to C-methyl in XIX, in agreement with acetate being the true precursor of the 7 position.

Important conclusions are to be drawn from the penicillic acid study (31,59). Provided conditions can be found to maximise uptake of [^{3}H]precursors and minimise incidental exchange-loss of label, then it is to be concluded (a) that tritium can at least usefully complement work with carbon isotopes, (b) that tritium may (as here) be the more powerful probe, (c) that only isotopic hydrogen will distinguish sp^2 (vinylic) methylene positions or prochiral sp^3 sites, and that tritium rather than deuterium is preferable for NMR observation, particularly where coupling constants to vicinal hydrogen will delineate the stereochemistry of the label, and (d) that the incidental (or enzymic) exchange-loss of tritium can provide detailed mechanistic information not accessible through use of carbon isotopes.

The particular advantages of ^{3}H NMR were put to outstanding use in the investigation, by Altman et al. (40), of the stereochemistry of cyclopropane ring formation, during the biosynthesis of cycloartenol (XXIV) from oxidosqualene (XXIII) via a carbonium ion intermediate. These are steps in the biosynthesis of steroids by photosynthetic organisms. An ingenious synthesis from D-malic acid via XXI gave oxidosqualene (XXIII) bearing a chiral (R)-CHDT methyl group. The precursor (XXII) gave rise to a 1:2:2:1 multiplet in its ^{1}H decoupled ^{3}H NMR spectrum in accord with its constitution. The geminal deuteron, with I = 1, splits the triton signal into a 1:1:1 triplet but this signal appears twice, and there happens to be overlap, to give the observed four-line signal. The doubling results from the compound (XXII) being a diastereotopic pair, the site adjacent to the triton being racemic in an otherwise chiral molecule. With ^{1}H coupling, the ^{3}H NMR signal appeared as a 13-line multiplet (Figure 18): this arises from a large doublet splitting of the basic 1:2:2:1 multiplet by the proton of the methyl group together with a lesser doublet splitting by the vicinal proton. Line overlap (accidently caused by the magnitudes of the splittings) results in the observed pattern, instead of a possible quartet of quartets.

The labeled oxidosqualene (XXIII) was incorporated very well by a microsomal fraction from the chosen organism <u>Ochromonas malhamensis</u> to give in good yield a labeled cycloartenol (XXIV).

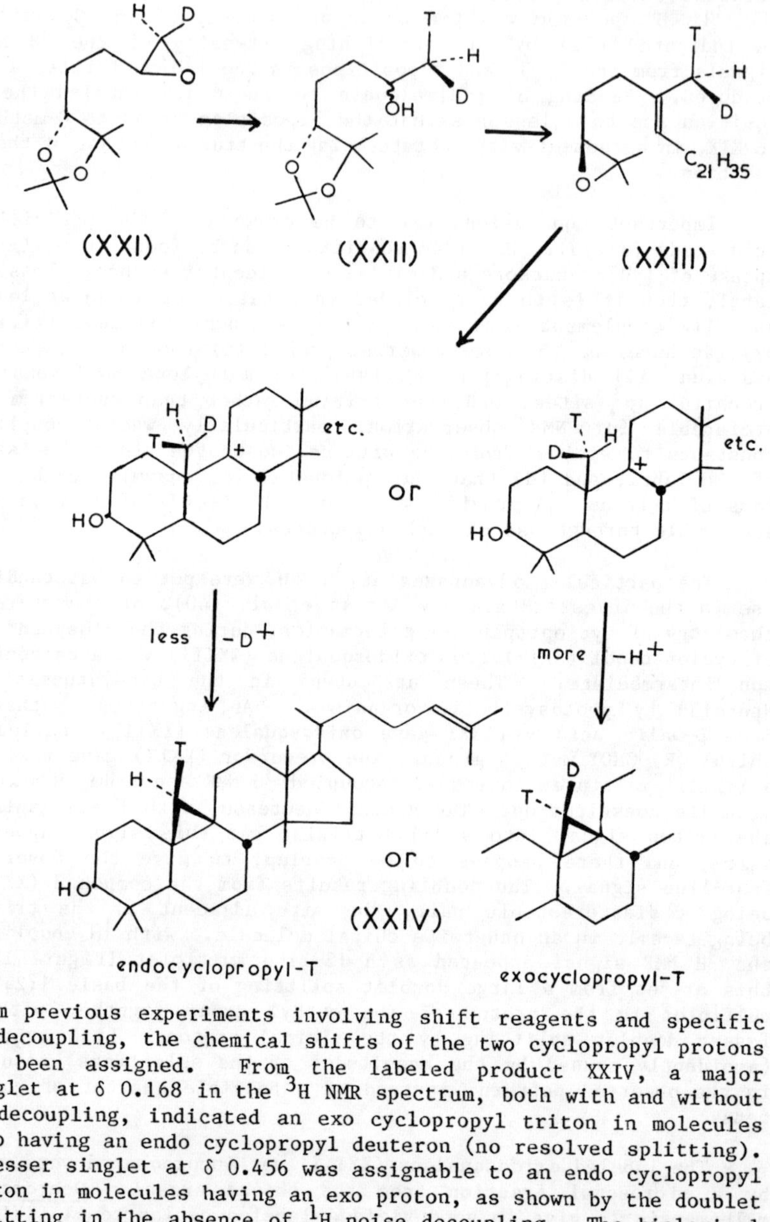

(XXI) (XXII) (XXIII)

(XXIV)

endocyclopropyl-T   exocyclopropyl-T

From previous experiments involving shift reagents and specific ^{1}H decoupling, the chemical shifts of the two cyclopropyl protons had been assigned. From the labeled product XXIV, a major singlet at δ 0.168 in the ^{3}H NMR spectrum, both with and without ^{1}H decoupling, indicated an exo cyclopropyl triton in molecules also having an endo cyclopropyl deuteron (no resolved splitting). A lesser singlet at δ 0.456 was assignable to an endo cyclopropyl triton in molecules having an exo proton, as shown by the doublet splitting in the absence of ^{1}H noise decoupling. The biological

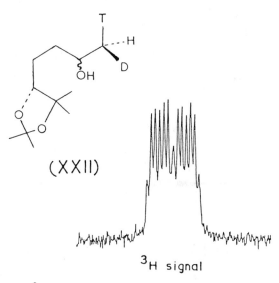

Figure 18. ^{3}H NMR spectrum of intermediate (XXII).

cyclisation had thus occurred with retention of configuration (see scheme) and not with inversion (40).

More recently, Aberhart and Tann (41) have used [^{3}H]labeling and ^{3}H NMR to elucidate the stereochemistry of the catabolism of L-valine (XXV), which proceeds by dehydrogenation of derived isobutyryl coenzyme A (XXVI) to methacrylyl CoA (XXVII). From pertinent spectra, in which signal separation was increased by addition of the shift reagent Eu(fod)$_3$, it was deduced that the elimination by Pseudomonas putida requires an antiperiplanar orientation of the two hydrogen atoms concerned, one of which resides on the 2 position, and the other on the 2-pro-S methyl group in XXVI. The 3-hydroxyisobutyric acid (XXVIII) which accumulates in the medium, as a result of hydration of the methacrylate (XXVII) and ester hydrolysis, was isolated. Since the stereochemistry of the hydration had earlier been established, and the prochiral 3 hydrogens in XXVIII identified by means of the Eu(fod)$_3$ shifted ^{1}H NMR spectrum of a simple derivative (60), the distribution of tritium in these 3 positions in XXVIII, resulting from chiral CHDT methyl groups in isobutyrate (XXVI), could be deduced by ^{3}H NMR.

Some use of ^{3}H NMR has also been made in elucidating the stereochemistry of the hydrogenation by Wilkinson's catalyst of the deuterium-labeled N-acetylisodehydrovaline (XXIX) with a hydrogen tritide--hydrogen mixture (61). The two diastereoiso-

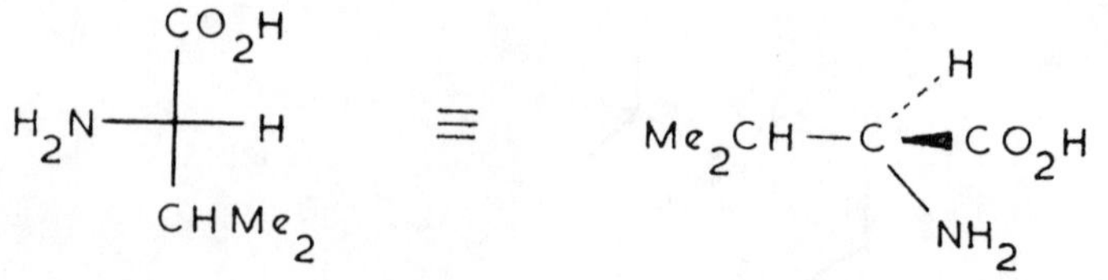

(XXV)

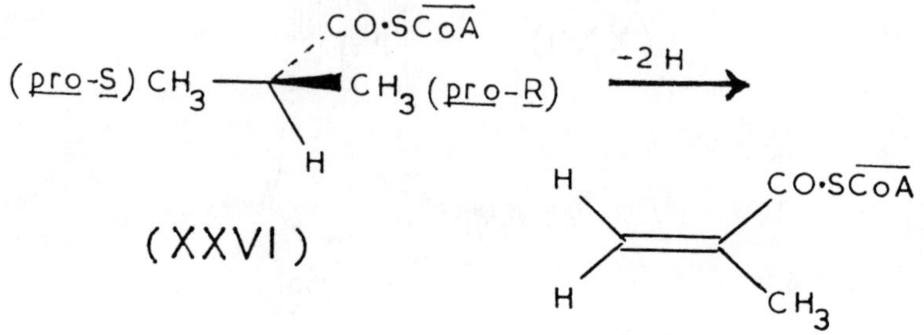

(XXVI)

(XXVII)

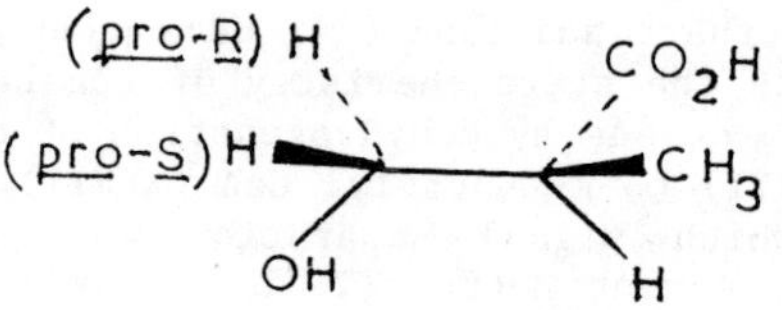

(XXVIII)

meric pairs of products from addition of HT in the 3,4-sense were formed in unequal amounts. This showed there was a preference for 3-re,4-si-addition to the S component of XXIX and 3-si,4-re-addition to the R component.

CONCLUSIONS AND PROSPECTS

The examples and spectra quoted in the foregoing account will have indicated the excellent NMR properties of the tritium nucleus for high resolution studies. Tritium NMR spectroscopy provides a rapid and nondestructive technique for locating the positions and amounts of the isotope in [^{3}H]compounds, and for

[Scheme showing compound (XXIX): two enantiomeric dehydroamino acid structures with CO₂H, AcNH, H, Me, D substituents at positions 3 and 4, undergoing 3-re,4-si addition and 3-si,4-re addition respectively, yielding CHDT (R) and CHDT (S) products.]

determination of the stereochemistry of labeling. No longer is it essential to employ chemical degradation procedures, which are inevitably very tedious and fraught with various uncertainties. There is no general need in quantitative ^{3}H NMR to use the rather time-consuming special pulse sequences for full suppression of nuclear Overhauser effects. Noise decoupling of the protons causes little variation in Overhauser enhancement of tritium signals given by the average, partially tritiated tracer compound, so the straightforward integration of relatively rapidly acquired ^{1}H-decoupled ^{3}H NMR spectra gives highly acceptable results.

As regards the relative merits of deuterium and tritium as isotopes of hydrogen (protium) for tracer and mechanistic studies, it must at once be stated that these isotopes are complementary. Each has advantages for specific applications. When molecular motion in macromolecules, lipid layers, membranes, or enzymes is to be studied, then ^{2}H labeling and ^{2}H NMR will provide the information by virtue of ^{2}H being a quadrupole nucleus. When simple chemical shift and intensity measurements will suffice to solve problems, then ^{2}H labeling at appropriate abundance, with either ^{1}H or ^{2}H NMR, will be the approach of choice. However, where spectral dispersion is all important, and particularly when stereochemical problems need to be resolved, then it is virtually essential to use tritium as the tracer with ^{3}H NMR as the method of observation. The sensitivity of the NMR technique whilst often satisfactory does not meet every requirement in biochemical tracer work as normally conducted. However, there is no special hazard in working with tritiated compounds with total radioactivities of the order of 10 μCi to 10 mCi or more: thin glass and the skin provide complete protection against the weakly penetrating β-radiation (mean range in air, 4.5 mm). The main precautions are to avoid contact, inhalation, and spillage, so good techniques and proper housekeeping are the essentials. Work should only be done over spill trays, using gloves and protective clothing, and in properly ventilated laboratories and fume-cupboards, equipped with washing facilities having foot-operated taps. The monitoring of workers who routinely prepare or handle tritiated compounds is best conducted by scintillation counting of their urine samples at weekly intervals. (A contaminated person would rest from work and drink adequately (1).) Only for the manipulation of tritium gas as in hydrogenations, or of other volatile highly radioactive samples, are special facilities absolutely essential, i.e., a gas line in a special hood.

In examining possible methods for increasing the sensitivity of NMR, particularly for the examination of [^{3}H]tracers, Behrendt (62) has made some theoretical proposals. These appear to have been made, in part, because of an impression that ^{3}H NMR using conventional spectrometers necessitates [^{3}H]labeling at levels sufficient to cause noticeable self-radiolysis of the compounds. This is not so. The average labeled compound (of moderate specific radioactivity) in a dilute solution containing a total of perhaps 10 or even up to 50 mCi of ^{3}H isotope does not suffer noticeable self-radiolysis over weeks or months. Indeed we early set up a variety of [^{3}H]compounds with the intention of monitoring their radiolytic decomposition by ^{3}H NMR. In the event, the intended programme failed because the samples remained unaffected, with virtually only the single exception of very highly tritiated uridine at 31 Ci mmol^{-1}, which undergoes hydroxylation and addition of water quite rapidly (7).

Finally a brief account is given of Behrendt's interesting suggestions for increasing the sensitivity of ^{3}H NMR spectroscopy. At a temperature T near to absolute zero, the NMR signal-to-noise will improve very markedly as $T^{-3/2}$. Samples at that temperature will necessarily be solid. However, satisfactorily resolved NMR spectra can now be obtained from solid samples by "magic angle" spinning, i.e., by rapid rotation of the sample in a turbine inclined at 54° 44' to the magnetic field axis. The frequency of rotation needs to exceed the frequency width of the broadest NMR signal given by the static solid sample in the magnetic field, if line broadening by dipolar interactions and by chemical shift anisotropy are both to be suppressed. Hence the frequency of rotation of the sample might need to be in the range 10-100 kHz, and this exceeds constructional possibilities with nonmagnetic materials. Behrendt therefore proposes that extremely good sensitivity for NMR examination of, e.g., [^{3}H]-tracers should be achieved using frozen samples in a probe cooled continuously at 0.01 K by a helium dilution refrigerator, with the applied magnetic field correctly rotated by electrical means. Because the relaxation time for the NMR nuclei would be too long near absolute zero for normal NMR observation, multipulse sequences designed to hasten relaxation would be necessary or other expedients such as paramagnetic solvents for the sample. However, the necessary instrumentation for such experiments has yet to be developed. Possibly, more immediately helpful, might be the design of a pulse sequence to enhance ^{3}H NMR signal strength as a result of cross polarisation of abundant spins in the sample (63), or merely to use a DEFT sequence to reduce acquisition time and so improve the effectiveness of spectral accumulation.

REFERENCES

(1) E. A. Evans, "Tritium and Its Compounds," 2nd ed., Butterworths, London, 1974.
(2) E.g., Chem. Abstracts (Tritium, Biological Studies-Reactions: 1981, 94, pp. 6415CS-6416CS, 6417CS. 1980, 93, pp. 6675CS-6677CS. 1980, 92, pp. 5890CS-5892CS. 1979, 91, pp. 5833CS-5835CS. 1979, 90, pp. 5994CS-5996CS. 1978, 89, pp. 5594CS-5595CS, 1978, 88, pp. 5205CS-5207CS, etc.
(3) Y. Osawa and D. G. Spaeth, Biochem., 10, pp. 66-71 (1971).
(4) J. Fishman and H. Guzik, J. Am. Chem. Soc., 91, pp. 2805-2806 (1969).
(5) J. A. Elvidge, J. R. Jones, R. B. Mane, V. M. A. Chambers, E. A. Evans, and D. C. Warrell, J. Label. Compounds Radiopharmaceuticals, 15, pp. 141-151 (1978).
(6) V. M. A. Chambers, E. A. Evans, J. A. Elvidge, and J. R. Jones, Review 19, The Radiochemical Centre, Amersham, pp. 1-68, 1978.
(7) J. P. Bloxsidge, J. A. Elvidge, J. R. Jones, and E. A. Evans, Org. Magn. Reson., 3, pp. 127-138 (1971).
(8) J. A. Elvidge, "^{3}H NMR Spectroscopy," in "Isotopes: Essential Chemistry and Applications," J. A. Elvidge and J. R. Jones, Eds., The Chemical Society, London, 1980, pp. 152-194.

(9) J. M. A. Al-Rawi, J. A. Elvidge, J. R. Jones, V. M. A. Chambers, and E. A. Evans, J. Label. Compounds Radiopharmaceuticals, 12, pp. 265-273 (1976).

(10) J. M. A. Al-Rawi, J. P. Bloxsidge, J. A. Elvidge, J. R. Jones, V. E. M. Chambers, V. M. A. Chambers, and E. A. Evans, Steroids, 28, pp. 359-375 (1976).

(11) J. P. Bloxsidge, J. A. Elvidge, M. Gower, J. R. Jones, E. A. Evans, J. P. Kitcher, and D. C. Warrell, J. Label. Compounds Radiopharmaceuticals, 18, pp. 1141-1165 (1981).

(12) R. R. Ernst, Advan. Magn. Reson., 2, pp. 1-135 (1966).

(13) J. A. Pople, W. G. Schneider, and H. J. Bernstein, "High Resolution Nuclear Magnetic Resonance," McGraw-Hill, New York, 1959, pp. 1-501.

(14) J. M. A. Al-Rawi, J. A. Elvidge, J. R. Jones, and E. A. Evans, J. Chem. Soc. Perkin Trans. 2, pp. 449-452 (1975).

(15) J. M. A. Al-Rawi and J. P. Bloxsidge, Org. Magn. Reson. 10, pp. 261-262 (1977).

(16) J. M. A. Al-Rawi, J. P. Bloxsidge, C. O'Brien, D. E. Caddy, J. A. Elvidge, J. R. Jones, and E. A. Evans, J. Chem. Soc. Perkin Trans. 2, pp. 1635-1638 (1974).

(17) J. P. Bloxsidge, J. A. Elvidge, J. R. Jones, R. B. Mane, and M. Saljoughian, Org. Magn. Reson., 12, pp. 574-578 (1979).

(18) J. P. Bloxsidge, J. A. Elvidge, J. R. Jones, E. A. Evans, J. P. Kitcher, and D. C. Warrell, Org. Magn. Reson., 15, pp. 214-217 (1981).

(19) B. Clin, J. de Bony, P. Lalanne, J. Biais, and B. Lemanceau, J. Magn. Reson., 33, pp. 457-463 (1979).

(20) M. A. Long and C. A. Lukey, Org. Magn. Reson., 12, pp. 440-441 (1979).

(21) L. J. Altman and N. Silberman, Anal. Biochem., 79, pp. 302-309 (1977).

(22) L. J. Altman and N. Silberman, Steroids, 29, pp. 557-565 (1977).

(23) L. J. Altman and L. Thomas, Anal. Chem., 52, pp. 992-995 (1980).

(24) R. M. Lynden-Bell and R. K. Harris, "Nuclear Magnetic Resonance Spectroscopy," Nelson, London, 1969, pp. 1-160.

(25) W. Duffy, Phys. Rev., 115, pp. 1012-1014 (1959).

(26) J. H. Noggle and R. E. Schirmer, "The Nuclear Overhauser Effect," Academic Press, New York, 1971, pp. 1-259.

(27) J. P. Bloxsidge, J. A. Elvidge, J. R. Jones, R. B. Mane, and E. A. Evans, J. Chem. Res. (S), pp. 258-259 (1977).

(28) J. M. A. Al-Rawi, J. P. Bloxsidge, J. A. Elvidge, J. R. Jones, V. M. A. Chambers, and E. A. Evans, J. Label. Compounds Radiopharmaceuticals, 12, pp. 293-306 (1976).

(29) D. Canet, J. Magn. Reson., 23, pp. 361-364 (1976).

(30) R. K. Harris and R. H. Newman, J. Magn. Reson., 24, pp. 449-456 (1976).

(31) J. A. Elvidge, D. K. Jaiswal, J. R. Jones, and R. Thomas, J. Chem. Soc. Perkin Trans. 1, pp. 1080-1083 (1977).

(32) J. A. Elvidge, J. R. Jones, R. M. Lenk, Y. S. Tang, E. A. Evans, G. L. Guilford, and D. C. Warrell, J. Chem. Res. (S), pp. 82-83 (1982).
(33) E. A. Evans, J. P. Kitcher, D. C. Warrell, J. A. Elvidge, J. R. Jones, and R. Lenk, J. Label. Compounds Radiopharmaceuticals, 16, pp. 697-710 (1979).
(34) J. P. Bloxsidge, J. A. Elvidge, J. R. Jones, R. B. Mane, V. M. A. Chambers, E. A. Evans, and D. Greenslade, J. Chem. Res. (S), pp. 42-43 (1977).
(35) J. A. Elvidge, J. R. Jones, M. A. Long, and R. B. Mane, Tetrahedron Lett., pp. 4349-4350 (1977).
(36) M. A. Long, J. L. Garnett, and J. C. West, Tetrahedron Lett., pp. 4171-4174 (1978).
(37) J. A. Elvidge, J. R. Jones, R. B. Mane, and M. Saljoughian, J. Chem. Soc. Perkin Trans. 1, pp. 1191-1194 (1978).
(38) J. A. Elvidge, J. R. Jones, M. S. Saieed, E. A. Evans, and D. C. Warrell, J. Chem. Res. (S), pp. 288-289 (1981).
(39) J. M. A. Al-Rawi, J. A. Elvidge, R. Thomas, and B. J. Wright, J. Chem. Soc. Chem. Commun., pp. 1031-1032 (1974).
(40) L. J. Altmann, C. Y. Han, A. Bertolino, G. Handy, D. Laungani, W. Muller, S. Schwartz, D. Shauker, W. H. de Wolf, and F. Yang, J. Am. Chem. Soc., 100, pp. 3235-3237 (1978).
(41) D. J. Aberhart and C-H. Tann, J. Am. Chem. Soc., 102, pp. 6377-6380 (1980).
(42) E. A. Evans, D. C. Warrell, J. A. Elvidge, and J. R. Jones, J. Radioanal. Chem., 64, pp. 41-45 (1981).
(43) E. Buncel, J. A. Elvidge, J. R. Jones, and K. T. Walkin, J. Chem. Res. (S), pp. 272-273 (1980).
(44) E. Buncel, A. R. Norris, J. A. Elvidge, J. R. Jones, and K. T. Walkin, J. Chem. Res. (S), pp. 326-327 (1980).
(45) J. A. Elvidge, J. R. Jones, and M. Saljoughian, J. Pharm. Pharmacol., 31, pp. 508-511 (1979).
(46) J. L. Garnett, M. A. Long, and A. L. Odell, Chem. in Australia, 47, pp. 215-220 (1980).
(47) M. A. Long and P. G. Williams, "Tritium NMR Spectroscopy of Compounds Labelled by Exchange over Zeolite and Related Catalysts," in "Synthesis and Applications of Isotopically Labeled Compounds," Proc. Int. Symp. (Kansas City), A. Susan, Ed., Elsevier, Amsterdam, 1982.
(48) J. A. Elvidge, "Tritium NMR Spectroscopy and Its Applications," in "Synthesis and Applications of Isotopically Labeled Compounds," Proc. Int. Symp. (Kansas City), A. Susan, Ed., Elsevier, Amsterdam, 1982.
(49) T. A. Victor, F. E. Hruska, K. Hikichi, S. S. Danyluk, and C. L. Bell, Nature, 223, pp. 302-303 (1969).
(50) Sadtler Standard Spectra, 6757 M.
(51) A. De Bruyn and M. Anteunis, Bull. Soc. Chim. Belges, 84, pp. 1201-1209 (1975).
(52) C. W. Haigh and R. B. Mallion, Mol. Phys., 18, pp. 737-750 (1970).

(53) J. A. Elvidge, J. R. Jones, R. B. Mane, and J. M. A. Al-Rawi, J. Chem. Soc. Perkin Trans. 2, pp. 386-388 (1979).
(54) J. C. Carter, G. W. Luther, and T. C. Long, J. Magn. Reson., 15, pp. 122-131 (1974).
(55) A. J. Birch, G. E. Blance, and H. Smith, J. Chem. Soc., pp. 4582-4583 (1958).
(56) K. Mosbach, Acta Chem. Scand., 14, pp. 457-464 (1960).
(57) R. Bentley and J. G. Keil, J. Biol. Chem., 237, pp. 867-873 (1962).
(58) H. Seto, L. W. Cary, and M. Tanabe, J. Antibiotics, 27, pp. 558-559 (1974).
(59) J. M. A. Al-Rawi, J. A. Elvidge, D. K. Jaiswal, J. R. Jones, and R. Thomas, J. Chem. Soc. Chem. Commun., pp. 220-221 (1974).
(60) D. J. Aberhart and C-H. Tann, J. Chem. Soc. Perkin Trans. 1, pp. 939-942 (1979).
(61) D. H. G. Crout, M. Lutstorf, P. J. Morgan, R. M. Adlington, J. E. Baldwin, and M. J. Crimmin, J. Chem. Soc. Chem. Commun., pp. 1175-1176 (1981).
(62) S. Behrendt, J. Radioanal. Chem., 29, pp. 335-342 (1976).
(63) W. McFarlane and D. S. Rycroft, "Multiple Resonance," in "Nuclear Magnetic Resonance," The Royal Society of Chemistry, London, 10, pp. 162-187 (1981); A. Pines, M. C. Gibby, and J. S. Waugh, J. Chem. Phys., 56, pp. 1776-1777 (1972); E. D. Becker, J. A. Ferretti, and T. C. Farrar, J. Am. Chem. Soc., 91, pp. 7784-7786 (1969).

CHAPTER 10

NITROGEN NUCLEAR MAGNETIC RESONANCE SPECTROSCOPY

Robert L. Lichter

Department of Chemistry, Hunter College of CUNY,
New York, NY 10021

ABSTRACT

Experimental aspects associated with the two magnetic isotopes of nitrogen, ^{14}N (natural abundance = 99.6%, spin quantum number = 1) and ^{15}N (natural abundance = 0.4%, spin quantum number = 1/2) are discussed briefly. General relationships of molecular structure to nitrogen chemical shifts and spin-spin coupling constants involving ^{15}N are presented. The utility of nitrogen NMR for examining molecular dynamics is described. Illustrations are provided that are chosen largely from organic and biochemistry.

INTRODUCTION AND SCOPE

Because of the widespread occurrence of nitrogen in organic and biological systems, it is not surprising that nitrogen NMR has evoked considerable interest among chemists. Antedating applications of proton NMR, first experiments using ^{14}N NMR for structure elucidation go back twenty-five years (1); these were followed by extensive characterization of ^{14}N chemical shifts and some relaxation properties (2-6). Only with the appearance of methods for sensitivity enhancement, however, did it become practical to expand nitrogen NMR studies to the less abundant but more readily resolvable ^{15}N isotope. Beginning in the early 1970's, this development, coupled with some knowledge of ^{15}N nuclear relaxation properties and related nuclear Overhauser effects (NOE), led to a burgeoning of research with ^{15}N in relatively small molecules. More recently, commercial introduction of high-field superconducting magnets, together with the existence of sophisticated polarization transfer pulse sequences, have allowed spectra of biological and polymeric materials at reasonable concentrations to be recorded within practical time periods. As a result, while experimental problems still arise, ^{15}N has effectively supplanted ^{14}N as the nucleus of choice for most studies of molecular structure by high resolution nitrogen NMR.

This chapter will provide a general introduction and overview of nitrogen NMR. It will not be a comprehensive survey of the current state of the field or a literature review. For these purposes readers are referred to available treatises (7-9), as well as to very recent reviews of inorganic (10) and biological (11) applications. Although the magnetic and relaxation properties of the two isotopes will be compared, most of the ensuing discussion will be based on ^{15}N results, for the reasons given above, and reflecting the author's experience.

COMPARISON OF ^{15}N AND ^{14}N

Some appreciation for the nitrogen nuclear characteristics may be obtained from Table 1. Because the relative sensitivities are comparable, the approximately 300-fold higher natural abundance of ^{14}N would seem to make it the nucleus of choice. It is even more sensitive than ^{13}C at natural abundance. However, ^{14}N, like all nuclei with spin quantum number $I > 1/2$, possesses an electric quadrupole moment that arises from a nonspherical electric charge distribution in the nucleus itself. When placed in an electric field gradient, such as that characteristic of most molecular electron distributions, a quadrupolar nucleus experiences random fluctuating electric fields. The characteristic frequencies of these motions have components at the ^{14}N resonance frequency and hence afford an efficient relaxation mechanism. As a result, spin-lattice relaxation times (T_1) are very short, 0.1-10 ms. Because $T_1 = T_2$ for ^{14}N in most molecules that are freely mobile in solution, ^{14}N linewidths are correspondingly broad, ranging from tens to thousands of hertz. Consequently, while ^{14}N resonances are readily detectable, small chemical shift differences in general are not resolvable, and hence the usefulness of the method for structure elucidation of complex molecules containing similar nitrogen atoms is limited. A few comparative values of ^{14}N and ^{15}N relaxation times are given in Table 2.

On the other hand, because $T_1(^{14}N)$ equals the quadrupole relaxation time $T_q(^{14}N)$, molecular motions may be investigated

Table 1. Comparison of ^{15}N and ^{14}N Nuclear Properties

	^{14}N	^{15}N
Natural abundance	99.64%	0.36%
Spin quantum number	1	1/2
NMR frequency at 4.7 Tesla[a]	14.46 MHz	20.26 MHz
Sensitivity relative to ^1H for equal number of nuclei	0.00101	0.00194
Sensitivity relative to ^{13}C at natural isotopic abundance	17.22	0.0214

[a]Magnetic field for proton resonance at 200 MHz.

Table 2. ^{14}N and ^{15}N Relaxation Times

Compound	$T_1{}^{14}N$, msec[a]	$T_1{}^{15}N$, sec[b]
Pyridine	1.6-2.3	85
Pyrrole	2	40
Pyrrolidine	1.2 (T_2)	58
Acetonitrile	4-5	90
Nitrobenzene	3.5	170

[a]From Ref. 12. [b]From Ref. 7.

directly by determination of $T_1(^{14}N)$ or $T_q(^{14}N)$ (eq. 1). In

$$\frac{1}{T_q} = \frac{3}{8}\left(1 + \frac{\eta^2}{3}\right)\left(\frac{e^2qQ}{\hbar}\right)^2 \tau_c \tag{1}$$

eq. 1, η is an asymmetry parameter that represents the distortion of the electric field from axial symmetry, eq is the electric field gradient, eQ is the nuclear electric quadrupole moment, and τ_c is an orientational correlation time characteristic of the molecular motion. Eq. 1 is in fact a special case for isotropic motion; more complex expressions arise when motions are anisotropic (12). Eq. 1 shows that quadrupolar broadening is zero when the electric field is spherically symmetric (eq = 0). This situation arises in only a small number of special cases: ammonium ion, a few tetralkylammonium ions, and isonitriles. For these compounds, ^{14}N and ^{15}N spectra are qualitatively similar, although the ^{14}N relaxation is still quadrupolar (10).

An additional consequence of ^{14}N quadrupolar relaxation is that spin-spin coupling to other nuclei, which is readily observable if ^{15}N is present, is usually eliminated. Hence a valuable source of structural information is lost.

Because I = 1/2 for ^{15}N, corresponding resonance lines are inherently sharp, and line resolution of 1 Hz or less is achievable. Detection of resonances is more difficult than for ^{14}N, not only because of the lower natural abundance, but also because T_1's are several orders of magnitude longer (see Table 2). Furthermore, the NOE, which can be highly advantageous for ^{13}C spectra, can in fact be a detriment for ^{15}N, because the magnetic moment of ^{15}N is negative. As a result, a maximum NOE, arising from exclusive dipole-dipole relaxation, gives rise to an inverted signal, the intensity of which is enhanced by a factor of four. When relaxation mechanisms other than the dipole-dipole mechanism intervene, the NOE is reduced. Depending on the relative contributions of non-dipole-dipole mechanisms (spin-rotation, chemical shielding anisotropy, electron-nuclear, scalar), the intensity may be reduced to less than that of the non-proton-irradiated signal, and in fact may be nulled. Thus, successful observation of a ^{15}N resonance requires some anticipation of relaxation behavior (7). The problem may be circumvented in a number of ways. Paramagnetic relaxation reagents have been used to shorten T_1's and to replace dipole-dipole relaxation with electron-nuclear relaxation. Gated decoupling techniques with NOE suppression have also been applied. Most recently, the INEPT polarization transfer sequence (13) has been used successfully. This method allows the ^{15}N nucleus to relax on a time scale determined by the T_1's of spin-spin coupled protons, whose values are commonly an order of magnitude shorter. Faster repetition rates combine with an inherent intensity enhancement by a factor of $\gamma_H/\gamma_{15N} \simeq 10$ to afford spectra significantly more quickly, or at significantly lower concentrations, than previously attainable. Spin-spin coupling may also be retained (14). Judicious choice of conditions can allow quantitative data to be obtained from ^{15}N spectra (15).

^{15}N CHEMICAL SHIFTS

Like those of other nonproton nuclei, nitrogen chemical shifts span a range of several hundred ppm, and nitrogen's nuclear screening is subject to structural and electronic influences similar to those of carbon, whose chemical shift behavior it often resembles. Additive substituent parameters (see below) exist that have some predictive value. In addition, nitrogen has some features that differentiate it from carbon, which result from the lone pair of electrons present in neutral molecules. As a result, a nitrogen resonance line position may change in either direction when the nitrogen is protonated, and these changes may be characteristic of the type of nitrogen.

A qualitative theoretical discussion of nitrogen chemical shift behavior has been given elsewhere (7). In general, nitrogen is deshielded when the atom possesses a lone electron pair that can be delocalized into an adjacent π system (e.g., aniline), or when it is itself part of an unsaturated system (e.g., azines, pyridines). In the latter case the deshielding is associated with the presence of low-lying n → π* electronic

states. Table 3 reveals some of these trends. Aniline lies to lower shielding compared to cyclohexylamine because of the lone-pair delocalization possible in the former. Pyridine is substantially deshielded relative to both its saturated analogue, piperidine, and aniline. Azobenzene lies to even lower shielding. Protonation of the saturated amines induces a deshielding, the magnitude of which depends on structure and solvent. However, protonation of the anilines and azines removes the deshielding contribution of the lone pair in those compounds and thus effects a shielding that compensates or outweighs the influence of positive charge generation. Qualitatively, a trend exists between the chemical shift of these compounds and the extent of shielding induced by protonation. Thus, an aniline-like nitrogen is easily distinguishable from a pyridine-like one.

Chemical Shift Referencing

As of this writing, no commonly accepted chemical shift standard exists. The difficulties associated with a variety of internal and external standards have been discussed (7-9). For its reproducibility and convenience, nitromethane in an external capillary has been suggested (7,8,16). With the widely--but not universally--adopted convention which specifies that chemical shifts of resonance lines lying to lower shielding of a reference be positive, use of the highly deshielded nitromethane nucleus requires that most values be negative. To circumvent this inconvenience, nitromethane has been calibrated carefully with respect to pure anhydrous liquid ammonia at 25°C (16b), such that $\delta(NH_3)$ = $\delta(CH_3NO_2)$ + 380.2 ppm. Values presented here are based on this reporting standard.

Table 3. Illustrative Nitrogen Chemical Shifts[a]

Compound	δ_N, ppm[b]
Ammonia	0
Methylamine	2
Ammonium chloride	25
Methylammonium chloride	28
Anilinium chloride	48
Cyclohexylamine, piperidine	39
Cyclohexylammonium ion	45
Aniline	52
Pyridinium ion	215
Diphenylketimine hydrochloride	168
Diphenylketimine	308
Pyridine	317
Protonated trans-azobenzene	358
trans-Azobenzene	508

[a]Taken from Ref. 7.
[b]Downfield from external anhydrous ammonia.

Assignment of Resonances

For many structures, assignment of resonances causes little difficulty, because only a small number of nitrogens are present in the molecule, and chemical shifts can be widely separated. When this is not the case, the arsenal of methods useful for ^{13}C NMR is available for nitrogen (especially ^{15}N) as well. This is particularly relevant for biological and synthetic macromolecules, which may contain larger numbers of nitrogens in similar environments. The techniques include use of additive substituent parameters, comparison with model compounds, determination of coupling constants and spin-spin splitting patterns (gated and off-resonance decoupling), selective decoupling, determination of T_1's and NOE's, and use of specific relaxation reagents. The differential sensitivity of nitrogen shifts to solvent effects, and the characteristic behavior on protonation described above, may also be exploited. Most recently, two-dimensional heteronuclear correlation spectroscopy, which has evoked enthusiasm for assignment of ^{13}C spectra, may also prove to be applicable.

Amines and Ammonium Ions

Like hydrocarbons in ^{13}C NMR spectroscopy, amines form the basis for much of the discussion of nitrogen chemical shifts. For many acyclic and some cyclic molecules, trends in nitrogen chemical shifts have been found to parallel trends in ^{13}C shifts of hydrocarbons derived by replacing nitrogen with carbon, e.g., $CH_3CH_2NHCH_2CH_3$ vs. $CH_3CH_2CH_2CH_2CH_3$. Indeed, for a large number of aliphatic amines and their hydrochlorides, statistically significant correlations of this type exist between ^{15}N and ^{13}C chemical shifts, that span a nitrogen range of >70 ppm (7,17). This observation has resulted in the derivation of substituent parameters similar to those that have been so successful for ^{13}C NMR. Values based on chemical shifts for primary, secondary, and tertiary amines are given in Table 4, and may be used in eq. 2,

$$\delta^j = B^j + n_\beta \beta^j + n_\gamma \gamma^j + n_b (cor\ \beta^j + cor\ \gamma) \qquad (2)$$

in which j denotes whether the parameters are for primary (p), secondary (s), or tertiary (t) amines, B^j is a base value, β^j and γ^j are for carbons two and three bonds away from the nitrogen, respectively, cor β^j is a correction for branching at the α carbon, and cor γ is a further correction for branching at the α carbon if a γ carbon is present in the same residue. The quantities n_β, n_γ, and n_b are the numbers of β carbons, γ carbons, and branches at α carbons, respectively. Calculated values are reported to display standard deviations <1.5 ppm, depending on solvent.

Table 4. Additive Substituent Effect Parameters for Aliphatic Amines[a]

Parameter	Value (ppm)		
	C_6H_{12}	CH_3OH	Hydrochloride CH_3OH
B^p	1.4 (1.6)	3.0 (0.7)	18.4 (0.4)
B^s	8.9 (1.3)	10.4 (0.8)	23.6 (0.4)
B^t	12.5 (0.9)	18.1 (0.6)	31.0 (0.3)
β^p	22.6 (1.4)	21.3 (0.8)	15.2 (0.4)
β^s	19.0 (0.8)	18.3 (0.5)	13.5 (0.3)
β^t	11.7 (0.5)	10.5 (0.4)	9.1 (0.2)
$\gamma^{p,s}$	-3.8 (0.4)	-3.4 (0.3)	-2.2 (0.1)
γ^t	-2.2 (0.3)	-1.8 (0.2)	-0.9 (0.1)
cor $\beta^{p,s}$	-4.9 (1.4)	-5.0 (0.9)	-3.6 (0.5)
cor β^t	-8.2 (0.7)	-5.7 (0.5)	-2.2 (0.3)
cor γ	-1.8 (0.6)	-0.6 (0.4)	0.4 (0.2)
Standard deviation[c]	1.4	1.0	0.5

[a] Ref. 7 and 17.
[b] Standard deviations in each parameter are given in parentheses.
[c] Standard deviation of all calculated shifts.

Representative amine chemical shifts are given in Table 5. Qualitatively, an axial $-NH_2$ group is shielded more than an equatorial one. A β substituent deshields a nitrogen more than it does a carbon of the corresponding acyclic hydrocarbon, although the magnitude of the effect depends on the degree of nitrogen substitution. While γ effects tend to be shielding, their magnitudes, too, depend on nitrogen substitution. In contrast to ^{13}C behavior, α substitution by methyl can lead to a marked shielding of the nucleus. Such shielding appears to be largest when the original α carbon is branched (cf., $(CH_3)_3CNH_2 \rightarrow (CH_3)_3CN(CH_3)_2$).

Table 5. Nitrogen Chemical Shifts of Acyclic Aliphatic Amines[a]

Compound	δ, ppm	Compound	δ, ppm
Primary			
CH_3NH_2	1.3[b]	$(CH_3)_3CCH_2NH_2$	11.5
$CH_3(CH_2)_3NH_2$	19.8	cyclopropylamine	29.3[b]
$CH_3CH_2CH(CH_3)NH_2$	37.8	trans-4-t-butyl-cyclohexylamine	39.4
$(CH_3)_3CNH_2$	57.8	cis-4-t-butylcyclo-hexylamine	30.6
Secondary			
$(CH_3)_2NH$	6.7[b]	$CH_3CH_2NHCH_3$	27.4
$CH_3(CH_2)_3NHCH_3$	24.1	$[(CH_3)_2CH]_2NH$	75.1
$(CH_3CH_2CH_2)_2NH$	38.0	$[(CH_3)_2CHCH_2]NH$	34.8
Tertiary			
$(CH_3)_3N$	13.0[b]	$(CH_3)_3CN(CH_3)_2$	31.4
$(CH_3CH_2)_3N$	46.6	$(CH_3)_3CCH_2N(CH_3)_2$	15.3
$CH_3CH_2CH_2N(CH_3)_2$	21.8	cyclohexyldi-methylamine	26.3

[a]Ref. 7, in cyclohexane. [b]Pure liquid.

Similar trends arise for protonated amines, although β effects and branching corrections are smaller than for the amines. Protonation shifts of primary amines appear to decrease with branching at the α carbon.

Alicyclic amines behave broadly in a manner similar to the acyclic molecules, but here geometric influences may be discerned. From studies of methyl-substituted piperidines (18a) and trans-decahydroquinolines (18b), the methyl substituent effects shown have been adduced. Furthermore, interaction between α and β substituents appears to effect an additional shielding that depends on geometry (18b). Configurationally isomeric acyclic amines (meso vs. rac forms) can also display chemical shift differences of ca. 1 ppm (18c).

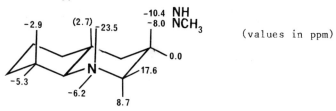

(values in ppm)

Aromatic Amines

Delocalization of an aromatic nitrogen lone pair deshields the nitrogen, and the magnitude of this deshielding appears to be related to independent measures of substituent electronic effects (e.g., Hammett σ values). Representative data are given in Table 6. Protonation has an apparent leveling effect on the shifts.

Table 6. Nitrogen Chemical Shifts (ppm) of para-Substituted Anilines[a]

Substituent	Aniline[b]	Anilinium[c]
NO_2	70.3	47.8
I	56.1	46.1
H	55.4	46.7
F	52.0	45.1
OCH_3	52.0	
CH_3	53.9	49.6[d]

[a]Ref. 7. [b]1 molal in acetone. [c]1 molal in $HFSO_3$. [d]2M in CF_3COOH.

N-Methylation of aniline (α substitution) leads to successive shielding of nitrogen by up to 12 ppm, that is not related to steric inhibition of delocalization. That change, however, is responsible for the large shielding (almost 30 ppm) induced by ortho methyl substitution in N,N-dimethylanilines

(19a). An extreme example of this behavior is displayed by 2,4,6-tri-tert-butyl-N,N-dimethylaniline, whose chemical shift is very close to that of liquid ammonia (19b). Data are given with the structures. Interestingly, the shifts of para-substituted

C$_6$H$_5$NH$_2$	C$_6$H$_5$NHCH$_3$	C$_6$H$_5$N(CH$_3$)$_2$	2,6-(CH$_3$)$_2$-C$_6$H$_3$N(CH$_3$)$_2$	2-CH$_3$-6-t-Bu-C$_6$H$_3$N(CH$_3$)$_2$	2,6-(t-Bu)$_2$-C$_6$H$_3$N(CH$_3$)$_2$
56.5	52.8	44.6	17.1		1.0

(chemical shifts in ppm)

N-phenylaziridines are much less influenced by substituents, and methyl substitution at the ortho positions changes the nitrogen shift by <2 ppm (19c).

Evidence for a moderate <u>deshielding</u> steric effect across the peri positions of 8-substituted-1-naphthylamines has been presented (20).

Amides and Other Carbonyl Derivatives

Selected data for amides are given in Table 7. By delocalizing the lone pair, the carbonyl group deshields nitrogen to a substantial extent compared with amines. Resonance positions can be very much influenced by solvent (20). As with anilines, N-methyl substitution shields nitrogen; this behavior appears to be characteristic of nitrogens whose lone pairs are delocalized into an adjacent π system. Where Z and E isomers are detectable, nitrogens of the Z forms in general lie at higher shielding (1-3 ppm) than the E forms (21). Attempts have been made to relate nitrogen shifts of amides and other compounds showing hindered rotation quantitatively to rotational barriers, although these procedures have severe limitations (22).

Lactams display values similar to those of simple amides, but ureas are consistently shielded compared with corresponding amides (7). By contrast, the nitrogens of thioamides are

H$_2$NCNH$_2$ (C=O) C$_6$H$_5$NHCNH$_2$ (C=O)

75.0 ppm   105.7 ppm   77.5 ppm

deshielded.

Table 7. Nitrogen Chemical Shifts of Amides[a]

Compound	δ, ppm	Solvent
$HCONH_2$	112.4	none
	108.5	CH_3NO_2
CH_3CONH_2	103.4	$CHCl_3$
$C_6H_5CONH_2$	105.4	$(CD_3)_2SO$
$HCONHCH_3$	109.9(E), 108.1(Z)	none
$HCONHC_6H_5$	141.0(E), 138.4(Z)	$CDCl_3$
$HCON(CH_3)_2$	103.8	none
	104.4	CH_3NO_2
$CH_3CON(CH_3)_2$	98.3	none
$C_6H_5CON(CH_3)_2$	100.5	none
$HCON(C_2H_5)_2$	131.5	none
$CH_3CON(C_2H_5)_2$	130.4	none

[a]Ref. 7.

Nitriles and Isonitriles

Like ^{13}C in alkynes, nitrogen in nitriles is shielded compared to that in their double-bond congeners, imines, as illustrated by benzonitrile and benzalmethylamine. For the reasons

$C_6H_5C\equiv N$ $C_6H_5CH=N-CH_3$

258.7 ppm   318.1 ppm

discussed earlier, protonation, alkylation, or N-oxide formation induces a substantial shielding, of which the -99 ppm change displayed by protonated acetonitrile is typical. Similarly, the nitrogen of the linear isonitrile group, $-N\equiv C$, which has no lone pair, is shielded (e.g., $CH_3-N\equiv C$, 162 ppm). The ^{14}N lines of these compounds are relatively narrow.

Heterocyclic Aromatic Compounds

Heterocyclic aromatic nitrogens are conveniently divided into two categories, pyrrole-like and pyridine-like. The lone pair of the former is part of a π system, and thus resembles aniline. That on the latter nitrogen is orthogonal to the π system,

pyrrole-like   pyridine-like

and hence contributes an additional deshielding. Proton tautomerism can obscure these differences.

Representative values for pyrroles and indoles are given in Table 8, from which effects of substituents in electronically conjugating positions can be seen. It should also be noted that the progressive shielding in the series pyrrole → indole → carbazole (112.7 ppm in acetone) is consistent with decreasing electron delocalization in this series.

Table 8. Nitrogen Chemical Shifts of Pyrroles and Indoles[a]

Compound	δ, ppm	Compound	δ, ppm
Pyrrole	145.1	Indole	132.9
2-Nitropyrrole	146.5	3-Acetylindole	142.3
3-Nitropyrrole	149.9	5-Aminoindole	129.8

[a]Ref. 7, in CDCl$_3$. Values in DMSO are deshielded by ~8-9 ppm.

The chemical shift region of pyrroles is also maintained in molecules (e.g., imidazole) which contain both pyrrole-like and pyridine-like nitrogens, although additional nitrogens can markedly affect resonance positions. Where some overlap of

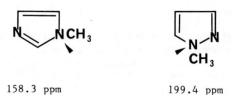

158.3 ppm   199.4 ppm

regions arises, protonation changes can afford an assignment mechanism.

Pyridine and pyridine-like nitrogens are among the most thoroughly investigated structural units experimentally and theoretically. These studies have been summarized very extensively (7-9). Chemical shifts depend strongly on solvent, and are related to the extent of the lone-pair contribution to the n → π* electronic state. Hydrogen bonding to a pyridine nitrogen has a particularly pronounced shielding effect that can be used to identify such interactions in complex systems.

Conjugating or electron polarizing substituents in ortho and para positions shield and deshield a pyridine nitrogen respectively, according to their qualitatively expected electron-donating or withdrawing properties. As discussed earlier (Table 3) protonation shields the nitrogen substantially, as does N-oxidation. The effect of the latter change is reduced somewhat because of the electronegativity of oxygen. Pyridine-like nitrogens in polyazaheterocycles retain their characteristic chemical shifts; Table 9 gives representative values. Several recent compilations of data for these types of compounds have appeared (23,24).

Table 9. Nitrogen Chemical Shifts of Pyridines[a]

Compound	δ, ppm
Pyridine	317.3
4-Picoline	309.3
2-Fluoropyridine	275.7
4-Aminopyridine	276.5[b]
4-Cyanopyridine	327.9
2-Carbethoxypyridine	316.1[b]
4-Carbethoxypyridine	312.7
Quinoline	316.2
Pyrimidine	294.8
Pyrazine	334.9[c]
N-Methylimidazole	252.2
N-Methylpyrazole	301.7

[a]Ref. 7. [b]S. Tobias, P. Schmitt, and H. Günther, in press, and private communication. [c]Ref. 24a.

Nitro and Nitroso Compounds

Nitro groups lie in the range ~350-380 ppm. Available ^{14}N data on a limited number of aliphatic nitroalkanes reveal alkyl substituent effects similar to those for amines, and a small solvent effect on the resonance positions (25). The ^{15}N chemical shifts of meta- and para-substituted nitrobenzenes not only are much less influenced by substituents than are corresponding anilines and pyridines, but a reverse effect is displayed: electron-withdrawing substituents induce shielding. A π-polarization mechanism to account for this effect has been suggested, in which the N=O bonds are polarized through space by the substituent (26a). In general, the shielding of aromatic nitro nitrogens relative to aliphatic analogues is attenuated when the nitro group is twisted out of coplanarity with the benzene ring by ortho substituents. This effect is opposite to that displayed by N,N-dimethylanilines. The nitro group of nitramines is shielded relative to nitrobenzene, consistent with the effect of delocalization noted for the aromatics (26b). Typical values are given in Table 10.

Table 10. Nitrogen Chemical Shifts of Nitro Compounds[a]

Compound	δ, ppm	Compound	δ, ppm
CH_3NO_2	380.23	p-Acetylnitrobenzene	370.2
$CH_3CH_2NO_2$	390.5[b]	p-Trifluoromethyl-	
$CH_3(CH_2)_2NO_2$	387.9[b]	nitrobenzene	367.3[c]
$(CH_3)_2CHNO_2$	399.7[b]	2-Nitro-m-xylene	379.3
Nitrocyclohexane	398.4	Nitromesitylene	379.7
CH_3NHNO_2	358.0[e]	Nitrodurene	382.6
Nitrobenzene	370.3	m-Acetylnitrobenzene	369.6
p-Aminonitrobenzene	371.0	m-Trifluoromethyl-	
p-Dinitrobenzene	366.1	nitrobenzene	366.6[c]
p-Fluoronitrobenzene	368.5	m-Dinitrobenzene	365.4[c]
p-Methoxynitro-benzene	369.9	1,8-Dinitronaphthalene	372.3[d]

[a]Ref. 7. [b]Ref. 25. [c]Ref. 26a. [d]Ref. 21a. [e]Ref. 26b.

Nitroso groups are very highly deshielded owing to the availability of low-lying n → π* excited states (the compounds are green-blue), and their chemical shifts are 800-900 ppm. When a nitroso group serves as an electron acceptor for a lone pair (e.g., nitrosamines), its nitrogen is shielded by several hundred ppm. A few values are given below (7,27). Attempts have been

C_6H_5NO	p-$(CH_3)_2N-C_6H_4-NO$	$(CH_3)_2NNO$
913	804	535

(chemical shifts in ppm)

made to correlate ^{15}N chemical shifts of nitrosamines and related compounds with barriers to N-N rotation, but these efforts rest on several assumptions that make the results problematic (27).

Imines, Hydrazones, Azo and Diazo Compounds

The pyridine-like nitrogens in imines and related compounds, as expected, lie at low shielding, and protonation induces a change to higher shielding. Chemical shift values increase algebraically (i.e., move to lower shielding) roughly in the order imines ≤ hydrazones < oximes < terminal diazo < azo. Where detectable, cis and trans azo compounds display different chemical shifts. The internal nitrogens of diazo compounds are shielded relative to the terminal ones, consistent with the absence of lone-pair character on the internal nitrogen. Differences in resonance positions of azo and hydrazone nitrogens have been used to determine azo-hydrazone tautomerism (28). Some representative data are given in Table 11, and more recent data have been reported that are consistent with these values (27,29).

Detailed values for other classes of compounds not discussed here may be found elsewhere (7-9).

Table 11. Nitrogen Chemical Shifts of Imine Nitrogens[a]

Compound	δ, ppm
$C_6H_5CH=NCH_3$	318.1
$C_6H_5CH=N-C_6H_5$	326.1
$C_6H_5CH=N^1-N^2HC_6H_5$	326.2 (N-1)
	143.2 (N-2)
$C_6H_5CH=N-OH$	353.9
$CH_2=\overset{+}{N}=\overset{-}{\underset{..}{N}}:$	286 (N-1)
	397 (N-2)
$C_6H_5CH=\overset{+}{N}=\overset{-}{\underset{..}{N}}:$	297.2 (N-1)
	436.5 (N-2)
$C_6H_5-\overset{+}{N}\equiv N:$	223.8 (N-1)
	316.8 (N-2)
$C_6H_5N=N-C_6H_5$	508.7
$C_6H_5N^1=N^2-C(CH_3)_2C_6H_5$	N_1: 512 (trans), 541 (cis)
	N_2: 550 (trans), 574 (cis)
$C_6H_5N=\overset{+}{N}=\bar{N}$ [b]	91.7 (N-1)
	243.5 (N-2)
	232.8 (N-3)

[a]Ref. 7. [b]Ref. 29.

SPIN-SPIN COUPLING TO NITROGEN

Indirect nuclear spin-spin coupling involving nitrogen, especially ^{15}N, has shown itself to be quite useful for structure elucidation. Because of nitrogen's small magnetic moment, μ, magnitudes of nitrogen couplings are smaller than are values for ^{13}C, ^{19}F, or ^{1}H in analogous structures, and variations with structure and bonding are less pronounced. Also, because μ_{15N} is negative, signs of J_{15N} couplings to a given nucleus are opposite to those for ^{13}C or ^{14}N. Thus, for comparison of nuclides, a reduced coupling constant K is used frequently, in which the influence of μ has been factored out. Here, however, J values will be reported, as these are more commonly used for structure elucidation. Coupling constants between nitrogen and other nuclei X, separated by n bonds, are denoted as $^{n}J_{NX}$, and only measured ^{15}N values, or those converted from the few available ^{14}N ones, are reported. Signs are given explicitly where experimentally determined, or in parentheses where they may be inferred by analogy. Detailed theoretical discussions may be found elsewhere (7-9) (see also Chapter 3).

Experimental determination of both magnitudes and signs of ^{15}N couplings, which generally has been carried out with ^{15}N-enriched materials, has been facilitated by the development of multiple pulse techniques, including selective population transfer (SPT) (30) and INEPT (13) procedures. The ability to detect these couplings at the natural abundance level of ^{15}N has afforded a much larger range of compounds for study.

Spin-Spin Coupling to Proton

One-bond Couplings. Values for $^{1}J_{NH}$ are among the largest in magnitude for ^{15}N couplings, and fall into the regions of ~75, ~90, and ~135 Hz for pyramidal, trigonal, and linear bonding to nitrogen, respectively (Table 12). Based on this observation, eq. 3, which relates $^{1}J_{NH}$ to s character at nitrogen, has been

$$\text{Percent s} = 0.43\ ^{1}J_{NH} - 6 \qquad (3)$$

suggested (31). It should be emphasized that eq. 3 holds only for that part of the coupling which derives from the Fermi contact mechanism, which in turn depends not only on s character but also on both effective nuclear charge and internuclear distances. Because of this and other reasons (7), it is inappropriate to use eq. 3 and similar equations (see eq. 5, below) to interpret small differences in J values for structurally related compounds in a numerically exact way. Other contributors to small variations in $^{1}J_{NH}$ include solvent and geometry. Changes arising from the latter source may be useful in making configurational and conformational assignments.

Table 12. One-Bond ^{15}N-H Coupling Constants[a]

Compound	$^1J_{NH}$, Hz
NH_3	(-) 61.2
CH_3NH_2	- 64.5
NH_4Cl	(-) 73.3
$C_6H_5NH_2$	(-) 78.5 (C_6D_6 solvent)
	(-) 82.6 ((CH_3)$_2$SO solvent)
$p-O_2N-C_6H_4NH_2$	(-) 92.6
$(CF_3)_2PNH_2$	(-) 85.6
$CH_3C\equiv N^+-H$	(-)136
F_4PNH_2	(-) 90.3

[a]Ref. 7.

The above considerations notwithstanding, eq. 1 can provide some useful insight into structure. Thus, $^1J_{NH}$ for aniline implies nearly pyramidal geometry, consistent with microwave results. Conjugative electron-withdrawing substituents increase $^1J_{NH}$ by inducing conversion of the amino group to trigonal geometry. The same geometrical inference applies to amides, values for which lie in the range 90-100 Hz. Here, small differences are displayed within a molecule according to geometry (cf., N-methylformamide (30b)). Heteroatoms directly bonded to

$$\begin{array}{c} H \\ \diagdown \\ O \end{array} C-N \begin{array}{c} H \\ \diagup \\ CH_3 \end{array} \quad -90.35 \qquad \begin{array}{c} H \\ \diagdown \\ O \end{array} C-N \begin{array}{c} CH_3 \\ \diagup \\ H \end{array} \quad -87.80$$

nitrogen increase the magnitude of $^1J_{NH}$ (i.e., algebraically decrease the coupling), presumably by increasing the "effective nuclear charge" at nitrogen.

Two-bond Couplings. Couplings across one intervening atom tend to be smaller than $^1J_{NH}$ values, but variations in structure have a larger proportional effect. Both positive and negative values are displayed. It is here that one of the unique characteristics of nitrogen couplings becomes evident: protons that are close in space to a lone pair orbital on a nitrogen that is geometrically fixed give rise to a large and negative value for $^2J_{NH}$, while spin-spin coupling to the remote proton is smaller and positive. Formaldehyde oxime (32) is the archetypical

$$\begin{array}{c} -11.8 \\ +2.0 \end{array} \quad \underset{H}{\overset{H}{>}}C=N\underset{OH}{}$$

example, but the same trend, with roughly the same magnitudes, holds for pyridines and other azaheterocycles, imines, hydrazones, and saturated compounds such as aziridines and oxaziridines (33). Available data show that removal of the lone pair by protonation or N-oxide formation substantially reduces the magnitude of $^2J_{NH}$. Representative data are in Table 13 (7,34).

Table 13. Two-Bond ^{15}N-H Coupling Constants[a]

pyridine, −10.76	pyridinium, −3.01	pyrimidine, (−)12.31 / (−)13.41
CH$_3$(H)C=N−OH, +3.0	CH$_3$(H)C=N−OH, −15.9	CH$_3$NH$_2$, −1.0
H−C≡N, 8.7	(CH$_3$)$_2$N−C(=O)H, +1.1 / −15.6	

[a]Ref. 7. [b]Ref. 34.

Three-bond Couplings. Determination of three-bond (vicinal) N-H couplings affords the potential for use in conformational analysis. Several Karplus-type relationships have been derived, of which eq. 4 is typical (35) (θ is the N-(CO)-C-H dihedral

$$^3J_{NH} = -4.6 \cos^2 \theta + 3.0 \cos \theta + 0.8 \qquad (4)$$

angle specifically for a peptide linkage, $R_2N(CO)-CHXY$). However, $^3J_{NH}$ values are small, the order of 6-7 Hz, so that variations with structure may not always be distinguishable from other factors. In any event, the small differences observed must be determined very precisely. Furthermore, spectra that appear to be first order may in fact require second-order spectral analysis in order to obtain correct values (30b). Conformational aspects of the peptide hormone oxytocin have been inferred using these methods (36). In rigid molecules, a lone pair dependence analogous to that for $^2J_{NH}$ also appears to exist (33). Some values are shown below (7.32a). In general, these quantities are negative, although positive values have been reported for some isothiazoles (30c).

$H_2N-C(=O)CH_3$ 2-X-aniline   pyridine   pyridinium

−1.0   1.5−1.8   −1.53   −3.98

$CH_3C\equiv N$ $CH_3C\equiv NH^+$ $(CH_3)(H)C=N-OH$ $(CH_3)(H)C=N-OH$

−1.7   2.8   (−)4.0   (−)2.4

Spin-Spin Coupling to Carbon

The effectively routine acquisition of ^{13}C spectra has allowed spin-spin coupling to that nucleus to be intensively characterized, and ^{15}N has not escaped inclusion in these efforts. Despite tremendous improvements in sensitivity, most determinations of ^{13}C-^{15}N couplings still require ^{15}N enrichment; hence, availability of data is limited by synthetic (and economic) constraints. Nonetheless, detailed trends are quite evident. A comprehensive review of ^{13}C coupling constants devotes substantial attention to J_{NC} (38).

One-bond Couplings. As expected from the Fermi contact mechanism, $^1J_{NC}$ is negative for most molecules, and for many, eq. 5, which relates s character at the bonded atoms to $^1J_{NC}$,

$$\text{Percent } s_C \cdot \text{Percent } s_N = 80\ J_{NC} \qquad (5)$$

appears to be valid (31). The same cautions apply here as for eq. 3. Values range from ca. −5 to ca. −20 Hz (Table 14). As is the case with $^1J_{NH}$, $^1J_{NC}$ also appears to be influenced by lone pair orientation, as illustrated by values for the two isomers of acetaldoxime. This observation has been treated theoretically in

$$CH_3\diagdown C=N \diagdown OH \qquad CH_3 \diagdown C=N \diagup OH$$
$$H \diagup \qquad\qquad\qquad H \diagup$$

(−) 4.0 Hz (E)   (−) 2.3 Hz (Z)

terms of a separate <u>positive</u> contribution to the Fermi contact term arising directly from the lone pair (39). The lone pair contribution in turn depends on the s character of the lone pair orbital, variations in which may exist as a function of structure (e.g., different steric interactions). The role of the lone pair is confirmed when it is removed by protonation or oxide formation: $^1J_{NC}$ values are comparable to those in, e.g., amides, and are broadly consistent with eq. 5.

Table 14. One-bond $^{15}N-^{13}C$ Coupling Constants[a]

Compound	$^1J_{NC}$, Hz	Compound	$^1J_{NC}$, Hz
CH_3NH_2	−4.5	Pyridine	+0.62
Quinuclidine	(−)2.5	Pyridinium	−11.85
Quinuclidinium	(−)4.8	Quinoline	1.4(C_2), 2.1(C_9)
Aniline	−11.43	Acetophenone	(−)3.8(E),
Nitrobenzene	−14.5	oxime	(−)3.1(Z)
p-Nitroaniline	(−)14.9	Benzonitrile	(−)77.5
Acetamide	−14.4[b]	oxide	
HC≡N	−18.5[c]	CH_3NHNO_2	(−)8.5[d]
$^-C≡N$	5.4		

[a]Ref. 7.   [b]Ref. 37a.   [c]Ref. 40.   [d]Ref. 26b.

<u>Two-bond Couplings</u> (Table 15). In general, $^2J_{NC}$ values are small and can be both positive and negative. A dependence on lone pair orientation analogous to that displayed by $^2J_{NH}$ also exists: a lone pair close in space to carbon makes a negative contribution to the coupling.

<u>Three-bond Couplings</u> (Table 16). As with $^3J_{NH}$, the possibility of using $^3J_{NC}$ for conformational analysis has been particularly attractive. However, values are small and no pronounced dihedral angle dependence has yet been adduced experimentally, although several have been calcualted (41): $^3J_{NC}$(cis) is predicted to be only slightly greater than $^3J_{NC}$(trans) in a peptide linkage.

Table 15. Two-bond $^{15}N-^{13}C$ Coupling Constants[a]

Compound	$^2J_{NC}$, Hz	Compound	$^2J_{NC}$, Hz
$CH_3CH_2CH_2NH_2$	1.2	Aniline	-2.68
CH_3CONH_2	-8.5[b]	Nitrobenzene	-1.67
$CH_3C\equiv N$	+3.0	Propionaldoxime	(-)7.3(E)
			(+)11.4(Z)
Quinoline	-9.3(C_8)	$C_2H_5OOC-\bar{C}H-\overset{+}{N}_1\equiv N_2$	3.7(C,N_2)
	+2.7(C_3)		1.2(C,N_1)
Quinolinium	1.0(C_8)		
	(+)1.0(C_3)		

[a]Ref. 7. [b]Ref. 37c.

Table 16. Three-bond $^{15}N-^{13}C$ Coupling Constants[a]

Compound	$^3J_{NC}$, Hz	Compound	$^3J_{NC}$, Hz
$CH_3CH_2CH_2NH_2$	1.4	Quinoline	(-)3.5(C_4)
Quinuclidine	2.8		3.9(C_7)
Quinuclidinium	6.7	Quinolinium	(-)4.6(C_4)
Aniline	-1.3		2.7(C_7)
Anilinium	2.1	Benzaldoxime	2.8
		Mesitaldoxime	1.2

[a]Ref. 7.

In geometrically fixed systems $^3J_{NC}$ depends on lone pair orientation in a manner very much like $^3J_{NH}$.

Spin-Spin Coupling to Nitrogen (Table 17)

Values of J_{NN}, most of which are directly bonded couplings, range from 4-21 Hz, but relationships with structure have not been clearly defined. In hydrazines, a dependence on both dihedral angle and lone pair s character has been predicted; the latter is expected to make a negative contribution to $^1J_{NN}$ because of the negative μ (42). Qualitatively, $^1J_{NN}$ appears to increase in magnitude with s character at nitrogen. The difference between $^1J_{NN}$ in azo compounds and hydrazones has been used to examine azo-hydrazone tautomerism (43).

Table 17. ^{15}N-^{15}N Coupling Constants[a]

Compound	J_{NN}, Hz	Compound	J_{NN}, Hz
$C_6H_5NHNH_2$	6.7	CH_3NHNO_2	4.9
p-$O_2NC_6H_4CH$=$NNHC_6H_5$	10.7	$CH_3N(NO_2)_2$	12.2
C_6H_5N=N-$C(CH_3)_2C_6H_5$	17(E), 21(Z)	CH_3N=$\overset{+}{N}(O^-)OCH_3$	12.8(Z), 14.0(E)
$(C_6H_5CH_2)_2N$-N=O	19.0	CH_3-$\bar{N}_1$-$\overset{+}{N}_2$≡N_3[b]	14.4(1,2), 8.2(2,3)

[a]Ref. 7. [b]Ref. 29.

Spin-Spin Coupling to Fluorine (Table 18)

Compared with other couplings, relatively few N-F couplings have been reported. Many of these have been summarized and compared with INDO-calculated values, which show scattered agreement with the experimental ones (44). That the nitrogen lone pair plays a role is evident in the difference between $^1J_{NF}$ in 2-fluoropyridine and in its ion. The five-bond coupling in 4-fluoroanilines is the only observable splitting in any of the fluoroanilines.

Table 18. ^{15}N-^{19}F Coupling Constants[a]

Compound	J_{NF}, Hz	Compound	J_{NF}, Hz
F-N=N-F (trans)	190(^1J), 102(^2J)	8-Fluoroquinoline	2.9
F-N=N-F (cis)	203(^1J), 52(^2J)	4-Fluoroaniline	1.5
2-Fluoropyridine	-52.64[b]	2-Fluorobenzamide	-7.0 ($CDCl_3$) -3.2 ((CD_3)$_2$SO)
2-Fluoropyridinium	(-)23.1	$HCF_2\overset{+}{N}H_3$	13.8[c]

[a]Ref. 7. [b]Ref. 45a, b. [c]Ref. 45c.

Spin-Spin Coupling to Phosphorus (Table 19)

Values for J_{NP} depend, as expected, on coordination state, bond angle, substituent electronegativity, and lone pair orientation. Although $^1J_{NP}$ decreases when the coordination number increases from 3 to 4, other systematic behavior is difficult to discern. The largest values frequently arise when phosphorus is bonded to electronegative atoms such as oxygen or nitrogen. In diazaphospholanes the exo coupling is ca. twice as large as the endo, which may reflect differences in the P-N conformation (46).

Table 19. $^{15}N-^{31}P$ Coupling Constants[a]

Compound	J_{NP}	Compound	J_{NP}
$C_6H_5NHP(CH_3)_2$	+53.0	diazaphospholane with CH$_3$N, NCH$_3$, P, N(CH$_3$)$_2$	52(endo)[b] 90(exo)
$C_6H_5NHP(O)(CH_3)_2$	-0.5	$C_6H_5N=P(OCH_3)_3$	(-)42.5[c]
$C_6H_5NHP(O)(\underline{t}-C_4H_9)_2$	+11.5	$Cl_2PN(CH_3)_2$	89.4
$[(CH_3)_2N]_3PO$	-26.9		

[a]Ref. 7. [b]Ref. 46a. [c]Ref. 46b.

RELAXATION OF NITROGEN NUCLEI

This section presents a qualitative overview of nitrogen relaxation, principally that of ^{15}N. Theoretical background may be obtained from other sources (47), and details pertaining to nitrogen specifically may be found in available texts (7,9).

As discussed above, ^{14}N relaxation is dominated by quadrupole interactions, even when eq ~ 0. The corresponding correlation time τ_c (eq. 1) reflects rotational contributions to molecular motions, which arise from various sources. These may include internal rotational diffusion about bond axes and overall molecular diffusion. The latter can be influenced markedly by solvent, temperature, and viscosity. Despite these complexities, theoretical models are available for relatively straightforward extraction of motional parameters from measured ^{14}N relaxation data (12). Furthermore, because T_1's are short, spectra may be accumulated quite rapidly (10-100 ms repetition times), thus allowing adequate signal strength to be attained within reasonable total times. At the low ^{14}N resonance frequencies, probe design is important in order to avoid artifacts such as acoustic ringing, the time decay of which can correspond to that for the ^{14}N free induction decay.

Changes in ^{14}N relaxation rates also arise from changes in the electric field gradient in the vicinity of the nucleus. Such variations can be induced by intermolecular aggregation, solvation, or binding of added species (e.g., metal ions to basic binding sites). Phase changes (isotropic → liquid crystal → plastic crystal) can also be probed via ^{14}N relaxation and lineshape studies (8).

In contrast to ^{14}N, relaxation of ^{15}N is determined by the same array of mechanisms as ^{13}C, and has a few peculiar features of its own. Unlike ^{14}N, values of T_1 do not equal T_2. However, in _most_ cases, ^{15}N linewidths (i.e., T_2^*) are determined by instrumental parameters such as proton decoupling power and magnetic field inhomogeneities. Thus, although T_2 can be obtained from multiple pulse sequences, T_1 is the parameter that has received most attention.

Significant contributions to ^{15}N spin-lattice relaxation are dipole-dipole interactions (DD), chemical shielding anisotropy (CSA), spin-rotation interaction (SR), scalar coupling (SC), and electron-nuclear interactions (EN). Contributions from each mechanism to the relaxation rate $R_1 = 1/T_1$ are additive (eq. 6).

$$R_1{}^{obs} = \frac{1}{T_1 obs} = \frac{1}{T_1(DD)} + \frac{1}{T_1(CSA)} + \frac{1}{T_1(SR)} +$$

$$\frac{1}{T_1(SC)} + \frac{1}{T_1(EN)} \tag{6}$$

Because each mechanism has different motional, temperature, and field dependences, it is important to identify the relative contributions in order to draw proper inferences about molecular dynamics from T_1 measurements. Operational procedures for this purpose have been outlined in some detail (7).

The relative contribution of T_1(DD) can be inferred from the measured NOE arising on saturation of proton resonances, using eq. 7, which indicates that relaxation mechanisms other than the

$$\frac{NOE(obs)}{NOE(max)} = \frac{T_1(obs)}{T_1(DD)} \tag{7}$$

DD mechanism act to shorten T_1. It is for this reason that determination of NOE's requires scrupulously careful sample preparation, for example, thorough degassing and removal of trace amounts of paramagnetics.

The CSA mechanism depends on the square of the value of B_0, the applied magnetic field. Thus, T_1 measurements at two fields serve to identify $T_1(CSA)$.

The $T_1(SR)$ decreases with temperature, while $T_1(DD)$ and $T_1(CSA)$ increase. Thus, while quantitation can be difficult, a qualitative indication of a SR mechanism may be adduced.

The SC mechanism arises from interaction between ^{15}N and a spin-spin coupled nucleus X, and depends not only on J_{NX} but also on ω_X, the X resonance frequency; ω_X and ω_N must be similar in value. While examples exist in ^{13}C NMR, none have been reported for ^{15}N.

The EN mechanism, arising from the presence of paramagnetic materials, has complex dependences on temperature and magnetic field. Operationally, relaxation mechanisms not accounted for by the above considerations commonly are ascribed to EN interaction. Concentrations of paramagnetics as low as 10^{-7} M can contribute substantially to overall ^{15}N relaxation, owing to the ability of nitrogen to complex with metal ligands. Conversely, this mechanism can be made straightforwardly to dominate ^{15}N relaxation, and the resulting T_1's can be used as a probe for nitrogen-ligand interaction. The $T_1(EN)$ values are then determined from the difference between the observed T_1's in the presence and absence of added paramagnetic ligand. In the latter case, contributions to T_1 from residual DD interaction must be corrected for by determination of a NOE. For example, although T_1 for the amino nitrogen of 2-aminopyridine is reduced from 13 to 4 seconds on addition of 10^{-3} M $Gd(dpm)_3$, the signal still displays a NOEF of -1.5, corresponding to only ~70% replacement of diamagnetic by paramagnetic relaxation (15).

Finally, the maximum NOE described above is attainable only under conditions of extreme narrowing, when molecular motions are vey fast compared to the spectrometer frequency, i.e., $\omega_N^2 \tau_c^2 \ll 1$. When this condition is not fulfilled, particularly when $\tau_c > 10^{-9}$ s, the NOE is reduced in magnitude, ultimately to zero. As a result, even under conditions where relaxation is exclusively dipolar, a ^{15}N signal may not only go unenhanced, but may also be nulled. This situation can arise most commonly with biomolecules.

STRUCTURAL APPLICATIONS OF NITROGEN NMR

This section presents selected examples of the use of nitrogen NMR in chemical and biochemical problems. Examples are meant to be illustrative rather than exhaustive. Additional applications may be found in previously cited sources (7,9,11).

Organic Structure and Mechanism

A 1,1-diazene, proposed as an intermediate in a variety of reactions, has been detected at low temperature by ^{15}N NMR. Oxidation of 1 (doubly enriched) at -78°C afforded a deep purple solution which gave rise to a ^{15}N spectrum consisting of four doublets at 321.4 ppm (J = 15.5 Hz), 917.0 ppm (J = 15.5 Hz), 419.5 ppm (J = 6.4 Hz), and 164.6 ppm (J = 6.4 Hz). The last two were assigned to tetrazene 3 by independent synthesis, and specific resonance assignments were confirmed by spectral determination of 2 and 3 derived from 1 labeled only at N_2. The remarkably deshielded N_2 represents the most highly deshielded neutral nitrogen known (49). Interestingly, N_1 in both 2 and 3 displayed maximum NOE's.

chemical shifts in ppm

The unstable heteroaromatic benz(cd)indazole (4) has been characterized in part from its ^{15}N shift, obtained on enriched material by photolysis of a diazide. Although lying at somewhat higher shielding, the resonance position is consistent with that expected for azo compounds (50).

The structure of the streptidine part of the antibiotic streptomyin (5) was shown by natural abundance ^{15}N NMR to be the same as that of its dihydro analogue 6 (51). The ^{15}N chemical shifts of both compounds are essentially identical and consistent with those expected for protonated amines (35 ppm) and for guanidines (74, 88 ppm) (52). Thus, a previously suggested four-ring structure involving condensation between the methylamino and nascent aldehyde groups is excluded.

Conformational Analysis

Nitrogen inversion in a series of bicyclic hydrazines has been characterized from low temperature ^{15}N lineshapes obtained at the natural abundance level (53). Two sets of resonances separated by 1-2 ppm coalesced into singlets in the range 90-100 ppm over the temperature range 250-280 K. Activation free energy inversion barriers of 12.7-14.8 kcal/mol were consistent with those obtained from proton and ^{13}C experiments. Similarly, conformational equilibria in cis-decahydroquinolines have been elucidated by comparison of ^{15}N shifts in rigid compounds with those in conformationally mobile ones (54).

Conformations of polypeptides may be elucidated by ^{15}N NMR and details of useful techniques, including comparison with model compounds, hydrogen-deuterium exchange, and solvent effects, have been described (7,55). As one example, the conformations of the tetrapeptide Tyr^1-Gly^2-Gly^3-Phe^4, an enkephalin model, have been investigated as a function of pH and solvent. The nitrogen of Phe[4] in the zwitterion is deshielded by 5.5 ppm in going from $(CH_3)_2SO$ to water as solvent, while the corresponding resonance positions in both the cation and anion change by 2.7-2.9 ppm. Furthermore, the chemical shift of the Phe[4] ^{15}N signal in $(CH_3)_2SO$ is intermediate between that for the cation and anion, while in water the value is the same as that in the anion. These results are interpreted in terms of a folded, intramolecularly hydrogen-bonded conformation for the zwitterion in $(CH_3)_2SO$, but an extended conformation for all species in water, and for the cation and anion in DMSO. These conclusions are based on the observation that solvation of an amide carbonyl by a hydrogen-bonding donor deshields a nitrogen more than does interaction between an amide N-H and a hydrogen-bonding acceptor (56).

Solvent effects have also been used to distinguish diastereomers of oligopeptides of alanine, phenylalanine, valine, and glycine. The ^{15}N nucleus appears to be much more sensitive to diastereomerism than do ^{13}C and proton, although enrichment may be required (57).

Vicinal nitrogen-proton couplings have been used in conjunction with carbon-proton and proton-proton couplings to characterize torsional angles in the cystine bridge of oxytocin (36). Measured values were compared with those calculated from rotational isomer populations and angular dependences of vicinal couplings, the latter taking a form similar to that of eq. 4. Eclipsed conformations with nitrogen and sulfur at -120°C were suggested.

Molecular Dynamics

The ^{15}N T_1 data for poly(vinylamine) were combined with ^{13}C values to elucidate segmental and group internal motions as a function of pH. In the manner described above, the DD components were extracted from NOE measurements. The DD mechanism is a function of both the effective correlation time, τ_{eff}, and the distance r between the nucleus and the n relaxing protons, according to eq. 8. If τ_{eff} is the same for each nucleus in the

$$\frac{1}{T_1(DD)} = n_H \gamma_H^2 \gamma_X^2 \hbar^2 r^{-6} \tau_{eff} \tag{8}$$

molecule under study, then eq. 9 holds. Significant increases

$$\frac{nT_1(DD)(^{15}N)}{nT_1(DD)(^{13}C)} = \left(\frac{r_{NH}}{r_{CH}}\right)^6 \cdot \frac{\gamma_C^2}{\gamma_N^2} = 4.4 \tag{9}$$

over this value imply significant rotation of the amino group. In fact, at low pH the nT_1 ratio was 13-30, while at neutral and alkaline pH value the ratio was close to theoretical. This change is consistent with restricted motion at the higher pH values because of formation of intramolecular hydrogen-bonded species such as 7, which cannot be formed under acidic condi-

7

tions. In all cases T_1 never exceeded 1.3 sec, a value consistent with restricted overall motion.

The ^{15}N spectrum of 90-95% enriched double-stranded DNA displays resonances assignable to nitrogens of guanine, adenine, thymine, and cytosine bases. A sample containing both single- and double-stranded DNA, run under conditions designed to retain some NOE, showed both positive and negative signals. Comparison with NOE curves calculated as a function of isotropic jump angles and orientational correlation times allowed assignment of negative peaks to protonated nitrogens on single-stranded DNA that have jump times <5 ns. Corresponding peaks in double-stranded material appear to have jump times >5 ns, and are shifted to lower shielding because of hydrogen bonding. Nonprotonated nitrogens may be relaxing in part via the CSA mechanism.

Biomolecules

Because pyridine-like and pyrrole-like nitrogens differ markedly in chemical shift, and because the former experience substantial shielding when they are protonated, nitrogen NMR is very useful for characterizing complex compounds containing these kinds of nitrogens. Such studies have been concerned largely with nucleic acid bases: their identification, their sites of protonation, and their involvement in hydrogen-bonding, metal-ion, or base-pair stacking interactions. Earlier results have been summarized (7-9,11).

When 2',3',5'-tri-O-benzoyluridine-$^{15}N_3$ (8) in $CDCl_3$ is mixed with varying mole fractions of 5'-O-acetyl-2',3'-O-isopropylideneadenosine (9), the ^{15}N-3 chemical shift of 8 moves progressively to lower shielding from 153.7 ppm to 158.5 ppm (58). The magnitude of this change parallels that of the N-3 proton ($\Delta\delta$ = 3.5 ppm), and both changes are consistent with formation of an A-U base pair. This result contrasts with the apparent absence of base pairing in dimethyl sulfoxide.

The ^{15}N resonances from individual bases in uniformly enriched DNA isolated from E. coli have been resolved (48). The N-3 of cytosine is shifted ~10 ppm to higher shielding relative to the monomer, indicating that site to be a hydrogen-bond acceptor (59). Based on the knowledge that N-1 of adenine is the only other acceptor, a resonance about 4-7 ppm to higher shielding from its position in the monomer is ascribable to that nitrogen. Reproducible multiple resonances in some regions suggest locally heterogeneous chemical environments.

The mechanism of α-lytic protease-catalyzed cleavage of peptide bonds has been addressed in a particularly elegant way using nitrogen NMR (60). The enzyme contains a "catalytic triad" consisting of serine, histidine, and aspartine residues, all of which are required for the cleavage. Mechanistic questions hinged on the actual pH of the histidine residue at the catalytic site. Titration of the enzyme enriched in the imidazole ring of the histidyl residue gave rise to substantial chemical shift changes, from which a pK_a of 7.0 was derived. This value contrasts with that reported earlier (pK_a < 4.0) based on ^{13}C chemical shifts. In addition, of the two possible tautomeric forms of imidazole, that with the proton on the nitrogen nearest the aspartyl residue predominates, ostensibly because of hydrogen-bonding stabilization by the carboxylate group. On the basis of this evidence a mechanism was proposed involving cooperative transfer of the serine proton to the proximate imidazole nitrogen (N-1) to give an imidazolium cation stabilized by both electrostatic and hydrogen bonding to the aspartyl carboxylate. Transfer of the N-1 proton to the peptide nitrogen would then produce the corresponding amine and a serine ester.

The binding of zinc in a modified carboxypeptidase A containing a ^{15}N-enriched arsanilazo group at tyrosine-248 has been investigated by ^{15}N NMR (61). The modified enzyme retains its catalytic properties while affording a specific label near the proposed catalytic site. With arsanilazo-N-acetyltyrosine (10)

<pre>
        CH₃CONHCHCO₂⁻
               |
              CH₂
               |
   10   [benzene ring]
               |
              OH   N=N
                     α β   [benzene ring]—AsO₃H₂
</pre>

as a model, chemical shifts were determined as a function of pH both with and without zinc. At pH 7, the N_α and N_β chemical shifts for the enzyme in water, 503.8 and 452.0 ppm, respectively, are close to those for 10, 501.9 and 456.1 pm. At pH 8.8, that for N_β moves to higher shielding by -26.9 ppm, while N_α is shielded by only -8.1 ppm. Finally, at pH 10.3 both nitrogens resume values characteristic of free azo compound. The model compound 10 undergoes changes of <2 ppm over the pH range investigated. These results were interpreted in terms of specific Zn coordination with N_β at pH 8, which is disrupted by hydroxide at higher pH values. Indeed, the azo resonance positions of the zinc-free apoenzyme at pH 8.8 are the same as those for 10.

A particularly exciting development is the detection of ^{15}N resonances of metabolites in vivo derived from Neurospora crassa cultured in the presence of ^{15}NH₄Cl (62). Over 17 components were identified, largely as side chain nitrogens of free amino acids or as protein residues. Amide nitrogens could be detected only with NOE suppression; evidently, the NOE is ~0 owing to restricted backbone motion.

Other biochemical applications include studies of purine (63) and imidazole (64) tautomerism, studies of the effects of redox conditions on NAD and NADH (65), and characterization of unstable porphyrin intermediates (66).

Macromolecules

The ubiquity of nitrogen in polyamides has naturally stimulated intense efforts to apply nitrogen NMR to their characterization. Many of these attempts have been described in a recent comprehensive review (67), in which a variety of complex criteria for sequencing polyamides have been summarized. Substituent effects comparable to those in small molecules are observed, but resonance positions and relaxation properties are also influenced by solvent (57,68).

The high molecular weight and resulting low concentrations of polyamides in solution considerably exacerbates an already unfavorable signal intensity problem. That polarization transfer methods described above can be expected to aid in overcoming this difficulty has been demonstrated in the characterization of nylons (69). Using the J cross polarization (JCP) technique (70), a signal enhancement of $\gamma_H/\gamma_N = 10$, corresponding to a time saving of ~100, can be attained, which may be increased by repeating the pulse sequence at a rate determined by the faster proton T_1 values, which are smaller than the ^{15}N values. These techniques allow adequately strong signals to be attained in one hour or less (depending on sample size), and permit the effects of temperature, concentration, and solvent to be studied over wide ranges. For example, chemical shifts for a variety of nylons move to lower shielding with increasing solvent acidity, presumably because of extensive hydrogen bonding to the amide carbonyl. At high acid/amide ratios, the amide N-H proton begins to exchange more rapidly and hence the signal intensity decreases.

The natural abundance ^{15}N spectrum of the Bunte (sulfonate) salt of bovine insulin A displays 20 of the 22 possible amide nitrogens in dimethyl sulfoxide as solvent. In water, 18 negative signals are obtained, although relative intensities vary. In DMSO, continuous proton irradiation results in signal nulling, so that NOE suppression was required to observe signals. Signals in the latter solvent are shielded by 3-4 ppm compared to their counterparts in water. Interestingly, varying irradiation times between two spectra resulted in slight chemical shift differences, owing to differences in sample heating from the decoupler. Seven of the resonances could be assigned tentatively by comparison with model oligopeptides and from qualitative T_1's. The latter are longer for side chain nitrogens (71).

High Resolution Nitrogen NMR of Solids

The techniques of cross polarization and magic angle spinning (CP-MAS) that have been applied successfully to ^{13}C NMR of solids (72) are broadly applicable to ^{15}N as well. By this means nitrogen metabolism in plants was studied in intact seed and dried extracts of soybeans grown on ^{15}N-enriched ammonium nitrate (73a) and ^{15}N-enriched glycine (73b). Resonances from protein amide nitrogens, amino acids, and histidine and guanidino nitrogens could be detected. Similarly, metabolism of ^{15}N nitrate in

Neurospora crassa was characterized, and the nature of ^{15}N assimilation was related to the method of culturing (74).

CP-MAS spectra of ^{15}N-enriched histidine and imidazole as lyophilized powders show two signals for the former material at 155 ppm and 222 ppm (relative to $(NH_4)_2SO_4$), relative intensities of which depend on whether the pH of the species from which the powders were obtained is low or high (75). In contrast to the time-averaged single peak displayed in solution, imidazole itself gives rise to two peaks at 150 and 222 ppm. Clearly, these lines are assignable to the pyrrole-like and pyridine-like nitrogens, respectively, and the average of these values is equal to the solution value. Because the ring nitrogens of histidine even in solution are nonequivalent, the pH dependence is more complex. Consideration of details of the equilibria among the five

possible species allowed the relative importance of the various tautomers to be assessed. The N_τ-H form appears to predominate in the neutral and anionic forms, and other evidence suggests the presence of an intramolecular hydrogen bond between the NH_3 group and N_π in the neutral form of the solid, just as in solution.

Finally, the first example of solid state ^{15}N spectra at the natural abundance level has been reported, in which intractable polymeric material from the reaction between HCN and ammonia have been partially characterized (76). Although specific nitrogen-containing compounds could not be ascertained, the types of nitrogens present, including nitrile, aromatic, amide, urea, and ammonia-like nitrogens, were readily identified.

CONCLUDING COMMENTS

There is little question that nitrogen NMR, using both isotopes, is an informative and practicable spectroscopic technique. Earlier concerns about sensitivity and detectability of resonances, while by no means eliminated, nonetheless have been alleviated by availability of modern technology, development of elegant and clever pulse sequences, and better understanding of required experimental techniques. Very strikingly, most of the predictions made in 1979 (7) about the potential utility of nitrogen NMR have already been realized, at least to a moderate extent. There is consequently little reason for a chemist to hesitate in thinking about how nitrogen NMR can be useful in addressing research problems, and every reason to expect that these approaches will be successful.

ACKNOWLEDGMENT

The author is grateful to several colleagues who provided unpublished data, and to Exxon Research and Engineering Company for its hospitality and support during preparation of this manuscript. Acknowledgment is also made to the National Science Foundation for Science Faculty Professional Development Award No. SPI-8165053.

REFERENCES

(1) J. D. Ray and R. A. Ogg, Jr., J. Chem. Phys. 26, pp. 1452-1454 (1957).
(2) B. E. Holder and M. P. Klein, J. Chem. Phys., 23, p. 1956 (1955).
(3) B. M. Schmidt, L. C. Brown, and D. Williams, J. Mol. Spectrosc., 2, pp. 539-550, 551-558 (1958); J. Mol. Spectrosc., 3, pp. 30-35 (1959).
(4) D. H. Evans and R. E. Richards, Mol. Phys., 8, pp. 19-31 (1964).
(5) M. Bose, N. Das, and N. Chatterjee, J. Mol. Spectrosc., 18, pp. 32-40 (1965).
(6) M. Witanowski, T. Urbański, and L. Stefaniak, J. Am. Chem. Soc., 86, pp. 2569-2570 (1964).
(7) G. C. Levy and R. L. Lichter, "Nitrogen-15 Nuclear Magnetic Resonance Spectroscopy," Wiley-Interscience, New York, 1979.
(8) M. Witanowski, L. Stefaniak, and G. A. Webb, Ann. Rep. NMR Spectrosc., 11B, (1981).
(9) G. J. Martin, M. C. Martin, and J. P. Gouesnard, NMR-Basic Prin. Progr., 18, (1981).
(10) J. Mason, Chem. Rev., 81, pp. 205-227 (1981).
(11) K. Kanamori and J. D. Roberts, Acc. Chem. Res., in press.
(12) J. M. Lehn and J. P. Kintzinger, in "Nitrogen NMR," M. Witanowski and G. A. Webb, Eds., Plenum Press, London, 1973, pp. 79-161.
(13) G. A. Morris, J. Am. Chem. Soc., 102, pp. 428-429. (1980).
(14) W. Städeli, P. Bigler, and W. von Philipsborn, Org. Magn. Reson., 16, pp. 170-172 (1981).
(15) G. C. Levy, T. Pehk, and P. R. Srinivasan, Org. Magn. Reson., 14, pp. 129-132 (1980).

(16) (a) M. Witanowski, L. Stefaniak, S. Szymanski, and H. Januszewski, J. Magn. Reson., 28, pp. 217-226 (1977).
 (b) P. R. Srinivasan and R. L. Lichter, ibid., 28, pp. 227-231 (1977).
(17) R. O. Duthaler and J. D. Roberts, J. Am. Chem. Soc., 100, pp. 3889-3895 (1980).
(18) (a) R. O. Duthaler, K. L. Williamson, D. D. Giannini, W. H. Bearden, and J. D. Roberts, J. Am. Chem. Soc., 99, pp. 8406-8414 (1977).
 (b) F. W. Vierhapper, G. T. Furst, S. N. Y. Fanso-Free, E. L. Eliel, and R. L. Lichter, J. Am. Chem. Soc., 103, pp. 5629-5633 (1981).
 (c) G. C. Levy, T. Pehk, and E. Lippmaa, Org. Magn. Reson., 14, pp. 214-219 (1980).
(19) (a) M. P. Sibi and R. L. Lichter, J. Org. Chem., 42, pp. 2999-3004 (1977).
 (b) E. Prince, J. LeMelle, and R. L. Lichter, unpublished results.
 (c) K. Crimaldi, R. L. Lichter, and A. D. Baker, J. Org. Chem., 42, pp. 3524-3528 (1982).
(20) (a) G. J. Martin, T. Bertrand, D. LeBotlan, and J. M. LeTourneux, J. Chem. Res. (S), pp. 408-409 (1979).
 (b) M. J. Kamlet, C. Dickinson, and R. W. Taft, J. Chem. Soc. Perkin 2, pp. 353-355 (1981).
(21) (a) I. I. Schuster and J. D. Roberts, J. Org. Chem., 45, pp. 284-287 (1980).
 (b) O. W. Sørensen, S. Scheibye, S. O. Lawesson, and H. J. Jakobsen, Org. Magn. Reson., 16, pp. 322-324 (1981).
(22) (a) G. J. Martin, J. P. Gouesnard, J. Dorie, C. Rabiller, and M. L. Martin, J. Am. Chem. Soc., 99, pp. 1381-1384 (1977).
 (b) J. Dorie, J. P. Gouesnard, C. Rabiller, B. Mechin, N. A. Naulet, and G. J. Martin, Org. Magn. Reson., 13, pp. 126-131 (1980).
(23) M. Witanowski, L. Stefaniak, and G. A. Webb, Org. Magn. Reson., 16, pp. 309-311 (1981).
(24) (a) W. Städeli and W. von Philipsborn, Org. Magn. Reson., 15, pp. 106-109 (1981).
 (b) W. Städeli, W. von Philipsborn, A. Wick, and I. Kompis, Helv. Chim. Acta, 63, pp. 504-522 (1980).
(25) M. Witanowski, L. Stefaniak, B. N. Lamphun, and G. A. Webb, Org. Magn. Reson., 16, pp. 57-59 (1981).
(26) (a) D. J. Craik, G. C. Levy, and R. T. C. Brownlee, submitted for publication and private communication.
 (b) S. Bulusu, T. Axenrod, and J. Autera, Org. Magn. Reson., 16, pp. 52-56 (1981).
(27) (a) J. P. Gouesnard and G. J. Martin, Org. Magn. Reson., 12, pp. 263-270 (1979).
 (b) N. Naulet and G. J. Martin, Tetrahedron Lett., pp. 1493-1496 (1979).

(28) A. Lyčka, D. Šnobl, V. Machaček, and M. Večeřa, Org. Magn. Reson., 16, pp. 17-19 (1981).
(29) D. M. Kanjia, J. Mason, I. A. Stenhouse, R. E. Banks, and N. D. Venayak, J. Chem. Soc. Perkin 2, pp. 975-979 (1981).
(30) (a) H. J. Jakobsen and W. S. Brey, J. Am. Chem. Soc., 101, pp. 774-775 (1979).
 (b) O. W. Sørensen, S. Scheibye, S.-O. Lawesson, and H. J. Jakobsen, Org. Magn. Reson., 16, pp. 322-324 (1981).
 (c) H. J. Jakobsen and S. Deshmukh, J. Magn. Reson., 42, pp. 337-340 (1981).
(31) G. Binsch, J. B. Lambert, B. W. Roberts, and J. D. Roberts, J. Am. Chem. Soc., 86, pp. 5564-5568 (1964).
(32) (a) D. Crepaux and J. M. Lehn, Mol. Phys., 14, pp. 547-550 (1968).
 (b) A. Danoff, M. Franzen-Sieveking, R. Lichter, and S. N. Y. Fanso-Free, Org. Magn. Reson., 12, pp. 83-86 (1979).
(33) D. R. Boyd, M. E. Stubbs, N. J. Thompson, H. J. C. Yeh, D. M. Jerina, and R. E. Wasylishen, Org. Magn. Reson., 14, pp. 528-533 (1980).
(34) W. Städeli, P. Bigler, and W. von Philipsborn, Org. Magn. Reson., 16, pp. 170-172 (1981).
(35) V. F. Bystrov, Y. D. Favrilov, and V. N. Solkan, J. Magn. Reson., 19, pp. 123-126 (1975).
(36) A. J. Fishman, D. H. Live, H. R. Wyssbrod, W. C. Agosta, and D. Cowburn, J. Am. Chem. Soc., 102, pp. 2533-2539 (1980).
(37) A. DeMarco and M. Llinas, Org. Magn. Reson., 12, pp. 454-460 (1979).
(38) R. E. Wasylishen, Ann. Rep. NMR Spectrosc., 7, pp. 245-291 (1977).
(39) J. M. Schulman and T. Venanzi, J. Am. Chem. Soc., 98, pp. 4701-4705, 6739-6741 (1976).
(40) R. Siegel, K. Crimaldi, R. Lichter, and J. Schulman, J. Phys. Chem., 85, p. 4157 (1981).
(41) (a) V. N. Solkan and V. F. Bystrov, Bull. Acad. Sci. USSR, Div. Chem. Sci., 23, pp. 95-97 (1975).
 (b) R. E. London, T. E. Walker, T. W. Whaley, and N. A. Matwiyoff, Org. Magn. Reson., 9, pp. 598-602 (1978).
(42) J. M. Schulman, J. Ruggio, and T. J. Venanzi, J. Am. Chem. Soc., 99, pp. 2045-2048 (1977).
(43) A. Lyčka and D. Šnobl, Coll. Czech. Chem. Comm., 46, pp. 892-897 (1981).
(44) T. Khin, S. Duangthai, and G. A. Webb, Org. Magn. Reson., 13, pp. 240-243 (1980).
(45) (a) H. J. Jakobsen and W. S. Brey, J. Chem. Soc. Chem. Commun., pp. 478-479 (1979).
 (b) M. P. Sibi and R. L. Lichter, Org. Magn. Reson., 14, pp. 494-498 (1980).
 (c) R. J. Gillespie and R. Hulme, J. Chem. Soc. Dalton Trans., pp. 1261-1267 (1973).

(46) (a) J. H. Hargis, W. B. Jennings, S. D. Worley, and M. S. Tolley, J. Am. Chem. Soc., 102, pp. 13-17 (1980).
(b) G. W. Buchanan, F. G. Morin, and R. R. Fraser, Can. J. Chem., 58, pp. 2442-2446 (1980).
(47) J. R. Lyerla, Jr., and G. C. Levy, Top. Carbon-13 NMR Spectrosc. 1, pp. 79-148 (1974).
(48) T. L. James and A. Lapidot, J. Am. Chem. Soc., 103, pp. 6748-6750 (1981).
(49) P. B. Dervan, M. E. Squillacote, P. M. Lahti, A. P. Sylvester, and J. D. Roberts, J. Am. Chem. Soc., 103, pp. 1120-1122 (1981).
(50) H. Nakanishi, A. Yabe, and K. Honda, J. Chem. Soc. Chem. Commun., pp. 86-87 (1982).
(51) R. E. Botto, J. H. Schwartz, and J. D. Roberts, Proc. Natl. Acad. Sci. USA, 77, pp. 23-25 (1980).
(52) W. E. Hull and H. R. Kricheldorf, Liebigs Ann. Chem., pp. 158-164 (1980).
(53) Y. Nomuro and Y. Takeuchi, J. Chem. Soc. Chem. Commun., pp. 295-296 (1979).
(54) F. W. Vierhapper, G. T. Furst, and R. L. Lichter, Org. Magn. Reson., 17, pp. 127-130 (1981).
(55) G. E. Hawkes, E. W. Randall, W. E. Hull, and O. Convert, Biopolymers, 19, pp. 1815-1826 (1980).
(56) C. Garbay-Jaureguiberry, J. Baudet, D. Florentin, and B. P. Roques, FEBS Letters, 115, pp. 315-318 (1980).
(57) H. R. Kricheldorf and W. E. Hull, Org. Magn. Reson., 12, pp. 607-611 (1979).
(58) C. D. Poulter and C. L. Livingston, Tetrahedron Lett., pp. 755-758 (1979).
(59) C. Dyllick-Brenzinger, G. R. Sullivan, P. P. Pang, and J. D. Roberts, Proc. Natl. Acad. Sci. USA, 77, pp. 5580-5582 (1980).
(60) W. Bachovchin and J. D. Roberts, J. Am. Chem. Soc., 100, pp. 8041-8048 (1978).
(61) W. Bachovchin, K. Kanamori, B. L. Vallee, and J. D. Roberts, Biochemistry, 21, pp. 2885-2892 (1982).
(62) T. L. Legerton, K. Kanamori, R. L. Weiss, and J. D. Roberts, Proc. Natl. Acad. USA, 78, pp. 1495-1498 (1981).
(63) N. C. Gonnella and J. D. Roberts, J. Am. Chem. Soc., 104, pp. 3162-3164 (1982).
(64) M. A. Alei, Jr., L. O. Morgan, W. E. Wageman, and T. W. Whaley, J. Am. Chem. Soc., 102, 2881-2887 (1980).
(65) N. J. Oppenheimer and R. M. Davidson, Org. Magn. Reson., 13, pp. 14-16 (1980).
(66) (a) G. Burton, P. E. Fagerness, P. M. Jordan, and A. I. Scott, Tetrahedron, 36, pp. 2721-2725 (1980).
(b) A. I. Scott, G. Burton, P. M. Jordan, H. Matsumoto, P. E. Fagerness, and L. M. Pryde, J. Chem. Soc. Chem. Commun., pp. 384-387 (1980).
(67) H. R. Kricheldorf, Pure Appl. Chem., 54, pp. 467-481 (1982).

(68) (a) H. R. Kricheldorf and W. E. Hull, Makromol. Chem., 182, pp. 1177-1196 (1981).
  (b) Macromolecules, 13, pp. 87-95 (1980).
(69) B. S. Holmes, G. C. Chingas, W. B. Moniz, and R. E. Ferguson, Macromolecules, 14, pp. 1785-1787 (1981).
(70) (a) R. D. Bertrand, W. B. Moniz, A. N. Garroway, and G. C. Chingas, J. Am. Chem. Soc., 100, pp. 5227-5229 (1978).
  (b) J. Magn. Reson., 32, pp. 465-467 (1978).
(71) W. E. Hull, E. Bullesbach, H.-J. Wieneke, H. Zahn, and H. R. Kricheldorf, Org. Magn. Reson., 17, pp. 92-96 (1981).
(72) J. Schaefer and E. O. Stejskal, Top. Carbon-13 NMR Spectrosc., 3, pp. 283-342 (1979).
(73) (a) J. Schaefer, E. O. Stejskal, and R. A. McKay, Biochem. Biophys. Res. Commun., 88, pp. 274-280 (1979).
  (b) T. A. Skokut, J. E. Varner, J. Schaefer, E. O. Stejskal, and R. A. McKay, Plant Physiol., 69, pp. 314-316 (1982).
(74) G. C. Jacob, J. Schaefer, E. O. Stejskal, and R. A. McKay, Biochem. Biophys. Res. Commun., 97, pp. 1176-1182 (1980).
(75) M. Munowitz, W. W. Bachovchin, J. Herzfeld, C. M. Dobson, and R. G. Griffin, J. Am. Chem. Soc., 104, pp. 1192-1196 (1982).
(76) J. Schaefer, E. O. Stejskal, G. S. Jacob, and R. A. McKay, Appl. Spectrosc., 36, pp. 179-182 (1982).

CHAPTER 11

APPLICATION OF ^{17}O NMR SPECTROSCOPY TO STRUCTURAL PROBLEMS

Walter G. Klemperer

School of Chemical Sciences
University of Illinois
Urbana, Illinois 61801 USA

ABSTRACT

As a low-abundance, quadrupolar nucleus, ^{17}O has often been overlooked as a viable NMR nucleus for structure determination. The ^{17}O nucleus has, however, proved to be a surprisingly convenient nucleus for study by NMR techniques due to the availability of water and dioxygen enriched in ^{17}O and the relatively small magnitude of the ^{17}O quadrupole moment. This chapter focuses on two practical aspects of ^{17}O NMR spectroscopy. First, methods for optimizing sensitivity and resolution are outlined. Correlations between chemical shift and structural environment are then presented in the context of examples selected from the organometallic, organic, inorganic, and biochemical literature.

RESOLUTION AND SENSITIVITY

Before the factors that influence spectral resolution and sensitivity are examined, it is important to establish that well resolved ^{17}O NMR spectra can and have been obtained for many classes of chemical compounds. This will be achieved through a discussion of three systems which have been studied in recent years.

Organic sulfones demonstrate well the sensitivity of ^{17}O chemical shifts to structural environment. The oxygens in phenyl 1-phenylethyl sulfone, 1, are diastereotopic. Resonances for both of these oxygens have been measured from stereospecifically labeled compounds, and appear 134 and 139 ppm downfield from pure water when measured in CHCl$_3$ solution (1). This Δδ value of 5 ppm for nonequivalent sulfone oxygens is of course magnified in conformationally rigid systems. Compound 2, for example, displays resonances at δ 210 and δ 243 (2), yielding a Δδ value of 33 ppm. When the symmetry-breaking substituents appear not on α carbons but only on β carbons as in 3a and 3b, smaller Δδ values are observed. For 3a (δ 182, 198) and 3b (δ 184, 187), Δδ values have been reported (2) to be 16 and 3 ppm, respectively.

1

2

3a R = C$_6$H$_5$S
3b R = OH

The H$_2$O molecule, its derivatives, and its adducts illustrate the potentially general usefulness of spin-spin coupling in ^{17}O NMR spectroscopy. Water in aprotic solvents displays the expected triplet shown in Figure 1a. The effect of H/D substitution is evident from the ^{17}O NMR spectrum of H$_2$O-D$_2$O in CH$_2$Cl$_2$ (see Figure 1b) (3). Although the H$_2$O triplet and HDO doublet of triplets are clearly resolved in Figure 1b, the D$_2$O quintet can only be resolved in the proton-decoupled spectrum shown in Figure 1c. Acidification of H$_2$O with HF-SbF$_5$ in SO$_2$ yields an adduct readily identified as H$_3$O$^+$ from its ^{17}O NMR spectrum shown in Figure 1d (3,4). Note that all four species, H$_2$O, HDO, D$_2$O, and H$_3$O$^+$, can be identified from coupling constants, without reference to any chemical shift data.

The degree of resolution obtained in spectra of water and related species just mentioned can frequently be observed in spectra of larger species. The spectrum shown in Figure 2, obtained from a mixture of three [(OC)$_3$Re(Nb$_2$W$_4$O$_{19}$)]$^{3-}$ diastereomers also shown in Figure 2, provides a good example. All four OC (carbonyl) oxygen resonances shown in the inset have linewidths of 13-15 Hz. The neighboring OW$_2$ (doubly bridging between tungsten) oxygen resonances are also quite narrow, in contrast with the ONbW and ONb$_2$ resonances which appear slightly further downfield. These resonances illustrate how spin-spin coupling can dramatically lower spectral resolution. The resonances for all oxygens in the [(OC)$_3$Re(Nb$_2$W$_4$O$_{19}$)]$^{3-}$ isomers which are bonded to at least one Nb are split into components by ^{17}O-^{93}Nb coupling. Since the 100% abundant, spin 9/2 ^{93}Nb nucleus undergoes rapid, quadrupolar relaxation, each of these

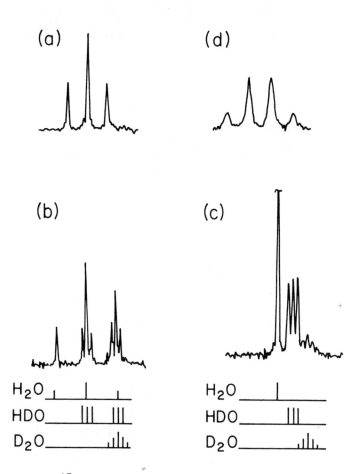

Figure 1. ^{17}O FT NMR spectra (3,4) of (a) H_2O in CH_2Cl_2 (J_{O-H} = 79 Hz), (b) H_2O-D_2O in CH_2Cl_2 (J_{OH} = 79 Hz, J_{OD} = 12 Hz) measured at 27 MHz, (c) H_2O-D_2O in CH_2Cl_2 measured at 27 MHz with proton decoupling, and (d) H_3O^+ in SO_2 (J_{O-H} = 107 Hz).

components has been broadened or has coalesced with other components. The net result is very low spectral resolution.

The remainder of this chapter will develop and generalize upon the points illustrated in the examples just discussed. It must be emphasized, however, that well resolved ^{17}O NMR spectra are usually not measured as easily as is the case in ^1H, ^{13}C,

^{19}F, and ^{31}P NMR spectroscopy due to the quadrupole moment and low natural abundance of the ^{17}O nucleus. A brief review of standard methods (6) for optimizing linewidth and signal-to-noise is therefore in order.

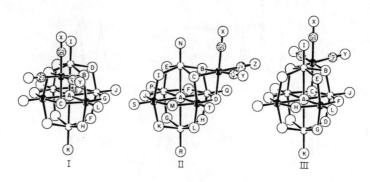

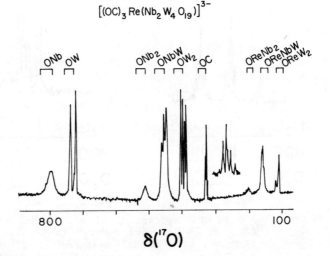

Figure 2. 33.9 MHz ^{17}O FT NMR spectrum of an acetonitrile solution containing the three $[(OC)_3Re(Nb_2W_4O_{19})]^{3-}$ isomers shown at the top of the figure (5): temperature = 79°C, total $[(OC)_3Re(Nb_2W_4O_{19})][(C_4H_9)_4N]_4$ concentration = 0.06 M, % ^{17}O = 10, number of accumulations = 115,000.

Poor spectral resolution resulting from quadrupolar broadening can be improved by increasing the separation (in Hz) between chemically shifted resonances and decreasing the linewidth (in Hz) of individual resonances. The former is achieved by measuring spectra in a high static magnetic field. The latter is achieved by lowering the rate of quadrupolar relaxation. If it is assumed that one is operating under the "usual" conditions of rapid molecular motion (rotational correlation time $\ll$ reciprocal of the ^{17}O Larmor frequency), quadrupolar relaxation rates are lowered by raising solution temperature and lowering solution viscosity. Low solution viscosity can be maintained using low solute concentrations in low viscosity solvents. The importance of sample temperature, solute concentration, and solvent viscosity cannot be overemphasized. High solute concentrations in high viscosity solvents at low temperatures almost invariably lead to very poor resolution ^{17}O NMR spectra.

Signal-to-noise ratios in ^{17}O NMR spectra can be enhanced in two ways. First, high magnetic field spectrometers and large volume sample tubes can be used. Large sample volume is especially important since low sample concentration is in general essential (see above). Second, the ^{17}O content can be enriched above its natural 0.037 atom per cent abundance. Since ^{17}O is relatively inexpensive and readily available in the form of water or dioxygen, ^{17}O-enriched compounds can usually be obtained inexpensively by either direct exchange or from a synthetic precursor enriched in ^{17}O. The 10% ^{17}O-enriched, 0.58 g sample used to measure the spectrum shown in Figure 2b, for example, contains less than $2.00 worth of ^{17}O. Treating this as a representative case, it is evident that even low yield preparations of ^{17}O-enriched compounds are not prohibitively expensive.

CHEMICAL SHIFTS

Since ^{17}O NMR was recently reviewed in a comprehensive fashion (7), no attempt will be made here to cover all areas of application. Instead, material has been selected from recent literature in an effort to include diverse systems of current organometallic, organic, inorganic, and biological interest.

Chemical shifts are useful to the structural chemist if they can be associated with specific chemical environments. In the case of ^{17}O NMR, this must be achieved element-by-element since oxygen combines with all the elements except helium, neon, and argon to form an extremely diverse family of compounds. For each element, chemical shifts for prototypical parent compounds, their derivatives, and their adducts must be measured in order to establish a chemical shift scale. Chemical shifts for incompletely characterized compounds can then be assigned to oxygens in specific structural environments by referring to the chemical shift scale. Carbon chemistry provides a good starting point since vast numbers of chemical shifts have been measured in this area.

Carbon Compounds

Figure 3 shows a fairly complete chemical shift scale for organic carbon compounds. Inspection of the scale reveals a dominant trend of increasing downfield chemical shifts resulting from increasing oxygen-carbon π bonding. For doubly bridging (two-coordinate) oxygens, comparison of chemical shifts for alcohols, ethers, and acetals with those for carbonates, esters, and anhydrides shows the effect of π bonding introduced by π acceptor substituents. Similar comparison of chemical shifts for aldehydes and ketones with those for acyl derivatives and carbonates shows the effect of π donor substituents on chemical shifts of carbonyl oxygens through reduction of oxygen-carbon π bond order. These trends have been discussed elsewhere (6-8). It is important to note, however, that there exists no overall correlation between ^{17}O chemical shift and electronic charge on oxygen. For terminal (carbonyl) oxygens in organic compounds, reduced oxygen-carbon π bond order is accompanied by increased negative charge on oxygen. There is therefore a relationship between increasing negative charge and upfield chemical shift (9-10). Precisely the opposite trend is observed for doubly bridging oxygens in saturated ethers and alcohols: increasing negative charge leads to downfield chemical shifts (9-10).

Comparisons of ^{17}O chemical shifts with ^{13}C chemical shifts for isostructural compounds often reveal extremely good linear correlation (7). The plot shown in Figure 4 of ^{17}O chemical shifts for aliphatic ethers ROR' against ^{13}C chemical shifts for RCH_2R' methylene groups illustrates this well (11). Since additive substituent parameters often can predict ^{13}C chemical shifts accurately (12), such a linear relationship allows corresponding additive substituent parameters to be calculated for oxygen analogues.

Oxygen-17 NMR provides a convenient means for monitoring perturbations of carbonyl groups which reduce carbon-oxygen π bond order. Protonated ketones R_2COH^+ show the most dramatic effect. As is to be expected from the resonance forms 4 and 5,

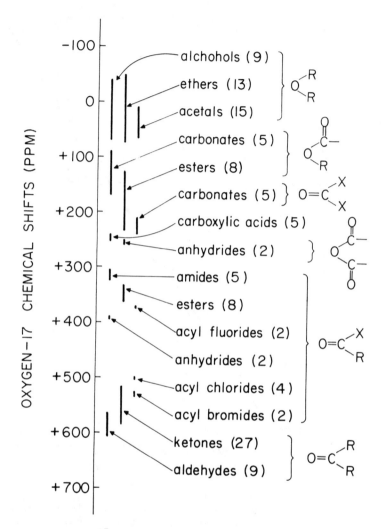

Figure 3. An ^{17}O chemical shift scale for carbon compounds. The parenthesized numbers indicate the numbers of chemical shifts used to define the chemical shift range in question. The substituents R are restricted to saturated hydrocarbons and hydrogen. All data, measured from neat liquid samples, were taken from Ref. 7. To simplify the chart, ketals, orthoesters, and orthocarbonates are classified as acetals.

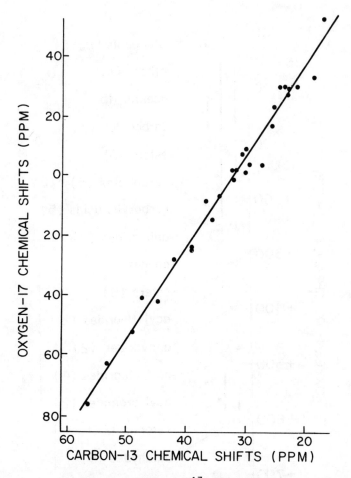

Figure 4. Correlation between ^{17}O NMR chemical shifts for aliphatic ethers R-O-R' and ^{13}C NMR chemical shifts for alkanes R-CH$_2$-R'. See Ref. 11 for numerical data. Reprinted with permission from Ref. 11, copyright 1978 Verlag Helvetica Chimica Acta.

these species display ^{17}O chemical shifts which approximate the average of values observed for ketones and alcohols: δ 310.3 for protonated acetone and δ 287-306 for protonated cyclic ketones $(CH_2)_nCO$, n = 3-7 (13). Lesser but parallel effects result from hydrogen bonding interactions (6-8). This effect has been exploited in comparative studies of isopropylidine-uridine 6 in

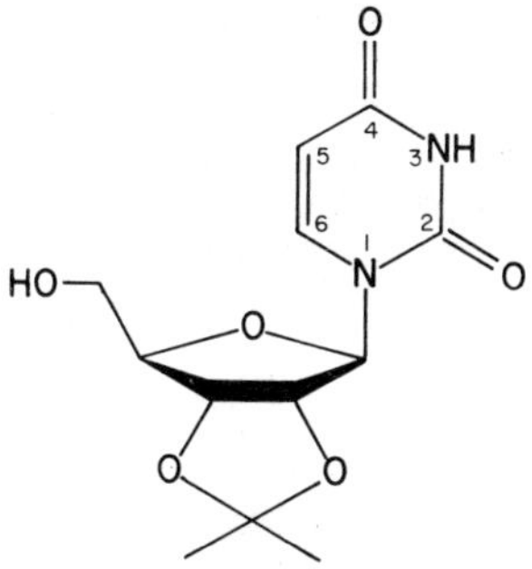

6

acetonitrile and water (14). In acetonitrile, the O_2 and O_4 resonances appear at δ 263 and 336, respectively. In water, hydrogen bonding interactions displace the O_2 resonance upfield by 8 ppm and the O_4 resonance by 32 ppm, reflecting preferential hydrogen bonding at O_4. Since the C_2 carbonyl is more polar than the C_4 carbonyl group, this preference reflects steric as opposed to electronic factors (15).

In general ^{17}O NMR has found few applications in carbon chemistry due in large part to the vast array of alternative spectroscopic techniques which are available for the study of carbon compounds. One class of experiments, however, is particularly suited for ^{17}O NMR: selective labeling experiments. The recently reported mechanistic study of homolytic O-O bond cleavage in 7 is representative (16). When compound 7 is labeled specifically at the carbonyl oxygen site and decomposed, the decomposition product 8 incorporates ^{17}O preferentially at its

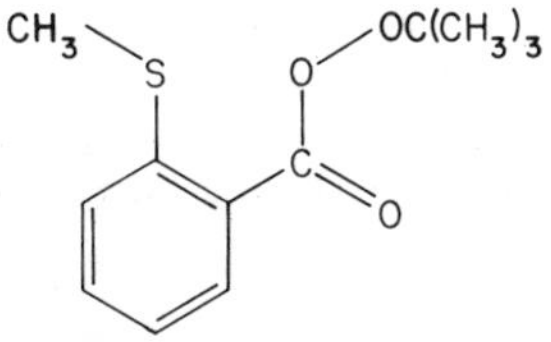

7 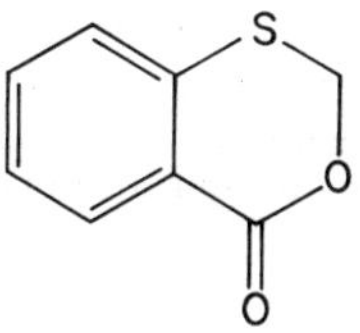   8

ketone oxygen site as evidenced by the relative intensities of its carbonyl (δ 356) and ester (δ 159) oxygen resonances.

Although ^{13}C NMR is a powerful technique for the study of metal carbonyl complexes, its utility is marginal when quadrupolar metal nuclei are involved due to metal-carbon spin-spin coupling which broadens ^{13}C resonances and distorts their intensities. Since two bond metal-oxygen couplings do not interfere in a similar fashion, ^{17}O NMR alone can provide definitive structural and dynamic information for many quadrupolar metal carbonyls (17-21). The most complex molecule of this type which has been studied to date is the C_{3v} HFeCo$_3$(CO)$_{12}$ cluster (20), __9__. At

__9__

-11°C, ^{17}O resonances for the four nonequivalent carbonyl oxygens can be resolved. As the sample temperature is raised, selective line broadenings and coalescences are observed which can be interpreted in terms of three distinct intramolecular rearrangement processes.

Phosphorus Compounds

As is the case with carbon compounds, ^{17}O chemical shifts for boron (21) and nitrogen (6,7) compounds cover a wide range and can in general be related to oxygen π bond order. Phosphorus compounds, like many of the heavier main group elements, behave differently and yield only a limited range of ^{17}O chemical shifts (δ 20-260 ppm), and these chemical shifts cannot be related to oxygen-phosphorus bond order in a straightforward fashion (6,7). Recently reported studies (22-24) show, however, that ^{17}O resonances for nonequivalent oxygens in phosphates and polyphosphates can be resolved and utilized in stereochemical and binding site studies.

The ability to resolve nonequivalent terminal oxygens in cyclic phosphates parallels the sulfone observations mentioned above: terminal phosphate oxygens in stereospecifically ^{17}O enriched cyclic 2'-deoxyadenosine 3',5'-monophosphate, __10__, have chemical shifts δ 92.8 (axial) and δ 91.2 (equatorial), $\Delta\delta$ = 1.66 (22).

10

The effect of terminal oxygen protonation on ^{17}O chemical shifts in phosphates and polyphosphates has also been noted (23,24). The suggestion (24) that the observed upfield displacements might be extended into a general relationship between increasing charge density on oxygen and upfield chemical shift should be carefully considered in light of the comments made above concerning charge and chemical shift in carbon compounds and the observation that protonation of methanol shifts the methanol ^{17}O resonance downfield (13). Caution must also be exercised when interpreting chemical shift displacements which occur upon protonation of a single oxygen in a set of equivalent oxygens under conditions of rapid proton exchange. The ^{17}O chemical shift for VO_4^{3-}, for example, is displaced downfield upon protonation in aqueous solution (25). This result does not necessarily imply that the HOV resonance for HVO_4^{3-} lies downfield from the OV resonance for VO_4^{3-}, since the chemical shift observed for aqueous HVO_4^{3-} under conditions of fast proton exchange is a weighted average of the chemical shifts for OV and HOV in HVO_4. In fact, if VO_4^{3-} behaves like $V_2W_4O_{19}^{4-}$ (26) and $V_{10}O_{28}^{6-}$ (27), the OV resonance in HVO_4 lies downfield relative to the VO_4^{3-} resonance and the HOV resonance lies upfield.

Early Transition Metal Compounds

Two general features of ^{17}O NMR chemical shifts for d^0, early transition metal compounds are evident from the Mo^{VI} data presented in Figure 5. First, there is a very wide range of chemical shift values. Second, there is a poor but real correlation between ^{17}O chemical shifts and Mo-O bond lengths. The chemical shifts can in fact be correlated with metal-oxygen π bond character in the same manner described above for carbon-oxygen compounds. Chemical shift scales analogous to the one shown in Figure 3 can be prepared for Mo^{VI}, W^{VI}, and V^V compounds, but since the structural units involved are probably unfamiliar to most readers, these scales will not be presented. Instead, only a single structural unit will be investigated in detail with regard to substituent effects and effects of cation addition.

The structural unit in question, 11, contains a hexavalent tungsten center doubly bonded to one oxygen ($d_{W=O}$ = ~1.7 Å), singly bonded to four doubly bridging oxygens (d_{W-O} = ~1.9 Å),

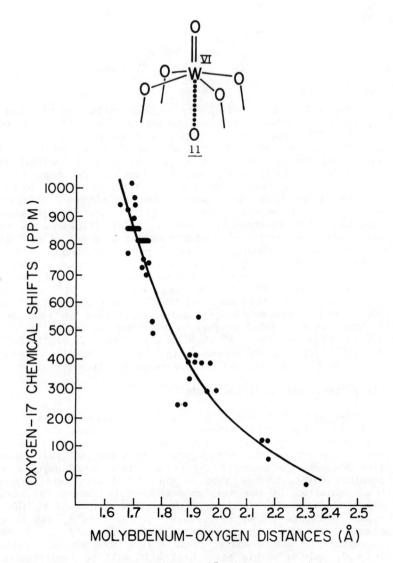

Figure 5. Correlation between ^{17}O NMR chemical shifts and the shortest O-Mo distance for oxo ligands in selected oxomolybdenum(VI) compounds. See Ref. 28 for numerical data. Reprinted with permission from Ref. 28, copyright 1982 American Chemical Society.

and very weakly bound to a sixth oxygen ($d_{W\cdots O} > 2.2$ Å). Note that in orbital terms, the W=O bond is a three component bond and the W-O bonds are two component bonds: the terminal oxygen's p orbitals can form two p-d π bonds with tungsten d_{xz} and d_{yz} orbitals; the bridging oxygens' p orbitals can also overlap in a π fashion with the tungsten d_{xy} orbital. The extreme length of the W····O bond precludes significant π bonding.

The $W_6^{VI}O_{19}^{2-}$ ion shown in Figure 6a incorporates six symmetry equivalent structural units 11. The chemical shifts observed for the three types of oxygen (see Figure 6c) are ordered qualitatively according to π bond order. Replacement of a hexavalent W^{VI} center in the $W_6O_{19}^{2-}$ dianion by a pentavalent V^V center generates the $V^V W_5^{VI} O_{19}^{3-}$ trianion (see Figure 6b). The $VW_5O_{19}^{3-}$ ^{17}O NMR spectrum shows two new resonances corresponding to the two new types of oxygen bonded to vanadium: OV terminal and OVW bridging oxygens. Note the unusual lineshapes which result from spin-spin coupling to the 99.8% abundant, spin 7/2 ^{51}V nucleus. Close comparison of the $W_6O_{19}^{2-}$ and $VW_5O_{19}^{3-}$ spectra reveals another key feature: upfield displacement of OW and OW_2 chemical shifts for the substituted trianion relative to the parent dianion. These presumably reflect diminished oxygen-tungsten π bonding resulting from increased negative charge on the anion. The analogy between O=C and O=W bonds is clear.

Another related effect is observed upon addition of a $Rh^{III}(C_5Me_5)^{2+}$ to the $(C_5H_5)Ti^{IV}(W_5^{VI}O_{18})^{3-}$ ion (see Figure 7) (29). Comparison of the $(C_5H_5)TiW_5O_{18}^{3-}$ and $(C_5H_5)Ti(W_5O_{18})Rh$-$(C_5Me_5)^-$ spectra is informative in two respects. First, the resonances for the three $ORhW_2$ oxygens which constitute the $Rh(C_5Me_5)^{2+}$ binding site are displaced upfield dramatically relative to their positions in the $CpTiW_5O_{18}^{3-}$ anion. This ca. 250 ppm displacement reflects substantial loss of W-O π bond character. The second point for comparison is the downfield shift of all OW, OTiW, and OW_2 resonances upon $Rh(C_5Me_5)^{2+}$ addition. This downfield shift upon negative charge reduction mirrors the trend of upfield shift upon increase of negative charge just discussed for the $W_6O_{19}^{2-}/VW_5O_{19}^{3-}$ case.

REFERENCES

(1) K. Kobayashi, T. Sugawara, and H. Iwamura, J. C. S. Chem. Commun., pp. 479-480 (1981).
(2) E. Block, A. B. Bazzi, J. B. Lambert, S. M. Wharry, K. K. Andersen, D. C. Dittmer, B. H. Patwardhan, and D. J. H. Smith, J. Org. Chem., 45, pp. 4807-4810 (1980).
(3) G. D. Mateescu, private communication.
(4) G. D. Mateescu and G. M. Benedikt, J. Am. Chem. Soc., 101, pp. 3959-3960 (1979).
(5) C. J. Besecker and W. G. Klemperer, manuscript in preparation.

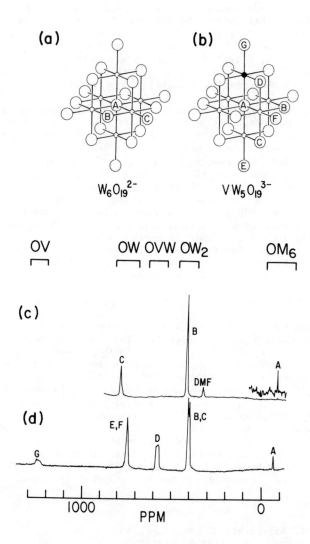

Figure 6. 13.5 MHz ^{17}O FT NMR spectra of $W_6O_{19}[(C_4H_9)_4N]_2$ and $VW_5O_{19}[(C_4H_9)_4N]_3$ are shown in (c) and (d), respectively. Note that the structures in (a) and (b) are idealized and show strictly octahedral metal coordination. For numerical data see Ref. 26. Reprinted with permission from Ref. 26, copyright 1978 American Chemical Society.

APPLICATION OF ^{17}O NMR SPECTROSCOPY TO STRUCTURAL PROBLEMS

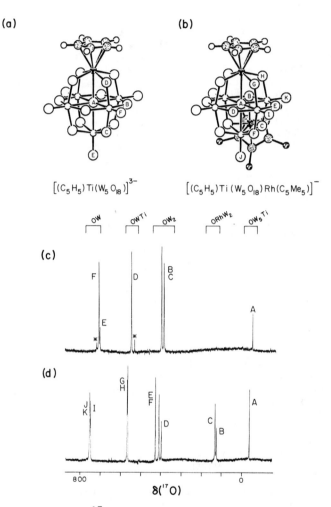

Figure 7. 33.9 MHz ^{17}O FT NMR spectra of $[(C_5H_5)Ti(W_5O_{18})]$-$[(C_4H_9)_4N]_3$ and $[(C_5H_5)Ti(W_5O_{18})Rh(C_5Me_5)]$-$[(C_4H_9)_4N]$ in CH_3CN are shown in (c) and (d), respectively (29). Temperature = 79°C, concentration = 0.013 ± 0.003 M, % ^{17}O = 6, number of accumulations = 200,000.

(6) W. G. Klemperer, Angew. Chem., Int. Ed. Engl., 17, pp. 246-254 (1978).
(7) J.-P. Kintzinger, NMR Basic Principles and Progress, 17, pp. 1-64 (1981).
(8) C. Rodger, N. Sheppard, C. McFarlane, and W. McFarlane, "NMR and the Periodic Table," R. K. Harris and B. E. Mann, Eds., Academic Press, 1978, pp. 383-419.
(9) T. Sugawara, Y. Kawada, M. Katoh, and H. Iwamura, Bull. Chem. Soc. Japan, 52, pp. 3391-3396 (1979).
(10) M.-T. Beraldin, E. Vauthier, and S. Fliszar, Can. J. Chem., 60, pp. 106-110 (1981).
(11) C. Delseth and J.-P. Kintzinger, Helv. Chim. Acta, 61, pp. 1327-1334 (1978).
(12) F. W. Wehrli and T. Wirthlin, "Interpretation of Carbon-13 NMR Spectra," Heyden & Son, 1978, pp. 35-48.
(13) G. A. Olah, A. L. Beiner, and G. K. S. Prakash, J. Am. Chem. Soc., 104, pp. 2373-2376 (1982).
(14) H. M. Schwartz, M. MacCoss, and S. S. Danyluk, Tetrahedron Lett., 21, pp. 3837-3840 (1980).
(15) H. M. Schwartz, private communication.
(16) W. Nakanishi, T. Jo, K. Miura, Y. Ikeda, T. Sugawara, Y. Kawada, and H. Iwamura, Chem. Lett., pp. 387-390 (1981).
(17) J. P. Hickey, I. M. Baibich, I. S. Butler, and L. J. Todd, Spectrosc. Lett., 11, pp. 671-680 (1978).
(18) S. Aime, L. Milone, D. Osella, G. E. Hawkes, and E. W. Randall, J. Organometal. Chem., 178, pp. 171-175 (1979).
(19) J. P. Hickey, J. R. Wilkinson, and L. J. Todd, J. Organometal. Chem., 179, pp. 159-168 (1979).
(20) S. Aime, D. Osella, L. Milone, G. E. Hawkes, and E. W. Randall, J. Am. Chem. Soc., 103, pp. 5920-5922 (1981).
(21) W. Biffar, H. Nöth, H. Pommerening, and B. Wrackmeyer, Chem. Ber., 113, pp. 333-341 (1980).
(22) J. A. Coderre, S. Mehdi, P. C. Demou, R. Weber, D. D. Traficante, and J. A. Gerlt, J. Am. Chem. Soc., 103, pp. 1870-1872 (1981).
(23) I. P. Gerothanassis and N. Sheppard, J. Magn. Reson., 46, pp. 423-439 (1982).
(24) J. A. Gerlt, P. C. Demou, and S. Mehdi, J. Am. Chem. Soc., 104, pp. 2848-2856 (1982).
(25) E. Heath and O. Howarth, J. C. S. Dalton, pp. 1105-1110 (1981).
(26) W. Klemperer and W. Shum, J. Am. Chem. Soc., 100, pp. 4891-4893 (1978).
(27) W. G. Klemperer and W. Shum, J. Am. Chem. Soc., 99, p. 3544 (1977).
(28) M. A. Freeman, F. A. Schultz, and C. N. Reilley, Inorg. Chem., 21, pp. 567-576 (1982).
(29) L. C. Francesconi and W. G. Klemperer, manuscript in preparation.

CHAPTER 12

THE ALKALI METALS

Pierre Laszlo

Institut de Chimie Organique et de Biochimie
Université de Liège au Sart-Tilman
4000 Liège, Belgium

ABSTRACT

A review of nuclear magnetic properties of alkali metal nuclei is followed by a brief survey of solvation of alkali metal cations, their hydration especially. The chemistry of alkali metal anions is then evoked. Chemical shifts and relaxation rates will be described with emphasis on the predominant factors contributing to these observables. A brief history of alkali metal NMR will be followed by selected applications to ion pairing phenomena, inclusion complexes, preferential solvation, the chelate effect, and polyelectrolytes.

NUCLEAR PROPERTIES

The nuclides to be considered in this chapter are ^{7}Li, ^{23}Na, ^{39}K, ^{87}Rb, and ^{133}Cs. These are all stable isotopes with <u>odd</u> atomic numbers and <u>even</u> numbers of protons: hence, one should expect natural abundance to be high.

Indeed, like other examples of this rule already familiar to NMR spectroscopists--such as ^{19}F, ^{31}P, and ^{59}Co--sodium-23 and cesium-133 are graced with 100% natural abundance. The other alkali metal nuclei all have a predominant odd/even isotope (1):

lithium	^{6}Li = 7.42%	potassium	^{41}K =	6.88%
	^{7}Li = 92.58%	rubidium	^{85}Rb =	72.15%
potassium	^{39}K = 93.1%		^{87}Rb =	27.85%
	^{40}K = 0.012%			

Spin Quantum Number

Table 1 lists the spin quantum numbers I for all these various nuclides. Only lithium-6 has an integer spin, all the others being half-integers; a majority of these nuclei have spins of 3/2.

Table 1. Spin Quantum Numbers I for the Alkali Metal Nuclei

Nucleus	$I/\hbar$	Nucleus	$I/\hbar$
^{6}Li	1	^{40}K	4
^{7}Li	3/2	^{41}K	3/2
^{23}Na	3/2	^{85}Rb	5/2
^{39}K	3/2	^{87}Rb	3/2
		^{133}Cs	7/2

Gyromagnetic Ratio

We list below (Table 2) the values of the gyromagnetic ratios γ (ratio of magnetic moment μ to angular momentum in $\hbar$) for the most important alkali metal nuclei. These gyromagnetic ratios span a large range from low and somewhat unwieldy values (because of the possibility of perturbations from acoustic resonances) for the potassium isotopes to quite high values for ^{87}Rb and ^{7}Li. Note also that for the three elements lithium, potassium, and rubidium represented by two magnetic nuclei, the gyromagnetic ratios differ quite substantially, thus permitting unambiguous tests between various pathways for nuclear relaxation.

Table 2. Gyromagnetic Ratios for the Alkali Metal Nuclei

Nucleus	$\gamma/10^7$ rad T^{-1}s^{-1}	Nucleus	$\gamma/10^7$ rad T^{-1}s^{-1}
^{6}Li	3.9366	^{40}K	1.5521
^{7}Li	10.3964	^{41}K	0.6851
^{23}Na	7.0761	^{85}Rb	2.5828
^{39}K	1.2483	^{87}Rb	8.7532
		^{133}Cs	3.5087

Receptivity

The NMR receptivities, expressed with reference to ^{13}C (= 1.00), are listed in Table 3. With the exception of potassium, the isotopes of which have poor to horrible receptivities, the alkali metal nuclei are endowed with quite good sensitivity. Hence, they are easily observed, provided that the quadrupolar interaction does not broaden their signal unduly.

Table 3. Receptivities of the Various Alkali Metal Nuclei

Nucleus	Receptivity/^{13}C	Nucleus	Receptivity/^{13}C
^6Li	3.58	^{40}K	3.51×10^{-3}
^7Li	1.54×10^3	^{41}K	3.28×10^{-2}
^{23}Na	525	^{85}Rb	43
^{39}K	2.69	^{87}Rb	277
		^{133}Cs	269

Quadrupole Moments

The quadrupole moments Q of the alkali metal nuclei are gathered in Table 4, together with the corresponding IQs. Two of these nuclei, ^6Li and ^{133}Cs, have quadrupole moments Q so small as to make them almost negligible; in many ways, these two nuclei are "honorary" spin 1/2 particles, and quadrupolar effects contribute but little to their properties.

Table 4. Quadrupole Moments and IQ Values ($IQ = 10^3 Q^2 (2I+3)/I^2(2I-1)$) for the Alkali Metal Nuclei

Nucleus	$Q/10^{-28}m^2$	IQ	Nucleus	$Q/10^{-28}m^2$	IQ
^6Li	-8×10^{-4}	3.2×10^{-3}	^{40}K	-7×10^{-1}	48.1
^7Li	-4.5×10^{-2}	2.7	^{41}K	6.7×10^{-2}	6.0
^{23}Na	0.12	19.2	^{85}Rb	0.25	20
^{39}K	5.5×10^{-2}	40.3	^{87}Rb	0.12	19.2
			^{133}Cs	-3×10^{-3}	12×10^{-4}

The same conclusion follows from consideration of the IQ values (Table 4); in the relaxation equations, relaxation rates are governed by the magnitude of the product Q^2 by the $(2I + 3)/I^2(2I-1)$ term. <u>All</u> the other alkali metal nuclei have moderate

IQ values, which makes them sensitive, but not unduly sensitive, to the presence of electrostatic field gradients (efg). Notice that ^{23}Na and ^{87}Rb could be termed <u>isoquads</u> since they share accidentally identical (or nearly identical) Q and IQ values.

Sternheimer Antishielding Factors

In practice, one does not examine naked nuclei, but nuclei in atoms. Because of the intervening electrons, the nuclei are submitted usually to a magnified efg (an $\ell = 2$ perturbation). This screening due to the atomic electrons leads to an antishielding effect which was first calculated by Sternheimer (2-5). Since the original work of Sternheimer (2-5) ca. 30 years of work, in numerous papers, have been devoted to this topic. Fortunately, a recent review is available (6). The Sternheimer antishielding factors γ_∞ have been calculated recently by a fully self-consistent Hartree-Fock treatment for the alkali metal cations, and we shall quote these results (7) (Table 5).

Table 5. Sternheimer Antishielding Factors for Alkali Metal Ions with the Rare Gas Electronic Configurations (7)

Ion	γ_∞
Li$^+$	0.249
Na$^+$	-5.26
K$^+$	-19.96
Rb$^+$	-47.66
Cs$^+$	-95.16

A Note of Caution

The example of ^{23}Na illustrates the problem in locating in the literature accurate values of the quadrupole moment Q and of the Sternheimer antishielding factor γ_∞. After undertaking a critical survey, I feel that values of Q close to 0.15 barn (0.15 x 10^{-28} m^2) are erroneous. Taking the experimental values obtained for the atom in the 3p or 4p state (13,14) and correcting for the antishielding contribution yields values of Q = 0.10 x 10^{-28} m^2 (15), in excellent accord with results from calculation (16).

THE ALKALI METALS

SOLVATION OF ALKALI METAL CATIONS

Geometric Considerations

The number of solvent molecules to be packed around and next to an ion of radius r, as a first shell, increases with the size of the ion. For instance (and we should not worry at this stage about the details of the method, nor about the accuracy of these numbers), Fabricand used an NMR method in 1964 (17) to determine the hydration numbers n of the alkali metal cations, listed below with their ionic radii r (Table 6). Evidently, the hydration number n, i.e., the number of nearest neighbor (coordinating) solvent molecules goes up as r^2, in proportion to the size of the outer surface of the ion. Pauling was the first, in the 1930s, to draw attention to this factor.

Table 6. Correspondence between Ionic Radii and Hydration Numbers for the Alkali Metal Cations, According to Fabricand (17)

Ion	Radius r, Å	Hydration number n
Li^+	0.59	1
Na^+	0.99	3.6 ± 1
K^+	1.38	6
Rb^+	1.49	9 ± 2
Cs^+	1.70	14.6 ± 2

Gas-Phase Solvation

Not only can the first shell be approached from packing considerations, as above, it can be visualized also from the changes in thermodynamics upon successive attachment of 0,1,2,...,n solvent molecules to a given ion. For instance, when Davidson and Kebarle (18) determined with their mass spectrometric method the van't Hoff plots of equilibrium constants for the reactions of Na^+ with 0-5 acetonitrile molecules, they found the results given in Table 7. Not only is there a large decrease of stability for the solvate having five acetonitrile molecules ($\Delta H° = 14.9 \to \Delta H° = 12.7$) but also "the fifth molecule going into the first shell imposes constraints on the freedom of all molecules. This is reflected in the particularly unfavorable entropy change. For example, for sodium $\Delta S°_{4,5} = -41.2$ e.u. is much larger than $\Delta S°_{3,4} = -27.9$ e.u." (18).

Table 7. Enthalpy and Entropy Changes for Gas-Phase Solvation of Na^+ by 2, 3, 4, or 5 Molecules of CH_3CN

Reaction	$\Delta H°$/kcal mol^{-1}	$\Delta S°$/e.u.
1,2	24.4 ± 0.3	22.7 + 0.5
2,3	20.6 ± 0.5	27.5 ± 0.9
3,4	14.9 ± 0.2	27.9 ± 0.6
4,5	12.7 ± 0.2	41.2 ± 0.7

For alkali metal cations--with the one exception of the Li^+ ion, whose solvation is a mix between electrostatic and covalent interactions--classical electrostatic calculations can account quite well for the observed enthalpies of gas-phase solvation. For instance, hydration can be split up into energy contributions from ion-dipole (E_{dip}), induced dipole (E_{ind}), dispersion (E_{dis}), and repulsive (E_{rep}) contributions (18) (Table 8). Clearly, the dominant term is the Coulombic attraction between the cation and the water dipoles, which decreases in magnitude as the ion gets bigger.

Table 8. Calculated and Experimental Solvation Energies for Some of the Alkali Metal Cations (kcal mol^{-1})

Ion M^+	Oxygen-to-ion distance R,Å	$-E_{dip}$	$-E_{ind}$	$-E_{dis}$	$-E_{rep}$	$-E_t$	$-\Delta H_{exp}$
Na^+	2.17	24.73	6.39	1.40	8.51	24.00	24.0
K^+	2.59	17.79	3.36	1.72	5.95	16.92	16.9
Rb^+	2.75	15.89	2.70	2.39	5.90	15.08	15.9
Cs^+	2.84	14.95	2.40	2.96	5.98	14.46	13.7

Indeed, the enthalpy of solvation decreases regularly from Li^+ to Cs^+ for each successive solvent molecule attached (19), as illustrated in Table 9 for water molecules.

If one turns to the ranking of solvent molecules in their solvating abilities towards a given ion (Table 10), dipolar solvents appear as remarkably similar to one another. Noteworthy is the absence of a chelate effect in the gas phase; addition to K^+ of two methylamine molecules is estimated (20) to release some 35 to 36 kcal mol^{-1}, much more than the observed $-\Delta H°_{0,1}$ = 25.7 kcal mol^{-1} for the bidentate ethylenediamine molecule

(Table 10). The explanation (20) is that the two unidentate ligands can minimize their dipolar repulsion by going to opposite ends of the potassium ion surface. Such a relief of the dipolar repulsion is forbidden to the bidentate ligand. In addition, when the ethylenediamine attaches its first ligator atom (N_1) to

Table 9. Hydration Enthalpies for the Alkali Metal Cations

Ion M^+ $\Delta H°$/kcal mol^{-1}:	0,1	1,2	2,3	3,4	4,5	5,6
Li^+	34	26	21	16	14	12
Na^+	24	20	16	14	12	11
K^+	16.9	16	13	12	11	10
Rb^+	16	14	12	11	10	-
Cs^+	14	12	11	10.6	-	-

Table 10. Enthalpies and Entropies (18,21,22) for the Reaction $M^+ + S \rightleftharpoons M^+S$

Solvent S	$\Delta H°_{0,1}$/kcal mol^{-1}	$\Delta S°_{0,1}$/e.u.
ETHERS		
H_2O	16.9	19.9
CH_3OCH_3	22.2	26.8
$C_2H_5OC_2H_5$	22.3	24.7
$(CH_3OCH_2)_2$	30.8	26.8
AMINES		
NH_3	17.9	20.0
CH_3NH_2	19.1	21.8
$(CH_3)_2NH$	19.5	21.4
$(CH_3)_3N$	20.0	23.4
$CH_3CH_2NH_2$	21.8	25.5
$(CH_2NH_2)_2$	25.7	22.3
$C_6H_5NH_2$	22.8	23.7
C_5H_5N	20.7	18.6
MISCELLANEOUS		
C_6H_6	18.3	22.4
CH_3CN	24.4	21.5

the cation, it undergoes a considerable internal polarization which draws electronic charge away from the second nitrogen (N_2), and makes it relatively electron deficient. As a consequence, the enthalpy change with the two unidentate ligands is so much larger that it overcomes the unfavorable entropy change (actually, for some unknown reason, the entropy loss with bidentate ligands (Table 10) turns out to be surprisingly close to that for the corresponding unidentate ligands (20)).

Quantum Calculations

The results of calculations of ion-molecule binding energies are made very reliable by fortunate cancellation of the electron correlation errors (23-24). The influence of Li^+, Na^+, and NH_4^+ ions on hydrogen bonding of near and distant water molecules has been studied at the CNDO/2 level (25). Clementi et al. (23) showed that the main contribution to the heat of hydration with a single water molecule for Li^+, Na^+, K^+ comes from Hartree-Fock binding, varying from about 35 kcal mol^{-1} (Li^+-H_2O), to 24 kcal mol^{-1} (Na^+-H_2O), and to 17 kcal mol^{-1} (K^+-H_2O). Ab initio SCF calculations on the Li^+-H_2O and Li^+-H_2CO systems (26) showed modest amounts of charge transfer and complicated electron density rearrangements. Pseudopotential calculations on Li^+, Na^+, and K^+ complexes with H_2O and H_2CO were found to provide reasonably correct description of binding energies and intermolecular distances (27). Ab initio calculations using an extensive basis set were performed on Li^+-NH_3 and Li^+-H_2O, indicating that, if the Li^+-base bond is basically electrostatic, simple models using charge-dipole and charge-polarizability interactions are not adequate; the Li^+ affinity is due primarily to an attractive electrostatic potential and a repulsive interaction resulting from the Pauli principle (28). Using the self-consistent field Hartree-Fock approximation, in the so-called electric-field-variant with Gaussian-type orbitals, the Sadlejs could compute the IR and Raman vibrational intensity changes due to the Li^+-H_2O interaction (29). Ab initio SCF calculations with a "double zeta" basis were performed of the K^+ affinities of a series of bases (30); dispersive interactions between the rather polarizable potassium cation and its ligands may explain that the relative K^+ affinities of methylated amines run parallel to their H^+ affinities, but not their Li^+ affinities:

$H^+, K^+ : NH_3 > CH_3NH_2 > (CH_3)_2NH > (CH_3)_3N$

$Li^+ : (CH_3)_2NH > (CH_3)_3N > CH_3NH_2 > NH_3$.

Ab initio SCF calculations served to determine (31) the relative Li^+ affinities of the carbonyl ligands R_2CO: (R) $NH_2 > CH_3 > OH > H > F$. Similar calculations on the H^+ and Li^+ ions interacting with H_2CO and OH^- showed predominant electrostatic Li-O interactions, although the Li-O interaction in LiOH has considerable covalent character (32). The merit of the MESQUAC (Mixed Electrostatic Quantum Chemical) MO method pioneered by Rode's group (see for instance (33)) is to include also the second and third hydration shells. Ab initio SCF-LCAO-MO calculations (for instance in (34)) have served to scan the potential energy hypersurface for alkali metal ions in the field of ligand molecules (ethers, amides, etc.) and to define transferable atom pair potentials.

Selected Experimental Studies

We list in Table 11 the Gibbs free energies for hydration (35) $\Delta G°$ and hydration numbers n for the alkali metal cations (the values for n are a combination of experimental determinations (17,36,37) and of personal choice between conflicting results!). Just like the gas-phase value (Table 8, for instance) the $\Delta G°$ values vary regularly with the ionic radius, the smallest ions being more strongly hydrated. Lithium ion owes to its very small radius an unusually small coordination sphere, consisting of only four water molecules in the primary hydration shell (37).

Table 11. Gibbs Free Enthalpies of Hydration and Mean Hydration Numbers for the Alkali Metal Cations

Cation	$-\Delta G°$/kcal mol^{-1}	n
Li$^+$	122.1	4
Na$^+$	98.4	6
K$^+$	80.6	6
Rb$^+$	75.5	9
Cs$^+$	67.8	12-15

A related matter is the electrostriction suffered by these water molecules in the first solvation sphere. One of the ways to approach it is through determination of the partial molal volumes of the ions (with various assumptions we will not go into here). These ionic partial molal volumes $\bar{V}_i°$ can be split, using the semi-empirical Hepler equation (eq. 1), into a geometric

$$\bar{V}_i° = Ar_i^3 - B/r_i \qquad (1)$$

contribution Ar_i^3 and the electrostriction contribution $-B/r_i$, in which r_i is the Pauling ionic radius. For alkali metal cations in water and in organic solvents, the A and B terms are given (37) in Table 12. Electrostriction is thus quite strong in all these solvents. With the solvents whose molecular sizes markedly exceed that of water (ethanol, DMF, DMSO), all alkali metal cations are probably surrounded by four solvent molecules in the first solvation sphere, and the electrostriction will decrease as the ionic radius increases, as indicated by the Hepler equation.

Table 12.

Solvent	A	B
Water	4.7	10 ± 2
Methanol	3.5	17 ± 2
Ethanol	3.3	12 ± 2
Dimethylformamide	2.5	18 ± 4
Dimethyl sulfoxide	3.6	12 ± 3

What about restriction in the motion of coordinated solvent molecules? A distinction has been made on this basis, for the alkali metal cations, between structure-making and structure-breaking ions. Samoilov (38) determined the change in the activation energy (ΔE) for self-diffusion in water (E) in the presence of an ion. The exchange rate of water from the coordination sphere of alkali ions is given by Frenkel's equation (eq. 2), in which τ_0 is the time characteristic of libration of a

$$\tau_i = \tau_0 \exp((E + \Delta E)/RT) \qquad (2)$$

water molecule at a given equilibrium site. For the smaller ions (Li^+, Na^+: structure-makers) in water solution, ΔE is positive. For the larger ions (K^+, Rb^+, and Cs^+: structure-breakers), also in water solution, ΔE is negative. This conclusion was confirmed by Fabricand et al. (39) (Table 13). There has been some controversy as to whether K^+ is structure-forming or structure-breaking in water; it is clearly a borderline case. For instance, Arakawa et al. (41), using a scaled particle theory, have ascribed an excess entropy of hydration δS to the ion-water interaction (Table 14); from these data, it would appear that K^+ is a very weak structure former. In methanol solution, the ions Li^+, Na^+, and K^+ are structure forming (40); the ratios τ^+/τ_0 are 7.9, 4.9 and 3.8, respectively, assuming tetracoordination (40).

Table 13. Ratio of Correlation Times for Water Near an Ion And Pure Water (τ^+/τ_0) and Activation Energy for Bound → Free Water Exchange (ΔE) (39)

Cation	τ^+/τ_0	ΔE/cal mol^{-1}
Li^+	8.29	1253 ± 200
Na^+	1.89	376 ± 150
K^+	0.94	−40
Rb^+	0.85	−95 ± 35
Cs^+	0.87	−81 ± 35

Table 14. Entropies of Hydration ($\Delta S°$) and the Excess Term due to Ion–Water Interaction (δS) for the Alkali Metal Ions (41)

Cation	$-\Delta S°$/e.u.	δS/e.u.
Li^+	33.7	−23.9
Na^+	26.2	−13.3
K^+	17.7	−1.3
Rb^+	14.8	2.9
Cs^+	14.1	6.0

Recently, neutron diffraction has been applied to the problem of ionic hydration and solvation. For instance, the Li^+ ion has hydration numbers of 3.3 ± 0.5 and 5.5 ± 0.3, respectively, for 9.95 and 3.57 molal solutions of LiCl in D_2O (42). The orientation of the water molecules at 3.57 molal in the coordination sphere is intermediate between the dipolar (a $\phi = 0°$ angle between the HOH plane and the Li–O axis) and the "lone pair" (with a $\phi = 55°$ angle) configuration (42).

By surveying data for salt hydrates, Friedman and Lewis (43) had been led to conclude, for the rare earth series, that water molecules next to a univalent cation are tetrahedral, i.e., coordinated to the metal ion, accepting a hydrogen bond from another water molecule, and donating two hydrogen bonds. Hence, for the alkali metal cations, it would appear that $Li^+(H_2O)_4$ has a mix of tetrahedral and "trigonal" water molecules, while the octahedral solvates $Na^+(H_2O)_6$ and $K^+(H_2O)_6$ have predominantly trigonal water molecules bound to the metal.

Molecular Dynamics and Monte Carlo Calculations

Hydration numbers for the alkali metal cations have been obtained from molecular dynamics simulations. Rao and Berne (44) have studied the interaction between a generalized spherical particle with 0, +1, -1, or +2 electronic charge and 215 H_2O molecules, pictured as consisting of four point charges placed within a single Lennard-Jones sphere centered at the oxygen atom. They found a hydration number of 6 for univalent cations, with the water molecules turning their electrical dipoles away from the cation. Heinzinger et al. have reviewed the evidence from molecular dynamics and from experiment and concluded that the hydration number is 4 (Li^+), 6.6-7.3 (Na^+, depending upon the salt concentration), 4-6 (K^+), and 6-12 (Cs^+) (45). Recently, the Heinzinger group at Mainz has focused especially on the Li^+ ion; their molecular dynamics studies have shown an octahedral first shell of six H_2O molecules, with 12 H_2O molecules in the second shell; the oxygen lone pair is directed towards the ion, deviating only very little (~6°) from linearity (46-49). Of the water molecules, 52% remain in the first hydration sphere for at least 10 ps, and another 31% during a minimum of 9 ps (48). Molecular dynamics also indicate, more generally for the alkali metal cations Li^+, Na^+, and Cs^+, that H_2O molecules in the first shell orient a lone pair towards the cation (50).

Clementi has been answering similar questions using Monte Carlo techniques. For instance, a cluster of 200 H_2O molecules around a Li^+F^- pair has been studied. Lithium hydration numbers are 4, 6, and 6 for anion-cation distances of 2, 6, and 10 Å, respectively, at 298 K (51). The same group has studied water solvation of Na^+-B-DNA at 300 K. The interesting findings include a stretched COO^- ··· Na^+ ion pair, by about 1 Å, allowing development of a more complete solvation shell both for the COO^- group and the Na^+ ion which, however, remains bound to the COO^- group, with no interstitial water molecules in between (52).

ALKALI METAL ANIONS

Alkali metal anions have been known to exist in the gas phase for over 30 years, and they have been proposed as major contributing species in metal-ammonia solutions since 1965 (53,54). By using cryptand C222, Dye and his co-workers were able in 1974 to isolate and to characterize fully--including an X-ray structure--a crystalline salt of the sodium anion Na^-, in which the counterion is the sodium cation Na^+ spirited away within the cavity of the cryptand (55,56). The Dye group has since prepared analogous crystalline solids for all the other alkali "metals" except lithium, and mixed systems in which the cryptated sodium cation coexists with the potassium, the rubidium, or the cesium anion.

The interest here is on the NMR properties: solution spectra of Na^+ C222, Na^- show separate ^{23}Na peaks for the sodium anion and the sodium cation (57) (Table 15). Noteworthy is the absence of a solvent shift for the cation resonance, encapsulated within the cryptand cavity and thus shielded from contact with solvent molecules. However, the real interest is the magnitude of the shift undergone by the anionic resonance, very close to the calculated diamagnetic shift of -2.6 for Na^- compared with Na(g). By reference, the paramagnetic shifts for Na^+ in water, in methylamine, and as the cryptate complex are 60.5, 72.2, and 50.4 ppm, respectively. The absence of any appreciable paramagnetic shift for Na^- in solution proves its structure; the solvent molecules are prevented from interacting with 2p orbitals by the presence of the outer spin-paired 2s electrons. The sodium anion in solution is thus truly "gas-like."

Table 15. Sodium-23 Chemical Shifts for the Cryptated Sodium Cation and for the Sodium Anion

Solvent $\delta(ppm):^a$	Na^+ C222	Na^-
Methylamine	+49.5	-1.4
Ethylamine	+49.5	-1.6
Tetrahydrofuran	+49.5	-2.3

^{a}With reference to Na(g).

More recently, Edwards et al. (58) have reported observation of the ^{23}Na resonance for Na^- simply by dissolving sodium metal in hexamethylphosphorictriamide (HMPT): they could not observe the companion Na^+ resonance, possibly due to the broadening undergone in the paramagnetic solution. They observed the Na^- signal some 62 ppm to high field of the aqueous NaCl reference, i.e., at a position extremely close to that observed in methylamine (-62.6 ppm), ethylamine (-62.8 ppm), and to that calculated for the isolated species in the gas phase (-63.8 ppm). As befits an unsolvated centrosymmetric entity, the linewidth of 10 Hz is very small. By contrast, the sodium cation Na^+, obtained by dissolving NaCl in HMPT under the same conditions, shows a resonance at +3.9 ppm, with a 30 Hz linewidth, suggesting that the bulky HMPT molecules cannot achieve fully symmetrical (T_d) tetracoordination of the sodium cation.

The ^{87}Rb resonances for Rb$^-$ are found at δ = +26.2 (ethylamine) and δ = 14.4 (tetrahydrofuran), versus δ = +213 (water) for Rb$^+$. The ^{133}Cs resonance for Cs$^-$ is at δ = +52.3 (tetrahydrofuran); that for Cs$^+$ (water) is at δ = +348. These values are all expressed with respect to the gaseous alkali metal atom (57).

THE CHEMICAL SHIFT FOR THE ALKALI NUCLIDES

While for the lithium nuclides, dia- and paramagnetic contributions to the chemical shift are of the same order of magnitude, the paramagnetic term is predominant for the other alkali metal nuclei.

Correlation of ^{23}Na Chemical Shifts and Linewidths

Theoretical work, dating back to 1959 (59), and to which Deverell's name is chiefly associated (60), is intuitively appealing; it finds the origin of <u>both</u> the paramagnetic part of the chemical shift and of the electrostatic field gradient for a sodium cation in deviations from a fully symmetrical T_d or O_h solvate. These deviations could be due to collisions with solvent molecules (Deverell), to vacancies in the Na$^+$ coordination shell (Valiev), or to asymmetric Na$^+$ coordination, either instantaneous or time-averaged (61).

The key feature of this "Deverell approach" is proportionality of the quadrupolar coupling constant $\chi = e^2qQ/\hbar$ to the paramagnetic part of the shielding constant σ^p. Hence, in the extreme narrowing limit, the square root of linewidths (reduced to unit viscosity) should be linear with respect to the <u>paramagnetic</u> part of the shielding constant, all other factors being assumed constant (eq. 3). This equation was indeed found to pre-

$$(\Delta v^*_{1/2})^{1/2} = C \, \Delta E \left(\frac{\tau_c}{\eta}\right)^{1/2} (\sigma - \sigma^d) \tag{3}$$

dict the experimental behavior for a number of ^{23}Na$^+$ solvates, with THF (t) or amines (a) coordinated to the cation (61). From the nonzero focal point of the linear regressions observed with the four different solvates (Na$^+$)$_{t_3a}$, (Na$^+$)$_{t_2a_2}$, (Na$^+$)$_{ta_3}$, and (Na$^+$)$_{a_4}$, it was possible to determine experimentally that the diamagnetic term σ^d for Na$^+$ ions dissolved in these binary mixtures of THF and amines differs from that of the hexaquo ion by about -11 ppm (61) (this may reflect the change from octahedral to tetrahedral coordination).

Shielding of Hydrated Alkali Metal Cations

Comparison of the magnetic moments for alkali metal cations purely surrounded by water molecules with the results of atomic beam magnetic resonance experiments yields the shielding of M^+ nuclei by water molecules around the ions. This shielding, which for instance for K^+ ions is worth -0.1052×10^{-3}, is a nearly linear function of the atomic number (62).

The shielding constant for Cs^+ ions, by reference to the free Cs atom, is -0.344×10^{-3} in water solution (63): about three times as large as that for hydrated K^+ ions (62) but only half as large as that of ^{133}Cs in solid CsI, which has the largest cesium shielding (63).

Ranges of Chemical Shifts for the Alkali Metal Cations and Empirical Correlations

Erlich and Popov (64) have reported a very useful correlation of $^{23}Na^+$ chemical shifts with the Gutmann donor number of solvents. As the solvent becomes a better electron donor, the resonance moves downfield. The regression is quite good, especially if one removes from it the point for water, a legitimate deletion since this point corresponds to octahedral as compared with tetrahedral coordination in all the others. Likewise, with $^{39}K^+$ there is a definite correlation between the magnitude of the downfield shift and the donor number (Table 16) (65). By comparison, with $^{133}Cs^+$ there is no longer a nice correlation but only a tendency towards downfield shifts as the donor number of the solvent increases (Table 17) (66).

To give an idea of the various ranges, the infinite dilution chemical shifts vary (i) for $^7Li^+$, from +2.80 ppm in acetonitrile to -2.54 ppm in pyridine (64,67), (ii) for $^{23}Na^+$, from +16.1 ppm in acetonitrile to -2.58 ppm in hexamethylphosphorictriamide (64), (iii) for $^{39}K^+$, from +21.1 ppm in nitromethane to -23.62 ppm in ethylenediamine (65), and (iv) for ^{133}Cs, from +59.8 ppm in acetonitrile to -68.0 ppm in dimethyl sulfoxide (66). The difference between chemical shifts of the respective nuclei in nitromethane and dimethyl sulfoxide is ~15.5 ppm for ^{23}Na, ~29 ppm for ^{39}K, and ~130 ppm for ^{133}Cs.

RELAXATION PROPERTIES OF ALKALI METAL CATIONS

All of the alkali metal nuclides possess an electric quadrupole $\overline{Q}$. As indicated above, ^6Li and ^{133}Cs have exceptionally low quadrupole moments Q, and therefore relax also by mechanisms other than the coupling between the nuclear quadrupole and the efg at the nucleus. For the other alkali metal nuclei, quadru-

Table 16. Potassium-39 Infinite Dilution Chemical Shifts for K^+ and Donor Numbers of the Corresponding Solvents

Solvent	δ/ppm	Donor Number
Nitromethane	21.10	2.7
Formic acid	11.60	17.0
Propylene carbonate	11.48	15.0
Acetone	10.48	17.0
Methanol	10.05	25.7
Formamide	4.55	24.7
Dimethylformamide	2.77	26.6
Acetonitrile	0.41	14.1
Water	0.00	33.0
Pyridine	-0.82	33.1
Dimethyl sulfoxide	-7.77	29.8
Ethylenediamine	-23.62	55.0

Table 17. Cesium-133 Infinite Dilution Chemical Shifts for Cs^+ and Donor Numbers for Various Solvents

Solvent	δ/ppm	Donor Number
Nitromethane	59.8	2.7
Acetonitrile	-32.0	14.1
Propylene carbonate	35.2	15.1
Acetone	26.8	17.0
Formamide	27.9	24.7
Methanol	45.2	25.7
Dimethylformamide	0.5	26.6
Dimethyl sulfoxide	-68.0	29.8
Water	0.0	33.0
Pyridine	-29.4	33.1

polar relaxation is often the dominant, if not the exclusive relaxation process.

Quadrupolar Relaxation Under Extreme Narrowing Conditions

Let us start by considering the general expression for the rate of relaxation for a quadrupolar nucleus under extreme narrowing conditions (eq. 4). We have discussed already the

$$\frac{1}{T_1} = \frac{1}{T_2} = \frac{3}{40} \frac{2I+3}{I^2(2I-1)} \left(\frac{PeQ(1+\gamma_\infty)}{\hbar} \right)^2 \left(\frac{\partial^2 V}{\partial z^2} \right)^2 \left(1 + \frac{\eta^2}{3} \right)^2 \tau_c \quad (4)$$

impact of the product of the two terms $(2I+3)/I^2(2I-1)$ and e^2Q^2 -- the so-called IQ factor. We shall consider now separately the meaning of the other terms in the above equation, as applied to alkali metal <u>cations</u>, in solution. P is a polarization factor, which can be expressed as a function of the dielectric constant ε of the solvent: $P = (2\varepsilon + 3)/5\varepsilon$. This is an approximation, all the more so that it assumes that solvent molecules in the first shell around an ion can be treated as a macroscopic continuum for which a dielectric constant is a well defined entity. We have already covered the Sternheimer antishielding term $(1 + \gamma_\infty)$. The electrostatic field gradient at the nucleus (efg) $q = \partial^2 V/\partial z^2$ intervenes with its second power. The asymmetry parameter η describes for an axially symmetric case (a symmetric top) the deviations from spherical symmetry. The asymmetry factor η varies between 0 and 1, so that setting arbitrarily $\eta = 0$ when one is not in a position to evaluate this factor leads to a maximum error of 30%.

Often, the relaxation expression is simplified by lumping together all these terms into a quadrupolar coupling constant χ (eq. 5). Such a procedure is incorrect since, strictly

$$\frac{1}{T_1} = \frac{1}{T_2} = \frac{3}{40} \frac{2I+3}{I^2(2I-1)} \chi^2 \cdot \tau_c \quad (5)$$

speaking, $\chi = (e^2qQ/\hbar)$.

Since, in the extreme narrowing limit, the rate of quadrupolar relaxation $R_{1,2} = T_{1,2}^{-1}$ is proportional to the product $(\chi^2 \tau_c)$, it is impossible to determine separately from one observable $(R_{1,2})$ the two unknowns χ and τ_c.

One can get an idea of the magnitude of the quadrupolar coupling constant χ from solid state studies (Table 18). The very small values of the quadrupolar coupling constant χ for $^{133}Cs^+$ is in line with the tiny quadrupole moment of this nucleus. For the other nuclei listed in Table 18, values of the quadrupolar coupling constants <u>in a pure solvent</u> are probably somewhat smaller than the solid state values, because symmetrical solvation of the alkali metal cation should reduce the magnitude of the efg q. Hence, the values in Table 18 should be taken as upper limits. For instance, solution studies imply ^{23}Na quadrupolar coupling constants in the range 0.2-2 MHz (72).

Table 18. Magnitudes of the Quadrupolar Coupling Constants χ and of the Asymmetry Factors η for Alkali Metal Cations in the Solid State

Nuclear species	Salt	χ/MHz	η	Reference
^{7}Li$^+$	LiI 6 M (H$_2$O)[a]	0.035	–	68
^{23}Na$^+$	NaOH	3.73	0.4	69
	NaOH·H$_2$O	2.27	0.6	69
	NaOH·4H$_2$O	1.29	0.6	69
	NaCN	1.28	0.37	70
^{39}K$^+$	KNO$_3$	<0.7	–	71
^{87}Rb$^+$	RbNO$_3$	0.88[b]	0.17[b]	71
	RbNO$_3$	0.90[b]	0.48[b]	71
	RbNO$_3$	1.1[b]	0.91[b]	71
^{133}Cs	CsNO$_3$	<0.03	–	71
	Cs$_2$SO$_4$	0.038	–	63

[a]Supercooled solution.
[b]There are three crystallographically distinct sites.

Origin of the EFG for Alkali Metal Cations in Pure Solvents

We come now to what may apppear, at first sight, as a paradoxical situation. If ions such as sodium cations Na$^+$ relax exclusively by a quadrupolar mechanism, and if they are considered in a pure solvent in which they exist as octahedral (water) or tetrahedral (organic solvents) species, then by reason of symmetry (O_h or T_d) the efg vanishes at the nucleus, and the relaxation rate ought to be zero. Yet, it is nonzero, and finite linewidths of a few Hz are observed. What is the explanation?

A considerable body of literature has been devoted to this question. We have mentioned already the collision model of Deverell (60): an efg occurs at each ion-solvent molecule collision, with attendant loss of spherical symmetry of the ion. In the solvent exchange model of the Russian school (73), migration of a solvent molecule from the first solvation sphere into the bulk solution creates a vacancy; the efg arises from coordinative unsaturation. In the electrostatic model of Hertz, the efg at the nuclear site is caused by the electric point dipoles of the surrounding solvent molecules and the point charges of other ions present in solution (74-76).

Yet another two models, which hold considerable attraction to this reviewer, should be mentioned. Let me first quote from a 1976 article by Friedman and Lewis (43): "In a complex ion $M(H_2O)_6^{+n}$ or $M(H_2O)_4^{+n}$ which is symmetrical enough, there is no electric field gradient at the center. In such a complex all the water molecules are trigonal. Trigonal-tetrahedral fluctuations in the orientations of the water molecules break the symmetry and give rise to electric field gradients at the sites of the metal ions. One may speculate whether this is the detailed mechanism for the quadrupole-driven NMR relaxation in the case of ions such as $^{23}Na^+$ (aq.) and for the zero-field-splitting-driven EPR relaxation of ions such as Ni^{2+} (aq.). The rate expected for the trigonal-tetrahedral fluctuations is about right..." Friedman showed (77) that such a libration model provides an interpretation of the relaxation process of many spin greater than one-half metal ions (Table 19). The order of magnitude agreement between the calculated and the experimental values (Table 19) for $^{23}Na^+$, $^{39}K^+$, $^{85}Rb^+$, and even $^{133}Cs^+$ looks good, and it is tempting to conclude that the libration or water "wagging" model of Friedman (43,77) explains the nonzero linewidths observed for the quadrupolar alkali metal cations in aqueous solutions.

Table 19. Comparison of Experimental and Calculated Rates, According to the Water Libration Model, for Alkali Metal Cations in Aqueous Solutions (77)

Ions	I	T_1^{-1}: Calc.	Exp./(Hz)
$^7Li^+$	3/2	0.31	0.027
$^{23}Na^+$	3/2	23	16
$^{39}K^+$	3/2	53	24
$^{85}Rb^+$	5/2	680	420
$^{133}Cs^+$	7/2	0.14	0.075

Another model is provided by the key finding, from Monte Carlo simulations of Li^+ + n H_2O clusters (n = 6, 50, and 150) (78), that the first hydration shell has an average hydration number just above five, which implies a noncubic average arrangement of the water oxygens. The Monte Carlo results ruled out the earlier models: in the Deverell model, the efg originates in overlap between the wave functions of the ion and a colliding water molecule. In contrast, the Swedish workers found that the contributions of water molecules in the first shell to the efg are all about the same. The Hertz electrostatic model (73-76) gives a good fit with experimental data, but it contains many adjustable parameters. The Friedman libration model (43,77), when tested in a Monte Carlo simulation, provides a fluctuation which was about 1/10 of the value found in a simulation without any restrictions.

We have addressed similar questions and we have come to a similar conclusion (61). When we measured the transverse relaxation rate (or, equivalently, the linewidth) for the tetracoordinated solvates $(^{23}Na^+)_{4L}$ in a series of six unidentate amines L (aniline, pyridine, piperidine, pyrrolidine, 1-propylamine, and 2-propylamine), we found that, far from being constant (as would befit T_d symmetry), the value of R_2 depended on the nature of the four identical amine ligands L. This "experimental observation of nonzero and variable linewidth thus indicates a deviation from the ideal T_d symmetry. It could occur as a permanent distortion, i.e., with a lifetime significantly greater than τ_c: due to steric congestion in the coordination sphere, the four ligands twist themselves into a lower symmetry arrangement, just as tetraarylmethanes also adopt a conformation of lowest energy which is not the T_d conformation of greatest symmetry. As an alternative to such a permanent distortion, deviation from the ideal T_d symmetry could occur within a time scale much shorter than τ_c: it would suffice, for the vibrations of the four ligands (rotations and elongations) to be strongly anharmonic, in order to produce nonzero average" (61).

Weingärtner (68), by studying ^7Li relaxation in supercooled aqueous solutions of LiI down to ca. 170 K, has been able to test the various available models. His conclusion was that the collision model of Deverell, the coordinative unsaturation model of Mishustin and Kessler (73), and the water wagging model of Friedman (43,77) all fail to account for the experimental data. Only the electrostatic model of Hertz, he found, "gave an acceptable explanation for the observed dynamic behavior" (76). Weingartner, however, did not test an unsymmetrical solvate model, which the Monte Carlo studies of Engstrom and Jonsson (78) have made an almost unescapable fact to contend with henceforth.

Nonquadrupolar Relaxation Mechanisms for Alkali Metal Nuclei

Lithium-6. Wehrli (79) has investigated carefully the relaxation of natural abundance ^6Li in aqueous LiCl (3.9 M). At 25°C, 86% of the relaxation rate is dipolar ($^6Li-^1H$), falling off to 48% at 100°C. The nondipolar contribution to ^6Li relaxation exhibits a maximum at ca. 40°C, pointing to the presence of spin-rotation relaxation. Thus, by estimating upper limits for the quadrupolar, for the chemical shielding anisotropy (negligible), and for the dipole-dipole ($^6Li-^7Li$), Wehrli could provide the individual contributions (80) (Table 20). Wehrli also studied alkyllithium compounds, which are known to form oligomers in solution. There also quadrupolar relaxation is relatively inefficient, contributing 3, 17 and 15%, respectively, of the total relaxation rates for methyl-, butyl-, and phenyllithium at 28°C; versus 20, 35, and 17% dipole-dipole (^{6}Li-^{1}H) contributions, at the same temperature (80) (Table 21). Wehrli could write therefore that "^{6}Li possesses most virtues of a spin 1/2 nucleus" (79).

Table 20. Individual Contributions to the Experimental ^{6}Li Spin-Lattice Relaxation Rate in Naturally Abundant Aqueous ^{6}LiCl, at 40°C

Contribution	T_1^{-1}/Hz
Dipole-dipole (^{6}Li-^{1}H)	84
Spin-rotation	8
Dipole-dipole (^{6}Li-^{7}Li)	<1
Quadrupolar	<0.1
Unidentified	7

Table 21. Individual Contributions to the Experimental ^{6}Li Longitudinal Relaxation Rate in Methyllithium, Butyllithium, and Phenyllithium (80), at 28°C

Contribution	T_1^{-1}/Hz:	CH$_3$Li	n-C$_4$H$_9$Li	C$_6$H$_5$Li
Dipole-dipole (^{6}Li-^{1}H)		20	35	17
Quadrupolar		3	17	15
Dipole-dipole (^{6}Li-^{7}Li)		<6	48	68
Unidentified		<61		

Lithium-7. According to the data of Hertz for ^{7}Li relaxation in LiCl-H$_2$O, ca. 50% of the total relaxation rate is dipolar induced (79). For ^{7}LiI, also in aqueous solutions, the Weingartner study (68) gave a comparable figure, approximately 30% of the total relaxation rate being dipolar and 70% quadrupolar. Likewise, in alkali silicate, borate, and phosphate glasses, ^{7}Li longitudinal relaxation is governed by a mixture of dipolar and quadrupolar effects (81).

Sodium, potassium, and rubidium nuclides. Because of their much larger quadrupole moments Q (Table 4), ^{23}Na, ^{39}K, and ^{87}Rb—which are all spin 3/2 particles—will have much more efficient quadrupolar relaxation. It can be safely assumed thus that this is the predominant relaxation mechanism for $^{23}Na^+$, $^{39}K^+$, and $^{87}Rb^+$ in most cases.

<u>Cesium-133</u>. This nuclide, however, has very small Q and IQ values (Table 4). Nevertheless, Wehrli (80) found, both for CsCl 0.2 M in water and for CsI 0.1 M in dimethylformamide that the relaxation of the Cs^+ cation is exclusively quadrupolar.

Slow Modulation of Spin 3/2 Relaxation

A spin 3/2 nucleus (e.g., ^7Li, ^{23}Na, ^{39}K, ^{41}K, or ^{87}Rb) has the spin states +3/2, +1/2, -1/2, and -3/2. Hence, there are three single quantum coherences: +3/2 → +1/2, +1/2 → -1/2, and -1/2 → -3/2. In the absence of an efg at the nucleus, these three SQ transitions are degenerate. Presence of a nonvanishing efg lifts the degeneracy and, if there is local order, three separate lines can be seen. However, in isotropic liquids, these are time-averaged into a single resonance. Nevertheless, each relaxation mode, whether longitudinal or transverse, will consist of <u>two</u> distinct relaxation processes: a slow relaxation, corresponding to the +1/2 → -1/2 transition; and a fast relaxation, associated with the ±3/2 → ± 1/2 transitions.

Let me interject, in passing, that a methyl group with three equivalent spin 1/2 protons behaves as a spin 3/2 composite particle, and that the simplified description to follow is also completely applicable to a methyl group bound to a slowly reorienting macromolecules, a protein for example (82-83).

For instance, the longitudinal magnetization of the I = 3/2 nucleus is given by eq. 6 (the P_i terms are populations of the

$$\langle I_z \rangle = \frac{3}{2}(P_{3/2} - P_{1/2}) + \frac{4}{2}(P_{1/2} - P_{-1/2}) + \frac{3}{2}(P_{-1/2} - P_{3/2}) \tag{6}$$

various spin states). In this description (84), there is longitudinal magnetization associated with a <u>broad</u> component (eq. 7)

$$\langle I_z^b \rangle = 3/2(P_{3/2} - P_{1/2} + P_{-1/2} - P_{-3/2}) \tag{7}$$

and there is longitudinal magnetization associated with a <u>narrow</u> component (eq. 8).

$$\langle I_z^n \rangle = 2(P_{1/2} - P_{-1/2}) \tag{8}$$

A similar situation holds for the transverse magnetization which, under certain simplifying assumptions, consists of 60% of a broad component and 40% of a narrow component. Hence, the resonance consists of two overlapping lorentzians which differ both in their widths, and in the corresponding spectral densities (eqs. 9, 10). While the narrow component has a spectral density

$$T_{2b}^{-1} = \frac{\pi^2}{5} (\frac{e^2qQ}{\hbar})^2 [\tau_c + \frac{\tau_c}{1 + \omega^2\tau_c^2}] \qquad (9)$$

$$T_{2n}^{-1} = \frac{\pi^2}{5} (\frac{e^2qQ}{\hbar})^2 [\frac{\tau_c}{1 + 4\omega^2\tau_c^2} + \frac{\tau_c}{1 + \omega^2\tau_c^2}] \qquad (10)$$

at Larmor (J_1) and twice the Larmor frequencies (J_2), it differs from the broad component whose spectral density depends upon zero (J_0) and Larmor frequencies (J_1).

In many cases, when the dynamic shift (see below) is negligible, deconvolution of the observed nonlorentzian lineshape--which can be performed digitally, even using a small hand calculator (85)--will provide these two separate lorentzian components (it should be checked then that they are indeed isochronous, and that the relative intensities are 60:40). Then, since the above expressions for the transverse relaxation rates T_{2n}^{-1} and T_{2b}^{-1} have the same functional dependence upon the square of the quadrupolar coupling constant $\chi^2 = ((e^2qQ)/\hbar)^2$, their ratio will be, at a given (operating) Larmor frequency ω, a measure of the correlation time τ_c. In other words, these two observables T_{2n}^{-1} and T_{2b}^{-1} will allow separate determination of the two quantities χ^2--in fact of the product $p_B \cdot \chi^2$, where p_B is the mole fraction of nuclei in the bound, slowly modulated state--and of the correlation time τ_c in the bound state.

Furthermore, the whole procedure can be cross-checked for internal consistency by working with another spectrometer on the identical sample, at quite another frequency. For instance, for ^{23}Na, if one measures the ratio of the widths of the narrow component at 23.81 and 62.86 MHz, this ratio changes by ca. 700% when the correlation time τ_c increases from 10^{-10} to 10^{-8} s. In the transition range, i.e., for correlation times τ_c = 1-50 ns, these correlation times can be quite accurately determined from this experimental quantity (85).

One should be wary, however, of the possibility of second-order dynamic frequency effects, shifting apart these two components (84). Perhaps, the first example of such a dynamic shift for ^{23}Na has been reported recently (86): aqueous solutions of 40 mM sodium laurate, 28 mM lauric acid, and 100 mM NaCl display cleanly such an effect.

Nearly Exponential Quadrupolar Relaxation for Higher Spin

For nuclei with half-integer spins, the magnetization decays as a weighted sum of $I + 1/2$ exponentials: for $I = 3/2$, these two exponentials are the Fourier inverts of the two lorentzians, narrow and broad, with characteristic times T_{2n} and T_{2b}, as we have just described. For the $I = 3/2$ case, there exist analytical expressions, which have been given above in a simplified form. For higher spin numbers, expressions in closed form do not exist. Fortunately, there is a recent perturbational treatment by Halle and Wennerstrom (87), applicable to the $I = 5/2$ (e.g., ^{85}Rb) and to the $I = 7/2$ (e.g., ^{133}Cs) cases. The relaxation behavior is well described as a simple biexponential decay, to first order (eqs. 11,12).

$$R_{11}^{(1)} = \frac{3}{40} \left(\frac{eQ}{\hbar}\right)^2 \frac{2I+3}{I^2(2I-1)} (2J_1 + 8J_2) \qquad (11)$$

$$R_{21}^{(1)} = \frac{3}{40} \left(\frac{eQ}{\hbar}\right)^2 \frac{2I+3}{I^2(2I-1)} (3J_0 + 5J_1 + 2J_2) \qquad (12)$$

The same group, in Lund, has calculated very recently (88) the corresponding dynamic frequency shifts. The spectral densities $J(\omega)$ and $Q(\omega)$ are defined by eqs. 13 and 14, i.e., with

$$J(\omega) = \int_0^\infty \langle V_0(0) \cdot V_0(\tau) \rangle \cos \omega\tau \, d\tau \qquad (13)$$

$$Q(\omega) = \int_0^\infty \langle V_0(0) \cdot V_0(\tau) \rangle \sin \omega\tau \, d\tau \qquad (14)$$

$$J(\omega) \propto 0.3 \, V_{zz}^2 (\tau_c/(1 + \omega^2 \tau_c^2)) \text{ and}$$

$$Q(\omega) \propto 0.3 V_{zz}^2 (\omega\tau_c^2/(1 + \omega^2 \tau_c^2)).$$

Then, in the slow motion limit, where $J_0 \gg J_1, J_2$, the dynamic shift undergone by the narrow $+1/2 \to -1/2$ line is $(-1/25)(eQ/\hbar) [Q_1 - Q_2]$, and $-(10/588) (eQ/\hbar)[Q_1 - Q_2]$ for a spin 5/2 and for a spin 7/2 nucleus, respectively (88). Near the extreme narrowing range, where $J_0 > J_1 > J_2$ approximately, the dynamic shifts are $+(1/125) (eQ/\hbar)^2 [Q_1 - 2Q_2]$ and $(1/294) (eQ/\hbar)^2 [Q_1 - 2Q_2]$, again for spins 5/2 and 7/2, respectively (88).

HIGH RESOLUTION SPECTRA IN SOLIDS

With noninteger quadrupolar nuclei, there are (at least) two strategies for obtaining high resolution spectra in solids. I shall describe these briefly for the example of spin 3/2 nuclei.

A first possibility, exploited by Eric Oldfield's group at the University of Illinois (89,90), takes advantage of the narrow central $+1/2 \rightarrow -1/2$ transition: it is only broadened by dipolar, chemical shielding (or Knight shift) anisotropy (CSA), and by second-order quadrupolar effects. Thus, the $+1/2 \rightarrow -1/2$ line has often a width significantly smaller than the quadrupolar coupling constant. With quadrupolar coupling constants of up to 2-3 MHz (Table 18), the alkali metal nuclei are thus especially good cases for such high resolution studies, which are performed with high-speed sample spinning techniques, and at as high a magnetic field as is convenient. The second-order broadenings of the $(+1/2, -1/2)$ line are inversely proportional to the magnitude of B_0. However, CSA and second-order quadrupolar interactions also increase with field strength. The influence of the applied magnetic field B_0 can be gauged from the attendant reduction of the linewidth for a static sample: from ca. 20-25 kHz to 10-12 kHz upon increasing B_0 from 3.5 to 8.5 T. As for the spinning, the optimum angle depends on the case at hand: for ^{23}Na samples with asymmetry parameters $\eta = 0$, angles of about ~80° and ~40° were found to be the best (90). The highest spinning velocities (at least 2 kHz) should be used. In this manner, high resolution spectra have been obtained for a number of ^{23}Na salts: NaCl, NaBrO$_3$, Na$_2$MoO$_4$, and Na$_2$SO$_4$ (89).

A second possibility is to resort to multiple quantum NMR. Consider again a spin 3/2 case. The quadrupolar Hamiltonian is of the form $\hat{H}^{(1)} = 1/3\ \omega_Q[3\ I_z^2 - I(I+1)]$. Hence, the Zeeman levels are modified as follows: $m = \pm 3/2$ is raised by ω_Q and $m = \pm 1/2$ is lowered by ω_Q. Therefore, while the three single quantum coherences are given (at first order) by $(3/2, 1/2) = \omega_0 - \omega_Q$; $(1/2, -1/2) = \omega_0$; $(-1/2, -3/2) = \omega_0 + \omega_Q$, where ω_Q is the magnitude of the quadrupolar interaction ($e^2qQ/\hbar$), the triple quantum transition, which is devoid of the quadrupolar interaction at first order, leads to much narrower lines. The first triple quantum FT spectra of I = 3/2 nuclei have been published recently (91). The work was performed on an oriented single crystal of sodium ammonium tartrate tetrahydrate, using ^{23}Na NMR. In this orientation of the single crystal there are two magnetically nonequivalent sodiums. They have quadrupolar coupling constants of 470 and 387.5 kHz, leading to second-order shifts: $\omega^{(2)}(3/2) = -0.9$ and 1.3 kHz and $\omega^{(2)}(1/2) = -1.4$ and -3.4 kHz.

SELECTED APPLICATIONS OF ALKALI METAL NMR

Development of alkali metal NMR has proceeded at a very unequal pace, depending on the nucleus considered. If consideration is limited to chemical and biochemical applications, thus excluding more physical studies, lithium-7 NMR was dominant in the 1960's and sodium-23 NMR became predominant in the 1970's. As compared with these two nuclei, the others, potassium, rubidium, and cesium, have curried less favor, and have led only to a handful of applications.

Lithium-7

Structure and bonding in alkyllithium compounds have been investigated by using, among other observables, the presence of scalar couplings to the ^{7}Li nucleus. Brown and Ladd reported in 1964 the collapse of the ^{7}Li-^{1}H scalar coupling in ethyllithium when temperature of the toluene solution increases, due to slow intermolecular exchange and reported quadrupolar coupling constants $\chi \leq 0.57$ and 0.14 MHz for C_2H_5Li and $(CH_3)_3CLi$, under these conditions (92), respectively. By looking at the ^{7}Li NMR of polycrystalline samples, Lucken reported an upper limit of 16 kHz for the quadrupolar coupling constant of methyllithium, and $\chi = 98$ kHz ($\eta = 0.62$) for ethyllithium (93). The first report of the presence of ^{13}C-^{7}Li scalar coupling in an alkyllithium compound appeared in 1968: the CH_3Li tetramer displays a 14.5 Hz coupling in THF solution (94). This is a crucial observation, since the multiplet pattern of seven lines for ^{13}C enriched CH_3Li demonstrates the tetrameric structure, in which lithium atoms occupy the four vertices, and methyl groups the four faces. The

[structure diagram of methyllithium tetramer with Li atoms at vertices and CH$_3$ groups at faces]

four faces. The seven lines, with intensity ratios 0.05:0.31:0.76:1.00:0.76:0.31:0.05 arise in the following manner: the dominant species (singlet) is due to ^{7}Li nuclei having only three ^{12}C nearest neighbors; the next species, abundance-wise, corresponds to a (^{13}C, ^{12}C, ^{12}C) configuration (doublet); and so on (95). The methyllithium tetramer, however, while showing ^{13}C-^{7}Li coupling, lacks (96) ^{6}Li-^{7}Li coupling, despite a close Li-Li distance of 2.68 Å (97).

The absence of $^{13}C-^{7}Li$ scalar coupling, for butyllithium in hydrocarbon solvents is proof of rapid <u>inter</u>aggregate exchange for this hexameric species (98).

Interaggregate exchange can also be explored from the ^{7}Li chemical shifts. Let us consider an equimolar binary mixture of two <u>tetrameric</u> alkyllithium compounds LiR and LiR', with significantly different ^{7}Li chemical shifts. With rapid intermolecular exchange, only a single, weighted-average resonance will be seen. Conversely, in the limit of infinitely slow chemical exchange, one should see distinct resonances for the following species (the relative amounts are given in parentheses): Li_4R_4 (1), Li_4R_3R' (4), $Li_4R_2R_2'$ (6), Li_4RR_3' (4), Li_4R_4' (1). However, let us recall that each lithium has only <u>three</u> nearest neighbor alkyl groups. Hence there are only four types of local environment for each lithium: 3R/(2R,1R')/(1R,2R')/3R', with <u>i</u>ntensity ratios 1:3:3:1. Such a prediction is borne out by the ^{7}Li spectrum for the mixture of t-butyllithium and trimethylsilylmethyllithium, after it has had time to equilibrate (99). Lithium-7 NMR was also used therefore to measure the rate of intermolecular exchange in this $(CH_3)_3CLi + (CH_3)_3SiCH_2Li$ system. It is rate controlled by dissociation of t-butyllithium tetramers, with a first-order rate constant in cyclopentane of 10^{-5} s^{-1} at 293 K (100).

The ^{7}Li chemical shift for benzyllithium is solvent dependent: with reference to internal butyllithium, it has the values 1.06 (THF), 1.47 (Et_2O), and 2.07 (C_6H_6). This upfield shift, from tetrahydrofuran to benzene, goes with a reduction in the one-bond $^{13}C-^{1}H$ scalar coupling, from 132 to 116 Hz: this was interpreted as "due to a substantial increase in anion-cation interaction with transfer of electron density from the benzyl moiety to lithium ••• : the greater the s-character of the C-H orbitals, the more p-character in the C-Li orbital" (101).

The mixed complexes formed between methyllithium and lithium bromide or iodide, in diethyl ether between 180 and 220 K, were studied using ^{7}Li NMR: with the bromide, there is coexistence of tetramers $Li_4(CH_3)_4$, $Li_4(CH_3)_3Br$, $Li_4(CH_3)_2Br_2$, and of oligomers $(LiBr)_n$. With the iodide system, only $Li_4(CH_3)_4$, $Li_4(CH_3)_3I$, and $(LiI)_n$ are present (102).

Cox et al. have investigated the structure of some aromatic ion pairs, by taking advantage of the ASIS (<u>a</u>romatic <u>s</u>olvent <u>i</u>nduced <u>s</u>hifts) (103). Comparison of the ^{7}Li chemical shifts obtained with four aromatic carbanions in various solvents (Table 22) shows differences. Note especially the large solvent-dependent shifts for the fluorenyllithium system, interpreted in terms of a predominant <u>tight</u> ion pair in ether, at van der Waals contact (~2.7 Å) being progressively stretched into a predominant loose, solvent-separated ion pair in the higher dielectric hexamethylphosphorictriamide (103).

Table 22. ^{7}Li Chemical Shifts for Some Aromatic Anion Systems (103)

Anion/Solvent	δ/ppm
Cyclopentadienyl/THF	8.37
Cyclopentadienyl/DME	8.66
Cyclopentadienyl/1,4-dioxane	8.68
Indenyl/THF	6.17
Indenyl/DME	6.62
Phenylallyl/THF	0.61
Phenylallyl/DME	0.96
Fluorenyl/Et$_2$O	6.95
Fluorenyl/DME	3.04
Fluorenyl/THF	2.07
Fluorenyl/HMPA	0.73

Sodium-23

Sodium-23 NMR has also been exploited for investigating the structure of ion pairs: in the series of solvents tetrahydrofuran, glyme, diglyme, triglyme--with increasing cation-complexing ability in this order--there is a gradual shift to high field of the ^{23}Na resonance for NaI, as the equilibrium is displaced from a tight ion pair in THF to a predominantly loose ion pair in triglyme. By contrast, NaClO$_4$ exists as the loose ion pair in all four of these ether and polyether solvents (104).

Preferential solvation has also been much studied using NMR of the (solvated) alkali metal ion: for instance ^{87}Rb$^+$ in water-methanol mixtures (105), ^{23}Na$^+$ in water-methanol mixtures (106) and aqueous mixtures of simple amides (107), ^{23}Na$^+$ and ^{87}Rb$^+$ in binary mixtures of water (H$_2$O and D$_2$O) and methanol, formamide, or dimethyl sulfoxide (108), etc.

Our approach to preferential solvation, in the Liège group, has been to consider jointly the two observables, the chemical shift and the relaxation rate, for the alkali metal cation (109-110). We have developed a model based upon transfer to this area of preferential solvation of the Hill formalism (111) widely used in biochemical studies of such phenomena as dioxygen binding to the four equivalent subunits of hemoglobin (112). The two problems are isomorphous, mathematically speaking: binding of a ligand to one of four equivalent sites in a biomolecule and preferential solvation of a tetracoordinated ion constitute the same problem (113,114). If the two members of a binary solvent mixture which compete for Na^+ cation solvation share the same solvation number n, (n+1) complexes can be formed in succession, related by the equilibria of eq. 15. Inserting the appropriate

$$nL_A + Na^+(L_B)_n \underset{}{\overset{K_1}{\rightleftarrows}} (n-1)L_A + L_B + Na^+(L_B)_{n-1}(L_A)$$

$$\overset{K_2}{\rightleftarrows} \cdots \overset{K_{n-1}}{\rightleftarrows} L_A + (n-1)L_B + Na^+(L_B)(L_A)_{n-1}$$

$$\overset{K_n}{\rightleftarrows} nL_B + Na^+(L_A)_n \qquad (15)$$

statistical factors leads to the <u>intrinsic</u> equilibrium constants k_i, defined from $k_i = (n-i+1)K_i/i$. With tetracoordination (n = 4) this leads to (114) eq. 16, in which Y is the fraction of

$$Y = \sum_{i-1}^{4} i\alpha_i/r = \sum_{i-1}^{4} \frac{i\beta_i X^i}{4D} \qquad (16)$$

sites on the Na^+ cation occupied by a ligand L_A, $X = [L_A]/[L_B]$, and $D = \Sigma(i=0$ to 4) $\beta_0 = 1$, $\beta_i = K_1 \cdots K_i$, and $\alpha_i = \beta_i X^i/D$. The Hill plot is a representation of $\ln(Y/1-Y)$ as a function of $\ln X$. In the case of equality of all the intrinsic constants (k_i), i.e., under the absence of cooperativity, the Hill plots are linear and with unit slope: $\ln(Y/1-Y) = \ln K + \ln X$. This equation is identical with existing formalisms, which had assumed equality of each intrinsic equilibrium constant (113) or, in the case of a nonstatistical distribution, had resorted to the introduction of an adjustable parameter (115). Thus preferential solvation data have been derived from Covington's statistical distribution of solvate species theory (116,117) for Na^+.

We have extended application of the Hill formalism to polydentate ligands, polyamines competing with tetrahydrofuran for solvation of the sodium cation (118). The Hill plots obtained are no longer linear, denoting cooperativity. Entries of the first and of the second diamine molecules into the sodium coordination shell are independent and equiprobable steps: $K_1 = K_3$ and $K_2 = K_4$. This permits also a <u>direct</u> determination of the chelate effect (118).

Many other types of applications of ^{23}Na NMR can be found in the literature which, up to 1976, has been reviewed elsewhere (72).

Potassium-39

For various reasons (a mediocre receptivity; a rather low resonance frequency; and (probably) lack of familiarity on the part of many NMR spectroscopists), ^{39}K has been unduly neglected, if one considers its considerable importance for biological applications. Besides the studies already quoted in this chapter, the reviewer wishes to bring attention to studies of complex formation between K$^+$ and various organic ligands (119), of self-ordering of the 5'-GMP nucleotide induced by the potassium cation (120,121), and of K$^+$ binding to gels of polysaccharides (122).

Rubidium-85

The binding of ^{85}Rb$^+$ to humic acid (a complex mixture of oxygen-containing organic molecules found in the soil) has been studied, in competition with various other cations (123). Its role in the self-assembly of a nucleotide has also been ascertained from rubidium NMR (121).

Cesium-133

Changes in the ^{133}Cs quadrupole splittings and chemical shielding anisotropy, for CsCl dissolved in the lamellar phase of liquid crystals, for aqueous solutions, have been related to the local ordering (124). The interaction of ^{133}Cs$^+$ with various organic ligands, such as 2,2-bipyridines (129), crown ethers (125-126), and cryptands (127-128), in nonaqueous solvents, has been studied. The thermodynamics (119-125, 127) and the kinetics (126,128-129) of exchange between free and bound Cs$^+$ ions have been determined. Cesium NMR has the advantage of narrow lines, a useful attribute in consideration of the poor solubility of most cesium salts in the usual solvents.

CONCLUSIONS

We have surveyed rapidly a family of nuclides which, by and large, have decent sensitivities and whose quadrupolar responses to their local environment prove quite useful. This family, chemically quite homogeneous, illustrates nevertheless most of the features of NMR, the two extremes, ^{6}Li and ^{133}Cs, being nuclei almost devoid of a quadrupolar moment.

Since their NMR characteristics are favorable, their chemical and biological importance--since one no longer believes, as Humphry Davy did, that volcanism originates in water seeping into an earth core made of sodium and potassium metal (!)--maintains these nuclei in the forefront. NMR imaging is a prime example;

at the time of writing, the first ^{23}Na pictures of the human body are being taken.

ACKNOWLEDGMENTS

I thank my co-workers whose names are cited in the references for their contributions as a team, and Fonds de la Recherche Fondamentale Collective, Brussels, for its constant support of our work.

REFERENCES

(1) For consistency, all the quoted values have been taken from C. Brevard and P. Granger, "Handbook of High Resolution Multinuclear NMR," Wiley, New York, 1981.
(2) R. Sternheimer, Phys. Rev., 84, pp. 244-253 (1951).
(3) R. Sternheimer, Phys. Rev., 95, pp. 736-751 (1954).
(4) R. Sternheimer, Phys. Rev., 96, pp. 951-968 (1954).
(5) R. Sternheimer, Phys. Rev., 105, p. 157 (1957).
(6) E. N. Kaufmann and R. J. Vianden, Rev. Mod. Phys., 51, pp. 161-214 (1979).
(7) P. C. Schmidt, K. D. Sen, T. P. Das, and A. Weiss, Phys. Rev. B, 22, pp. 4167-4179 (1980).
(8) H. Pfeifer, in P. Diehl, NMR Basic Principles and Progress, 7, pp. 53 (1972).
(9) J. R. Huizinga, P. F. Grieger, and F. T. Wall, J. Am. Chem. Soc., 72, pp. 4228-4232 (1950).
(10) H. Ackerman, Z. Physik, 194, pp. 253-269 (1966).
(11) M. Baumann et al., Z. Physik, 194, pp. 270-279 (1966).
(12) K. Lee and W. Anderson, Table of Nuclear Properties, Varian Associates, Palo Alto, California (1967).
(13) M. Baumann, Z. Naturforsch, A23, pp. 620-622 (1968).
(14) D. Schönberner and D. Zimmermann, Z. Physik, 216, pp. 172-182 (1968).
(15) R. M. Sternheimer and R. F. Peierls, Phys. Rev., A3, pp. 837-848 (1971).
(16) S. Garpman, et al., Phys. Rev. A11, pp. 758-781 (1975).
(17) B. P. Fabricand, Mol. Phys., 7, pp. 425-432 (1964).
(18) W. R. Davidson and P. Kebarle, J. Am. Chem. Soc., 98, pp. 6125-6133 (1976).
(19) P. Kebarle, Ann. Rev. Phys. Chem., 28, pp. 445-476 (1977).
(20) W. R. Davidson and P. Kebarle, Can. J. Chem., 54, pp. 2594-2599 (1976).
(21) W. R. Davidson and P. Kebarle, J. Am. Chem. Soc., 98, pp. 6133-6138 (1976).
(22) J. Sunner, K. Nishizawa, and P. Kebarle, J. Phys. Chem., 85, pp. 1814-1820 (1981).

(23) H. Kistenmacher, H. Popkie, and E. Clementi, J. Chem. Phys., 59, pp. 5842-5848 (1973).
(24) H. Kistenmacher, H. Popkie, and E. Clementi, J. Chem. Phys., 61, pp. 799-815 (1974).
(25) R. E. Burton and J. Daly, Trans. Faraday Soc., 57, pp. 1219-1225 (1971).
(26) P. Schuster, W. Marius, A. Pullman, and H. Berthod, Theor. Chim. Acta (Berl.), 40, pp. 323-341, (1975).
(27) W. Marius and P. Schuster, Theor. Chim. Acta (Berl.), 42, pp. 5-11 (1976).
(28) R. L. Woodin, F. A. Houle, and W. A. Goddard, III, Chem. Phys., 14, pp. 461-468 (1976).
(29) J. Sadlej and A. J. Sadlej, Acta Phys. Polonica, A53, pp. 747-759 (1978).
(30) P. Kollman, Chem. Phys. Lett., 55, pp. 555-559 (1978).
(31) J. E. Del Bene, Chem. Phys., 40, pp. 329-335 (1979).
(32) J. E. Del Bene, Chem. Phys. Lett., 64, pp. 227-229 (1979).
(33) B. M. Rode, G. J. Reibnegger, and S. Fujiwara, J. Chem. Soc., 76, pp. 1268-1274 (1980).
(34) G. Corongiu, E. Clementi, E. Pretsch, and W. Simon, J. Chem. Phys., 72, pp. 3096-3102 (1980).
(35) R. M. Noyes, J. Am. Chem. Soc., 84, pp. 513-522 (1962).
(36) G. Geier, U. Karlen, and A. V. Zelewsky, Helv. Chim. Acta, 52, pp. 1967-1975 (1969).
(37) K. M. Kale and R. Zana, J. Sol. Chem., 6, pp. 733-746 (1977).
(38) O. Ya. Samoilov, Disc. Faraday Soc., 24, pp. 141-157 (1957).
(39) B. P. Fabricand, S. S. Goldberg, R. Leifer, and S. G. Ungar, Mol. Phys., 7, pp. 425-432 (1964).
(40) Z. Pajak and L. Latanowicz, Acta Phys. Polon., A53, pp. 555-569 (1978).
(41) K. Arakawa, K. Tokiwano, N. Ohtomo, and H. Vedaira, Bull. Chem. Soc. Japan, 52, pp. 2483-2488 (1979).
(42) J. R. Newsome, G. W. Neilson, and J. E. Enderby, J. Phys. C. Solid St. Phys., 13, pp. L923-926 (1980).
(43) H. L. Friedman and L. Lewis, J. Sol. Chem., 5, pp. 445-455 (1976).
(44) M. Rao and B. J. Berne, J. Phys. Chem., 85, pp. 1498-1505 (1981).
(45) P. Bopp, K. Heinzinger, and G. Jancso, Z. Naturfosch, 32a, pp. 620-623 (1977).
(46) T. Radnai, G. Palinkas, Gy. I. Szasz, and K. Heinzinger, Z. Naturforsch, 36a, pp. 1076-1082 (1981).
(47) Gy. I. Szasz, K. Heinzinger, and W. O. Riede, Z. Naturforsch, 36a, pp. 1067-1075 (1981).
(48) Gy. I. Szasz, K. Heinzinger, and W. O. Riede, Ber. Bunsenges. Phys. Chem., 85, pp. 1056-1059 (1981).
(49) Gy. I. Szasz, K. Heinzinger, and G. Palinkas, Chem. Phys. Lett., 78, pp. 194-196 (1981).
(50) K. Heinzinger, Z. Naturforsch, 31a, pp. 1073-1076 (1976).

(51) J. Fromm, E. Clementi, and R. O. Watts, J. Chem. Phys., 62, pp. 1388-1398 (1975).
(52) G. Corongiu and E. Clementi, Biopolymers, 20, pp. 2427-2483 (1981).
(53) S. Golden, C. Guttman, and T. R. Tuttle, Jr., J. Am. Chem. Soc., 87, pp. 135-136 (1965).
(54) S. Golden, C. Guttman, and T. R. Tuttle, Jr., J. Chem. Phys., 44, pp. 3791-3796 (1966).
(55) J. L. Dye, J. M. Ceraso, M. T. Lok, B. L. Barnett, and F. J. Tehan, J. Am. Chem. Soc., 96, pp. 608-609 (1974).
(56) F. J. Tehan, B. L. Barnett, and J. L. Dye, J. Am. Chem. Soc., 96, pp. 7203-7208 (1974).
(57) For instance, see J. L. Dye, Angew. Chem., Int. Ed. Engl., 18, pp. 587-598 (1979).
(58) P. P. Edwards, S. C. Guy, D. M. Holton, and W. McFarlane, J. Chem. Soc. Chem. Commun., pp. 1185-1186 (1981).
(59) J. Kondo and J. Yamashita, J. Phys. Chem. Solids, 10, pp. 245-253 (1959).
(60) C. Deverell, Progr. NMR Spectrosc., 4, pp. 235-334 (1969); Mol. Phys., 16, pp. 491-500 (1969).
(61) A. Delville, C. Detellier, A. Gerstmans, and P. Laszlo, J. Magn. Reson., 42, pp. 14-27 (1981).
(62) W. Sahm and A. Schwenk, Z. Naturforsch., 29a, pp. 1754-1762 (1974).
(63) A. R. Haase, M. A. Kerber, D. Kessler, J. Kronenbitter, H. Krüger, O. Lutz, M. Müller, and A. Nolle, Z. Naturforsch, 32a, pp. 952-956 (1977).
(64) R. H. Erlich and A. I. Popov, J. Am. Chem. Soc., 92, pp. 4989-4990 (1970).
(65) J. S. Shih and A. I. Popov, Inorg. Nucl. Chem. Lett., 13, pp. 105-110 (1977).
(66) W. J. De Witte, L. Liu, E. Mei, J. L. Dye, and A. I. Popov, J. Sol. Chem., 6, pp. 337-348 (1977).
(67) A. I. Popov, in "Characterization of Solutes in Nonaqueous Solvents," G. Mamantov, Ed., Plenum Press, New York, 1978, pp. 47-63.
(68) H. Weingärtner, J. Magn. Reson., 41, pp. 74-87 (1980).
(69) D. T. Edmonds and J. P. G. Mailer, J. Magn. Reson., 36, pp. 411-418 (1979).
(70) D. E. O'Reilly, E. M. Peterson, C. E. Scheie, and P. K. Kadaba, J. Chem. Phys., 58, pp. 3018-3022 (1973).
(71) S. L. Segel, J. Chem. Phys., 73, pp. 4146-4147 (1980).
(72) P. Laszlo, Angew. Chem., Int. Ed. Engl., 17, pp. 254-266 (1978).
(73) A. I. Mishustin and Yu. M. Kessler, J. Sol. Chem., 4, pp. 779 (1975).
(74) C. A. Melendres and H. G. Hertz, J. Chem. Phys., 61, pp. 4156-4162 (1974).
(75) H. G. Hertz, M. Holz, G. Keller, H. Versmold, and C. Yoon, Ber. Bunsenges. Phys. Chem., 78, pp. 493-509 (1974).

(76) H. Weingärtner and H. G. Hertz, Ber. Bunsenges. Phys. Chem., 81, pp. 1204-1221 (1977).
(77) H. Friedman, in "Protons and Ions Involved in Fast Dynamics Phenomena," P. Laszlo, Ed., Elsevier, Amsterdam, 1978, pp. 27-42.
(78) S. Engström and B. Jönsson, Mol. Phys., 43, pp. 1235-1253 (1981); see also S. Engström, "On the Interpretation of NMR Spectra of Quadrupolar Nuclei. Quantum Chemical and Statistical Mechanical Calculations," Ph.D. Dissertation, University of Lund (1980).
(79) F. Wehrli, J. Magn. Reson., 23, pp. 527-532 (1976).
(80) F. Wehrli, J. Magn. Reson., 30, pp. 193-209 (1978).
(81) E. Göbel, W. Müller-Warmuth, H. Olyschläger, and H. Dutz, J. Magn. Reson., 36, pp. 371-387 (1979).
(82) L. G. Werbelow and D. M. Grant, J. Magn. Reson., 29, pp. 603-605 (1978).
(83) L. G. Werbelow, A. Thevand, and G. Pouzard, J. Chem. Soc. Faraday Trans. 2, 75, pp. 971-974 (1979).
(84) L. G. Werbelow and A. G. Marshall, J. Magn. Reson., 43, pp. 443-448 (1981).
(85) A. Delville, C. Detellier, and P. Laszlo, J. Magn. Reson., 34, pp. 301-315 (1979).
(86) A. G. Marshall, T. C. Lin, and C. E. Cottrell, to be published; poster at the 43rd Experimental NMR Conference, Madison, Wisconsin, May, 1982.
(87) B. Halle and H. Wennerström, J. Magn. Reson., 44, pp. 89-100 (1981).
(88) S. Forsén, private communication, 1982.
(89) E. Oldfield, S. Schramm, M. D. Meadows, K. A. Smith, R. A. Kinsey, and J. Ackerman, J. Am. Chem. Soc., 104, pp. 919-920 (1982); see also M. D. Meadows, K. A. Smith, R. A. Kinsey, T. M. Rothgeb, R. P. Skarjune, and E. Oldfield, Proc. Nat. Acad. Sci. USA, 79, pp. 1351-1355 (1982).
(90) E. Oldfield, lecture at the 23rd Experimental NMR Conference, Madison, Wisconsin, May, 1982.
(91) S. Vega and Y Naor, J. Chem. Phys., 75, pp. 75-86 (1981).
(92) T. L. Brown and J. A. Ladd, J. Organometal. Chem., 2, pp. 373-379 (1964).
(93) E. A. C. Lucken, J. Organometal. Chem., 4, pp. 252-254 (1965).
(94) L. D. McKeever, R. Waack, M. A. Doran, and E. B. Baker, J. Am. Chem. Soc., 90, p. 3244 (1968).
(95) L. D. McKeever, R. Waack, M. A. Doran, and E. B. Baker, J. Am. Chem. Soc., 91, pp. 1057-1061 (1969).
(96) T. L. Brown, L. M. Seitz, and B. Y. Kimura, J. Am. Chem. Soc., 90, p. 3245 (1968).
(97) E. Weiss and G. Hencken, J. Organometal. Chem., 21, pp. 265-268 (1970).
(98) L. D. McKeever and R. Waack, J. Chem. Soc. Chem. Commun., pp. 750-751 (1969).

(99) T. L. Brown, Acc. Chem. Res., 1, pp. 23-32 (1968).
(100) M. Y. Darensbourg, B. Y. Kimura, G. E. Hartwell, and T. L. Brown, J. Am. Chem. Soc., 92, pp. 1236-1237 (1970).
(101) R. Waack, L. D. McKeever, and M. A. Doran, J. Chem. Soc. Chem. Commun., pp. 117-118 (1969).
(102) D. P. Novak and T. L. Brown, J. Am. Chem. Soc., 94, pp. 3793-3798 (1972).
(103) R. H. Cox, H. W. Terry, Jr., and L. W. Harrison, J. Am. Chem. Soc., 93, pp. 3297-3298 (1971).
(104) C. Detellier and P. Laszlo, Bull. Soc. Chim. Belg., 84, pp. 1081-1086 (1975).
(105) P. Neggia, M. Holz, and G. H. Hertz, J. Chim. Phys., 71, pp. 56-60 (1974).
(106) M. Holz, H. Weingärtner, and H. G. Hertz, J. Chem. Soc. Faraday Trans. 1, 73, pp. 71-83 (1977).
(107) M. Holz, H. Weingärtner, and H. G. Hertz, J. Sol. Chem., 7, pp. 705-719 (1978).
(108) M. Holz, J. Chem. Soc. Faraday Trans. 1., 74, pp. 644-656 (1978).
(109) C. Detellier and P. Laszlo, Helv. Chim. Acta, 59, pp. 1333-1345 (1976).
(110) C. Detellier and P. Laszlo, Helv. Chim. Acta, 59, pp. 1346-1351 (1976).
(111) A. V. Hill, J. Physiol., 40, pp. iv-xi (1910).
(112) S. J. Gill, H. T. Gaud, J. Wyman, and B. G. Barisas, Biophys. Chem., 8, pp. 53-59 (1978).
(113) A. Delville, J. Grandjean, P. Laszlo, C. Gerday, Z. Grabarek, and W. Drabikowski, Eur. J. Biochem., 105, pp. 289-295 (1980).
(114) A. Delville, C. Detellier, A. Gerstmans, and P. Laszlo, Helv. Chim. Acta, 64, pp. 547-555 (1981).
(115) A. K. Covington and K. E. Newman, in "Thermodynamic Behavior of Electrolytes in Mixed Solvents," W. Furter, Ed., ACS Chemistry Series, 155, 1976, pp. 153-196.
(116) R. H. Erlich, M. S. Greenberg, and A. I. Popov, Spectrochim. Acta, 29A, pp. 543-549 (1973).
(117) M. S. Greenberg and A. I. Popov, Spectrochim. Acta, 31A, pp. 697-705 (1975).
(118) A. Delville, C. Detellier, A. Gerstmans, and P. Laszlo, Helv. Chim. Acta, 64, pp. 556-567 (1981).
(119) E. Schmidt, A. Hourdakis, and A. I. Popov, Inorg. Chim. Acta, 52, pp. 91-95 (1981).
(120) C. Detellier, A. Paris, and P. Laszlo, Compt. Rend. Ac. Sci. Paris, 286D, pp. 781-783 (1978).
(121) C. Detellier and P. Laszlo, J. Am. Chem. Soc., 102, pp. 1135-1141 (1980).
(122) V. J. Morris and P. S. Belton, J. Chem. Soc. Chem. Commun., pp. 983-984 (1980).
(123) B. Lindman and I. Lindqvist, Chem. Scripta, 1, pp. 195-196 (1971).

(124) N. O. Persson and G. Lindblom, J. Phys. Chem., 83, pp. 3015-3019 (1979).

(125) M. Shamsipur, G. Rounaghi, and A. I. Popov, J. Sol. Chem., 9, 701-714 (1980).

(126) E. Mei, J. L. Dye, and A. I. Popov, J. Am. Chem. Soc., 99, pp. 5308-5311 (1977).

(127) E. Mei, L. Liu, J. L. Dye, and A. I. Popov, J. Sol. Chem., 6, pp. 771-778 (1977).

(128) E. Mei, A. I. Popov, and J. L. Dye, J. Am. Chem. Soc., 99, pp. 6532-6536 (1977).

(129) E. Mei, A. I. Popov, and J. L. Dye, J. Phys. Chem., 81, pp. 1677-1681 (1977).

CHAPTER 13

ALKALINE EARTH METALS

Otto Lutz

Physikalisches Institut der Universität
Tübingen, Bundesrepublik Deutschland

ABSTRACT

The NMR properties of the alkaline earth nuclides are discussed, together with experimental problems arising from the relatively low natural abundance, the low gyromagnetic ratios, and the quadrupole moments of most members of this series. Chemical shifts, coupling constants, and longitudinal relaxation times are discussed and illustrated.

INTRODUCTION

The IIa elements of the Periodic Table are very important candidates for NMR investigations, e.g., in biological samples. Unfortunately, with the exception of ^{9}Be, all the nuclei are not very favourable for NMR experiments.

In Table 1, NMR properties of the different isotopes are given (A1). Since the Larmor frequencies are relatively small, ringing problems often arise. Furthermore, the natural abundance is low for nuclei other than ^{9}Be, resulting in receptivities which are not very encouraging for the experimentalist.

Nevertheless, all these nuclei have been investigated in more detail. The very accurate magnetic moments given in Table 1 have been derived (Mg 1, Ca 1, Sr 1, Ba 1) from precise measure-

Table 1. NMR Properties of Alkaline Earth Metal Nuclei

Nucleus	Natural abundance (%)	Larmor frequency at 2.11 T (MHz)	Receptivity (^{1}H = 1)	Nuclear spin	Nuclear magnetic moment (μ_N)	Nuclear quadrupole moment ($10^{-28}m^2$)
^{9}Be	100	12.65	1.4×10^{-2}	3/2	-1.1776	+0.05
^{25}Mg	10.1	5.51	2.7×10^{-4}	5/2	-0.854813	+0.22
^{43}Ca	0.15	6.06	9.3×10^{-6}	7/2	-1.315675	-0.06
^{87}Sr	7.0	3.90	1.9×10^{-4}	9/2	-1.089273	+0.3
^{135}Ba	6.6	8.94	3.2×10^{-4}	3/2	+0.832339	+0.18
^{137}Ba	11.3	10.00	7.8×10^{-4}	3/2	+0.931075	+0.28

ments of their Larmor frequencies relative to a standard nucleus. The nuclear magnetic moments are given for the ions in infinitely diluted aqueous solutions without diamagnetic corrections.

From optical pumping experiments using free atoms or ions, relatively accurate magnetic moments also are available (see data given in Ref. Mg 1, Sr 1, Ba 1). Now it is possible to give shielding constants for some of the IIa elements, which are referred to the free atom by eq. 1. This yields a more general

$$\sigma^* = \frac{\mu_{atom} - \mu_{NMR}}{\mu_{atom}} \quad (1)$$

shielding scale than the arbitrary chemical shift scale. For instance for $\sigma^*(^{135}Ba) = -(8.2 \pm 0.2) \times 10^{-4}$ (Ba 1, Ba 2)) was found. Unfortunately the optical pumping data, in the case of the other IIa elements, are not very accurate, so that for the other nuclei only upper limits could be given: $\sigma^*(^9Be) = (3 \pm 7) \times 10^{-4}$ (A 2), $\sigma^*(^{43}Ca) = -(2 \pm 5) \times 10^{-4}$ (Mg 1), $\sigma^*(^{87}Sr) = -(6.3 \pm 6.0) \times 10^{-4}$ (Sr 1).

It is not possible to review in a short chapter all the results available on the IIa elements. Therefore some important topics have been selected which are connected with the author's interests. Review articles have been written by Forsén and Lindman (A2, A3). The article by Wehrli (A 4) on quadrupolar nuclei contains also a small chapter on the alkaline earth metals.

ALKALINE EARTH METALS IN AQUEOUS SOLUTION

In Table 2 chemical shifts and linewidths for typical aqueous solutions of the alkali earth metals are given. Usually the chemical shifts are referred to infinite dilution of the salts. For beryllium the shifts are smaller than 1 ppm and the linewidths are also reported as very narrow (Be 1). From a detailed study of T_1 as a function of temperature in beryllium nitrate a value of $T_1 \sim 9$ s can be taken from the figure given by Wehrli (Be 4). For ^{25}Mg as well, very small chemical shifts have been found and the linewidths for solutions of low concentration are in the range of a few Hertz (Mg 1, Mg 2). For ^{43}Ca significant chemical shifts are found and the lines are narrow (Ca 1, Mg 3). Although ^{87}Sr shows much broader lines, chemical shifts of several ppm are observable (Sr 1). This is not possible for the two isotopes of barium since in their cases linewidths in the range of a thousand Hertz are observed (Ba 1, Ba 2).

Table 2. Typical Experimental NMR Data for Alkaline Earth Metal Nuclei in Aqueous Solution

Nucleus	Sample	Chemical shift (δ)	Linewidth ($\Delta\nu$/Hz)	Spin-lattice relaxation time T_1 (s)	Ref.
^{9}Be	Be(NO$_3$)$_2$ 0.1 m in D$_2$O	–	1.5	–	A 5
^{9}Be	Be(NO$_3$)$_2$ 1 m in H$_2$O	0	–	~9	Be 1, Be 4
^{25}Mg	MgCl$_2$ 1 m in H$_2$O	0 (0.2)	4	0.178	Mg 1, Mg 2
^{43}Ca	CaCl$_2$ 4 m in H$_2$O	+2 (0.2)	2.7	–	Ca 1, Ca 2
^{87}Sr	SrBr$_2$ 1.8 m in H$_2$O	+2 (0.4)	101	–	Sr 1, Sr 2
^{135}Ba	BaCl$_2$ 0.5 m in H$_2$O	<5	650	–	Ba 1
^{137}Ba	BaCl$_2$ 0.5 m in H$_2$O	<5	1500	–	Ba 1

As an example of the signal-to-noise ratio an experimental curve of a ^{43}Ca NMR signal (Ca 2) is given in Figure 1. Using a high resolution probe with internal field stabilization by ^{2}H, the given signal-to-noise ratio was obtained in a 10 mm sample at 2.11 T within 10 minutes. Other typical NMR signals with experimental parameters of the diverse nuclei can be found in Ref. A 5.

Figure 2 is illustrative for all the IIa elements. The chemical shift of ^{43}Ca in aqueous solutions of CaCl$_2$ and Ca(NO$_3$)$_2$ is given. The data have been taken from the dissertation of A. Uhl (Mg 1, Mg 4). As found also with the alkali nuclei, chlorine shifts the resonance to higher frequencies and nitrate to lower ones compared with the resonance frequency found for infinite dilution, which frequency reflects the shielding effects due to the water molecules.

RATIO OF THE LARMOR FREQUENCIES OF ^{41}Ca AND ^{43}Ca

The element calcium possesses a radioactive isotope ^{41}Ca with $T_{1/2} = 1.1 \times 10^5$ years. Brun et al. (Ca 3) have given a ratio of the Larmor frequencies of ^{41}Ca, measured in saturated calcium nitrate solution with a "small amount of cobalt acetate" by comparison with the ^{2}H frequency measured in D$_2$O.

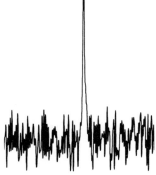

Figure 1. The ^{43}Ca NMR signal (Ca 2) of a solution of 4 molal CaCl$_2$ in D$_2$O at a Larmor frequency of about 6.06 MHz. A thousand free induction decays were accumulated within 10 min. The linewidth is 2.7 Hz (no exponential line broadening). A high resolution probe with ^{2}H-internal field stabilization, a Bruker SXP 4-100 pulse spectrometer with a high resolution iron magnet system B-E 38, and a B-NC-12 data system was used.

For an evaluation of the hyperfine structure anomaly in some atomic states, a reliable ratio of ^{41}Ca and ^{43}Ca should be available (G. zu Putlitz, private communication). Therefore first a ratio of the Larmor frequencies of ^{43}Ca and ^{2}H in a 4 molal solution of CaCl$_2$ in D$_2$O was measured with a high resolution probe internally stabilized by ^{2}H. The result of the measurements (Ca 2) is $\nu(^{43}\text{Ca})/\nu(^2\text{H}) = 0.438419737(77)$.

Using the chemical shift between this sample and a sample which is saturated in Ca(NO$_3$)$_2$, the following frequency ratio was found (Ca 4) $\nu(^{43}\text{Ca})$(saturated Ca(NO$_3$)$_2$)/$\nu(^2\text{H})$(D$_2$O) = 0.43841408(6). Now the influence of the paramagnetic cobalt acetate on the chemical shift of ^{43}Ca had to be studied (Ca 4). This was done with a 5 molal Ca(NO$_3$)$_2$ solution with various amounts of cobalt acetate. The results are given in Figure 2 (filled circles). The ^{43}Ca signal is shifted to lower frequencies with increasing Co^{2+} concentration up to 0.65 molal in cylindrical samples. Furthermore, the lines broaden to about 20 Hz.

To get comparable ratios of Larmor frequencies for ^{41}Ca and ^{43}Ca, the given ratio for the saturated nitrate solution was corrected by the largest shift occurring in the Co^{2+} containing

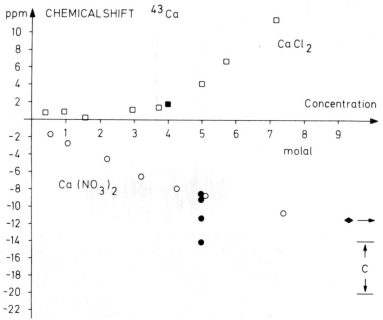

Figure 2. Chemical shift of ^{43}Ca (Mg 4, Ca 2, Ca 4) in aqueous solutions of $CaCl_2$ and $Ca(NO_3)_2$. The full circles are samples of 5 molal $Ca(NO_3)_2$ with added cobalt acetate having concentrations up to 0.65 molal. The chemical shift of saturated $Ca(NO_3)_2$ is given by the filled diamond. The meaning of c is given in the text.

samples (point c in Figure 2). As a reasonable error the whole range of the Co^{2+} induced shift has been used (see Figure 2). This procedure was necessary since the exact amount of Co^{2+} in Brun's solution (Ca 3) was not known. Our result for $\nu(^{43}Ca)/\nu(^{2}H)$ in a sample comparable to that of Brun et al. (Ca 3) is now $(^{43}Ca)_{corr}/\nu(^{2}H) = 0.4384116(13)$. Together with the value of Brun et al. one obtains $\nu(^{43}Ca)/\nu(^{41}Ca) = 0.826208(5)$, where the error of about 6.4×10^{-6} results mainly from the value of Brun et al. (Ca 3).

BERYLLIUM CHEMICAL SHIFTS

Beryllium with its marked tendency for involvement in covalent binding plays in a way an exceptional role within the IIa elements. Some material is available for chemical shifts and linewidths in organoberyllium compounds. These data have been collected in Table 3. As can be seen from the reference list,

Table 3. Chemical Shifts and Linewidths of Beryllium Compounds

No	Compound	Solvent	δ	Δν/Hz	Ref.
1	$C_5H_5BeBH_4$	C_6F_6	-22.1	31	Be 8
2	$C_5H_5BeB_5H_8$	pentane/CF_3Br	-21.0	28	Be 8
3	$C_5H_5BeCH_3$	C_6H_6	-20.6	10	Be 6
4	$C_5H_5BeCH_3$	$C_6H_5CH_3$	-20.5	9	Be 6
5	$C_5H_5BeCH_3$	pentane/CF_3Br	-20.4	8	Be 8
6	$C_5H_5BeCH_3$	cyclohexane	-20.1	7	Be 6
7	C_5H_5BeBr	C_6H_6	-19.5	4	Be 6
8	C_5H_5BeCl	C_6F_6	-19.1	3	Be 8
9	C_5H_5BeCl	C_6H_6	-18.8	3	Be 6
10	$(C_5H_5)_2Be$	$C_6H_5CH_3$	-18.5	–	Be 2
11	$(C_5H_5)_2Be$	methylcyclohexane	-18.3	–	Be 2
12[a]	$Be(HMPT)_4^{2+}$	$MeNO_2$	-3.3	–	Be 5
13[a]	$Be(TMPA)_4^{2+}$	$MeNO_2$	-3.2	–	Be 5
14[a]	$Be(DMMP)_4^{2+}$	$MeNO_2$	-2.9	–	Be 5
15	BeF_4^{2-}	H_2O	ca. -2	ca. 10	Be 1
16	$BeF_2(H_2O)_2$	H_2O	-0.8	–	Be 3
17	$Be(H_2O)_4^{2+}$	H_2O	0.0	0.15	Be 3, Be 8
18	$Be(NH_3)_4^{2+}$	NH_3	1.7	–	Be 3
19	$Be(B_3H_8)_2$	C_6F_6	2.9	27	Be 8
20	$(Be[N(CH_3)_2]_2)_3$, cent	C_6H_6	3.0	5	Be 3
21	$BeCl_2(Et_2O)_2$	Et_2O	3.1	–	Be 3
22	$(CH_3)_2Be[P(CH_3)_3]_2$	$P(CH_3)_3$	3.6	–	Be 3
23	$Be(B_3H_8)_2$	CH_2Cl_2/CF_3Br	4.0	28	Be 8
24	$CH_3BeCl[S(CH_3)_2]_2$	$(CH_3)_2S$	4.2	–	Be 3
25	$BeCl_2[S(CH_3)_2]_2$	$(CH_3)_2S$	5.5	–	Be 3
26	$(B_3H_8BeCH_3)_2$	CH_2Cl_2/CF_3Br	6.6	19	Be 8
27	$(Be_3H_8BeCH_3)_2$	C_6F_6	7.6	18	Be 8
28	$(Be[N(CH_3)_2]_2)_3$, end	C_6H_6	9.8	64	Be 3
29	$B_5H_{10}BeBH_4$	pentane/CF_3Br	11.4	26	Be 8
30	$(CH_3)_2BeS(CH_3)_2$	$S(CH_3)_2$	11.6	–	Be 3
31	$(CH_3BeBH_4)_2$	pentane/CF_3Br	11.7	7	Be 8
32	$(CH_3)_2Be[N(CH_3)_3]_2$	$N(CH_3)_3$	12.0	–	Be 3
33	$(CH_3BeBH_4)_2$	C_6F_6	12.2	12	Be 8
34	$B_5H_{10}BeBH_4$	C_6F_6	12.3	23	Be 8
35	$B_5H_{10}BeBr$	C_6F_6	14.3	30	Be 8
36	$B_5H_{10}BeCl$	C_6F_6	14.6	32	Be 8
37	$(CH_3)_2BeN(CH_3)_2$	C_6H_{12}	19.9	20	Be 3
38	$(CH_3)_2BeO(C_2H_5)_2$	Et_2O	20.8	–	Be 3

[a] HMPT: hexamethylphosphorotriamide, TMPT: trimethylphosphate, DMMP: dimethylmethylphosphonate.

only a few papers have been published on this subject (Be 1 - Be 3, Be 5, Be 6, Be 8). The whole range of the chemical shift is only about 40 ppm but due to the small linewidths the dispersion is quite good. As reference, usually the tetrahydrated Be^{2+} is used since the line is narrow and in aqueous solutions no concentration dependence has been observed (Be 1, Be 8).

In a recent paper from Gaines et al. (Be 8), some beryllaboranes have been studied. Further, Wehrli and Wehrli investigated aqueous acetonitrile species at high field (Be 9).

LONGITUDINAL RELAXATION TIME OF ^{25}Mg

From the earlier measurements of the linewidths of ^{25}Mg in aqueous solutions (Mg 1, Mg 2) it is obvious that for low concentration solutions the T_1's are expected to be long. Due to the high interest of ^{25}Mg in complexes some basic data on spin-lattice relaxation times in aqueous solutions have been investigated (Mg 3); T_1 as a function of the concentration of magnesium chloride and nitrate in water has been measured. In Figure 3 is a typical stacked plot, given for a sample of 5 molal magnesium chloride in H_2O (Mg 4). A strong difference for the relaxation rates between nitrate and chloride solutions has been found for higher concentrations. In 4 molal solutions $1/T_1$ equals 44.6 s^{-1} for the nitrate and 17.8 s^{-1} for the chloride. For vanishing concentrations the relaxation rates are the same, $(1/T_1)_0 = (4.5 \pm 0.2)$ s^{-1}. A very strong increase for $1/T_1$ was observed when to a 1 molal magnesium chloride solution ($1/T_1$ = 5.6 s^{-1}) sodium citrate (0.1 molal) was added: $1/T_1$ = 94.3 s^{-1}. From this fact the usefulness of relaxation rate measurements in biological samples may be imagined. Furthermore, the relaxation rates are in an experimental range in which it is not too difficult to work. Further information on this subject is developed in the next chapter, by Drakenberg and Forsén.

RELAXATION BEHAVIOUR AT INFINITE DILUTION

The relaxation behaviour at infinite dilution provides a good possibility for a comparison with theory. Since all the IIa elements have isotopes with quadrupole moments, the relaxation rates are governed by the interaction between the quadrupole moment and the electric field gradient. The relaxation rates can be quite well understood in terms of the electrostatic model developed by Hertz (A 6). In the extreme narrowing case, the condition which is usually fulfilled in aqueous solution, $1/T_1$ equals $1/T_2$. That means that also from the linewidths information on $1/T_1$ can be taken if magnetic field inhomogeneity effects can be neglected. Data for $1/T_1$ and $1/T_2$ at infinite dilution are collected for the alkaline earth nuclei in Table 4.

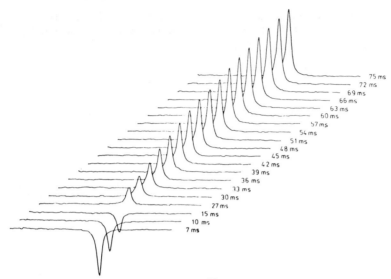

Figure 3. Stacked plot of a ^{25}Mg inversion recovery Fourier transform measurement (Mg 5) of the spin-lattice relaxation time T_1 in a 5 molal solution of $MgCl_2$ in H_2O. For each spectrum 300 free induction decays were accumulated. For the 20 spectra with different time intervals between the 180° and 90° pulses a measuring time of something less than 2 hours was necessary. The result is T_1 = (30.3 ± 1.5) ms. A high power probe with external ^{1}H stabilization was available for the Larmor frequency of 5.5 MHz (2.11 T).

An early test (A 7) of the theory of Hertz (A 6) suffered from the nonavailability of good experimental data. But one of the great successes of this paper was that the quadrupole moment of ^{43}Ca was put in question, since the very narrow NMR lines for ^{43}Ca lead to a strong discrepancy with the theory. In fact, the quadrupole moment of ^{43}Ca has now been found to be much smaller by three different groups (1 b = 10^{-28} m^2): Q = -0.065(20) b (Ref. Ca 5), Q = -0.048 (8) b (Ref. Ca 6), Q = -0.062(2) b (Ref. Ca 7).

In a very recent paper, Helm and Hertz (A 8) have investigated $1/T_1$ and $1/T_2$ at infinite dilution by isotopically enriched samples (see Table 4). The earlier experimental results have been confirmed and extended.

The infinite dilution values of the relaxation rates have been interpreted intensively in this paper (A 8) in terms of the

Table 4. Experimental Relaxation Rates $1/T_1$ and $1/T_2$ at Infinitely Diluted Aqueous Solutions Given in s^{-1}

	Natural abundance		Enriched material		Ref.
	$1/T_1$	$1/T_2$	$1/T_1$	$1/T_2$	
^{9}Be	0.07	–	–	–	Be 4
^{25}Mg	4.5 ± 0.2	–	4.1 ± 0.3	–	Mg 3, A 8
^{43}Ca	–	–	0.75	–	A 7
			0.80 ± 0.10	–	A 8
^{87}Sr	–	205 ± 40	–	170 ± 15	Sr 1, A 8
^{135}Ba	–	1700 ± 160	–	–	Ba 1, Ba 2
^{137}Ba	–	–	–	4000 ± 400	A8

electrostatic quadrupolar relaxation theory. The high symmetry quenching of the electric field gradients at the nuclear site reported for Mg^{2+} (Mg 3) was confirmed and also found for the other alkaline earth ions. A strongly increasing trend has been obtained going from Mg^{2+} to Ba^{2+}. Some conclusions on the hydration of the alkaline earth metal ions are also made by Helm and Hertz (A 8).

ACKNOWLEDGMENTS

I thank the Deutsche Forschungsgemeinschaft for financial support. I acknowledge gratefully the help of the following co-workers: Dr. habil. A. Nolle, Dr. H. Oehler, Dr. A. Uhl, E. Haid, S. Günther, D. Köhnlein, G. Nothaft, G. Kössler, W. Messner, K.-R. Mohn, W. Schich, P. Schrade, and N. Steinhauser.

REFERENCES

(A 1) G. H. Fuller, J. Phys. Chem. Ref. Data 5, p. 835 (1976).
(A 2) S. Forsén and B. Lindman, Ann. Reports NMR Spectrosc., 11A (1979).

(A 3) B. Lindman and S. Forsén, in "NMR and the Periodic Table," R. K. Harris and B. E. Mann, Eds., Academic Press, London, 1978.
(A 4) F. W. Wehrli, Ann. Reports NMR Spectrosc., 9 (1979).
(A 5) C. Brevard and P. Granger, "Handbook of High Resolution Multinuclear NMR," J. Wiley, New York, 1981.
(A 6) H. G. Hertz, Ber. Bunsenges. Phys. Chem., 77, p. 531 (1973).
(A 7) B. Lindman, S. Forsén, and H. Lilja, Chem. Script., 11, p. 91 (1977).
(A 8) L. Helm and H. G. Hertz, Z. Physik. Chemie, Neue Folge, 127, p. 23 (1981).
(Be 1) J. C. Kotz, R. Schaeffer, and A. Clouse, Inorg. Chem., 6, p. 620 (1967).
(Be 2) G. L. Morgan and G. B. McVicker, J. Am. Chem. Soc., 90, p. 2789 (1968).
(Be 3) R. A. Kovar and G. L. Morgan, J. Am. Chem. Soc., 92, p. 5067 (1970).
(Be 4) F. W. Wehrli, J. Magn. Reson., 23, p. 181 (1976).
(Be 5) J. J. Delpuech, A. Péguy, P. Rubini, and J. Steinmetz, Nouveau J. Chim., 1, p. 133 (1976).
(Be 6) D. A. Drew and G. L. Morgan, Inorg. Chem., 16, p. 1704 (1977).
(Be 7) F. W. Wehrli, J. Magn. Reson., 30, p. 193 (1978).
(Be 8) D. F. Gaines, K. M. Coleson, and D. F. Hillenbrand, J. Magn. Reson., 44, p. 84 (1981).
(Be 9) F. W. Wehrli and S. L. Wehrli, J. Magn. Reson., 47, p. 151 (1982).
(Mg 1) O. Lutz, A. Schwenk, and A. Uhl, Z. Naturforsch., 30a, p. 1122 (1975).
(Mg 2) L. Simeral and G. E. Maciel, J. Phys. Chem., 80, p. 552 (1976).
(Mg 3) S. Günther, O. Lutz, A. Nolle, and P.-G. Schrade, Z. Naturforsch., 34a, p. 944 (1979).
(Mg 4) A. Uhl, Dissertation, Tübingen (1975).
(Mg 5) S. Günther, Wissenschaftliche Arbeit, Tübingen (1978).
(Mg 6) M. Ellenberger, M. Villemin, and M. J. Lecomte, C. R. Acad. Sci. Paris, B 266, p. 1430 (1968).
(Mg 7) S. Forsén and B. Lindman, Chem. Brit., 14, p. 29 (1978).
(Mg 8) E. O. Bishop, S. J. Kimber, B. E. Smith, and P. J. Beynon, FEBS Lett., 101, p. 31 (1979).
(Mg 9) R. G. Bryant, J. Magn. Reson., 6, p. 159 (1972).
(Mg 10) J. A. Magnuson and A. A. Bothner-By, in "Magnetic Resonance in Biological Research," C. Franconi, Ed., Gordon and Breach Sci. Publ., London, 1971, p. 365.
(Mg 11) P.-H. Heubel and A. I. Popov, J. Sol. Chem., 8, p. 283 (1979).
(Ca 1) O. Lutz, A. Schwenk, and A. Uhl, Z. Naturforsch., 28a, p. 1534 (1973).

(Ca 2) N. Steinhauser, Wissenschaftliche Arbeit, Tübingen (1981).
(Ca 3) E. Brun, J. J. Kraushaar, W. L. Pierce, and W. J. Veigele, Phys. Rev. Lett., 9, p. 166 (1962).
(Ca 4) W. Schich, Wissenschaftliche Arbeit, Tübingen (1982).
(Ca 5) P. Grundevic, M. Gustavsson, I. Lindgren, G. Olsson, L. Robertson, A. Rosén, and S. Swanberg, Phys. Rev. Lett., 42, p. 1528 (1979).
(Ca 6) E. Bergmann, P. Bopp, Ch. Dorsch, J. Kowalski, and F. Träger, Z. Physik, A294, p. 319 (1980).
(Ca 7) R. Aydin, W. Ertmer, and U. Johann, Z. Phys., A306, p. 1 (1982).
(Sr 1) J. Banck and A. Schwenk, Z. Physik, 265, p. 165 (1973).
(Sr 2) D. Köhnlein, unpublished results.
(Ba 1) H. Krüger, O. Lutz, and H. Oehler, Phys. Lett., 62A, p. 131 (1977).
(Ba 2) O. Lutz and H. Oehler, Z. Physik., A288, p. 11 (1978).
(Ba 3) L. Helm and H. G. Hertz, Z. Physik. Ch., 127, p. 23 (1981).

CHAPTER 14

THE ALKALINE EARTH METALS--BIOLOGICAL APPLICATIONS

Torbjörn Drakenberg and Sture Forsén

Physical Chemistry 2, Chemical Center
University of Lund, S-220 07 Lund 7, Sweden

ABSTRACT

Both ^{25}Mg and ^{43}Ca are nuclei with low receptivity due to low magnetogyric ratios and, especially for ^{43}Ca, low natural abundance. It is therefore not surprising that the interest in these nuclei in biological studies by NMR is of quite recent date, despite the well documented importance of these nuclei in biological systems (1). The first ^{25}Mg and ^{43}Ca NMR studies with some biological implications appeared in 1969 (2,3); however, not until 1978 was the first study of metal binding to proteins published (4). Even though subsequently it has been made clear that both ^{25}Mg and ^{43}Ca NMR can be very useful tools in biological studies, very few groups are active in this area, as can be judged from the literature. This is most probably, at least for ^{43}Ca, due to the high cost for enriched samples. In this chapter we will show, with a few examples, how ^{25}Mg and ^{43}Ca can be used to obtain information about the metal binding to proteins and other biologically interesting molecules. Some of this information is hard, if not impossible, to get with other methods, e.g., metal exchange rates, whereas other problems can often be better solved by other methods, e.g., binding constants.

EXPERIMENTAL ASPECTS

In both ^{25}Mg and ^{43}Ca NMR sensitivity is of utmost importance and resolution is normally of very little importance. In order to gain the highest possible sensitivity, and possibly sacrifice some resolution, one therefore wants to use high fields and large sample volumes. However, due to the high cost for enriched ^{43}Ca the sample volumes cannot be made very large. In superconducting magnets the orientation of the magnetic field has made it necessary to use saddle-shaped Helmholtz coils for the transmitter/receiver when one wants to have vertically oriented sample tubes. It has been shown (5) that the Helmholtz coil results in a sensitivity that is a factor of ca. 2.5 less than for a solenoid. Therefore, in order to get the highest possible sensitivity for a superconducting magnet it is necessary to use horizontally arranged solenoids. It is not so convenient as with the regular probe design to change sample since the eject cannot be used. However, a saving of a factor of ca. 6 in time is well worth some inconvenience. We have in Lund been using such a system for several years now and it has been working to our satisfaction.

Another experimental problem when observing broad lines is that the spectrometer has a ring down, or dead, time after the rf pulse, during which no useful signal can be observed. The most serious problem in this respect seems to be acoustic ringing in the probe. This problem is especially serious at low frequencies. If we define the dead time as the time delay after the rf pulse needed to have no observable distortion of the first point in the FID after the accumulation of 10^5 transients, then we have on our Nicolet 360WB system a dead time of ca. 300 μs for ^{25}Mg using a home-built horizontal probe. With a pulse sequence suggested by Paul Ellis (personal communication) $(-(180_{(0)}-90_{(180)}+FID-180_{(0)}-90_{(0)}-FID-90_{(0)}+FID-90_{(180)}-FID)-)$ it was in our system possible to reduce the dead time to between 50 and 100 μs.

THEORETICAL CONSIDERATIONS

We will here briefly describe the theoretical background which is necessary in the treatment of the experimental data presented later on.

If the fluctuation, characterized by a correlation time τ_c, of the electric field gradient causing quadrupole relaxation is sufficiently rapid so that $\omega_0 \tau_c \ll 1$, where ω_0 is the Larmor frequency of the nucleus, the relaxation can be described by a simple exponential. This applies to both the longitudinal, R_1, and the transverse, R_2, relaxation rates, and furthermore in this limit $R_1 = R_2$. For nuclei with spins $I > 1$ under conditions such that $\omega_0 \tau_c > 1$, the relaxation processes are multiexponential and the observed NMR lineshapes are no longer Lorentzians (6). A perturbation treatment of quadrupolar relaxation of spin $I = 5/2$ and 7/2 nuclei shows, however, that the relaxation behaviour is well described by a simple exponential decay for $\omega_0\tau_c$ up to 1.5 (7). The effective rates, R_1 and R_2, are unequal and with some loose restrictions are given by eqs. 1 and 2, in which χ is the

$$R_1 = \frac{3\pi^2}{10} \cdot \chi^2 \cdot \frac{2I+3}{I^2(2I-1)} [0.2J(\omega_0) + 0.8J(2\omega_0)] \tag{1}$$

$$R_2 = \frac{3\pi^2}{10} \cdot \chi^2 \cdot \frac{2I+3}{I^2(2I-1)} [0.3J(0) + 0.5J(\omega_0) + 0.2J(2\omega_0)] \tag{2}$$

quadrupole coupling constant characterizing the coupling of the nuclear quadrupole and the electric field gradient and $J(\omega_0)$ is defined by eq. 3. A further complication for quadrupolar nuclei

$$J(\omega_0) = \frac{\tau_c}{1 + (\omega_0\tau_c)^2} \tag{3}$$

with spin $I > 1$ is that the quadrupolar interaction not only gives rise to relaxation but that there is also a second order (dynamic) shift of the observed NMR signals. The shift approaches zero in the extreme narrowing limit ($\omega_0\tau_c \ll 1$), and in the opposite limit, $\omega_0\tau_c \gg 1$, it approaches a constant value proportional to χ^2/ω_0 (8,9). A detailed analysis of the influence of second order shifts on the lineshape of spin $I = 5/2$ and $7/2$ nuclei under various conditions including chemical exchange has recently been carried out by Westlund and Wennerström (10). Some general results from this treatment are summarized in Figure 1. Of particular interest here is the lineshape in the presence of chemical exchange. Due to the rather complex interplay between relaxation and dynamic shifts, the influence of the exchange rate on the lineshape can take various forms depending on the parameters. Under conditions of nearly extreme narrowing and when $\chi^2/\omega_0 < 10^4$ Hz it is possible to treat the exchange effects using a combination of the perturbation expression of eqs. 1 and 2 and the modified Bloch equations due to McConnell (11).

The strategy employed at our department in Lund to evaluate exchange rate parameters can briefly be summarized as follows (12).

(i) The modified Bloch equations are used to derive a general expression for the lineshape, $G(\nu)$, for the N-site system under study.

(ii) The transverse relaxation rate at the macromolecular binding site is expressed by eq. (2) ($\omega_0\tau_c < 1.5$).

(iii) The exchange rate, $k_{ex} = 1/\tau_{ex}$, is given by eq. 4.

Dynamic frequency shifts - some general characteristics

Def. of spectral densities

$$\begin{cases} J(\omega) \equiv \int_0^\infty \langle v_0(0) v_0(\tau) \rangle \cos\omega\tau \, d\tau \implies 0.3 V_{zz}^2 \cdot \dfrac{\tau_c}{(1+\omega^2\tau_c^2)} \\ Q(\omega) \equiv \int_0^\infty \langle v_0(0) v_0(\tau) \rangle \sin\omega\tau \, d\tau \implies 0.3 V_{zz}^2 \cdot \dfrac{\omega\tau_c^2}{(1+\omega^2\tau_c^2)} \end{cases}$$

(A) <u>Slow motion limit</u> $(J(0) \gg J(\omega_0), J(2\omega_0))$:

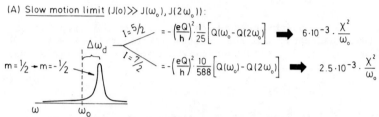

$I = 5/2$: $\quad = -\left(\dfrac{eQ}{h}\right)^2 \cdot \dfrac{1}{25}\left[Q(\omega_0) - Q(2\omega_0)\right] \implies 6\cdot 10^{-3} \cdot \dfrac{\chi^2}{\omega_0}$

$I = 7/2$: $\quad = -\left(\dfrac{eQ}{h}\right)^2 \cdot \dfrac{10}{588}\left[Q(\omega_0) - Q(2\omega_0)\right] \implies 2.5\cdot 10^{-3} \cdot \dfrac{\chi^2}{\omega_0}$

$m = 1/2 \rightarrow m = -1/2$

(B) <u>Near extreme narrowing</u> $(J(0) \gtrsim J(\omega) \gtrsim J(2\omega))$:

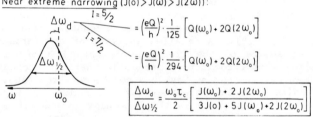

$I = 5/2$: $\quad = \left(\dfrac{eQ}{h}\right)^2 \cdot \dfrac{1}{125}\left[Q(\omega_0) + 2Q(2\omega_0)\right]$

$I = 7/2$: $\quad = \left(\dfrac{eQ}{h}\right)^2 \cdot \dfrac{1}{294}\left[Q(\omega_0) + 2Q(2\omega_0)\right]$

$$\boxed{\dfrac{\Delta\omega_d}{\Delta\omega_{1/2}} = \dfrac{\omega_0\tau_c}{2}\left[\dfrac{J(\omega_0) + 2J(2\omega_0)}{3J(0) + 5J(\omega_0) + 2J(2\omega_0)}\right]}$$

(C) <u>Extreme narrowing</u>

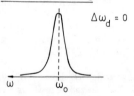

$\Delta\omega_d = 0$

Figure 1. General characteristics of dynamic frequency shifts.

$$k_{ex} = \frac{kT}{h} \exp(-\frac{\Delta G^{\ddagger}}{RT}) = \frac{kT}{h} \exp(\frac{\Delta S^{\ddagger}}{R} - \frac{\Delta H^{\ddagger}}{RT}) \qquad (4)$$

(iv) The correlation time, τ_c, is assumed to follow eq. 5. The

$$1/\tau_c = \frac{kT}{h} \exp(-\Delta G_c/RT) \qquad (5)$$

correlaton time at a given temperature is determined from measurements of the ratio between $\langle R_1 \rangle$ and $\langle R_2 \rangle$, which according to eqs. 1 and 2 is solely dependent on $\omega_0 \tau_c$ (eq. 6). From a knowledge of τ_c it is then possible to

$$\langle R_1 \rangle / \langle R_2 \rangle = \frac{0.2J(\omega) + 0.8J(2\omega)}{0.3J(0) + 0.5J(\omega) + 0.2J(2\omega)} \qquad (6)$$

calculate χ at this temperature.

(v) As a first approximation, the temperature dependence of the binding constant, K_B, is neglected. Since the NMR experiment is usually performed with ^{43}Ca or ^{25}Mg in considerable molar excess over the biological macromolecule, a modest change in K_B will have a negligible influence on the fraction of ions bound to the macromolecule as long as K_B is reasonably large.

(vi) Experimental NMR spectra obtained at ten or more different temperatures are used for an iterative nonlinear least squares computer fitting procedure, whereby ΔG_c, $\Delta G^{\ddagger}$ (or $\Delta S^{\ddagger}$ and $\Delta H^{\ddagger}$), as well as χ are varied.

METAL BINDING TO SMALL LIGANDS

Four early studies of ^{25}Mg and ^{43}Ca NMR were made with quite high ion concentration (ca. 1 M or more); nevertheless these studies fully demonstrated some of the possibilities with ion NMR. Magnuson and Bothner-By (2) studied ^{25}Mg linewidths as a function of added ligands like carboxylic acids and organic phosphates. By variable pH studies they were able to estimate binding constants as well as the relaxation rate of the complexed ^{25}Mg^{2+} ions. It was found that the complexed Mg^{2+} ion relaxes about 100 times faster than the free ion. Bryant (13) reported similar results and included also variable temperature studies. For the Mg^{2+}-ATP system Bryant found a considerable exchange contribution to the line broadening and also an unlikely large chemical shift difference between Mg-ATP and free Mg^{2+}.

More recently, using FT NMR, Robertson et al. (14) have investigated the binding of ^{25}Mg and ^{43}Ca ions to a peptide containing γ-carboxyglutamic acid. Figures 2 and 3 show their results for ^{25}Mg and ^{43}Ca experiments, respectively. For ^{25}Mg a broadening of the NMR signal is observed as a function of added peptide; however, no chemical shift change is observed. For ^{43}Ca on the other hand, chemical shift changes are reported, whereas the broadening is much less than in the ^{25}Mg experiment. Both these experiments have been used to calculate the dissociation constant for the metal complexes, resulting in K_d = 0.6 mM for both Mg^{2+} and Ca^{2+}.

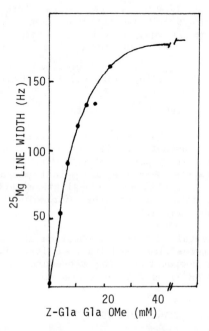

Figure 2. The dependence of the ^{25}Mg linewidth on the concentration of Z-D-GlaGlaOMe. Magnesium concentration varied from 21.3 to 18.7 mM, pH 6.5 to 6.6, temperature 27°C (from Ref. 14).

DIRECT OBSERVATION OF ^{25}Mg AND ^{43}Ca NMR SIGNALS FROM IONS BOUND TO PROTEINS AND OTHER LIGANDS

For ^{25}Mg, like the halogens, with its large quadrupole moment there is not much hope for the observation of an NMR signal from magnesium firmly bound to a macromolecule. As far as

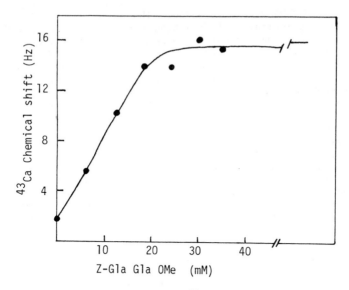

Figure 3. The dependence of the ^{43}Ca chemical shift on the concentration of Z-D-GlaGlaOMe. Calcium ion concentration varied from 19.5 to 17.4 mM, pH 6.6, temperature 27°C (from Ref. 14).

we are aware there is only one observation of the ^{25}Mg NMR signal from complexed magnesium (15). This was in the small complex with EDTA, and the signal was ca. 500 Hz broad. We have made the same observation in Lund, and we furthermore have found that the ^{25}Mg signal from the Mg-EDTA complex was even broader. In fact it was so broad that it could not be satisfactorily observed due to the dead time in the spectrometer of about 300 μs (Teleman et al. unpublished results.)

For ^{43}Ca the possibility to observe the NMR signal, even from calcium bound to a macromolecule, is much better than for magnesium due to its small quadrupole moment (0.05). Furthermore, the chemical shift for ^{43}Ca is expected to be more sensitive to changes in the environment. Chemical shifts of about ±40 ppm from the free (solvated) Ca^{2+} ion have been observed to date (Figure 4) (15-17). So far there is only one publication (18) in which the direct observation of ^{43}Ca NMR signals from calcium bound to proteins is reported. This is for the three homologous calcium binding proteins, parvalbumin, troponin C, and calmodulin. In all three cases only one broad signal is observed even though there are two strong sites with slow calcium exchange in each protein. However, as can be seen from Figure 5 two signals were observed for parvalbumin when more than two equivalents of ^{43}Ca had been added. The narrow one is due to calcium free in solution, or interacting weakly with other sites, and the broad one is due to calcium bound to the two strong sites. It is not surprising that only one signal can be observed for the bound ions when considering the small shift between free and bound ^{43}Ca^{2+} ions (ca. 10 ppm) and the broad lines from bound calcium, 500-800 Hz (at the used field 30 to 50 ppm). This demonstrates clearly that in this type of study it will be the relaxation rates and not the chemical shifts that will be the important parameters. The relaxation rates, R_1 and R_2, are not equal, showing that the extreme narrowing condition does not apply and therefore R_1 and R_2 can be used to calculate both the correlation time and the quadrupole coupling constant from eqs. 1, 2, and 6. The correlation times agree well with values calculated from the hydrodynamic radius, an indication that there is very little mobility within the binding sites. It might be interesting to note that although the sites in these molecules are assumed to have close to octahedral symmetry the quadrupole coupling constant (ca. 1.1 MHz) is about twice as large as for the Ca-EDTA complex, which has a much lower symmetry.

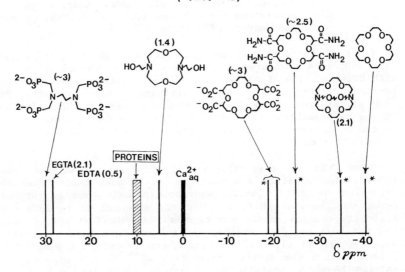

Figure 4. Chemical shifts of the ^{43}Ca^{2+} ion in complexes. Positive values to increasing frequency. Values in parenthesis are quadrupole coupling constants. Asterisks indicate methanol solution, otherwise water.

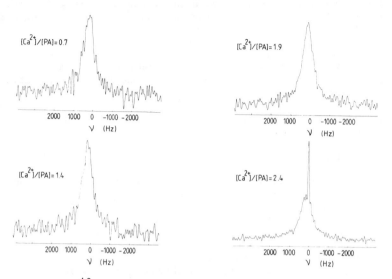

Figure 5. ^{43}Ca NMR spectra showing a calcium titration of the protein parvalbumin. Reprinted with permission from Ref. 18. Copyright 1982, American Chemical Society.

INTERMEDIATE TO FAST EXCHANGING METAL IONS

In systems where the metal exchange is not slow on the NMR time scale, information regarding the binding site has to be extracted from observed signals which are average signals for all sites, free as well as bound. Due to the low sensitivity of ^{25}Mg and ^{43}Ca NMR these experiments are often performed with an excess of metal ions over possible binding sites and, e.g., exchange rates can be evaluated as delineated in the theoretical section (vide supra). Also pH dependencies and binding constants can be determined. Both DNA and a few calcium-binding proteins have been studied with either ^{25}Mg or ^{43}Ca NMR, or both. We will in the following sections discuss some of these studies to point out what can be done with alkaline earth metal NMR in these systems.

γ-Carboxyglutamic Acid Containing Proteins

There is a group of proteins involved in blood coagulation, which contain the unusual amino acid γ-carboxyglutamic acid (Gla) (19). The Gla residues in these proteins are responsible for their Ca^{2+} binding properties. Robertson et al. (20) have used

^{25}Mg NMR to study metal binding to the prothrombin fragment 1, which is the N-terminal peptide containing residues 1 to 156, and all ten Gla residues are situated among its first 45 amino acid residues. The pH dependence of the ^{25}Mg NMR line broadening was studied for a solution which contained a 100-fold excess of Mg^{2+} over protein. This resulted in two pK_a's, 4.2 and 7.5. The lower pK_a agrees well with results obtained for small Gla-containing peptides and is interpreted as due to the protonation of the Gla carboxylates. The higher pK_a, however, has no counterpart in the small peptides and it is assumed that a functional group other than the carboxylate groups of Gla is involved in the magnesium binding.

Phospholipase A_2

Phospholipase A_2 is a pancreatic enzyme that catalyses the hydrolysis of the 2-acyl linkage in phospholipids (21). The enzyme is secreted by the pancreas in an inactive form, prophospholipase A_2 (PPLA2), which is activated by trypsin cleavage. It has been shown that both the enzyme (PLA2) and its zymogen bind Ca^{2+} ions with binding constants of 1-5 x 10^3 M^{-1} (22,23) and that this calcium binding is essential for the activity of the enzyme. Furthermore, it has been shown that PLA2 has a second, weaker Ca^{2+} site with a binding constant of $K_a \sim 50$ M^{-1} (23). These binding constants are in the range where ^{43}Ca NMR will be useful, $K_a < 10^4$ M^{-1}. Figures 6 and 7 show the dependence of ^{43}Ca NMR line broadening on the calcium concentration for PLA2 and PPLA2, respectively (34,25). The data in Figure 7 can be nicely fitted, assuming one calcium binding site with a binding constant $K_a = 2.5 \ 10^3$ M^{-1} and an exchange rate of $k_{off} = 3 \times 10^3$ s^{-1}.

It may not be obvious that also the exchange rate can be determined from the metal concentration dependence. However, the shape of the curve depends also on the exchange rate. In order to show this, one has to go into total band shape calculations based on the Bloch equations, because this dependence shows up only when the Swift-Connick approximation, $p_A \gg p_B$, breaks down. No good fit of calculated linewidths to the ones shown in Figure 6 could be obtained with the assumption of a single binding constant for Ca^{2+}, independent of the choice of K_a and k_{off}. Introducing a second Ca^{2+} site, however, with a lower binding constant and in fast exchange resulted in the calculated curve shown in Figure 6 with $K_a^W = 15$ M^{-1} and $K_a^S = 10^3$ M^{-1}. For every site introduced in the calculation there are three adjustable parameters: binding constant, K_a, exchange rate, k_{off}, and relaxation rate of bound ions, R_{2B}. In this case already two sites contain more adjustable parameters than one can reasonably estimate from Figure 6. Therefore as a first approximation the relaxation rates of the two bound sites were taken to be the

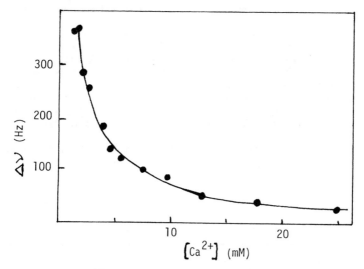

Figure 6. The ^{43}Ca NMR linewidth as a function of calcium concentration for a 2.0 mM PPLA2 solution (pH 7.5). The spectra were recorded at 24°C (from Ref. 24).

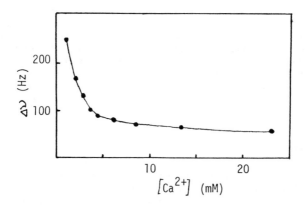

Figure 7. The ^{43}Ca NMR linewidth as a function of calcium concentration for a 0.74 mM PLA2 solution, pH 7.1, temperature = 23°C (from Ref. 25).

same, and the exchange rate of the weak site was taken to be so fast that a variation in it did not affect the calculation. The assumption of fast exchange for the weak site is probably reasonable for such a weak site. Sometimes it might be justified to assume that the relaxation rates of all bound ions are the same for all sites. However, for PLA2 there is no information available regarding the structure of the weak site, and there is therefore no basis for the assumption of the same symmetry of the two sites. It may well be that the larger observed quadrupole coupling constant for PLA2 than for PPLA2 is due to the fact that the two sites in PLA2 were assumed to have the same quadrupole coupling constant. The results are summarized in Table 1.

Table 1. A Comparison of the Thermodyanmic and Dynamic Properties of Calcium Binding to PLA2 and PPLA2

| | PLA2 | | PPLA2 |
	Active Site	Weak Site	Active Site
K (M^{-1})	1.0×10^3	15	2.5×10^3
pK_a	5.7	4.6	5.2
χ (MHz)	1.5	1.5	0.8
k_{off} (s^{-1})	1.3×10^3	fast	3.0×10^3
k_{on} (s^{-1})	1.3×10^6	–	7.5×10^6

Also the pH dependence of the line broadening of the ^{43}Ca NMR signal shows significant differences between PLA2 and PPLA2. The PPLA2 data can, even with a large excess of ^{43}Ca ions, be nicely fitted assuming one pK_a = 5.2. The PLA2 data on the other hand show a pronounced concentration dependence. At low calcium ion concentration the pH dependence shows mainly a pK_a = 5.7, whereas with increasing calcium concentration also a pK_a = 4.6 has to be included. This can be rationalized assuming that the strong site is sensitive to a pK_a = 5.7 and the weak site to a pK_a = 4.6.

Troponin C

The initial event in the contraction of skeletal muscles is generally assumed to be the binding of calcium ions to troponin C, one of three subunits in the protein complex, troponin, on the actin thin filament. Troponin C (MW ~ 18,000) contains 46 carboxylic groups out of a total of 159 amino acids (26). Based on primary structure analogies between parvalbumin, the crystal structure of which is known, and troponin C (TnC), Kretsinger and Barry have predicted the location and general structure of four calcium binding sites (27). Equilibrium dialysis and other types of biophysical studies have indicated that TnC indeed has four Ca^{2+} binding sites, two of which have a high affinity for Ca^{2+} ($K_{a_3} = 2.1 \times 10^7$ M^{-1}) and also bind Mg^{2+} competitively ($K_a = 5 \times 10^3$ M^{-1}) (28). These sites are usually called "the Ca^{2+}-Mg^{2+} sites." The other two sites have a lower affinity for Ca^{2+} ($K_a = 3.2 \times 10^5$ M^{-1}) and the Mg^{2+} affinity of these sites was too small to be measurable. These sites are called "the Ca^{2+}-specific sites."

The ^{43}Ca NMR excess linewidth in the presence of TnC as a function of the Ca^{2+} concentration is shown in Figure 8. Up to a [Ca^{2+}]/[TnC] ratio of about two, the linewidth of the observed signal is approximately constant and ca. 800 Hz. Only the high affinity sites are expected to be populated and, since the fraction of free Ca^{2+} under the experimental conditions is exceedingly low, the observed signal here corresponds to $^{43}Ca^{2+}$ bound to these sites (vide supra) (29).

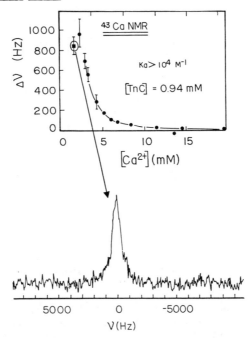

Figure 8. The ^{43}Ca NMR linewidth as a function of calcium concentration for a solution containing 0.94 mM TnC at pH 6.6 and 23°C. Lower trace shows the ^{43}Ca signal observed for the lowest calcium concentration.

At $[Ca^{2+}]/[TnC]$ ratios exceeding two, the observed ^{43}Ca NMR spectrum is a superposition of a broad signal from the Ca^{2+} ions bound to the two "Ca-Mg" sites and a narrower signal due to calcium ions exchanging between the low affinity sites and the free state. The results of Figure 8 are compatible with Ca^{2+} ions exchanging with two macromolecular sites with a binding constant $K_a > 10^4$ M^{-1}.

The temperature dependence of the ^{43}Ca NMR linewidth is shown in Figure 9. The curve is typical for a nucleus undergoing slow exchange at low temperatures and progressing through intermediate exchange rates to essentially fast exchange at higher temperatures. The results of Figure 9 strongly indicate that what we are observing is predominantly chemical exchange of $^{43}Ca^{2+}$ with the two weak Ca^{2+} specific sites of TnC. An analysis of the temperature dependence along the lines presented in the theoretical section yields a common rate constant for these two sites of $k_{off} = 1.0 \pm 0.1 \times 10^3$ s^{-1} at 23°C. The activation free energy, $\Delta G^{\ddagger}$, corresponding to this value is 57 kJ/mol (13.6 kcal/mol).

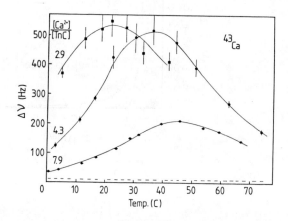

Figure 9. The temperature dependence of the ^{43}Ca linewidth for solutions containing various $[Ca^{2+}]/[TnC]$ ratios as indicated.

In the presence of 20 mM of Mg^{2+}, the temperature dependence of the ^{43}Ca NMR linewidth is flatter than in the absence of Mg^{2+}, as is shown in Figure 10. The data can no longer be fitted assuming two identical weak Ca^{2+} sites. The best fit is obtained when two different rate constants are used, $k_{off}^{(1)} = 10^3$ s^{-1} at 23°C and $k_{off}^{(2)} = 0.5 \times 10^2$ s^{-1} at 23°C. These results can be rationalized if one assumes the existence on TnC of secondary Mg^{2+} binding site(s) with an affinity of the order

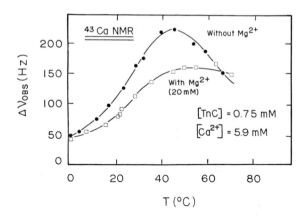

Figure 10. The temperature dependence of the ^{43}Ca linewidth for solutions containing TnC, without Mg^{2+} or with 20 mM Mg^{2+}.

of 10^2-10^3 M^{-1}. This assumption is further substantiated by a competition experiment where the ^{43}Ca NMR signal at constant $[Ca^{2+}]/[TnC]$ ratio is observed as a function of the Mg^{2+} concentration, Figure 11. It is seen that the ^{43}Ca linewidth decreases about 60% as the Mg^{2+} concentration is increased from zero to about 40-50 mM. This result is not consistent with a direct competition between Ca^{2+} and Mg^{2+} for the weak Ca^{2+} specific sites, since Mg^{2+} is known to bind very weakly to these sites (28). The results of Figure 11 are, however, consistent with the binding of Mg^{2+} to secondary sites that <u>indirectly</u> influence the mode of binding of Ca^{2+} to the Ca^{2+} specific sites. Further evidence for the presence of secondary Mg^{2+} and Ca^{2+} binding sites on TnC, apart from the "Ca^{2+}-Mg^{2+}" and "Ca^{2+} specific" sites, comes from the studies of Hincke, McCubbin, and Kay (30), cf. also Potter et al. (31).

There is still some uncertainty as to what is the concentration of "free" or "diffusible" Mg^{2+} in muscle cells. The concentration measured in different systems with various techniques is 2 to 6 mM (32). It appears from the present NMR results that under such conditions Mg^{2+} will have an obvious effect on the rate of exchange of Ca^{2+} ions from the Ca^{2+} specific sites. Since the latter are often regarded as the regulatory sites of TnC (31), this effect may have physiolgoical consequences. It would, for example, appear that primarily only <u>one</u> of the two Ca^{2+} specific sites will be alternatively filled and emptied during a cycle of muscle contraction and relaxation.

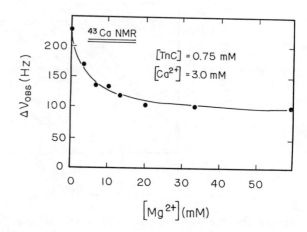

Figure 11. The effect on the ^{43}Ca NMR linewidth for a solution containing 3.0 mM Ca^{2+}, 0.75 mM TnC, and increasing amounts of Mg^{2+}, at pH 7.1 and 23°C.

It may be of some interest to show the effect of addition of Ca^{2+} ion on the ^{25}Mg NMR linewidth in the presence of TnC, Figure 12. Firstly we note that in the absence of Ca^{2+} the ^{25}Mg NMR excess linewidth is considerable. The broadening is primarily attributed to ^{25}Mg^{2+} exchange with the "Ca^{2+}-Mg^{2+}" sites. The steep decrease in the ^{25}Mg linewidth as Ca^{2+} is added is consistent with this; the binding constant of Ca^{2+} for these sites is reported to be nearly four orders of magnitude larger than that of Mg^{2+} (27). The break point in the ^{25}Mg linewidth curve at a ratio for (Ca^{2+})/(TnC) ~ 2 is also in accordance with this interpretation.

In order to determine the roles played in muscle contraction by the Ca^{2+}-Mg^{2+} sites and the Ca^{2+} specific sites of TnC, it is crucial to have good values for the "on" and "off" rates of Ca^{2+} and Mg^{2+} for the different sites. The Ca^{2+} specific sites have been argued to be the true regulatory sites (31). One of the arguments is that in a resting state the Ca^{2+}-Mg^{2+} sites would be mainly occupied by Mg^{2+}, and the off rate of Mg^{2+} from these sites (determined by a stopped flow technique using TnC (or troponin) labeled with a fluorescent probe) would be far too low to allow Ca^{2+} to bind appreciably during the few milliseconds that we know are required for triggering the contractile processes. The off rate for Mg^{2+} from the Ca^{2+}-Mg^{2+} sites determined by the stopped flow technique is 8 s^{-1}. From the tempera-

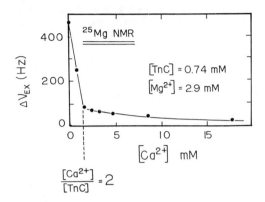

Figure 12. The ^{25}Mg NMR linewidth as a function of Ca^{2+} for a 2.9 mM Mg^{2+} solution (pH 7.1) in the presence of 0.74 mM TnC at 23°C (from Ref. 29).

ture dependence of the ^{25}Mg NMR signal, we have calculated a preliminary off rate from the same sites to be ca. 8×10^3 s^{-1}. If this value can be trusted, it would largely remove the above argument against the Ca^{2+}-Mg^{2+} sites being involved in the muscle triggering process. More experiments have to be performed until this question is satisfactorily resolved.

DNA

Double-stranded, helical DNA can be considered as a polyelectrolyte. According to Manning's ion condensation model (33) the amount of counterions bound to a linear polyelectrolyte is mainly given by the linear charge density. This quantity changes extensively at the DNA melting point. Figure 13 shows the ^{43}Ca linewidth as a function of temperature for a solution containing DNA with a phosphate concentration of 7.1 mM and 35 mM Ca^{2+} (34). The data in this figure indicate that the Ca^{2+} exchange between free and DNA-bound ions is slow up to ca. 80°C, where the DNA melts and the double helix is unfolded. At this point the broadening of the ^{43}Ca NMR signal drops to a very small value, indicative of no interaction between Ca^{2+} and unfolded DNA. This is in agreement with the ion condensation model since the charge density is much less for the unfolded than for the native DNA.

In another ^{25}Mg NMR experiment, Rose et al. (35) for the first time observed non-Lorentzian lineshapes for ^{25}Mg when Mg^{2+} ion is interacting with DNA. From a competition experiment using Na^+ they found that one part of the ^{25}Mg excess linewidth can be removed using a ca. 10-fold excess of Na^+ over Mg^{2+}. At very high Na^+ concentrations the ^{25}Mg linewidth is independent of the sodium concentration, Figure 14. This indicates that the ^{25}Mg broadening is due to two different types of binding, (i) site binding which is not affected by the addition of Na^+ and (ii) territorially bound Mg^{2+} ions that will be displaced by the addition of Na^+.

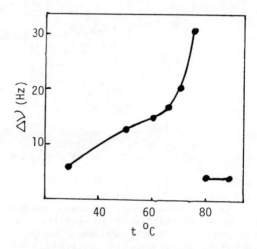

Figure 13. The dependence of the ^{43}Ca NMR linewidth on temperature for a solution containing 7.1 mM DNA-P, 35 mM Ca^{2+}, 57 mM Na^+ in 13 mM Tris-HCl at pH 5.2 (from Ref. 34).

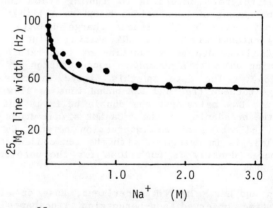

Figure 14. The ^{25}Mg NMR linewidth as a function of sodium concentration in a sample containing 14.8 mM DNA-P and 77.2 mM $MgCl_2$ (from Ref. 35).

REFERENCES

(1) R. J. P. Williams, Chem. Soc. Quart. Rev., 331 (1970).
(2) J. A. Magnuson and A. A. Bothner-By, in "Magnetic Resonance in Biological Research," A. Franconi, Ed., Gordon and Beach, London, 1969.
(3) R. G. Bryant, J. Am. Chem. Soc., 91, p. 1870 (1969).
(4) J. Parello, H. Lilja, A. Cavé, and B. Lindman, FEBS Lett., 89, p. 191 (1978).
(5) D. Hoult, Prog. NMR Spectrosc., 12, p. 45 (1978).
(6) S. Forsén and B. Lindman, in "Methods of Biochemical Analysis," Vol. 27, D. Glock, Ed., J. Wiley, New York, 1981.
(7) B. Halle and H. Wennerström, J. Magn. Reson., 44, p. 89 (1981).
(8) L. G. Werbelow, J. Chem. Phys., 70, p. 5381 (1979).
(9) L. G. Werbelow and A. G. Marshall, J. Magn. Reson., 43, p. 443 (1981).
(10) P. O. Westlund and H. Wennerström, J. Magn. Reson. in press.
(11) H. McConnell, J. Chem. Phys., 28, p. 430 (1958).
(12) T. Drakenberg, S. Forsén, and H. Lilja, manuscript in preparation.
(13) R. G. Bryant, J. Magn. Reson., 6, p. 159 (1972).
(14) J. P. Robertson, Jr., R. G. Hiskey, and K. A. Koehler, J. Biol. Chem., 253, p. 5880 (1978).
(15) E. Bouhoutsos-Brown, D. M. Rose, and R. G. Bryant, J. Inorg. Nucl. Chem., 43, p. 2247 (1981).
(16) T. Drakenberg, Acta. Chem. Scand., A36, p. 79 (1982).
(17) R. M. Farmer and A. I. Popov, Inorg. Nucl. Chem. Lett., 17, p. 51 (1981).
(18) T. Andersson, T. Drakenberg, S. Forsén, E. Thulin, and M. Swärd, J. Am. Chem. Soc., 104, p. 576 (1982).
(19) J. Stenflo and J. W. Suttie, Ann. Rev. Biochem., 46, p. 157 (1977).
(20) P. Robertson, Jr. K. A. Koehler, and R. G. Hiskey, Biochem. Biophys. Res. Commun., 86, p. 265 (1979).
(21) G. H. deHaas, N. M. Postema, W. Nieuwenhuizen, and L. L. M. van Deenen, Biochim. Biophys. Acta, 159, p. 103 (1968).
(22) W. A. Pieterson, J. J. Volwerk, and G. H. deHaas, Biochemistry, 13, p. 1439 (1974).
(23) A. J. Sloboom, E. H. J. M. Jansen, H. Vlijm, F. Pottus, P. Soares de Araujo, and G. H. deHaas, Biochemistry, 17 p. 4593 (1978).
(24) T. Andersson, T. Drakenberg, S. Forsén, T. Wielock, and M. Lindström, FEBS Lett., 123, p. 115 (1981).
(25) T. Andersson, "NMR Studies of Ion Binding to Proteins," Dissertation, Lund (1981).
(26) J. H. Collins, M. L. Grease, J. D. Potter, and M. J. Horn, J. Biol. Chem., 252, p. 6356 (1977).

(27) R. H. Kretsinger and C. D. Barry, Biochim. Biophys. Acta, 405, p. 40 (1975).
(28) J. D. Potter and J. Gergely, J. Biol. Chem., 250, p. 4628 (1975).
(29) T. Andersson, T. Drakenberg, S. Forsén, and E. Thulin, FEBS Lett., 125, p. 39 (1981).
(30) M. T. Hincke, W. D. McCubbin, and M. Kay, Can. J. Biochem., 56, p. 384 (1978).
(31) J. D. Potter, S. P. Robertson, and J. D. Johnson, Fed. Proc., 40, p. 2653 (1981).
(32) A. Scarpa and F. Brinley, Fed. Proc., 40, p. 2646 (1981).
(33) G. S. Manning, Quart. Rev. Biophys., 11, p. 179 (1978).
(34) P. Reimarsson, J. Parello, T. Drakenberg, H. Gustavsson, and B. Lindman, FEBS Lett., 108, p. 439 (1979).
(35) D. M. Rose, M. L. Bleam, M. T. Record, Jr., and R. G. Bryant, Proc. Natl. Acad. Sci. (USA), 77, p. 6289 (1980).

CHAPTER 15

GROUP III ATOM NMR SPECTROSCOPY

R. Garth Kidd

Faculty of Graduate Studies
The University of Western Ontario
London, Canada

ABSTRACT

The three factors determining the nuclear shielding of any atom are introduced and their relative importance discussed. The dependence of shiftability on the inverse cube radius $\langle 1/r^3 \rangle$ for valence shell electrons is identified for Group III atoms. The deshielding effects of ligand substitution and their correlation with the nephelauxetic effect are discussed. The dependence of relaxation rate and mechanisms available upon quadrupole moment is explored.

INTRODUCTION

Group III is of particular significance to NMR spectroscopists because exhibited among the chemical properties of its compounds are those three that allow us to recognize the pattern which yields a broad understanding of the whole chemical shift phenomenon.

(i) The molecules formed by Group III atoms are kinetically stable in solution. Rates of ligand exchange are slow and, as a result, chemical shifts are essentially independent of solvent and concentration.

(ii) For a single atom, two separate classes of molecule distinguished by the oxidation state of the atom are observed. Both Tl^{III} and Tl^{I} compounds are well characterized in solution. We will be looking at the meagre but significant ^{69}Ga results for Ga^{I} to compare with those for the more stable Ga^{III} compounds.

(iii) For a single atom, static environments of different <u>coordination symmetry</u> are encountered, the ones which correlate well with and appear to influence chemical shifts being trigonal, tetrahedral, and octahedral.

When we sit down at the spectrometer and measure the NMR spectra from a particular nucleus, it is ligand substitution that appears to be the most obvious cause of chemical shifts. Easily overlooked is the fact that changes in coordination which can be represented as ligand substitution can also represent change in oxidation state and change in coordination symmetry. Adopting the conventional scientific wisdom of considering major effects before worrying about minor ones has the result of relegating the most obvious effect to third place behind oxidation state and coordination symmetry, both of which represent larger perturbations than does the substitution of one ligand for another. (Table I).

Table I. Factors Determining Nuclear Shielding

Oxidation State	Lower oxidation states experience greater shielding than more highly oxidized ones.	$\delta(Tl^I) - \delta(Tl^{III}) \sim 3000$ ppm Fraction ^{205}Tl range = 55%
Coordination Symmetry	High symmetry environments experience greater shielding than low symmetry ones.	$\delta(B_{Trigonal}) - \delta(B_{Tetrahedral})$ ~ 80 ppm Fraction ^{11}B range = 35%
Ligand Substitution	Ligands high in the nephelauxetic series cause greater shielding than those that exert a low nepheleuxetic effect.	$\delta(AlI_3Cl^-) - \delta(AlI_4^-) = 49$ ppm Fraction ^{27}Al range = 8%

We will look at the details of these factors as they reveal themselves in the spectroscopies of each of the Group III atoms, but before doing so we must reflect upon the nature of molecular descriptions and pay our respects to the more successful theoretical ideas that convert this art into a science.

ON MOLECULAR DESCRIPTIONS

NMR spectroscopy has been referred to as that branch of science where physicists have seduced the chemist with instruments so elegantly conceived and chemically penetrating he forgets why he sat down at the spectrometer. Let us therefore begin by asking that fundamental question lying behind all chemical enquiry and research. Can a molecule be represented by the sum of its constituent parts, i.e, does molecular property = Σ(atomic property)$_i$? The reason for this fundamental motivation is fairly obvious. Given the mathematically limitless number of ways in which atoms might combine with one another, it provides an approach for reducing the infinite complexity of matter to an order and scale that the human mind can comprehend.

The organic chemist sees his molecule as a collection of functional groups and regards each functional group as having certain chemical properties that remain constant from molecule to molecule, independent of the carbon skeleton to which it is attached. The inorganic chemist sees molecules as central atoms (usually metals) in specific oxidation states that define the number of vacant orbitals available for bonding, and ligands which are Lewis bases with pairs of electrons available to form the chemical bonds. The inorganic chemist's ligand is analogous to the organic chemist's functional group in representing a molecular component viewed as transferable with properties constant from molecule to molecule.

Casting this bipartite view of a molecule into quantitative terms requires a central atom factor, α, and a ligand factor, β, allowing one to represent the value of any molecular property as the product given in eq. 1, the summation extending over all the

$$\text{Molecular property} = \alpha \sum_i \beta_i \tag{1}$$

ligands bonded to the central atom. Values for β are determined by analyzing for ligand dependence the spectral range observed for a particular nucleus. Values for α are determined by measuring for different central atoms the property difference experienced under ligand substitutions that represent a constant difference in ligand environment.

It is the task of the NMR spectroscopist to determine the extent to which this ideal view of the molecule is realized and, perhaps of even greater importance, to indicate where transferability of ligand properties breaks down.

THE IMPORTANCE OF $\langle 1/r^3 \rangle$

If a search for that intrinsic property of a ligand which determines the nuclear shielding of a metal and which is transferable without change from metal to metal were to be successful, the outcome would be a table of ligands with their individual shielding parameters, β, the value of which would be constant from molecule to molecule. This β parameter would be regarded as a substituent effect and would in combination with the β parameters for other ligands bound to a metal and multiplied by a metal parameter, α, describing its inherent "shiftability," yield the metal atom chemical shift for any ligand environment. Mathematically the chemical shift, δ, would be calculated from eq. 2, and this model would be a first-order

$$\delta_{Metal} = \delta_{Metal}(\beta_{Ligand\ 1} + \beta_{Ligand\ 2} + \ldots) \qquad (2)$$

model in that each ligand contributes (independently of the other ligands present) a constant increment to the shielding.

Whether or not the first-order substituent constant model is a valid one for metal atom chemical shifts (and we will see shortly that it is not), some metals will be more susceptible than others to the same change in ligand environment ($\Delta\Sigma\beta_i$). The broad pattern of variation in shiftability, α, was laid out in 1964 by Cynthia Jameson (1). An initial approximation to the NMR shiftability of an atom is the chemical shift range that it experiences. For Group III atoms these are ^{11}B, 220 ppm; ^{27}Al, 270 ppm; ^{71}Ga, 1400 ppm; ^{119}In, 2000 ppm; and ^{205}Tl, 5400 ppm. This pattern of comparable ranges for rows 1 and 2 followed by substantial increases in the 3rd, 4th, and 5th rows is typical of the main group elements, but is less pronounced among the transition metals where a partially filled d shell complicates the pattern (Figure 1). Jameson has shown that the inverse cube radius of the valence shell electrons for atoms, as deduced from the spin-orbit splittings observed in their atomic spectra, mirrors the pattern of chemical shift ranges. This correlation, taken together with the appearance of the $\langle 1/r^3 \rangle$ variable in Ramsey's generalized screening equation, leads us to attribute variations in shiftability α to variations in $\langle 1/r^3 \rangle_{np}$.

For a correlation as rough as this one necessarily must be, the parallel between $\langle 1/r^3 \rangle_{np}$ for the Group III atoms and the range of chemical shifts which they experience is truly remarkable (Table II). In an area where agreement within $\pm 100\%$ is good, the average for the ratio δ-range/$\langle a_o^3/r^3 \rangle_{np}$, excluding the ^{27}Al case, is 332 with a standard deviation of 7%. Even when we include ^{27}Al in the averaging process, the standard deviation is still only 22%.

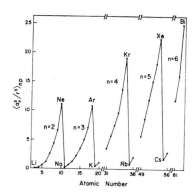

Figure 1. Variation with atomic number of the inverse cube radius of valence shell electrons. Reproduced from Ref. 1 by permission of the American Institute of Physics.

Table II. Shift Range and $\langle 1/r^3 \rangle$ for Group III Atoms

Isotope	n	δ range	$\langle a_o^3/r^3 \rangle_{np}$ [a]	Ratio
^{11}B	2	220 ppm	0.61	360
^{27}Al	3	270	1.27	210
^{71}Ga	4	1400	3.48	400
^{115}In	5	2000	5.71	350
^{205}Tl	6	4000[b]	11.81	340

[a] R. G. Barnes and W. V. Smith, Phys. Rev., 93, p. 95 (1954); a_o is the atomic unit of length, 5.3×10^{-10} m.
[b] Range for Tl^{III} compounds only.

When comparing chemical properties among the atoms of a Periodic Group, it is the first row element that is usually anomalous. With respect to nuclear shielding, however, it is the second row element that represents the anomaly. Whereas the trend is to increased shielding range with increasing atomic size down a Group, the ^{11}B and ^{27}Al ranges are nominally the same, the ^{13}C range exceeds the ^{29}Si range by a factor of 2, the ^{14}N range exceeds the ^{31}P range by a factor of 1.3, the ^{17}O range exceeds the ^{33}S range by a factor of 3, and the ^{19}F range of 1000 ppm is comparable to that of ^{35}Cl. Clearly these comparisons cannot be rationalized in terms of $\langle 1/r^3 \rangle$ alone. It is important to recognize, however, that it is the second row element that is anomalous, not the first, and the $\langle 1/r^3 \rangle$ correlation in Table II does this with conviction.

SHIELDING OF GROUP III ATOMS

The compounds formed by metal atoms of Periodic Groups I and II discussed in other chapters are characterized in solution by their high degree of kinetic lability. From an NMR standpoint this means that single time-averaged signals, whose chemical shifts exhibit a marked concentration dependence, are more commonly encountered than multi-line spectra containing discrete lines assignable to specific molecules in solution. On reaching Group III, however, we encounter covalently bonded metal environments that are kinetically stable and persist in solution without exchanging their ligands long enough for the hapless NMR spectroscopist with his slow-shuttered camera to obtain a static chemical shift measurement.

Except for thallium(I), which in some respects behaves like an alkali metal, thallium(III) and the other Group III atoms all provide NMR spectra the interpretation of which is open to little ambiguity (Table III). Boron-11 was among the first half dozen nuclei to be investigated during the earliest days of NMR and its literature is now extensive. Next in volume of data is ^{205}Tl, whose narrow lines and high receptivity have made it a favourite with James Hinton and his research group in Arkansas. Aluminum-27 with its medium-sized quadrupole moment has received moderate but regular attention first by O'Reilly in 1960 (2) and latterly by Akitt in Leeds (3). My own group at Western Ontario did the first work on the tetrahaloaluminate anions (4). Work on ^{71}Ga has been limited to about a dozen studies, while ^{115}In data appear in about half that number.

The chemical shift range charts for ^{11}B, ^{27}Al, ^{71}Ga, ^{115}In, and ^{205}Tl presented in Figures 2, 3, 4, and 5 illustrate the broad pattern of ligand shielding effects. In these charts the ligands have been arranged from top to bottom in diminishing order of their ability to shift the metal resonance to high frequencies, i.e., in diminishing order of their ability to deshield the metal. A rapid scan of these charts in such a way as to eliminate oxidation state and coordination symmetry effects provides

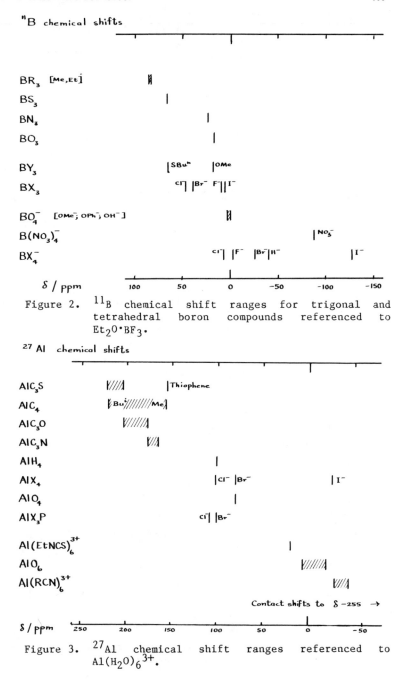

Figure 2. ^{11}B chemical shift ranges for trigonal and tetrahedral boron compounds referenced to $Et_2O \cdot BF_3$.

Figure 3. ^{27}Al chemical shift ranges referenced to $Al(H_2O)_6^{3+}$.

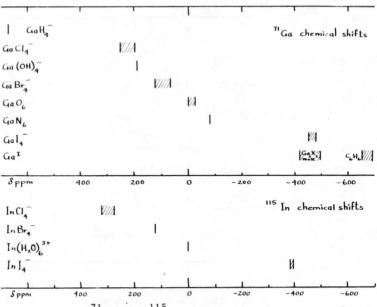

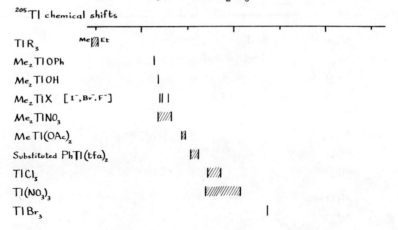

Figure 4. ^{71}Ga and ^{115}In chemical shift ranges referenced to the respective $M(H_2O)_6^{3+}$.

Figure 5. ^{205}Tl chemical shift ranges for Tl^{III} and Tl^{I} compounds referenced to Tl(aq)$^+$.

us with a reasonably consistent pattern of chemical shifts due solely to ligand substitution (Table IV).

Table III. Nuclear Properties of the Group III Isotopes

Nucleus	Spin	Abundance(%)	Quadrupole Moment $Q/10^{28}$ m^2	Usual shielding reference	Receptivity re ^{13}C
^{10}B	3	19.6	0.074	$Et_2O \cdot BF_3$	22
^{11}B	3/2	80.4	0.036	$Et_2O \cdot BF_3$	754
^{27}Al	5/2	100	0.15	$Al(H_2O)_6^{3+}$	1170
^{69}Ga	3/2	60.4	0.18	$Ga(H_2O)_6^{3+}$	237
^{71}Ga	3/2	39.6	0.11	$Ga(H_2O)_6^{3+}$	319
^{113}In	9/2	4.3	1.14	$In(H_2O)_6^{3+}$	84
^{115}In	9/2	95.7	1.16	$In(H_2O)_6^{3+}$	1890
^{203}Tl	1/2	29.5	----	$Tl(aq)^+$	313
^{205}Tl	1/2	70.5	----	$Tl(aq)^+$	769

Table IV. Deshielding Effects of Ligand Substitution

^{27}Al Shifts:	C ~ S > Cl ~ H > Br ~ O > N > I
^{71}Ga Shifts:	H > Cl > Br ~ O > N > I
^{115}In Shifts:	Cl > Br > I
^{205}Tl Shifts:	C > Cl ~ NO > Br

The most consistent element of this pattern is the extremely high shielding exerted by iodide, which has been remarked in the spectroscopies of all atoms that form iodide compounds. Taken together with the resonances for bromide and chloride compounds at successively lower shieldings with a (Cl-Br)/(Cl-I) ratio of about 0.3 which persists among most covalent halides, this pattern has come to be known as the normal halogen dependence (NHD), to distinguish it from the inverse halogen dependence (IHD) exhibited by a few metals such as ^{47,49}Ti and ^{93}Nb.

It was the ^{27}Al spectra for the AlX$_4^-$ (X = Cl, Br, I) compounds that in 1968 provided the first theoretical basis for the NHD shielding pattern (4). Since deshielding is proportional to $\langle 1/r^3 \rangle$ through the paramagnetic term of the Ramsey equation expressed in the form of eq. 3, the ligand whose binding causes

$$\sigma^P = -\frac{2}{3} (e^2\hbar^2/m^2c^2) \frac{1}{\Delta E} \langle 1/r^3 \rangle_p P_u \qquad (3)$$

the largest increase in r for the ^{27}Al valence shell electrons, other things being equal, will cause the smallest deshielding and the highest field ^{27}Al resonance. This model, while perfectly respectable on theoretical grounds, is of little value unless it can be empirically tested with an independent measure of the influence of ligand substitution upon $\langle 1/r^3 \rangle$. Fortunately this information is available in the form of the nephelauxetic series established by Klixbull Jorgenson (5) in which the halides are arranged Cl$^-$ < Br$^-$ << I in order of their abilities to "expand the cloud" of a metal atom's valence shell electrons.

For the other ligands appearing in the range charts, nephelauxetic parameters are unavailable. What can be said is that, based on chemical intuition, their positions in the deshielding sequences are not inconsistent with their expected influences on valence shell cloud expansion. It should also be noted that for every metal, the chloride to iodide chemical shift represents well over half the total shielding range for that metal, so shielding differences for other ligand substitutions are less significant than those accompanying halide substitutions.

RELAXATION OF GROUP III ATOMS

The nuclear magnetic relaxation characteristics of an atom and the chemical information that flows from this measurement are highly dependent upon the quadrupole moment of the nucleus under observation. By this criterion, four categories of nucleus are recognized (Table V).

The dependence of relaxation rate and hence linewidth upon quadrupole moment is shown in the relaxation equation which is now well established over a broad range of quadrupolar nuclei (6) (eq. 4). Relaxation rate is seen to be proportional to Q^2 and

$$R_2 = \frac{1}{T_2} = \pi \cdot \Delta\nu = \frac{3\pi^2}{10} \cdot \frac{2I+3}{I^2(2I-1)} \cdot \left(\frac{eqQ}{h}\right)^2 \cdot \tau_\theta \qquad (4)$$

Table V. Quadrupole Moment Dependence of Relaxation Rate

High Q nuclei	$Q > 1.0 \times 10^{-28}$ m^2	Signals broadened beyond recognition except for nuclei in highly symmetric tetrahedral or octahedral environments.
Medium Q nuclei	$\frac{0.1}{10^{28} \text{m}^{-2}} < Q < \frac{1.0}{10^{28} \text{m}^{-2}}$	Relaxation dominated by quadrupolar mechanism. Resonances sufficiently narrow to be observable in a representative range of chemical environments.
Low Q nuclei	$Q < 0.1 \times 10^{-28}$ m^2	Resonances relatively narrow even in environments of low symmetry. Mechanisms other than quadrupolar may contribute to total relaxation.
I=1/2 nuclei	Q = 0	Relaxation by spin-rotation, dipole-dipole, scalar coupling, or shielding anisotropy mechanisms.

for high-Q nuclei it can reach values of 10^6 s^{-1}. For medium-Q and low-Q nuclei, typical R_2 values of 10^4 s^{-1} and 10^2 s^{-1} are observed, the precise values being determined by the electric field gradient at the nuclear position and giving linewidths in the regions of 3000 Hz and 30 Hz respectively.

Table III containing the nuclear properties for Group III isotopes shows both isotopes of In to be in the high-Q category, accounting for the fact that few indium spectra have been reported. Those that have been observed all represent environments in which In experiences a relatively low electric field gradient. Some relief from the high-Q syndrome is obtained if the nucleus, like ^{115}In, also has high spin. The value of the spin factor in the relaxation equation is inversely proportional to I and varies by a factor of 18 between I = 3/2 and I = 9/2.

Both ^{71}Ga and ^{27}Al are medium-Q nuclei whose resonances can be observed in a variety of chemical environments and whose linewidths provide structural information about electric field gradients at these sites. Although ^{27}Al has a slightly higher Q than does ^{71}Ga, its spin factor is smaller by a factor of 4, and in tetrahedral environments linewidths as narrow as 10 Hz are observed. In C_{3v} environments such as $AlBr_3Cl^-$ and $AlBrCl_3^-$ linewidths of just under 20 Hz are observed and the widest ^{27}Al resonance in a tetrahaloaluminate is 58 Hz for $AlI_2Cl_2^-$.

Both of the magnetically active isotopes of boron are low-Q nuclei, and while the rate of quadrupolar relaxation is slow, it is still sufficient to adumbrate or totally vitiate spin-spin couplings to neighbouring nuclei. The quadrupolar relaxation rates observed are in the region of 50 s^{-1}, sufficiently slow that other mechanisms can effectively compete for some of the action. Total relaxation rates are still sufficiently slow that excessive linewidth is not a major problem in boron spectroscopy.

The only spin 1/2 atoms in Group III are the two isotopes of thallium, and in ^{205}Tl spectroscopy, the whole relaxation picture is completely different from that for the quadrupolar nuclei where excessive relaxation can be the greatest single obstacle to obtaining resolved spectra. Completely barred from the one relaxation path that is 3-4 orders of magnitude more effective than any of the others, ^{205}Tl is left in splendid thermal isolation searching for ways to divest itself of excess resonance energy. Failure to do so results in the signal-destroying phenomenon of rf saturation. Among the four relaxation paths that remain open to the spin 1/2 nucleus, it is dipole-dipole coupling to neighboring nuclei that carries most of the energy for ^{1}H, ^{13}C, ^{15}N, and ^{19}F. Provided the dipole-dipole separation remains small, this is an effective path, but its effectiveness varies as r^{-6} and for atoms beyond the first row whose covalent radii exceed 1 Å, it rapidly ceases to become an effective mechanism. As a concrete example, the size increase of 0.4 Å between ^{13}C and ^{29}Si reduces the effectiveness of dipole-dipole relaxation for the latter by a factor of 13.

With a covalent radius of 1.5 Å, ^{205}Tl is less effectively relaxed through this mechanism by a factor of 55. In common with other spin 1/2 nuclei from the lower two rows of the Periodic Table (^{119}Sn, ^{125}Te, ^{129}Xe, and ^{207}Pb), ^{205}Tl is generally agreed to undergo the bulk of its relaxation by the spin-rotation mechanism. A linewidth that <u>increases</u> with increasing temperature provides strong evidence for spin-rotation relaxation since the temperature dependence for R_2^{sr} operating through the correlation time factor is the inverse of that for R_2^{dd}. Where the two mechanisms are in competition, R_2^{Total} will be dominated by R_2^{sr} at high temperatures and by R_2^{dd} in the cold. In thallium molecules containing covalently bound $^{35,37}Cl$, $^{79,81}Br$, or ^{127}I, relaxation through scalar coupling, recognized by $R_2^{sc} > R_1^{sc}$, should be anticipated. Also a possibility with any atom having a chemical shift range as wide as thallium's is relaxation resulting from chemical shielding anisotropy. Values of $\sigma_\perp - \sigma_\parallel$ in excess of 1600 ppm are required before R_2^{csa} need be considered, and since the anisotropy parameter in frequency units is field dependent, its presence is recognized by the higher fraction of R_2^{csa} in spectra taken at higher field strengths. Particularly with spectra obtained using super-conducting solenoids, the spectroscopist should be on the lookout for an anisotropy component to the relaxation.

REFERENCES

(1) C. J. Jameson and H. S. Gutowsky, J. Chem. Phys., 40, p. 1714 (1964).
(2) D. E. O'Reilly, J. Chem. Phys., 32, p. 1007 (1960).
(3) J. W. Akitt, Ann. Reports NMR Spectrosc., 5A, p. 465 (1972).
(4) R. G. Kidd and D. R. Truax, J. Am. Chem. Soc., 90, p. 6867 (1968).
(5) C. K. Jorgenson, "Absorption Spectra and Chemical Bonding in Complexes," Pergamon Press, London, 1962, Chapter 7.
(6) A. Abragam, "The Principles of Nuclear Magnetism," Oxford Unversity Press, London, 1961, Chapter 8.

CHAPTER 16

SOLUTION-STATE NMR STUDIES OF GROUP IV ELEMENTS (OTHER THAN CARBON)

Robin K. Harris

University of East Anglia, Norwich, England

ABSTRACT

The spin properties of the magnetic nuclides of silicon, germanium, tin, and lead are described. Each has a useful spin 1/2 nuclide, except for germanium. Experimental methods are discussed, including those of importance because of the long relaxation times frequently encountered (especially for ^{29}Si). Data on relaxation times, chemical shifts, and coupling constants are briefly reviewed.

INTRODUCTION

Group IV is unique in the Periodic Table in the eyes of NMR spectroscopists because no fewer than four of the five elements have stable spin 1/2 isotopes. Tin is blessed with three such nuclides! Only germanium has a stable quadrupolar nucleus (but no spin 1/2 isotope). Fortunately, germanium is arguably the least chemically important element in Group IV. For the other elements the existence of spin 1/2 nuclides is invaluable, since the resulting sharp lines give excellent spectral dispersion, which is particularly necessary for carbon and silicon in view of the richness of the covalent chemistry involving these elements.

NMR work on silicon, germanium, tin, and lead has been reviewed (1, Chapter 10). Two more recent reviews on ^{29}Si NMR have appeared (2,3), as has one on ^{119}Sn chemical shifts (4). A summary of ^{73}Ge studies has been published (5) as part of a review on less-common quadrupolar nuclei. Chemical applications will therefore only be mentioned briefly here, and the emphasis will be placed on the NMR properties of the Group IV nuclei and on techniques used.

NUCLEAR SPIN PROPERTIES

The spin properties of the magnetic nuclides of Group IV elements are given in Table 1. All the nuclides (with the possible exception of ^{207}Pb) may be classified as dilute spin species by virtue of their low natural abundance. Thus, when spectra are obtained under conditions of proton noise decoupling, each chemically distinct environment to the nucleus in question will yield a single line (in the absence of ^{19}F, ^{31}P, etc.), making spectral interpretation relatively easy. All three magnetically active tin isotopes, together with ^{29}Si and ^{73}Ge, have negative magnetogyric ratios, so that continuous proton decoupling gives (in principle) a negative nuclear Overhauser effect. Thus peak intensities are sometimes reduced and frequently inverted for ^{29}Si and for the tin isotopes. Moreover, an appropriate balance between relaxation mechanisms may result in a near-zero peak intensity--the "null signal problem." This difficulty does not arise for ^{73}Ge because the relaxation is usually dominated by its quadrupolar properties.

For investigators of tin NMR, there is generally no point in studying ^{115}Sn or ^{117}Sn, since ^{119}Sn is the most receptive of the three nuclei. Moreover, ^{73}Ge provides an exception in every way to the spin properties of the other group IV elements, and ^{13}C, being so commonly studied, is excluded from this survey. Consequently, attention will be focused on the important trio ^{29}Si, ^{119}Sn, and ^{207}Pb, which all have substantial magnetogyric ratios, namely about one-fifth that of the proton for ^{29}Si and ^{207}Pb, rising to nearly two-fifths for ^{119}Sn. In addition, their receptivities are larger than that for ^{13}C--particularly for ^{119}Sn. As a result, all three are relatively easy to study, except where long relaxation times or the null signal problem make for difficulties. Therefore, the relevant NMR literature is extensive. In the three years 1978-80, ^{29}Si, ^{119}Sn, and ^{207}Pb proved to be the 7th, 18th, and 26th most popular nuclei, respectively, for chemical NMR studies, with at least 85, 28, and 9 publications, respectively (including solid state studies). There seems to have been only one publication (6) involving ^{73}Ge (measured via the INDOR technique) in the same period, and there do not appear to have been any in 1981.

One reason for the lack of popularity of ^{73}Ge (in addition to the low interest in its chemistry and its quadrupolar nature) is its small magnetogyric ratio, leading to a low (and hitherto

Table 1. Spin Properties of Group IV Nuclides

Isotope	Nuclear spin	Natural abundance C/%	Magnetic moment[a] μ/μ_N	Magnetogyric ratio $\gamma/10^7$ rad T^{-1}s^{-1}	NMR frequency[b] Ξ/MHz	Standard	Relative receptivity[c] D^P	D^C
^{13}C	1/2	1.108	1.2166	6.7283	25.145004	Me$_4$Si	1.76 × 10^{-4}	1.00
^{29}Si	1/2	4.70	−0.96174	−5.3188	19.867184	Me$_4$Si	3.69 × 10^{-4}	2.10
^{73}Ge	9/2	7.76	−0.97197	−0.93574	3.488315	Me$_4$Ge	1.10 × 10^{-4}	0.622
^{115}Sn	1/2	0.35	−1.590	−8.792	(32.86)[d]	Me$_4$Sn	1.24 × 10^{-4}	0.705
^{117}Sn	1/2	7.61	−1.732	−9.578	35.632295	Me$_4$Sn	3.49 × 10^{-3}	19.8
^{119}Sn	1/2	8.58	−1.8119	−10.021	37.290662	Me$_4$Sn	4.51 × 10^{-3}	25.6
^{207}Pb	1/2	22.6	1.002	5.540	20.920597	Me$_4$Pb	2.01 × 10^{-3}	11.4

[a] This is the vector value, $\gamma\hbar[I(I+1)]^{1/2}/\mu_N$, not the maximum component, and it is corrected for diamagnetic shielding.
[b] This is the value for the standard (listed in the next column) in a magnetic field such that the protons in Me$_4$Si resonate at exactly 100 MHz.
[c] Receptivity, $\gamma^3 CI(I+1)$, relative to that of the proton (penultimate column) or ^{13}C (final column).
[d] Approximate value for the bare nucleus.

relatively inaccessible) NMR frequency and a receptivity that is also not good. The quadrupole moment is moderate (0.18 × 10^{-28} m^2), giving a reasonably good linewidth factor $\ell = (2I + 3)Q^2/I^2(2I - 1) = 2.4 \times 10^{-59}$ m^4 (compare ^{17}O at 2.2×10^{-60} m^4 and ^{59}Co at 2.0×10^{-58} m^4)--the high spin quantum number of 9/2 is a considerable advantage in this respect. Moreover, the tetravalency, with tetrahedral coordination normal, is a considerable advantage in keeping linewidths moderate--values of less than 25 Hz are normal for tetraalkyl- and tetrahalogermanes. Indeed, Me$_4$Ge gives (7) a linewidth of 0.58 Hz.

NMR METHODS

The general methods of Fourier transform NMR, as commonly used for ^{13}C, are usually adequate for the other Group IV elements. Most of the compounds studied contain protons, usually in organic moieties, so proton noise decoupling is normal. Difficulties arising from long spin-lattice relaxation times (common for ^{29}Si) and the null signal problem may be overcome (8,9) by the use of suitable shiftless relaxation agents. Figure 1 shows how the ^{29}Si relaxation rate induced by Cr(acac)$_3$, T_{1e}^{-1}, depends on concentration for a sample of octamethylcyclotetrasiloxane (with 10% CH$_2$Cl$_2$ and 10% C$_6$D$_6$ by volume) (10). For this system $T_1(^{29}$Si) = 91.7 s and the nuclear Overhauser enhancement $\eta = -2.19$ in the absence of the paramagnetic species, so only 0.01 mol dm^{-3} of Cr(acac)$_3$ is necessary to reduce T_1 by over a factor of 10 (to 7.0 s) and to give a negligible NOE ($\eta = -0.05$). Table 2 shows (10) the efficiency of Cr(acac)$_3$ and Fe(acac)$_3$ for

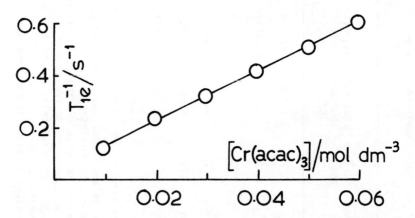

Figure 1. The effect of Cr(acac)$_3$ on $T_1(^{29}$Si) for [Me$_2$SiO]$_4$. The plot shows (10) the additional relaxation rate ascribed to the presence of the paramagnetic material.

inducing ^{29}Si relaxation in three siloxanes according to the equation $T_{1e}^{-1} = k[E]$, where E represents the paramagnetic species. The quantity d is the distance of closest approach of the paramagnetic centre to the relaxed nucleus, according to the model of Luz and Meiboom (11). The null signal problem may also be treated (12) using gated decoupling, and this is, of course, essential (13) if T_1 and the NOE are to be measured for such cases.

For ^{29}Si work the glass in the probe area can be a nuisance, since it gives a broad resonance ca. 110 ppm to low frequency of the signal due to Me_4Si. This is only really a difficulty when compounds containing silicon tetrahedrally bonded to four oxygens are being studied, since otherwise chemical shifts are sufficiently removed from the problem area. Use of plastic NMR tubes (and even plastic probe components) can reduce or eliminate the problem. Convolution-difference methods are also applicable in some cases, but not when the sample itself contains silica-type materials--in such cases a blank run must be employed (14), which reduces spectrometer efficiency substantially.

Table 2. Silicon-29 Relaxation Efficiencies for Paramagnetic Species (10)

Substrate	Cr(acac)$_3$		Fe(acac)$_3$	
	$k/dm^3mol^{-1}s^{-1}$	d/nm	$k/dm^3mol^{-1}s^{-1}$	d/nm
$(Me_3SiO)_2$	5.5	0.61	12.4	0.61
$(Me_2SiO)_4$	9.7	0.50	13.7	0.53
silicone oil[a]	11.3	0.48		

[a]Of average composition $Me_3SiO(Me_2SiO)_{50}SiMe_3$. The data are for the main-chain silicon nuclei.

Signal intensities can be greatly improved by the INEPT technique, and this has been demonstrated (15) for both ^{29}Si and ^{119}Sn. It has been shown that the relatively small (Si,H) and (Sn,H) coupling constants for Me_3Si- and Me_3Sn- groups, respectively, are suitable for INEPT transfer, but that the second delay time in the INEPT sequence is optimally set at 0.108 J_{SiH}^{-1} or 0.108 J_{SnH}^{-1}, respectively, in which case the enhancement is 9.42 for ^{29}Si and 5.00 for ^{119}Sn. The INEPT technique (or related selective-inversion methods) is particularly valuable (16,17) for measuring ^{29}Si relaxation times. Moreover, since repetition rates for pulsing in INEPT experiments depend on $T_1(^1H)$ rather than (for example) $T_1(^{29}Si)$, problems of inefficient relaxation (see below) are largely overcome.

Because of the low natural abundance of ^{29}Si, molecules containing more than one type of silicon do not give splittings due to (Si,Si) coupling on the principal peaks. However, "satellite" resonances can sometimes be detected (3,18,19), and yield values of J_{SiSi}. Such homonuclear satellite resonances are, of course, more pronounced for ^{119}Sn and ^{207}Pb in suitable molecules. For tin, double satellite resonances appear because of the existence of both ^{117}Sn and ^{119}Sn in comparable natural abundance. Whereas the satellites in ^{119}Sn due to the presence of ^{117}Sn are normally (but not always (20)) first order, the ^{119}Sn satellites in ^{119}Sn spectra are often second order.

Isotopic enrichment, which is common for ^{13}C and ^{15}N NMR, is much rarer for ^{29}Si because of the higher costs involved and the limitation on chemical species which can be purchased in the enriched form (in effect, only SiO_2 is available). However, Harris et al. have made considerable use (21-24) of the enrichment technique for unraveling the chemical complexities of aqueous silicate solutions. Isotopic enrichment for ^{119}Sn or ^{207}Pb NMR is unnecessary and has never been used.

In general, proton-decoupled ^{29}Si, ^{119}Sn, and ^{207}Pb spectra are simple, and consist of single lines for each type of chemical environment for the nucleus in question in the relevant sample. Satellite resonances can be more complicated, as mentioned above, but they can usually be readily distinguished from the main peaks, even for ^{119}Sn. When ^{19}F or ^{31}P is present, first-order splitting patterns are normal. Proton-coupled spectra are not often studied, so that second-order ^{29}Si, ^{119}Sn, or ^{207}Pb spectra have rarely been treated (see, however, 25). Enrichment with silicon-29 can lead to second-order features for symmetrical species, and an example (26) is shown in Figure 2. In such cases homonuclear decoupling is a desirable technique (22,24).

SPIN-LATTICE RELAXATION

Relaxation studies have concentrated on relatively small molecules, so that the validity of generalisations may be limited. For simple halogenated compounds of silicon, tin, and lead, scalar coupling to the relevant halogen provides (1,27) an important contribution to relaxation, particularly at low temperatures. The competing mechanism in these cases is spin-rotation, and this is also important for other small molecules. However, when the compound contains many protons the dipolar mechanism usually becomes effective. On the other hand, simple

SOLUTION-STATE NMR STUDIES OF GROUP IV ELEMENTS   349

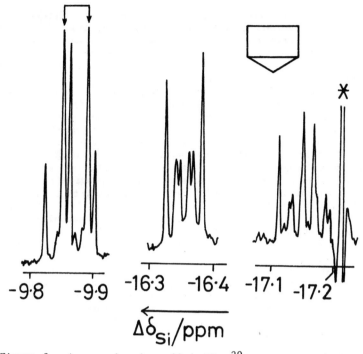

Figure 2. A second-order, 99.4 MHz ^{29}Si spectrum (26) for the bicyclic species indicated at the top right, in an alkaline solution of potassium silicate (0.65 M in SiO_2, with K:Si = 1.0). The spectrum is of the $[AM]_2X$ type. The doublet denoted by arrows is due to another species, as is most of the intensity in the peak marked by an asterisk (not taken to its full height). The chemical shift scale is with respect to the resonance of the orthosilicate ion. The simplified molecular diagram is such that Si atoms are at the corners, with oxygen atoms in between, plus such additional O^- or OH groups as are necessary for the tetravalency of silicon.

considerations show that the efficiency of dipolar relaxation is likely to be substantially less than in the case of ^{13}C. The dependence of $T_{1dd}(X)$ on $\gamma_X^2 r_{XH}^{-6}$ means that typical dipolar relaxation times from (^{29}Si,^{1}H) interactions are expected to be a factor of ca. 10 longer than from (^{13}C,^{1}H) interactions. Two factors reduce the efficiency of dipolar relaxation for ^{29}Si even further: (i) compounds with direct Si-H bonds are considerably less common in silicon chemistry than those with direct C-H bonds in carbon chemistry, and (ii) internal rotation about the Si-O bond is relatively easy (compared with that about the C-C bond). Consequently most measured spin-lattice relaxation times for ^{29}Si in solution-state conditions are in excess of 30 s. The consequences for routinely obtaining ^{29}Si spectra are obvious. The ease of internal rotation about the Si-O bond probably accounts for the surprisingly long values of $T_1(^{29}Si)$ for silicone groups even in very large molecules and in adsorbed species (see the discussion in 1, pp. 339-340). In fact it appears that spin-(internal-rotation) may be a competitive mechanism for most Me$_3$SiO groups. Such a mechanism has been suggested (28) as likely to account for the very low relaxation time (5.5 s) for phenylsilane. Aqueous silicate solutions provide the only other known examples of short ^{29}Si relaxation times in solution: typical values are 1-10 s, as is illustrated in Table 3. However, there is controversy about the relaxation mechanism since the samples appear to "age" when contained in glass tubes, and dissolution of paramagnetics from the glass is suspected (29). Shielding anisotropy is often small (30,31) for ^{29}Si, so that it is unlikely to provide a common relaxation mechanism. However, such a mechanism has been shown (32,33) to be important for ^{207}Pb in a few compounds. There appear to be relatively few studies of tin or lead relaxation in organometallic compounds, but reports suggest $T_1(^{119}Sn)$ to be ca. 0.2 to 7 s (34-37) and $T_1(^{207}Pb)$ to be ca. 0.1 to 2 s (38).

Table 3. Values (26) of $T_1(^{29}Si)$ for Various Silicate Structural Groupings[a]

Structural unit[b]	Q^0[c]	$\{Q^1\}$[d]	$\{Q^2\}$[e]	Q_3^2[f]	Q_6^3[g]	Q_8^3[h]
$T_1(^{29}Si)$/s	3.8	2.4	2.0	4.0	5.0	9.8

[a]Average values from three separate experiments on two similar potassium silicate solutions.
[b]The Q notation indicates silicon tetrahedrally coordinated by four oxygens. The superscript gives the number of coordinated oxygens which are part of siloxy bridges.
[c]The orthosilicate monomer.
[d]End groups.
[e]Middle groups.
[f]Cyclic trimer.
[g]Prismatic hexamer.
[h]Cubic octamer (in this case in a 2M tetramethylammonium silicate solution).

The maximum NOE values (with respect to proton irradiation) for ^{29}Si, ^{119}Sn, and ^{207}Pb are -1.51, -0.33, and 3.41, respectively, when the possibility of scalar relaxation is ignored. The fact that dipolar relaxation is relatively inefficient in many cases leads to less-than-maximum NOE observations. For ^{29}Si, in particular, this sometimes results in very low intensities, and the efficiency of NMR operation for both ^{29}Si and ^{119}Sn is often impaired.

CHEMICAL SHIFTS

The reference standards normally used are the relevant tetramethyl compounds, as indicated in Table 1, though objections can be raised to these on grounds of availability (apart from Me$_4$Si) and, in particular, inapplicability for internal use in aqueous solutions. Silicon-29 work on aqueous silicate solutions, for example, often uses the signal due to the orthosilicate ("monomer") species as a secondary reference, although the resonance frequency is known to be sensitive to pH, concentration, and gegenion. Unlike the situation for ^{13}C and ^{1}H studies, the ^{29}Si resonance of Me$_4$Si occurs towards the high frequency end of the chemical shift range, and it is often not well separated from signals due to other organosilanes containing Me$_3$Si groups. Recently (Me$_3$Si)$_4$C has been suggested (39) as an alternative reference for ^{29}Si NMR (as well as for ^{1}H and ^{13}C). Tetramethylstannane gives a tin signal near the middle of the range, as does Me$_4$Pb for lead NMR (when Pb(II) compounds are ignored).

Table 4 shows the shift ranges for the Group IV nuclei other than ^{13}C. The most shielded compounds are the tetraiodides (as is also the case for ^{13}C), except for ^{207}Pb, for which studies of PbI$_4$ have not been reported. This exemplifies the well recognised "heavy atom effect," which has been shown (47) to involve electron spin-orbital interactions on the heavy atom (i.e., iodine). When iodides are discounted, the most shielded cases are much less extreme--for silicon, for example, δ = -197.2 ppm provides (48) the next most shielded case.

As one might expect, the shielding range increases with atomic number for Group IV elements (though that for ^{29}Si is actually less than that for ^{13}C, probably largely because the mean distances between the nucleus in question and the valence p electrons are similar for carbon and silicon). This situation reflects, inter alia, the increasing numbers of electrons and the increasing atomic polarizabilities. Table 4 gives a separate range for ^{207}Pb resonance in solids, which is much larger than that hitherto observed for solutions.

Table 4. Shift Ranges for Group IV Elements Other than Carbon[a]

Element	Highest Shift			Lowest Shift			Range/ppm
	δ/ppm	Compound	Ref.	δ/ppm	Compound	Ref.	
Silicon	146.65	$Cl_2Si[Fe(CO)_2(C_5H_5)]_2$	40	−346.2 −351.7	SiI_4	41 42	~500
Germanium	153	$Ge(SMe)_4$	b	−1080.7 −1108.1	GeI_4	7	~1250
Tin	483	$MeSn[Co(CO)_4]_3$	43	−1701	SnI_4	44	~2180
Lead	472	Et_3PbCl	45	−2961.2	$Pb(NO_3)_2$	38	~3400
	11,150	Pb (solid)	1,46	−4750	$PbCl_2$ (solid)	1,46	15900

[a] Assuming no paramagnetic species are present.
[b] Ref. 1, page 341.

Reference 3 contains a comprehensive compilation of ^{29}Si shifts as at 1980. Figure 3 gives an indication of some shifts

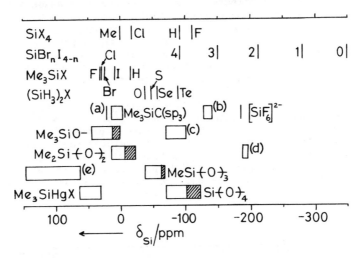

Figure 3. Some ^{29}Si chemical shifts and shift ranges for the systems indicated. Species (a) to (e) are shown below. The tropolonato complexes (b) and the silatranes (c) involve pentacoordinate silicon, while the diketone and ketoester systems (d) contain hexacoordinate silicon, as does $[SiF_6]^{2-}$. The shaded part of the ranges for $Me_nSi(-O-)_{4-n}$, n = 0-3, refers to compounds in which all the oxygens are involved in siloxane bridges.

and shift ranges for ^{29}Si in different chemical situations. It is rather more difficult to categorise ^{119}Sn chemical shifts because of the effects of solvation and auto-association, which are frequently pronounced (since they lead to changes in coordination). Thus in reporting ^{119}Sn NMR results it is vital to specify the medium conditions and the temperature. An extensive list of ^{119}Sn shifts appears in references 1 and 4. It is not proposed to give any detailed discussion of chemical shifts here, but a few points will be made briefly.

(i) Compounds of the type MX_nY_{4-n} (M ≡ ^{29}Si, ^{119}Sn, ^{207}Pb) usually show a characteristic "sagging pattern" when shielding is plotted as a function of n. Figure 4 gives a typical example. Such patterns have been shown to be related to the way in which p electron "imbalance" (49) affects the paramagnetic shielding term, and the case of ^{29}Si has been specifically discussed (50,51) as has that for Sn. Positive deviations from additivity are sometimes found for Sn, and these are also discussed in reference 51.

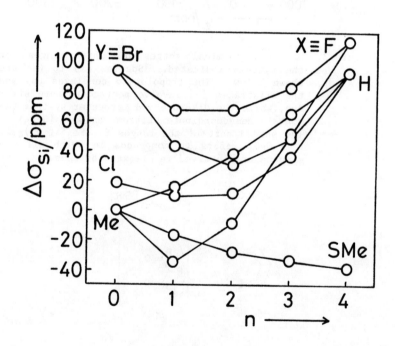

Figure 4. The "sagging pattern" for ^{29}Si shielding for several series of compounds of general formula SiX_nY_{4-n}.

(ii) There is a close parallel (52) between ^{119}Sn and ^{207}Pb shifts for corresponding four-coordinate compounds of tin and lead, as would be expected if the same mechanism were operating. The slope of the linear plot is, however, 3.0, whereas simple theory would suggest a value of 1.4 (being the ratio of inverse cubes of mean distances from the nucleus to valence p electrons).

(iii) Shielding is substantially affected by coordination. Even for ^{29}Si several arrangements are possible, the silatranes (for example), having effective five-coordinate geometry (which appears (53,54) to cause additional shielding by ca. 20 ppm in the silatranes). Hexacoordination also causes substantial shielding for ^{29}Si (for SiF_4 and SiF_6^{2-} chemical shifts are (40,55) δ_{Si} = -109.0 and -185.3 ppm, respectively). For tin, coordination numbers of 4, 5, and 6 are also well recognised, and are reflected (4) in the observed shifts. Data have also been reported (56) for seven coordinate organotin species, as have (57) results for Sn(II) halides. Frequently, however, stability constants of the more highly coordinated species are such that rapid, reversible equilibria occur in solution, with relative proportions of compounds sensitive to changes of concentration, temperature, etc. Kennedy and McFarlane (58) have shown how ^{119}Sn chemical shift measurements may be used to monitor the equilibria. For Pb(II) compounds substantial low frequency shifts from the typical values for Pb(IV) species are found (1), though relatively few solution-state data have been reported.

(iv) Data for alkylsilanes show (1, pages 316 and 317) the same type of dependence on the substitution pattern of the alkyl groups as has been made well known for ^{13}C by Grant and Paul (59). This parallelism includes the so-called γ shielding effect, which has a value (1) of 2.24 ppm for ^{29}Si compared with 2.49 ppm for ^{13}C.

(v) Dispersion of chemical shifts is excellent for certain groups of compounds. This is illustrated for ^{29}Si in Figure 5, which shows the spectrum (60) of a mixture of oligomers of type $[Me_2SiO]_n$. Distinctions are possible up to n = 16, probably due to restrictions in the allowed conformations. Clearly ^{29}Si can be used for discriminatory analysis of such oligomers.

COUPLING CONSTANTS

A considerable body of data on coupling to silicon, tin, and lead has been obtained from observations involving the other nucleus concerned, and has been in the literature for some while. The results have been reviewed for silicon by Schraml and Bellama (61) and by Marsmann (3). It appears that coupling only involving Group IV elements and hydrogen, or between Group IV

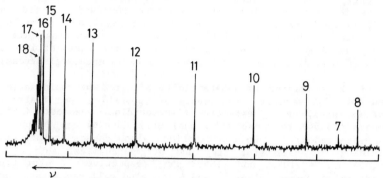

Figure 5. Silicon-29 spectrum, at 79.5 MHz, for a mixture of cyclic oligomers of general formula $[Me_2SiO]_n$. Resolution enhancement was used. The values of n are indicated on the figure. The markers on the chemical shift scale are at 10 Hz intervals. Spectra on a more expanded scale show separate peaks for n = 19 to 25.

elements themselves, behaves substantially as expected from the extensive literature on (C,H) and (C,C) coupling. The implication, at least for (Si,H), (Si,C), and (Si,Si) coupling, is that the Fermi contact term is dominant. For example, Kovačevič and Maksič show (62) that one-bond (Si,C) coupling depends on % s characters of the relevant hybrid orbitals on silicon and carbon (α_{Si}^2 and α_C^2 respectively) according to eq. 1.

$$\left| {}^1J_{SiC} \right| /Hz = 5.554 \times 10^{-2} \alpha_{Si}^2 \alpha_C^2 + 18.2 \tag{1}$$

Coupling constants involving tin or lead are generally larger in magnitude than the analogous ones for silicon, as would be expected from the increase in s electron density at the heavier nuclei. Indeed, one-bond (^{119}Sn,^{119}Sn) coupling constants can be up to ca. 5000 Hz in magnitude (1), and $^1J(^{195}Pt,^{119}Sn)$ has been reported (63) as 27,640 Hz for $[PtCl_2(SnCl_3)_2]^{2-}$. Two-bond (^{119}Sn,^{117}Sn) coupling constants as high as 21,248 Hz in magnitude have been found (20,64) in complexes for tin atoms trans to one another across metal atoms such as Ru. When the signs of coupling constants are being compared it is important to bear in mind the fact that the magnetogyric ratios of all the nuclei in Table 1 except ^{13}C and ^{207}Pb are negative.

For further discussion of coupling constants the reader is referred to the literature already cited.

CONCLUSION

Solution state ^{29}Si, ^{119}Sn, and ^{207}Pb NMR is flourishing and is yielding a substantial quantity of data of chemical significance. More work on relaxation phenomena, particularly for ^{119}Sn and ^{207}Pb, is desirable but there appear to be no outstanding problems as far as technical requirements for obtaining good spectra are concerned. Germanium-73 NMR, however, is languishing, and is unlikely to become important. Our understanding of chemical shifts and coupling constants for Group IV elements is relatively good compared with that for other groups in the Periodic Table.

REFERENCES

1. R. K. Harris, J. D. Kennedy, and W. McFarlane in "NMR and the Periodic Table", R. K. Harris and B. E. Mann, Eds., Academic Press, 1978, Chapter 10.
2. E. A. Williams and J. D. Cargioli, Ann. Repts. NMR Spectrosc., 9, p. 221 (1979).
3. H. C. Marsmann, NMR Basic Principles and Progr., 17, p. 65 (1981).
4. P. J. Smith and A. P. Tupčiauskas, Ann. Repts. NMR Spectrosc., 8, p. 291 (1981).
5. F. W. Wehrli, Ann. Repts. NMR Spectrosc., 9, p. 126 (1979).
6. V. A. Pestunovich, S. N. Tandura, B. Z. Shterenberg, N. Y. Khromova, T. K. Gar, and V. F. Mironov, Izv. Akad. Nauk SSSR (Khim.) 959 (1980): Chem. Abstr. 93, no. 94387.
7. J. Kaufmann, W. Sahm, and A. Schwenk, Z. Naturforsch., A26, p. 1384 (1971).
8. G. C. Levy, J. D. Cargioli, P. C. Juliano, and T. D. Mitchell, J. Am. Chem. Soc., 95, p. 3445 (1973).
9. R. K. Harris and B. J. Kimber, Appl. Spectrosc. Rev., 10, p. 117 (1975).
10. B. J. Kimber, Ph.D. thesis, University of East Anglia, 1974.
11. Z. Luz and S. Meiboom, J. Chem. Phys., 40, p. 2686 (1964).
12. R. K. Harris and B. J. Kimber, Org. Magn. Reson., 7, p. 460 (1975).
13. B. J. Kimber and R. K. Harris, J. Magn. Reson., 16, p. 354 (1974).
14. R. K. Harris and R. H. Newman, Org. Magn. Reson., 9, p. 426 (1977).
15. D. M. Doddrell, D. T. Pegg, W. Brooks, and M. R. Bendall, J. Am. Chem. Soc., 103, p. 727 (1981).
16. J. Kowalewski and G. A. Morris, J. Magn. Reson., 47, p. 331 (1982).
17. R. K. Harris and R. J. Morrow, unpublished work.
18. R. K. Harris and B. J. Kimber, J. Magn. Reson., 17, p. 174 (1975).

19. K. G. Sharp, P. A. Sutor, E. A. Williams, J. D. Cargioli, T. C. Farrar, and K. Ishibitsu, J. Am. Chem. Soc., 98, p. 1977 (1976).
20. C. R. Lassigne, E. J. Wells, L. J. Farrugia, and B. R. James, J. Magn. Reson., 43, p. 488 (1981).
21. R. K. Harris, J. Jones, C. T. G. Knight, and D. Pawson, J. Mol. Struct., 69, p. 95 (1980).
22. R. K. Harris, C. T. G. Knight, and W. Hull, J. Am. Chem. Soc., 103, p. 1577 (1981).
23. R. K. Harris and C. T. G. Knight, J. Mol. Struct., 78, p. 273 (1982).
24. R. K. Harris, C. T. G. Knight, and W. Hull, ACS Symp. Ser., 194 (1982).
25. S. Carr, R. Colton, and D. Dakternieks, J. Magn. Reson., 47, p. 156 (1982).
26. C. T. G. Knight, Ph.D. thesis, University of East Anglia, 1982.
27. A. Briguet, J. C. Duplan, and J. Delmau, J. Magn. Reson., 42, p. 141 (1981).
28. R. K. Harris and B. J. Kimber, Adv. Mol. Relaxation Processes, 8, p. 23 (1976).
29. R. K. Harris and R. H. Newman, J. Chem. Soc. Faraday 2, 73, p. 1204 (1977).
30. M. G. Gibby, A. Pines, and J. S. Waugh, J. Am. Chem. Soc., 94, p. 6231 (1972).
31. E. T. Lippmaa, M. A. Alla, T. J. Pehk, and G. Engelhardt, J. Am. Chem. Soc., 100, 1929 (1978).
32. R. M. Hawk and R. R. Sharp, J. Magn. Reson., 10, p. 385 (1973); J. Chem. Phys., 60, 1522 (1974).
33. G. R. Hays, D. G. Gillies, L. P. Blaauw, and A. D. H. Clague, J. Magn. Reson., 45, p. 102 (1981).
34. Y. K. Puskar, T. A. Saluvere, E. Lippmaa, A. B. Permin, and V. S. Petrosyan, Dokl. Akad. Nauk SSSR, 220, p. 112 (1975).
35. T. Hasebe, G. Soda, and H. Chihara, Bull. Chem. Soc. Japan, 49, p. 3684 (1976).
36. C. R. Lassigne and E. J. Wells, Can. J. Chem., 55, p. 927 (1977).
37. C. R. Lassigne and E. J. Wells, J. Magn. Reson., 26, p. 55 (1977).
38. G. E. Maciel and J. L. Dallas, J. Am. Chem. Soc., 95, p. 3039 (1973).
39. N. K. Zemlyanskii and O. K. Sokolikova, Zh. Anal. Khim., 36, p. 1990 (1981).
40. W. Malisch and W. Ries, Angew. Chem., 90, p. 140 (1978).
41. H. C. Marsmann and H. G. Horn, Chem. Ztg., 96, p. 456 (1972).
42. U. Niemann and H. C. Marsmann, Z. Naturforsch., B30, p. 202 (1975).
43. D. H. Harris, M. F. Lappert, J. S. Poland, and W. McFarlane, J. Chem. Soc. Dalton, 311 (1975).

44. J. J. Burke and P. C. Lauterbur, J. Am. Chem. Soc., 83, p. 326 (1961).
45. T. N. Mitchell, J. Gmehling, and F. Huber, J. Chem. Soc. Dalton, 960 (1978).
46. L. H. Piette and H. E. Weaver, J. Chem. Phys., 28, p. 735 (1958).
47. A. A. Cheremisin and P. V. Schastnev, J. Magn. Reson., 40, p. 459 (1980).
48. J. A. Cella, J. D. Cargioli, and E. A. Williams, J. Organomet. Chem., 186, p. 13 (1980).
49. C. J. Jameson and H. S. Gutowsky, J. Chem. Phys., 40, p. 1714 (1964).
50. G. Engelhardt, R. Radeglia, H. Jancke, E. Lippmaa, and M. Mägi, Org. Magn. Reson., 5, p. 561 (1973).
51. S. P. Ionov and V. S. Lyubimov, Russ. J. Phys. Chem., 54, p. 5 (1980).
52. J. D. Kennedy, W. McFarlane, and G. S. Pyne, J. Chem. Soc. Dalton, 2332 (1977).
53. R. K. Harris, J. Jones, and Soon Ng, J. Magn. Reson., 30, p. 521 (1978).
54. V. A. Pestunovich, S. N. Tandura, M. G. Voronkov, G. Engelhardt, E. Lippmaa, T. Pehk, V. F. Sidorkin, G. I. Zelchan, and V. P. Baryshok, Doklady Phys. Chem., 240, p. 516 (1978).
55. H. C. Marsmann and R. Löwer, Chem. Ztg., 97, p. 660 (1973).
56. J. Otera, T. Hinoishi, and R. Okawara, J. Organomet. Chem., 202, p. 93 (1980).
57. H. M. Yeh and R. A. Geanangel, Inorg. Chim. Acta, 52, p. 113 (1981).
58. J. D. Kennedy and W. McFarlane, Rev. Silicon, Germanium, Tin and Lead Compounds, 1, p. 235 (1973).
59. D. M. Grant and E. G. Paul, J. Am. Chem. Soc., 86, p. 2984 (1964).
60. D. J. Burton and R. K. Harris, unpublished work.
61. J. Schraml and J. M. Bellama, Determ. Org. Struct. Phys. Meth., 6, (1976).
62. K. Kovačevič and Z. B. Maksič, J. Mol. Struct. 17, p. 203 (1973).
63. J. H. Nelson, V. Cooper, and R. W. Rudolph, Inorg. Nucl. Chem. Lett., 16, p. 263 (1980).
64. L. J. Farrugia, B. R. James, C. R. Lassigne, and E. J. Wells, Inorg. Chim. Acta, 53, p. 261 (1981).

CHAPTER 17

HIGH RESOLUTION SOLID-STATE NMR STUDIES OF GROUP IV ELEMENTS

Robin K. Harris

University of East Anglia, Norwich, England

ABSTRACT

The practical limitations of the high-power-decoupling/cross-polarization/magic-angle-rotation suite of NMR techniques are discussed, with mention made of such matters as "dilution," mobility, and physical state. Various types of information available from such solid-state NMR studies which distinguish them from solution work are listed. A number of examples of applications for ^{13}C are reviewed, including questions of crystal symmetry, molecular conformation and motion, polymorphism, and tautomerism. Work on the silicon-29 nucleus, especially in studies of zeolites and related materials, is discussed and references given.

INTRODUCTION

The methods currently in use for obtaining high resolution NMR spectra of "dilute" spins in solids have been discussed in Chapter 6. These methods (high-power decoupling, cross polarization, and magic angle rotation) can be implemented on several currently available commercial spectrometers. Therefore, during the past two or three years there has been a rapid increase in the number of publications describing such work, and it is now possible to evaluate the usefulness of the techniques for chemical applications. Most publications to date have concerned the ^{13}C nucleus, but high resolution solid-state ^{29}Si NMR has proved to be invaluable for several problems of considerable interest to industry, and there have been a substantial number of reports of its use recently. It is the purpose of this article to describe some of the points of interest in chemical applications of both ^{13}C and ^{29}Si NMR with high resolution in solids. Quadrupolar nuclei give somewhat different problems, and though successful work is feasible in some cases, the high resolution techniques have not so far been applied to ^{73}Ge. Although the suite of techniques used for ^{13}C and ^{29}Si is fully applicable to all the magnetic tin isotopes and to ^{207}Pb, few reports have appeared on these nuclei to date. A publication (1) in 1978 described results for ^{119}Sn NMR of three organotin compounds. There was a surprisingly large (ca. 160 ppm) change in the isotropic shift for di-n-butyltin dichloride on going from CHCl$_3$ solution to the solid state. Magic angle rotation on its own has been shown (2) to suffice to obtain a narrow resonance (width ~80 Hz) for ^{207}Pb NMR of lead nitrate.

This article will be solely concerned with high resolution work under MAR conditions, so that studies of static solids, including single crystals and oriented systems, will not be discussed. The determination of shielding anisotropies is also beyond the scope of this review, even though such data may be obtained either by slow spinning about the magic angle or by rapid rotation off-angle.

GENERAL CONSIDERATIONS

It is important to know what limitations there are to the types of system that may be studied. The first problem is the question of what is meant by a "dilute" spin. The pragmatic answer is that a nuclide is sufficiently dilute when homonuclear dipolar interactions are sufficiently weak that MAR at readily attainable rates (say, ca. 3 kHz) suffices for averaging purposes. Fortunately, dilution can take three guises. (i) Spin dilution--as with ^{13}C, which has a natural abundance of 1.108%, leading to dipolar broadening of the order of only tens of Hz. (ii) Physical dilution--which may be achieved by matrix isolation of a compound containing the nuclide in question in a matrix of molecules in which that nuclide is absent. For instance, phosphines (^{31}P has a natural abundance of 100%) could be matrix-isolated in cyclohexane. (iii) Chemical dilution — which occurs for large molecules containing only one atom of the relevant element. An example would be provided by biomolecules containing a single atom of phosphorus.

In fact, there would appear to be few cases that are not suitable for the suite of techniques discussed here except for ^{1}H and ^{19}F NMR. Even in those cases one can envisage devising appropriate situations, e.g., heavily deuterating a sample to give proton spin dilution (deuterium decoupling and magic angle rotation would then be needed). Of course, the situation for nuclei other than ^{1}H is favoured by the dependence of dipolar interactions on both magnetogyric ratios and inverse cubes of distances. It may be noted that for ^{13}C specific enrichment at a particular site will, in general, still give a valid situation of chemical dilution. The matrix isolation technique together with enrichment, has been applied (3,4) successfully to ^{13}C, though without MAR. It is certainly clear that all the spin 1/2 nuclei of Group IV are suitable for the high resolution trio of techniques. Of course the gain in S/N obtained by cross polarization (CP) from ^{1}H to nucleus X depends on the ratio γ_H/γ_X. Thus the weaker the intrinsic sensitivity of X for NMR, the greater the advantage of CP, though the order of sensitivities is not inverted. In this sense, ^{29}Si is the Group IV nucleus for which CP shows the most benefit. However, such generalizations need to be treated with caution because of the effect of other factors, particularly relaxation times.

Mobility on the molecular scale can give limitations in the applicability of high resolution techniques. This occurs in two separate ways. Firstly mobility may render the proton relaxation times T_1 and $T_{1\rho}$ (particularly the latter) unfavourable for CP operation. Secondly, mobility at appropriate frequencies can modulate either decoupling power or MAR rate to yield a static component to dipolar interactions and hence provide a source of line broadening. Variable-temperature work can frequently overcome both these problems, since mobility is strongly temperature dependent.

Possible limitations due to physical state also require some discussion. The situation here is more favourable. Provided <u>molecular</u> mobility is not inappropriate (see above) the following situations may all be studied effectively: single crystals; microcrystalline material; powders; amorphous systems of various kinds; glasses; gels; rubbers; composites; and adsorbed species. Silicon chemistry, in particular, provides plenty of examples of glasses, gels, amorphous systems, and adsorption. Packing material into a rotor is also not normally a problem, though low-density samples may give rise to sensitivity difficulties if they cannot be effectively compressed. Pressures developed at the periphery of rotors can provide problems for the study of gels and related materials.

There are a number of limitations on resolution. Few of these will be discussed in this article, but they have been carefully described by Garroway, VanderHart, and Earl (5), with particular reference to ^{13}C NMR work.

AVAILABLE INFORMATION

It is, perhaps, the differences between solid-state and solution-state NMR that attract the most interest. Of course, ^{13}C chemical shifts can be obtained for solids and interpreted in terms of molecular structure, as has been customary for solution-state NMR. Moreover, heteronuclear scalar coupling of ^{13}C to ^{199}Hg (6), $^{203/205}Tl$ (7), ^{31}P (7), and ^{195}Pt (8) has been demonstrated for solids. The following features are, however, unique to solids or are of special importance for solids.

(i) <u>Spectra for insoluble materials</u>. Such spectra clearly represent considerable potential for NMR, and a wide variety of diverse materials (e.g., coals, wood, paper, cross-linked polymers, and simple insoluble organic compounds) have all been studied.

(ii) <u>Crystallographic nonequivalence</u>. Solution-state NMR spectra are interpreted in terms of the structure and symmetry of isolated molecules, but clearly this is inadequate for many solid-state situations. In particular, for crystalline materials the nature of the unit cell is important, and this topic will be further discussed below.

(iii) <u>Motional nonequivalences</u>. The mobility of molecules in solution has profound effects for NMR, and "chemical exchange" kinetics can be studied in detail where the rates are appropriate. However, much information is lost when rates are so fast that averaged spectra are observed at all accessible temperatures. Mobility is, however, naturally much restricted in solids and additional information is therefore available. Furthermore, in principle there is effectively no lower limit to the temperatures available for NMR studies of solids, whereas freezing and limitations on solubility place severe restrictions for solutions.

(iv) <u>Shielding effects specific to solids</u>. It may be anticipated that chemical shifts will differ between the solution (including neat liquids) and solid states. In fact such changes are extreme examples of medium effects, already well known in solution state NMR. In the absence of specific interactions (e.g., hydrogen bonding) the ^{13}C shift changes are of the order of a few ppm, as is shown below (7) for the type I polymorph of isotactic polybutene.

	$(\delta_d-\delta_c)$/ppm	$(\delta_c-\delta_b)$/ppm	$(\delta_b-\delta_a)$/ppm
solution	5.2	7.5	17.1
solid	6.9	4.8	14.5

CH_3 (a)
(b) CH_2
$-CH_2-CH-$
(d)   (c)

(v) **Linewidth variations**. For amorphous materials local environments may vary widely within the sample for a given chemical site, and this may lead to extensive line broadening. Thus for glassy polymers, ^{13}C linewidths are generally ca. 0.8 ppm. However, sites that are physically remote from neighbourhood variations, such as the quaternary sp^3 carbon in the grouping I,

$$\text{Ph}-\underset{\underset{\text{Me}}{|}}{\overset{\overset{\text{Me}}{|}}{C}}-\text{Ph}$$

I

which is present in a number of polymers (such as polycarbonate and cross-linked epoxides (5) based on bisphenol-A diglycidyl ether), give characteristically sharper lines, ca. 0.2 ppm wide. It may be noted that because linewidths in amorphous materials are governed by shielding variations rather than relaxation phenomena, the spectral dispersion is not usually improved by going to higher applied magnetic fields.

(vi) **Bandshapes due to ^{13}C bonded to nitrogen**. Resonances for carbon in C-N moieties can give rise to a variety of lineshapes, which are frequently characteristic and can be used for assignment purposes. Particularly at low applied fields unsymmetrical doublets (~2:1 in intensity) are often seen (9,10), as is illustrated in Figure 1. These shapes (which can have the more intense peak on either side of the less intense peak) arise, as is now well known (12,13) because MAR is unable to average completely the $(^{14}N, ^{13}C)$ dipolar interactions due to the competing quadrupolar effects of the ^{14}N. This situation only occurs when quadrupolar energies are not negligible with respect to Zeeman energies. Hence the splittings get smaller as the applied magnetic field is increased. They are not present in ^{15}N-enriched material (14).

(vii) **Relaxation measurements**. Unlike the case for the solution state the extreme narrowing situation does not normally obtain for solids. Under some circumstances variable temperature experiments should show $T_1(^{13}C)$ or $T_{1\rho}(^{13}C)$ minima arising from internal motions. Cross-polarization rates can also yield important information. The use of $T_{1\rho}(^{13}C)$ for individual carbons in polymers has been described by Schaefer et al. (15).

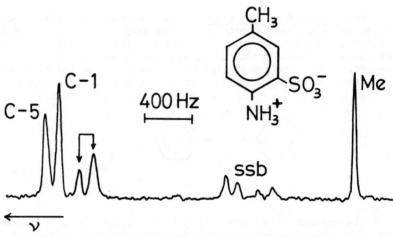

Figure 1. CP/MAR ^{13}C spectrum (11) at 22.6 MHz of 2-amino-5-methylbenzenesulphonic acid, present in the zwitterion form indicated. The spectrum shows clearly the unsymmetrical doublet (indicated by linked arrows) for C-2, caused by the (^{14}N, ^{13}C) dipolar interaction, which cannot be averaged by MAR because of the competing quadrupolar effects. The splitting is 114 Hz in this instance. The spectrum is simplified because the nonquaternary suppression (NQS) pulse sequence was used, and shows the value of this procedure.

CHEMICAL APPLICATIONS (CARBON-13 NMR OF SOLIDS)

At this point a range of chemical applications of solid-state high resolution ^{13}C NMR, including several examples from work at the University of East Anglia, will be discussed under a number of general headings. The number of publications is now so large that a proper review is beyond the scope of the present article, and the author makes no apology for the idiosyncratic choice of examples, but it is hoped that they provide a useful illustration of the types of information available.

(i) *Intermolecular symmetry*. It has already been asserted that crystallographic effects may be important in solid-state NMR. In particular, different crystallographic sites (unrelated by unit-cell symmetry) may be occupied by chemically identical molecules. There is no reason why corresponding atoms in such crystallographically inequivalent molecules should have the same shielding so, <u>a priori</u>, one expects to see splittings in spectra from such situations. Figure 2 shows the ^{13}C spectrum of solid barium acetate monohydrate as an example. The carboxyl region shows (16) four peaks, suggesting there are four different crystallographic sites (a somewhat unusual situation). Subsequent X-ray work (17) confirms this conclusion. It may be noted that the shift differences amount to ca. 4 ppm at the extreme in this case. Differences clearly also exist in the methyl region, but are not so well resolved.

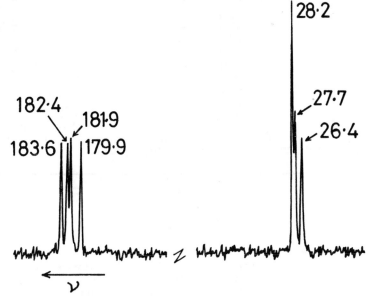

Figure 2. CP/MAR ^{13}C spectrum (16) at 22.6 MHz of barium acetate monohydrate, showing at the left the carboxyl carbon resonances and at the right the methyl resonances. The chemical shifts are given in ppm from the signal of liquid Me$_4$Si.

(ii) <u>Intramolecular symmetry</u>. Crystallogrpahic effects can have a second influence on chemical shift equivalence, viz., they can lower molecular symmetry, again producing extra splittings. Figure 3 shows (18) an example from dyestuff chemistry. In this case, the two-fold symmetry of the anthraquinone derivative (II) appears to be lifted in the crystal, causing doubling of many of the peaks. Again, subsequent X-ray work (17) has confirmed the NMR finding, and the conformations of the two amino substituents in each molecule are different. However, the whole molecule is the crystallographic asymmetric unit, i.e., there is only one type of crystallographic site for the molecules.

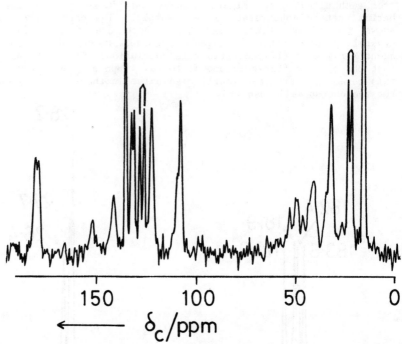

Figure 3. CP/MAR ^{13}C spectrum (18) at 22.6 MHz of 1,4-bis(n-butylamino)anthraquinone (II) showing splittings arising from the lack of molecular symmetry in the crystal. Splittings for C-5/C-8 and for C-13/C-17 are marked (and are 2.2 and 1.7 ppm, respectively), but others (in the carbonyl and methyl resonances for instance) are also visible.

(iii) <u>Molecular conformation</u>. This question is frequently related to that discussed under (ii). A clear example has been published by Bunn et al. (19). Spectra of solid syndiotactic polypropene show a splitting of ca. 8 ppm in the resonance assigned to the methylene carbons, whereas peaks due to CH_3 and CH peaks show no such effect. This can be traced to the curious figure-of-eight double-helix conformation adopted in the solid material, as demonstrated by earlier X-ray work (20). The magnitude of the splitting may be seen to be twice the well recognised γ-gauche shielding effect. It is clear that solid-state NMR work, combined with X-ray results, will enable more detailed investigations to be made of conformational influences on chemical shifts than are feasible with solution-state NMR.

(iv) <u>Intramolecular motion</u>. This topic is closely associated with the preceding one, since detailed conformational information is only available when intramolecular mobility is slow on the NMR time scale. Naturally, as pointed out earlier, the rigid order associated with the solid state tends to inhibit motions, and conditions of slow exchange are therefore more likely for a solid than for a solution of the same solid at the same temperature. Figure 4 shows (16) an example from organoplatinum chemistry. Reversal of the six-membered ring in III is rapid on the NMR time scale at ambient temperature for a solution, rendering two of the carbon atoms of the spiro-fused four-membered ring equivalent (21). In the solid, however, the motion is much slower, and Figure 4 shows four separate ^{13}C signals for the cyclobutane ring. Of course, as for solutions, variable temperature work for solids will reveal cases where intramolecular motion is at an appropriate rate for study by bandshape analysis. This is so, for instance, for a cyclooctaetraene complex of ruthenium studied by Lyerla et al. (22) and for hydrogen switching between carbonyl and OH groups in naphthazarin B (23), though bandshape fitting was not attempted in either case.

(v) <u>Intermolecular interactions</u>. Crystallographic effects generally result either from questions of the way molecules pack together or from chemical interactions between them. Blann et al. (24) have used solid-state NMR to study an effect of the latter type. They examined several systems which might contain π-π molecular complexes and showed (Table 1) that chemical shifts could be used as a measure of the strength of the interaction.

(vi) <u>Diastereoisomerism</u>. Crystallographic effects can, of course, lift equivalences between carbon atoms in a given diastereomeric form. Moreover, it is unlikely that different diastereoisomers of the same compound will be affected identically. Hence it is not, perhaps, surprising that Hill, Zens, and Jacobus (25) were able to differentiate between optically

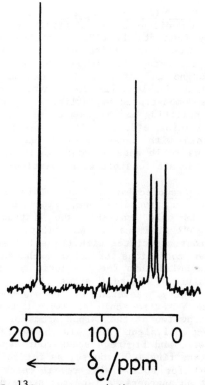

Figure 4. CP/MAR ^{13}C spectrum (16) at 22.6 MHz for III, <u>cis</u>-(diammino)(1,1-cyclobutanedicarboxylato)Pt-(II), showing four separate resonances for the cyclobutyl carbons.

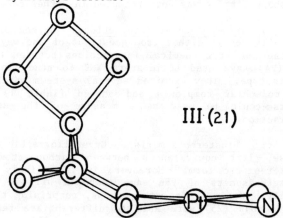

pure and meso forms of tartaric acid. The true dℓ racemate gave a spectrum that was different again, allowing distinction from a conglomerate. The data are given in Table 2. In all three samples there are two molecules per unit cell, but the two halves of a given molecule are not symmetry related, so each type of carbon gives two peaks in each case.

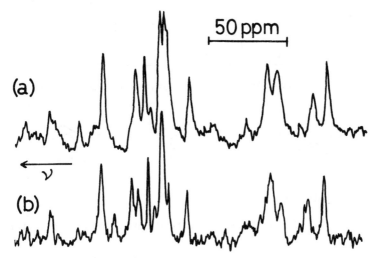

Figure 5. CP/MAR ^{13}C spectrum (26) at 22.6 MHz of two polymorphs of cephaloridine (VII): (a) δ form, (b) α form.

VII

(vii) <u>Polymorphism</u>. This is an extremely important topic in solid-state chemistry, particularly for the plastics and pharmaceuticals industries. Naturally, solution-state NMR is valueless in this context! Figure 5 shows (26) spectra of two polymorphs of cephaloridine (VII), and it is clear that the spectra are different. Such differences are valuable in themselves for diagnostic purposes, but one also expects that a

Table 1. Carbon-13 NMR Data for Solid π–π Complexes (24)

Acceptor molecule	Electron affinity/eV	Δδ/ppm	Electron transfer[a]/%
TCNE (IV)	2.2	2.2 ± 0.5	8.1
Chloroanil (V)	1.37	1.2 ± 0.6	4.8
TCNB (VI)	0.4	<1	<2

[a]Estimated from $\Delta\sigma = -(160 \pm 20)P$, where P is the fractional charge.

Table 2. Carbon-13 Solid-State Chemical Shifts of Tartaric Acids (25)

Form	Chemical Shifts, δ_C/ppm			
	Carbonyl		α carbon	
Optically pure (2R, 3R)	175.98	171.16	74.11	71.78
dℓ (racemate)	179.24	177.45	74.26	73.10
Meso	176.99	174.58	76.82	74.81

detailed examination of the two spectra will increase understanding of the effects of conformation and packing on chemical shifts. A combination of X-ray diffraction and solid-state NMR is extremely powerful.

(viii) <u>Tautomerism</u>. NMR has frequently been used for the study of tautomerism in the solution state. Clearly, the recent advances in technique allow such work to be extended to the solid state. For example, Figure 6 shows (11) the spectrum of a pigment (VIII). Observation of the peak at δ_C = 182 ppm, which must be assigned to a carbonyl group, clearly indicates that the molecule is in the hydrazo form (VIIIa) and not the azo form (VIIIb). The latter would give a signal at $\delta_C \sim$ 150 ppm due to the phenolic carbon.

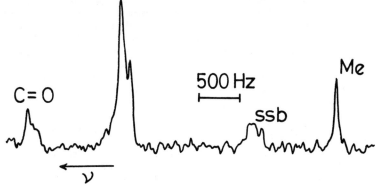

Figure 6. CP/MAR ^{13}C spectrum (11) at 22.6 MHz of C.I. Pigment Red 3 (VIII). This spectrum was obtained with the NQS sequence and therefore shows only peaks due to quaternary carbons, plus a residual methyl signal. The molecular diagrams are not drawn with realistic conformation, but the hydrazo form is presumably favoured by hydrogen bonding.

hydrazo form
VIIIa

azo form
VIIIb

(ix) <u>Heterogeneity</u>. The spectra discussed so far have all been relatively simple because the systems under discussion were physically homogeneous, even at the molecular level. However, solid-state problems frequently involve heterogeneous systems. These are many and diverse, so that each case requires special consideration. In favourable examples much information can be obtained, but it must be stressed that a single CP experiment is unlikely to be very useful. Rather, it is essential to combine proton T_1 and $T_{1\rho}$ measurements with ^{13}C spectra obtained under a variety of conditions. As a simple example, consider polyethene (polyethylene). This usually has "crystalline" and "amorphous" regions. Fyfe et al. (27) and Earl and VanderHart (28) showed, by varying the nature of the experiment, that the amorphous material gives a signal 2.4 ppm to low frequency of that of crystalline regions. Aujla et al. (29) have used different discriminatory techniques to distinguish spectra from crystalline and amorphous regions of polyethylene terephthalate.

(x) Adsorption. Adsorbed species clearly form a special case of heterogeneity (in common with interfacial regions of several types), but present particular problems because the concentration of the relevant species is usually low. However, successful investigations have already been carried out. For instance, Hays (30) has recorded ^{13}C spectra of a range of chemically modified silica gels which involve surface trialkylsiloxy groups. Such species are widely used as stationary phases in chromatographic columns. Similar work involving aminosilane coupling agents on silica surfaces has been reported by other workers (31).

(xi) Geochemical Applications. Solid-state NMR may, of course, be applied in virtually all areas of chemistry and biochemistry, but in most cases solution-state NMR has already been used extensively. The use of solid-state NMR for organic geochemistry is singled out for mention here because it is, in general, a novel area for NMR spectroscopists, although some solution-state NMR has been carried out on coal extracts. High resolution ^{13}C NMR has now been applied to solid coals and related materials, to oil shales, to soils, and to ancient buried woods. One recent piece of such research (32) has involved the study of amber samples from a variety of locations. Such materials are noncrystalline and have poor solubility, so they are not easy to characterise. However, the usual suite of techniques for high resolution NMR work on solids yielded diagnostic spectra with more detailed fine structure than is shown by coal samples. It proved possible to differentiate amber samples from different sources.

SILICON-29 NMR OF SOLIDS

There has been a quite remarkable spate of articles on ^{29}Si NMR of solids since 1979. Some of these have not involved MAR, but the majority have done so and have therefore been truly high resolution. The topics covered have been wide ranging, but there have been very few reports of work on simple organosilicon compounds. Indeed, interest has centred on four areas which are currently of considerable industrial interest, viz., (i) amorphous silicon/hydrogen systems, (ii) silica gel and other silica materials, (iii) zeolites and related systems, and (iv) silicones.

The first publication (1) on ^{29}Si MAR work, however, described results for two organosilanes, as well as two siloxanes and two silicates. Shielding anisotropies proved to be modest (20-80 ppm) so that MAR rates do not need to be very high. Lippmaa et al. extended their work (33) to a variety of simple silicates and found it was feasible to identify the nature of a silicate group (end, middle, branching, or cross-linking) by its chemical shift, in much the same way as had already been done for silicate solutions. Moreover, they found aluminosilicates could be similarly characterized, so that it was possible to determine the number of Si-O-Al bridges involving a given silicon--substitution of Si-O-Si by Si-O-Al causes deshielding of the ^{29}Si of ca. 6 ppm on average. Even finer distinctions were observed. Albite, Na[AlSi$_3$O$_8$], showed three peaks, two of them assigned to silicons with one Si-O-Al bridge each. The shift difference between these resonances due to similar silicons is as high as 7.5 ppm. These discoveries led to the first application of high resolution ^{29}Si NMR to a zeolite (34). This showed that NaA-type zeolites (Si:Al = 1) do not obey Loewenstein's rule, i.e., Al-O-Al connections are present, though this conclusion is still controversial. This report (34) caused considerable excitement to zeolite chemists, and a number of research groups immediately started work in this area, so that over a dozen papers have since appeared, and a wide variety of zeolites have been studied. Collaboration between Thomas in the UK and Fyfe in Canada has been particularly fruitful and has, for example (35), yielded detailed information on the effects of processing zeolites (calcining, leaching, etc.). It is interesting to note that the combination of ^{29}Si and ^{27}Al MAR NMR is proving particularly powerful in this context. However, even highly siliceous zeolites can yield useful information via characteristic chemical shifts, e.g., the region δ_{Si} = -112 to -114 ppm has been recognised (36) as characteristic of silicon atoms (with no Si-O-Al bridges) in five-membered rings.

One reason for the early application of high resolution ^{29}Si NMR to zeolites is that for many cases neither high-power decoupling nor cross polarization is necessary (protons having been excluded from the system). Therefore only MAR is needed, in conjunction with standard FT spectrometers. It seems obvious that use of the CP/decoupling techniques will greatly extend the applicability of the techniques, even for zeolite systems.

Already the literature on ^{29}Si NMR of solids is too large to discuss in full here. Consequently, just one more application wil be chosen for mention, and then one practical point will be discussed.

The application relates to the study of silica gel. Here, it is essential to use cross polarization and decoupling in most cases, particularly because one of the objectives is to determine the nature of hydroxyl groups. The use of cross polarization discriminates against the bulk of the silicon nuclei, which are not near the surface hydroxyls, so that Maciel and Sindorf (37) were able to distinguish dihydroxy and monohydroxy silicate groups (δ_{Si} = -90.6 and -99.8 ppm respectively) from silicons with four siloxy bridges (δ_{Si} = -109.3 ppm). Naturally, the lowest-frequency peak had a substantially longer cross polarization time (12.7 s) than the others.

The practical point concerns the choice of a suitable test sample to check such matters as magnet shimming, Hartmann-Hahn matching, and magic angle setting. For ^{13}C hexamethylbenzene is often used. For ^{29}Si, the trimethylsilyl ester of the octameric silicate, $Si_8O_{12}(OSiMe_3)_8$, has been suggested (38). This has silicons at the corner of a cube, joined by oxygens, and the fourth valency at each corner-silicon is occupied by a trimethylsiloxy group. The compound is often designated Q_8M_8. The Q units give rise to four sharp lines because of nonequivalences in the unit cell.

CONCLUSIONS

Work to date on high resolution ^{13}C NMR of solids has already yielded useful chemical information regarding a wide range of materials, and there is no doubt whatsoever that the next few years will see a further large growth in the number of relevant publications. Commercial instruments capable of such work are being ordered or installed at a significant number of locations around the world. Conceivably the time will come when NMR spectroscopists will normally try solid-state ^{13}C work before (or even instead of) obtaining solution-state ^{13}C spectra. Similar studies of ^{29}Si are already surprisingly advanced for some topics, but less so in others. Geochemistry is to the fore in both ^{13}C and ^{29}Si work, but organometallic chemistry currently lags behind. High resolution solid-state NMR studies of tin and lead have scarcely started, but there is substantial potential in these areas also.

REFERENCES

(1) E. T. Lippmaa, M. A. Alla, T. J. Pehk, and G. Engelhardt, J. Am. Chem. Soc., 100, p. 1929 (1978).
(2) D. J. Burton, R. K. Harris, and L. H. Merwin, J. Magn. Reson., 39, p. 159 (1980).
(3) K. W. Zilm, R. T. Conlin, D. M. Grant, and J. Michl, J. Am. Chem. Soc., 100, p. 8038 (1978).

(4) K. W. Zilm and D. M. Grant, J. Am. Chem. Soc., 103, p. 2913 (1981).
(5) A. N. Garroway, D. L. VanderHart, and W. L. Earl, Phil. Trans. Roy. Soc. A, 299, p. 609 (1981).
(6) R. D. Kendrick, C. S. Yannoni, R. Aikman, and R. J. Lagow, J. Magn. Reson., 37, p. 555 (1980).
(7) G. E. Balimann, C. J. Groombridge, R. K. Harris, K. J. Packer, B. J. Say, and S. F. Tanner, Phil. Trans. Roy. Soc. A, 299, p. 643 (1981).
(8) C. J. Groombridge, R. K. Harris, and K. J. Packer, unpublished work.
(9) M. H. Frey and S. J. Opella, J. C. S. Chem. Commun., 474 (1980).
(10) C. J. Groombridge, R. K. Harris, K. J. Packer, B. J. Say, and S. F. Tanner, J. C. S. Chem. Commun., 174 (1980).
(11) C. D. Campbell, R. K. Harris, P. Jonsen, and K. J. Packer, unpublished work.
(12) A. Naito, S. Ganapathy, and C. A. McDowell, J. Chem. Phys., 74, p. 5393 (1981).
(13) N. Zumbulyadis, P. M. Henrichs, and R. H. Young, J. Chem. Phys., 75, p. 1603 (1981).
(14) S. J. Opella, J. G. Hexem, M. H. Frey, and T. A. Cross, Phil. Trans. Roy. Soc. A, 299, p. 665 (1981).
(15) J. Schaefer, E. O. Stejskal, M. D. Sefcik, and R. A. McKay, Phil. Trans. Roy. Soc. A, 299, p. 593 (1981).
(16) C. J. Groombridge, R. K. Harris, and K. J. Packer, unpublished work.
(17) M. B. Hursthouse, unpublished work.
(18) R. A. Aujla, A. M. Chippendale, R. K. Harris, A. Mathias, and K. J. Packer, to be published.
(19) A. Bunn, M. E. A. Cudby, R. K. Harris, K. J. Packer, and B. J. Say, J. C. S. Chem. Commun., p. 15 (1981).
(20) P. Corradini, G. Natta, P. Ganis, and P. A. Temussi, J. Polym. Sci., C 16, p. 2477 (1967).
(21) S. Neidle, I. M. Ismail, and P. J. Sadler, J. Inorg. Biochem. 13, p. 205 (1980).
(22) J. R. Lyerla, C. A. Fyfe, and C. S. Yannoni, J. Am. Chem. Soc., 101, p. 1351 (1979).
(23) W.-I. Shiau, E. N. Duesler, I. C. Paul, D. Y. Curtin, W. G. Blann, and C. A. Fyfe, J. Am. Chem. Soc., 102, p. 4546 (1980).
(24) W. G. Blann, C. A. Fyfe, J. R. Lyerla, and C. S. Yannoni, J. Am. Chem. Soc., 103, p. 4030 (1981).
(25) H. D. W. Hill, A. P. Zens, and J. Jacobus, J. Am. Chem. Soc., 101, p. 7090 (1979).
(26) R. K. Harris, A. M. Kenwright, K. J. Packer, R. A. Fletton, and G. I. Gregory, unpublished work.
(27) C. A. Fyfe, J. R. Lyerla, W. Volksen, and C. S. Yannoni, Macromolecules, 12, p. 757 (1979).

(28) W. L. Earl and D. L. VanderHart, Macromolecules, 12, p. 762 (1979).
(29) R. S. Aujla, R. K. Harris, K. J. Packer, M. Parameswaran, B. J. Say, A. Bunn, and M. E. A. Cudby, Polymer Bull., 8, p. 253 (1982).
(30) G. R. Hays, The Analyst, 107, p. 241 (1982).
(31) C-H. Chiang, N-I. Liu, and J. L. Koenig, J. Colloid. Interfac. Sci., 86, p. 26 (1982).
(32) J. B. Lambert and J. S. Frye, Science, 217, p. 55 (1982).
(33) E. Lippmaa, M. Mägi, A. Samosan, G. Engelhardt, and A-R. Grimmer, J. Am. Chem. Soc., 102, p. 4889 (1980).
(34) G. Engelhardt, D. Zeigan, E. Lippmaa, and M. Mägi, Z. anorg. allgem. Chem., 468, p. 35 (1980).
(35) J. Klinowski, J. M. Thomas, C. A. Fyfe, and G. C. Gobbi, Nature, 296, p. 533 (1982).
(36) J. B. Nagy, J.-P. Gilson, and E. G. Derouane, J. C. S. Chem. Commun., p. 1129 (1981).
(37) G. E. Maciel and D. W. Sindorf, J. Am. Chem. Soc., 102, p. 7606 (1980).
(38) E. Lippmaa and A. Samosan, Bruker Report No. 1, p. 6 (1982).

CHAPTER 18

GROUP V ATOM NMR SPECTROSCOPY OTHER THAN NITROGEN

R. Garth Kidd

Faculty of Graduate Studies
The University of Western Ontario
London, Canada

ABSTRACT

Primary dependence of nuclear shielding on oxidation state observed throughout the earlier Periodic Groups is not evident in Group V. Symmetry considerations related to the lower oxidation state lone pair do, however, affect the spectroscopies of the quadrupolar ^{75}As and ^{121}Sb. Reversal effects in ^{75}As shielding upon methyl substitution are remarked. The precision of the pairwise additive model for ^{121}Sb shieldings is used to assign cis isomers in the $SbCl_nBr_{6-n}^-$ system.

INTRODUCTION

What can we say about Group V that will give it a flavour all its own? Once the NMR spectroscopist gets below ^{31}P he is venturing into terra incognita where few others have gone before. This region of unfamiliarity is not dissimilar from Group VI below ^{17}O and Group VII below ^{35,37}Cl. In this whole corner of the Periodic Table it is ironic that the rare gas Xe has received more attention than any other atom. The inordinate interest in Xe NMR is due in part to the flurry of synthetic activity that followed Neil Bartlett's (1) preparation of the first xenon compound just 20 years ago; also in part it is due to the isotopic characteristics of xenon providing the spectroscopist with a spin 1/2 isotope of high receptivity for measuring accurate chemical shifts and coupling constants, and a quadrupolar isotope of moderate receptivity to assay electric field gradients. Not many atoms provide the spectroscopist with this versatility of approach. One that does is the ^{15}N, ^{14}N isotope pair from Group V. Others, with the I = 1/2 isotope listed first in each case, are ^{1}H, ^{2}D; ^{129}Xe, ^{131}Xe; ^{187}Os, ^{189}Os; and ^{199}Hg, ^{201}Hg. It can be seen that the incidence of this feature is not high, and Group V is fortunate to have one.

Table I lists the nuclear properties of all magnetically active isotopes in Group V. Most features in the NMR history of these elements can be readily understood through a quick survey of the Table; one of them cannot.

Table I. Nuclear Properties of the Group V Isotopes

Nucleus	Spin	Abundance (%)	Quadrupole moment $Q/(10^{-28}m)$	Usual shielding reference	Receptivity re ^{13}C
^{14}N	1	99.63	0.016	NO_3^-(aq)	5.7
^{15}N	1/2	0.37	0	NO_3^-(aq)	0.022
^{31}P	1/2	100	0	85% H_3PO_4	377
^{75}As	3/2	100	0.3	AsF_6^-	143
^{121}Sb	5/2	57.25	-0.5	$SbCl_6^-$	520
^{123}Sb	7/2	42.75	-0.7	$SbCl_6^-$	111
^{209}Bi	9/2	100	-0.4		777

Chapter 10 summarized the large body of ^{14}N spectroscopy now available to us dating as far back as Proctor and Yu's work in 1951 (2). It is only since the signal enhance-ment capabilities of FT instrumentation came on the scene that ^{15}N spectroscopy became a routine procedure. When we look at the receptivity difference between the two isotopes, the reason is obvious. It is sobering to recall that all of the first ^{15}N spectroscopy carried out in J. D. Robert's lab was done on a CW instrument.

Consistent with its status as the only spin 1/2 nucleus in Group V having a forthcoming receptivity, ^{31}P together with ^{1}H and ^{19}F was among the very first nuclei to yield its secrets to the blandishments of the magnet, and, led by the group working with John Van Wazer at Monsanto, chemists both organic and inorganic have provided over a period of 30 years a rich NMR literature on this isotope (3). Beyond phosphorus in Group V, the NMR pickings become very thin. In "NMR and the Periodic Table," a scant three pages are devoted to arsenic, antimony, and bismuth (4). Pregosin (5) in Zurich has introduced us to the range of shieldings experienced by ^{75}As, and my group at Western Ontario has produced most of the ^{121}Sb spectroscopy available (6).

Bismuth-209 is a conundrum. It is 100% abundant. It has the highest receptivity of any Group V isotope. It has a medium Q quadrupole moment, the deleterious effects of which are ameliorated by a high spin of 9/2. Yet apart from a single measurement 30 years ago (2) used to establish its nuclear magnetic moment, nothing of an NMR nature has been done with ^{209}Bi.

OXIDATION STATE AND COORDINATION SYMMETRY CONSIDERATIONS

Every main group element forms a class of stable compounds in which its oxidation state coincides with the group number. This represents the highest state of oxidation that is chemically accessible to the element and in Group V is exemplified by nitrogen(V) compounds such as HNO_3 and phosphorus(V) compounds such as P_4O_{10}. As we move from the lighter to the heavier members in each of these groups, the higher oxidation state becomes less stable relative to the lower state, two units less than the group number. Thus the nitrogen(III) nitrites are good reducing agents on their way to nitrogen(V) nitrates, whereas bismuth(V) in $NaBiO_3$ is a powerful oxidizing agent looking for a less energetic existence as bismuth(III). Phosphorous(III) compounds are all reducing in character seeking to achieve the more stable phosphorous(V) state.

Intimately connected with oxidation state is the coordination symmetry for Group V atoms. In oxidation state V no lone pairs remain in the valence shell and tetrahedral or octahedral symmetry can be achieved through binding 4 or 6 donor ligands as happens in PCl_4^+ and PCl_6^-. In oxidation state III, however, tetrahedral symmetry is only achieved on those few occasions when the atom acts as a donor, as in NH_4^+. On all other occasions there is a lone pair in its valence shell and the symmetry is limited to C_{3v} or lower. It is the heavier members of the Group, for which oxidation state III is more stable, that show little tendency to act as electron donors. As a consequence their structures will be characterized by substantial electrical field gradients at the central atom, and if its nuclear quadrupole moment is sizable the spectroscopist can anticipate linewidth problems.

Phosphorus-31 has no quadrupole moment and so the lower oxidation state linewidth problem does not arise. All of the

^{75}As chemical shifts reported have been for arsenic(V) compounds, and we should not expect to encounter much NMR data from arsenic(III). Relaxation times for ^{75}As in AsH$_3$ are available (7) and at 25°C these translate into a linewidth of 3350 Hz. With ^{121}Sb the situation is similar. In spite of the fact that its higher spin number works in favour of slower relaxation, no chemical shifts for Sb^{III} have been reported and two authors actually admit failing to locate ^{121}Sb resonances in solutions containing Sb^{III}.

In 1951, Proctor and Yu obtained a ^{209}Bi signal from a 0.69 M aqueous solution of Bi(NO$_3$)$_3$. This experiment has been repeated by Brevard and Granger using modern FT instrumentation and they close out their Handbook (8) by reporting a 3200 Hz linewidth for Bi^{III}. Since it contains the symmetry-perturbing lone pair, Bi^V should give narrower signals and it is remarkable that none have been reported, even allowing for the instability of the oxidation state.

PHOSPHORUS-31

Figures 1 and 2 illustrate the essential highlights from the extensive literature on ^{31}P spectroscopy, and immediately apparent is the fact that it fits few of the shielding patterns established with the earlier Groups. Apart from the zero oxidation state P$_4$ marking the high field limit for ^{31}P chemical shifts, there is no oxidation state correlation. The 500 ppm range for phosphorus(III) compounds completely overlaps the more limited range for phosphorus(V), which lies at the <u>shielded</u> end of the larger P^{III} one rather than the deshielded end expected if oxidation state were a significant factor.

Whereas the Cl, Br, and I compounds of other atoms span at least 50% of the total shielding range, these halides of phosphorus(III) are separated by only 45 ppm, barely 6% of the total ^{31}P range of 720 ppm. Even more frustrating is the fact that there is no halogen sequence that can be used to establish either normal or inverse halogen dependence.

Van Wazer (3, 9) in a comprehensive series of papers plus a book on the subject has identified bond angle, substituent electronegativity, and π-bonding character of substituent as the three variables with which ^{31}P chemical shifts can be correlated. Provided only one variable undergoes significant change in the series of compounds being studied, then reasonable models to explain the trends are available. As more frequently happens when several variables are changing simultaneously, the theoretical models presently available are of little help and each person selects data to fit the vestigial patterns he perceives, as has been done in Figures 1 and 2.

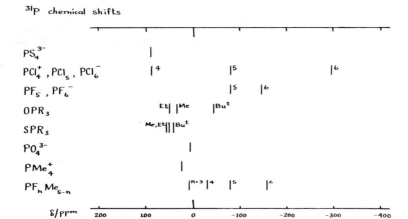

Figure 1. ^{31}P chemical shift ranges for phosphorus(V) compounds referenced to 85% H_3PO_4.

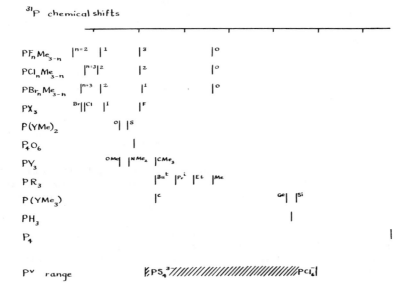

Figure 2. ^{31}P chemical shift ranges for phosphorus(III) compounds referenced to 85% H_3PO_4.

ARSENIC-75

All of the ^{75}As spectroscopy available deals with arsenic(V) compounds, and their chemical shifts are summarized in Figure 3. Although the amount of information is limited, we see some familiar patterns re-emerging. The tetrahedral oxyanion environment common among the transition metals occupies its familiar spot at the deshielded end of the range. The hydrido ligand, on the other hand, emerges as a strongly shielding substituent, the same influence that it brings to bear on the Group III and Group IV atoms.

^{75}As chemical shifts

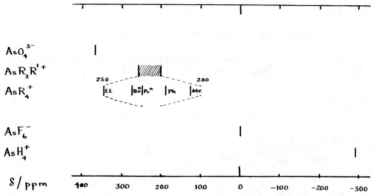

Figure 3. ^{75}As chemical shift ranges for arsenic(V) compounds referenced to AsF_6^-. Note x5 scale expansion for tetraalkylarsenates.

While there might appear to be no regularity among the shifts of the AsR_4^+ compounds, the ordering in terms of alkyl substituent is consistent with a pattern that is observed in most of the Group IV and Group V alkyls. If we label the atoms of the structural unit in the following way, $(As_\alpha - C_\beta - C_\gamma - C_\delta)$, then methyl substitution for one hydrogen at the β position ($\Delta\beta$) deshields by 11 ppm, $\Delta\gamma$ shields by 5 ppm, and $\Delta\delta$ shields by 1 ppm. This reversal in shielding effect as one proceeds down the chain is characteristic of methyl substitution.

ANTIMONY-121

Although ^{121}Sb spectroscopy is in roughly the same state of development as that of ^{75}As, rather more information is available in that chloride and bromide compounds have been studied. This allows us to establish that ^{121}Sb falls in the normal halogen dependence category and also to obtain from the $SbCl_nBr_{6-n}^-$

mixed halide compounds the pairwise additivity parameters that enable the assignment of cis and trans isomers.

Figure 4 shows a ^{121}Sb range of shifts that spans 3450 ppm. A comparison of the MR_4^+-MF_6^- differences for ^{121}Sb and ^{75}As indicates the shielding sensitivity of ^{121}Sb to be greater by a factor of 3.1; SbS_4^{3-} at the deshielded end of the range is consistent with the highly deshielded environment created for ^{95}Mo by tetrahedral sulfurs and observed by Otto Lutz (10).

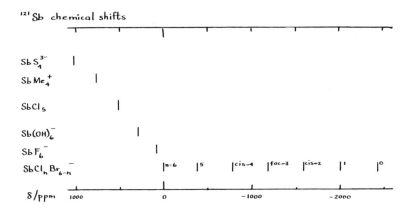

Figure 4. ^{121}Sb chemical shifts for antimony(V) compounds referenced to SbF_6^-.

The ligands in $SbCl_6^-$ and $SbBr_6^-$ are sufficiently labile in acetonitrile solution that a mixture of the two results through ligand exchange in a solution containing 7 different compounds and yielding 7 ^{121}Sb resonances. Table II lists the δ values for each of the 7, together with the δ values one calculates using a first-order substituent constant model and also those calculated using a second-order pairwise additive model.[6] Not only is the substituent constant model incapable of distinguishing between cis and trans isomers, but the agreement between calculated and observed shifts is poor, with worst-case discrepancies reaching 7%. The maximum discrepancy using the pairwise additive model is 3 ppm, and since calculated differences between cis and trans isomers are 11 ppm in each case, the precision of the model is sufficient to enable the assignment of geometrical isomers.

Table II. ^{121}Sb Chemical Shifts for $[SbCl_nBr_{6-n}]^-$ Anions in Acetonitrile Solution

Species	Coefficients, $C_{i,j}$			$\sigma C_{ij} n_{ij}$	$\delta_{calcd}{}^a$ $\delta_{obsd}{}^b$	$\delta_{calcd}{}^c$ Substituent additive
	Cl,Cl	Cl,Br	Br,Br			
$[SbCl_6]^-$	12	0	0	+1	0	(0)
$[SbCl_5Br]^-$	8	4	0	−383	−380	−405
$[SbCl_4Br_2]^-$						−810
Cis	5	6	1	−778	−780	
Trans	4	8	0	−767		
$[SbCl_3Br_3]^-$						−1215
Fac	3	6	3	−1183	−1180	
Mer	2	8	2	−1172		
$[SbCl_2Br_4]^-$						−1620
Cis	1	6	5	−1589	−1590	
Trans	0	8	4	−1578		
$[SbClBr_5]^-$	0	4	8	−2005	−2005	−2025
$[SbBr_6]^-$	0	0	12	−2432	−2430	(−2430)

[a] Pairwise additivity parameter values (ppm): $n_{Cl,Cl} = +0.1$; $n_{Cl,Br} = -95.9$; $n_{Br,Br} = -202.7$. (Uncertainty limits ± 0.2 in each case). [b] Negative shifts designate higher shielding. Experimental uncertainty ± 5 ppm. [c] Substituent constants (ppm): Cl = 0; Br = −405.

REFERENCES

(1) N. Bartlett, Proc. Chem. Soc. (London), p. 218 (1962).
(2) W. G. Proctor and F. C. Yu, Phys. Rev., 81, p. 20 (1951).
(3) M. M. Crutchfield, C. H. Dungan, and J. R. Van Wazer, Top. Phosphorous Chem. 5 (1967); J. Nixon and A. Pidcock, Ann. Reports NMR Spectrosc., 2, p. 345 (1969); G. Mavel, Ann. Reports NMR Spectrosc., 5B, p. 1 (1973).
(4) R. K. Harris in "NMR and the Periodic Table," R. K. Harris and B. E. Mann, Eds., Academic Press, London, 1978, Chapter 11.
(5) G. Balimann and P. S. Pregosin, J. Magn. Reson., 26, p. 283 (1977).
(6) R. G. Kidd and R. W. Matthews, J. Inorg. Nucl. Chem., 37, p. 661 (1975); R. G. Kidd and H. G. Spinney, Can. J. Chem., 59, p. 2940 (1981).
(7) L. J. Burnett and A. H. Zeltmann, J. Chem. Phys., 56, p. 4695 (1972).
(8) C. Brevard and P. Granger, "Handbook of High Resolution Multinuclear NMR," Wiley-Interscience, New York, 1981.
(9) J. H. Letcher and J. R. Van Wazer, J. Chem. Phys., 44, p. 815 (1966); 45, p. 2916 (1966); 45, p. 2926 (1966).
(10) O. Lutz, A. Nolle, and P. Kroneck, Z. Naturforsch, 32a, p. 505 (1977).

CHAPTER 19

GROUP VI ELEMENTS OTHER THAN OXYGEN

Otto Lutz

Physikalisches Institut der Universität
Tübingen, Bundesrepublik Deutschland

ABSTRACT

The NMR properties of ^{33}S, ^{77}Se, ^{123}Te, and ^{125}Te are discussed, together with experimental difficulties associated with low abundance and, for ^{33}S, a quadrupole moment. Chemical shifts, coupling constants, and relaxation times are discussed and illustrated when available.

INTRODUCTION

All the Group VI elements of the Periodic Table have isotopes with nuclear properties (A 1) which are not very favourable for NMR applications: ^{17}O and ^{33}S have nuclear quadrupole moments and very low natural abundance; ^{77}Se, ^{123}Te, and ^{125}Te are spin 1/2 nuclei with low (^{123}Te) or moderate natural abundance; ^{209}Po is radioactive with a half-life of 103 y. The nuclear properties of ^{209}Po have been examined only by hyperfine structure investigations (A 2): $I = 1/2$, $\mu = +0.76\ \mu_N$. Data which are important for NMR investigations are collected in Table 1. The magnetic moments given for the different nuclei with the exception of ^{209}Po are obtained from the ratios of the Larmor frequencies of the X nuclei and ^{2}H or ^{1}H in well defined species: ^{33}S: SO_4^{2-} at infinite dilution in D_2O (S 6); ^{77}Se: SeO_3^{2-} at infinite dilution in H_2O (Se 34); ^{123}Te and ^{125}Te: TeO_3^{2-} at infinite dilution in D_2O (Te 16). The magnetic moments are not corrected for diamagnetism and further are influenced by chemical shift effects. By comparing these accurate values with magnetic moments measured with high precision in free atoms, the nuclear screening could be evaluated.

Table 1. NMR Properties of Group VIa Elements

Nucleus	Natural abundance (%)	Larmor frequency at 2.11 T (MHz)	Receptivity ($^1H = 1$)	Nuclear spin	Nuclear magnetic moment[a] (μ_N)
^{33}S	0.76	6.91	1.7×10^{-5}	$3/2$[b]	+0.643103
^{77}Se	7.58	17.19	5.3×10^{-4}	$1/2$	+0.5332996
^{123}Te	0.87	23.57	1.6×10^{-4}	$1/2$	-0.732131
^{125}Te	6.99	28.44	2.2×10^{-3}	$1/2$	-0.882664

[a]Nuclear magnetic moments from NMR measurements: Refs. S 6, Se 34, Te 16.
[b]Nuclear quadrupole moment of ^{33}S: $Q = -0.055 \times 10^{-28}$ m^2 (A 1).

In addition to the chapter on Group VI elements in reference A 3, Wehrli has summarized ^{33}S data in his article on quadrupolar nuclei (A 4). General aspects of NMR of tellurium have been presented by Granger in a conference proceeding (Te 19).

SULPHUR-33

The problems of NMR studies with ^{33}S arise from the low receptivity, the low Larmor frequency, and the quadrupole moment. Very broad lines often beyond detectibility are expected.

As chemical shift reference the sulphate ion is proposed since the linewidth of ^{33}S, for instance in a 1 molal aqueous Cs_2SO_4 solution, is about 4 Hz (see Figure 1). Furthermore, the concentration and temperature dependence is small. Due to the relatively moderate T_1 (see later), many transients can be accumulated in a reasonable period of time.

It is therefore not surprising that only a small number of investigations have been published (S 1 - S 12). The rather limited chemical shift data now available are collected in Table 2 to our best knowledge. The whole range of chemical shifts is at this time about 1000 ppm. In addition to the less reliable earlier compounds, recently Faure et al. (S 11) have investigated some sulphones and sulphonic acids which had moderate linewidths. The whole spread of their chemical shifts was about 70 ppm. From Table 2 one finds for the linewidths a range of a few Hz to some kHz. Very impressive is the narrow ^{33}S signal in solid ZnS. This surely results from the cubic symmetry and the very magnetic dilution of this compound.

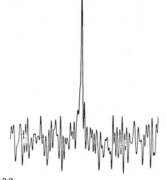

Figure 1. The ^{33}S NMR signal of a 1 molal solution of Cs_2SO_4 in D_2O (A 6); 8000 free induction decays were accumulated. The measuring time was 94 min. The linewidth is 4.1 Hz (no exponential line broadening). The signal was obtained with a high resolution probe at 6.91 MHz (2.11 T) with internal ^2H stabilization. A Bruker SXP 4-100 pulse spectrometer was used in connection with a high resolution magnet B-E 38 and a B-NC-12 data unit.

Table 2. Chemical Shifts and Linewidths of ^{33}S

Compound	Sample	Chemical shift (δ)	$\Delta\nu$ (Hz)	Ref.
Na_2S	H_2O	-592(-)	1600	S 5
Sphalerite (ZnS)	solid	-561(6)	65	S 5
ZnS	powder	-562(1)	53	S 10
$S_2(C_2H_5)_2$	neat	-499(88)	5200	S 5
Tetrahydrothiophene	neat	-420(38)	2600	S 5
CS_2	neat	-331(3)	360	S 6
3-Bromothiophene	neat	-197(-)	1600	S 5
2-Methylthiophene	neat	-153(9)	1300	S 5
3-Methylthiophene	neat	-134(26)	1600	S 5
Thiophene	90% in CS_2	-111(6)	620	S 5
H_2SO_4	conc.	-106(32)	2300	S 5
Dimethyl sulfoxide	neat	-98(20)	2600	S 5
$(CH_2=CH_2)_2SO_2$	DMSO-d_6	-26(1.5)	60	S 11
$(C_6H_5)_2SO_2$	DMSO-d_6	-23(2)	130	S 11
$C_6H_5SO_2CH_3$	DMSO-d_6	-20(2)	120	S 11
H_2SO_4	10 molal in H_2O	-17(11)	500	S 6
$CH_2=CHSO_3Na$	D_2O	-11(1.5)	70	S 11
p-$CH_3C_6H_5SO_3H$	D_2O	-10(1.5)	90	S 11
L-Cysteic acid	D_2O	-9(1.5)	80	S 11

Compound	Solvent/Conc.	δ (Δν)	Linewidth	Ref.
H_2SO_4	4 molal in H_2O	-8(4)	340	S 6
$(CH_3)_2SO_2$	DMSO-d_6	-7(1.5)	50	S 11
CH_3SO_3H	D_2O	-5(2.5)	150	S 11
Rb_2SO_4	0.5-1.5 molal D_2O	0(1)	70[a]	S 6
Cs_2SO_4	0.5-1.5 molal D_2O	0(1)	70[a]	S 6
$CdSO_4$	2.2 m D_2O	0(1)	130	S 6
Cs_2SO_4	1 m in D_2O	0	4	Ref.
$(n-C_3H_7)_2SO_2$	DMSO-d_6	7(2)	130	S 11
(sulfolene)	DMSO-d_6	32(1.5)	50	S 11
(3-amino sulfolane)	DMSO-d_6	33(1.5)	80	S 11
$(NH_4)_2S_2O_3$[b]	4 m in H_2O	34.5(0.2)	36	S 7
(3-hydroxy sulfolane)	DMSO-d_6	36(1.5)	100	S 11
(2,4-dimethyl sulfolane)	DMSO-d_6	37(1.5)	90	S 11
(3-methyl sulfolane)	DMSO-d_6	37(1.5)	60	S 11
(sulfolane)	DMSO-d_6	42(1.5)	50	S 11
$(NH_4)_2MoS_4$	0.1 m in H_2O	343(2)	40	S 9

[a] Steady state line broadening.
[b] Only one signal observed.

Due to the high sulphur concentration in liquid CS_2, the good signal-to-noise ratio allowed the accurate measurement of the linewidths of ^{33}S as a function of temperature (S 8). A quadrupole coupling constant of $e^2qQ/h = (14.9 \pm 0.3)$ MHz could be derived. No spin-lattice relaxation time measurements have been published. Just recently, in our laboratory an investigation of T_1 of central nuclei in oxoanions has been finished (S 12). Also some T_1 values in aqueous cesium sulphate solutions have been obtained. Figure 2 shows an example of such a measurement with experimental details. In the 4 molal Cs_2SO_4 solution $T_1 = (36 \pm 2)$ ms was found. The spin-lattice relaxation time increases with decreasing concentration. An extrapolated value for vanishing concentration of $(T_1)_0 = (110 \pm 5)$ ms can be given (S 12).

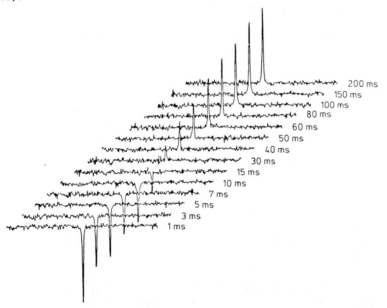

Figure 2. Stacked plot of an inversion recovery Fourier transform T_1 measurement of ^{33}S in a 4 molal solution of Cs_2SO_4 in H_2O (S12); 500 free induction decays were accumulated for each signal. A measuring time of 5.5 h was used for the 14 spectra. The delay times between the 180° and 90° pulses are given in the figure. A longitudinal relaxation time $T_1 = (36 \pm 2)$ ms was obtained. A high power probe with external ^1H field stabilization was used. Further experimental data as in Figure 1.

All the presented data have been measured with iron magnets at relatively low frequencies with all the ringing problems. With the measurements on superconducting magnets it seems not unreasonable that further useful information on ^{33}S can be obtained.

^{77}Se AND ^{123}Te IN AQUEOUS SOLUTIONS

For measuring the magnetic moments (see Table 1) detailed studies of the behaviour of ^{77}Se and ^{125}Te in SeO_3^{2-}, SeO_4^{2-} and TeO_3^{2-} ions in aqueous solutions with different cations have been performed (Se 34, Te 16).

The chemical shift of ^{77}Se has been studied as a function of concentration for the following compounds: Na_2SeO_4, Na_2SeO_3, H_2SeO_3, and $NaHSeO_3$. For some concentrations T_1 has been measured also. The change in chemical shifts as a function of concentration is small, about a few ppm for each salt. Large differences are observed for the different ions. In Table 3 chemical shift data of those samples are given, for which also T_1 has been measured.

Table 3. Typical Chemical Shifts and Spin-Lattice Relaxation Times of ^{77}Se in Aqueous Solution (Se 35)[a]

Compound	Concentration (molal)	Chemical shift (δ)	Relaxation time T_1 (s)	NOE
Na_2SeO_4	0.5, H_2O	-227.9(0.5)	10.2(0.8)	-
Na_2SeO_4	1.6, H_2O	-227.3(0.6)	10.1(0.7)	0
Na_2SeO_3	1.0, H_2O	1.5(0.5)	10.7(0.6)	-
	4.0, H_2O	2.7(0.5)	5.0(0.5)	0.4
H_2SeO_3	4.0, H_2O	28.9(0.5)	1.1(0.1)	0
	4.0, D_2O	-	1.4(0.1)	-
$NaHSeO_3$	4.0, H_2O	50.7(0.5)	0.34(5)	0

[a] Na_2SeO_3 at infinite dilution was used as $\delta = 0$, positive shifts are to higher frequencies.

For SeO_4^{2-} and SeO_3^{2-} ions at low concentrations, a T_1 of about 10 s is found; for SeO_3^{2-} at higher concentrations half of this value is obtained. The form $HSeO_3^-$ has a short T_1: the line broadens to about 10 Hz. To determine whether the intermolecular dipole-dipole interaction dominates the relaxation process, some NOE measurements have been performed. With the exception of SeO_3^{2-}, no NOE enhancement could be observed. That means that the dipole-dipole interaction plays only a minor role. Dawson et al. (Se 10) also observed a similar behaviour in organoselenic compounds.

For ^{125}Te in K_2TeO_3 solutions again the chemical shifts as a function of concentration are small (Te 16). For a 4 molal solution in D_2O, T_1 = 2.5 (0.1) s was measured. Linewidths indicated that $T_1 \neq T_2$. A direct measurement using the method of Kronenbitter and Schwenk (A 5) yielded T_1/T_2 = 8.2 ± 0.4, indicating strong exchange phenomena in tellurite solutions (Te 16).

CHEMICAL SHIFTS OF SELENIUM AND TELLURIUM COMPOUNDS

In the chapter on selenium and tellurium NMR in reference A 3, much material can be found on selenium, but for tellurium only a small number of investigated compounds are available. This trend has continued: for selenium a lot of papers have appeared in the meanwhile and therefore it is here not possible to list all the chemical shifts measured directly by ^{77}Se NMR. Therefore only a small table with typical chemical shifts is presented. From Table 4 one derives for instance that the chemical shift range is about 3000 ppm. Positive and negative chemical shifts occur when Me_2Se is chosen as reference.

Table 4. Selected Chemical Shifts of Selenium Compounds

No.	Compound	Solvent	Chemical shift (δ)	Ref.
1	$(Me_3Sn)_2Se$	liquid	-547	Se 37
2	SeCO	C_6D_6	-447	Se 20
3	$(MeO)_3PSe$	-	-396	Se 38
4	Me_3PSe	-	-235	Se 38
5	H_2Se	liquid	-226	Se 2
6	Me_2Se	liquid	0	Ref.
7	Et_2Se	liquid	+233	Se 31
8	$CF_3SeSeCF_3$	C_6F_6	+550	Te 15
9	Selenophene	$CDCl_3$	+605	Se 3
10	$(CF_3)_2Se$	liquid	+694	Se 2
11	$\begin{array}{c}OO\\ \diagdown\diagup\\ Se\\ \diagup\diagdown\\ OO\end{array}$	-	+1261	Se 18
12	H_2SeO_3	H_2O, sat.	+1282	Se 2
13	H_2SeCl_6	H_2O	+1451	Se 2
14	$SeC(t-Bu)_2$	C_6D_6	+2162	Se 20

The reference list on ^{77}Se papers is now very long (Se 1 - Se 38, Te 15), but ^{125}Te has been investigated to a much lesser extent. Here a list (Table 5) of chemical shifts is presented which may contain nearly all the published material (Te 1 - Te

25, Se 15, Se 18, Se 35). The chemical shifts are larger than in the case of selenium. The partly covalently bonded chalcogenides show negative shifts with respect to the accepted reference sample Me_2Te (see Table 6).

Table 5. Chemical Shifts and Linewidths of Tellurium Compounds

No.	Compound	Solvent	Chemical shift (δ)	$\Delta\nu$ (Hz)	Ref.
1	$(CH_3)_2Te$	neat	Reference	–	–
2	$(CH_3)_2Te$	$CDCl_3$	0	6.0	Te 10
3	(thienyl)$_2Te_2$	$CDCl_3$	264.1	–	Te 12
4	$(C_2H_5)_2Te_2$	$CDCl_3$	392.1	4.0	Te 10
5	$(C_6H_5)_2Te_2$	$CDCl_3$	420.8	–	Te 12
6	$(C_6H_5)_2Te_2$	CH_2Cl_2	422	–	Te 20
7	$(p-CH_3C_6H_4)_2Te_2$	$CDCl_3$	432.2	–	Te 12
8	$(p-ClC_6H_4)_2Te_2$	$CDCl_3$	451.8	–	Te 12
9	$(p-EtOC_6H_4)_2Te_2$	$CDCl_3$	456.0	–	Te 12
10	$(2,4,6-(CH_3)_3C_6H_3)_2Te$	$CDCl_3$	611.6	5.0	Te 10
11	$CF_3TeTeCF_3$	C_6D_6	686	–	Te 15
12	H_2TeO_6	D_2O	712	–	Te 20
13	H_2TeI_6	HI/D_2O	857	–	Te 20
14	$p-C_8H_{17}-C_6H_4-\overset{O}{\underset{\|}{C}}-Te-C_6H_4-p-C_5H_{11}$	–	905	–	Se 15
15	$p-CH_3O-C_6H_4-\overset{O}{\underset{\|}{C}}-Te-C_6H_4-p-CH_3$	–	910	–	Se 15
16	$C_6H_5-\overset{O}{\underset{\|}{C}}-Te-C_6H_4-p-CH_3$	–	934	–	Se 15
17	$(CF_3)_2TeCl_2$	CH_3CN/CD_3CN	1114	–	Te 15
18	$Te(OH)_5F$	40%HF, D_2O	1160	4	Te 10
19	$(CF_3)_2TeBr_2$	CH_3CN/CD_3CN	1180	–	Te 15
20	$(CF_3)_2TeF_2$	CH_3CN/CD_3CN	1187	–	Te 15
21	$Te(OH)_6$	40%HF, D_2O	1215	4	Te 10
22	$[NBu_4^+]_2TeCl_6^{2-}$	$(CD_3)_2CO$	1329	–	Te 20
23	$[NBu_4^+]_2TeBr_6^{2-}$	$(CD_3)_2CO$	1341	–	Te 20
24	catecholate Te	CCl_2H_2	1355	–	Se 18
25	H_2TeBr_6	HBr/D_2O	1356	–	Te 20

No.	Compound	Solvent	δ	Δν (Hz)	Ref.
26	CF$_3$TeCF$_3$	C$_6$D$_6$	1368	–	Te 15
27	Te[OCH(CF$_3$)$_2$]$_4$THF	neat	1394	–	Se 18
28	H$_2$TeCl$_6$	HCl/D$_2$O	1403	–	Te 20
29	Te(OCH$_2$CF$_3$)$_4$	neat	1463	–	Se 18
30	Te(OCH$_2$CH$_3$)$_4$	neat	1503	–	Se 18
31	(C$_6$H$_5$)$_2$TeBr$_2$	THF	1508	18	Te 10
32	Te(OCH$_3$)$_4$	neat	1510	–	Se 18
33	Te(OCH(CH$_3$)$_2$)$_4$	neat	1523	–	Se 18
34	Te(OCH$_2$C(CH$_3$)$_3$)$_4$	neat	1525	–	Se 18
35	[Te(OCH$_2$CH$_2$O)$_2$ cyclic]	–	1526	–	Se 18
36	CF$_3$TeCF$_2$Cl	C$_6$D$_6$	1566	–	Te 15
37	[Te(OCH$_2$CH$_2$O)$_2$ cyclic]	–	1601	–	Se 18
38	TeCl$_4$	THF	1725	–	Te 20
39	K$_2$TeO$_3$	D$_2$O	1732	–	Te 20
40	TeBr$_4$	THF	1962	–	Te 20

Table 6. Chemical Shifts and Linewidths of Tellurides

No.	Compound	Solvent	δ[a,b]	Δν (Hz)	Ref.
41	PbTe	powder	–1191	1500	Se 35
42	CdTe	powder	–1090	280	Se 35
43	ZnTe	powder	–890	200	Se 35
44	HgTe	powder	–874	820	Se 35

[a]With δ = 1732 (K$_2$TeO$_3$, 2 m in D$_2$O) from Goodfellow (Te 20) and δ ≈ 2 (infinite dilution of K$_2$TeO$_3$) from Ref. Te 16 and Ref. Se 35.
[b]Reference in Me$_2$Te at δ ≡ 0.0.

SOLID CHALCOGENIDES

The IIb and IV chalcogenides have found many interests, e.g., in optoelectronics. Due to the partial covalency of the bonding, large chemical shifts, chemical shielding anisotropies, and indirect spin-spin couplings are expected in favourable cases. Some papers have been published in this field (Te 5, S 10, Se 35, Te 17, Te 18, Te 21).

Experimental curves for ^{33}S, ^{77}Se, and ^{125}Te in the zinc compounds are given with the appropriate parameters in Figures 3, 4, and 5. In Table 7 experimental results which have been taken from references S 10 and Se 35 are listed for IIb chalcogen nuclei. For the chemical shifts, infinitely diluted aqueous solutions of SO_4^{2-}, SeO_3^{2-}, and TeO_3^{2-} have been used.

Table 7. ^{33}S, ^{77}Se, and ^{125}Te NMR Parameters in IIb Chalcogenides (Ref. S 10, Se 35)

Nucleus	Compound	Chemical shift (δ)	Typical linewidth $\Delta\nu$ (Hz)
^{33}S	ZnS	-562^a	53
^{77}Se	ZnSe	-1622^b	170
	CdSe	-1743^b	170
	HgSe	-1365^b	700
^{125}Te	ZnTe	-2620^c	200
	CdTe	-2820^c	280
	HgTe	-2604^c	820

^{a}Reference is infinitely dilute aqueous SO_4^{2-} at $\delta \equiv 0.0$.
^{b}Reference is infinitely dilute aqueous SeO_3^{2-} at $\delta \equiv 0.0$.
^{c}Reference is infinitely dilute aqueous TeO_3^{2-} at $\delta \equiv 0.0$.

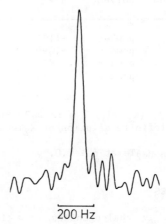

200 Hz

Figure 3. The ^{33}S NMR signal in powdered ZnS (S 13); 4800 free induction decays have been accumulated within 13.5 h. A linewidth of 53 Hz was found. Larmor frequency: 6.90 MHz.

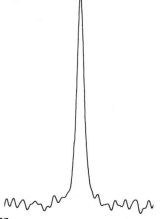

Figure 4. The ^{77}Se NMR signal of powdered ZnSe. Only 56 free induction decays could be accumulated within 15.5 h due to the long spin-lattice relaxation time. The linewidth is 150 Hz; Larmor frequency: 17.19 MHz.

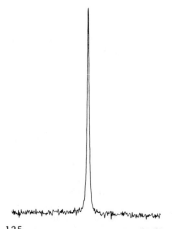

Figure 5. The ^{125}Te NMR signal of powdered ZnTe; 180 free induction decays were accumulated for this signal; a measuring time of only 30 min was necessary. The line is 280 Hz broad. Larmor frequency: 28.44 MHz.

The shifts are all to lower frequencies, and the amount of the shift increases strongly with the atomic numbers of the chalcogen nuclei. Since CdSe has no cubic symmetry, an anisotropy of the shielding of $\sigma_\parallel - \sigma_\perp = 44 \times 10^{-6}$ has been found (Se 35).

Due to the small linewidths of ^{125}Te an interesting line splitting in single crystals of CdTe and also in powdered samples of HgTe and PbTe was observed (Te 17, Te 18).

Detailed studies of this phenomenon revealed that this fine structure can be explained by an indirect spin-spin interaction between ^{125}Te and ^{111}Cd or ^{113}Cd, respectively. This fact is possibly due to the covalency of the binding in CdTe. Isotropic and anisotropic indirect spin-spin coupling constants have been evaluated from a very tiresome single crystal study of ^{125}Te in CdTe (Te 18) and also with less accuracy from powder samples in other compounds like HgTe and PbTe (Te 17).

As an example the following data from Nolles work (Te 18) on ^{125}Te in a CdTe single crystal are presented for the interaction of ^{125}Te with ^{113}Cd: direct coupling constant (calculated from distance and gyromagnetic ratios): D = 380(1) Hz, isotropic indirect coupling: J = 655(60) Hz, anisotropic indirect coupling: $J_\perp$ = 765(80) Hz), $J_\parallel$ = 435(120) Hz. This is one of the very few cases where indirect coupling has been observed directly in a solid.

ACKNOWLEDGMENTS

I thank the Deutsche Forschungsgemeinschaft for financial support. I acknowledge gratefully the help of the following co-workers: Dr. W. Koch, Dr. habil. A. Nolle, R. Balz, K. U. Buckler, M. Haller, W. Hertler, D. Köhnlein, G. Kössler, W. Messner, K. R. Mohn, and R. Schafitel. I thank Prof. P. Granger, Rouen, for helpful remarks on tellurium NMR.

REFERENCES

(A 1) G. H. Fuller, J. Phys. Chem. Ref. Data, 5, p. 835 (1976).
(A 2) G. W. Charles, J. Opt. Soc. Am., 5b, p. 1292 (1966).
(A 3) R. K. Harris and B. E. Mann, Eds., "NMR and the Periodic Table," Academic Press, London, 1978.
(A 4) F. W. Wehrli, Ann. Reports NMR Spectrosc., 9 (1979).
(A 5) J. Kronenbitter and A. Schwenk, J. Magn. Reson., 25, p. 147 (1977).
(A 6) D. Köhnlein, unpublished results.
(S 1) S. S. Dharmatti and H. E. Weaver, Jr., Phys. Rev., 83, p. 845 (1951).
(S 2) K. Lee, Phys. Rev., 172, p. 284 (1968).
(S 3) H. L. Retcofsky and R. A. Friedel, Appl. Spectrosc., 24, p. 379 (1970).
(S 4) H. D. Schultz, C. Carr, Jr., and G. D. Vickers, Appl. Spectrosc., 25, p. 363 (1971).

(S 5) H. L. Retcofsky and R. A. Friedel, J. Am. Chem. Soc., 94, p. 6579 (1972).
(S 6) O. Lutz, A. Nolle, and A. Schwenk, Z. Naturforsch., 28a, p. 1370 (1973).
(S 7) O. Lutz, W. Nepple, and A. Nolle, Z. Naturforsch., 31a, p. 978 (1976).
(S 8) R. R. Vold, S. W. Sparks, and R. L. Vold, J. Magn. Reson. 30, p. 497 (1978).
(S 9) P. Kroneck, O. Lutz, and A Nolle, Z. Naturforsch., 35a, p. 226 (1979).
(S 10) M. Haller, W. E. Hertler, O. Lutz, and A. Nolle, Solid State Comm., 33, p. 1051 (1980).
(S 11) R. Faure, E. J. Vincent, J. M. Ruiz, and L. Lena, Org. Magn. Reson., 15, p. 401 (1981).
(S 12) G. Kössler and O. Lutz, to be published.
(S 13) M. Haller, Wissenschaftliche Arbeit, Tübingen (1979).
(Se 1) W. McFarlane, J. Chem. Soc. (A), p. 670 (1969).
(Se 2) T. Birchall, R. J. Gillespie, and S. L. Vekris, Can. J. Chem., 43, p. 1672 (1965).
(Se 3) S. Gronowitz, I. Johnson, and A.-B. Hörnfeldt, Chem. Scripta, 3, p. 94 (1973).
(Se 4) S. Gronowitz, I. Johnson, and A.-B. Hörnfeldt, Chem. Scripta, 8, p. 8 (1975).
(Se 5) A. Fredga, S. Gronowitz, and A.-B. Hörnfeldt, Chem. Scripta, 8, p. 15 (1975).
(Se 6) L. Christiaens, J.-L. Piette, L. Laitem, M. Baiwir, J. Denoel, and G. Llabres, Org. Magn. Reson., 8, p. 354 (1976).
(Se 7) A. Fredga, S. Gronowitz, and A.-B. Hörnfeldt, Chem. Scripta, 11, p. 37 (1977).
(Se 8) W.-H. Pan and J. P. Fackler, Jr., J. Am. Chem. Soc., 100, p. 5783 (1978).
(Se 9) O. A. Gansow, W. D. Vernon, and J. J. Dechter, J. Magn. Reson., 32, p. 19 (1978).
(Se 10) W. H. Dawson and J. D. Odom, J. Am. Chem. Soc., 99, p. 8352 (1977).
(Se 11) S. Ueda and T. Shimizu, Phys. Stat. Sol. (b), 88, K1 (1978).
(Se 12) H. J. Jakobsen, R. S. Hansen, J. Magn. Reson., 30, p. 397 (1978).
(Se 13) J. D. Odom, W. H. Dawson, and P. D. Ellis, J. Am. Chem. Soc., 101, p. 5815 (1979).
(Se 14) P. A. W. Dean, Can. J. Chem., 57, p. 754 (1979).
(Se 15) B. Kohne, W. Lohner, K. Praefcke, H. J. Jakobsen, and B. Villadsen, J. Organomet. Chem., 166, p. 373 (1979).
(Se 16) H. J. Jakobsen, A. J. Zozulin, P. D. Ellis, and J. D. Odom, J. Magn. Reson., 38, p. 219 (1980).
(Se 17) E. R. Cullen, F. S. Guziec, Jr., C. J. Murphy, T. C. Wong, and K. K. Andersen, J. Am. Chem. Soc., 103, p. 7055 (1981).

(Se 18) D. B. Denney, D. Z. Denney, P. J. Hammond, and Y. F. Hsu, J. Am. Chem. Soc., 103, p. 2340 (1981).
(Se 19) M. Baiwir, G. Llabrès, L. Christiaens, and J.-L. Piette, Org. Magn. Reson., 16, p. 14 (1981).
(Se 20) W. Gombler, Z. Naturforsch., 36b, p. 1561 (1981).
(Se 21) M. Baiwir, G. Llabrès, L. Christiaens, and J.-L. Piette, Org. Magn. Reson., 18, p. 33 (1982).
(Se 22) N. P. Luthra, R. B. Dunlap, and J. D. Odom, J. Magn. Reson., 46, p. 152 (1982).
(Se 23) P. Granger, S. Chapelle, and C. Paulmier, Org. Magn. Reson., 14, p. 240 (1980).
(Se 24) R. Colton and D. Dakternieks, Austral. J. Chem., 33, p. 1463 (1980).
(Se 25) S. W. Carr and R. Colton, Austral. J. Chem., 34, p. 35 (1981).
(Se 26) M. Baiwir, G. Llabrès, J. L. Piette, and L. Christaens, Spectrochem. Acta, 36A, p. 819 (1980).
(Se 27) T. Drakenberg, A.-B. Hörnfeldt, S. Gronowitz, J.-M. Talbot, and J. L. Piette, Chem. Scripta, 13, p. 152 (1978).
(Se 28) S. Gronowitz, A. Konar, and A.-B. Hörnfeldt, Org. Magn. Reson., 9, p. 213 (1977).
(Se 29) M. Baiwir, G. Llabrès, J.-L. Piette, and L. Christiaens, Org. Magn. Reson., 14, p. 293 (1980).
(Se 30) G. A. Kalabin, D. F. Kushnarev, V. M. Bzesovsky, and G. A. Tschmutova, Org. Magn. Reson., 12, p. 598 (1979).
(Se 31) W. McFarlane and R. J. Wood, J. Chem. Soc. Dalton, 1397 (1972).
(Se 32) R. Dupree, W. W. Warren, Jr., and F. J. DiSalvo, Jr., Phys. Rev., B 16, p. 1001 (1977).
(Se 33) M. Lardon, J. Am. Chem. Soc., 92, p. 5063 (1970).
(Se 34) W. Koch, O. Lutz, and A. Nolle, Z. Naturforsch, 33a, p. 1025 (1978).
(Se 35) W. Koch, O. Lutz, and A. Nolle, Z. Physik, A289, p. 17 (1978).
(Se 36) R. Keat, D. S. Rycroft, and D. G. Thompson, Org. Magn. Reson., 12, p. 391 (1979).
(Se 37) J. D. Kennedy and W. McFarlane, J. Organomet. Chem., 94, p. 7 (1975).
(Se 38) W. McFarlane and D. S. Rycroft, J. C. S. Dalton, 2162 (1973).
(Te 1) G. Pfisterer, H. Dreeskamp, Ber. Bunsenges., 73, p. 654 (1967).
(Te 2) H. C. E. McFarlane and W. McFarlane, J. Chem. Soc. Dalton, 2416 (1973).
(Te 3) W. McFarlane, F. Berry, and B. C. Smith, J. Organomet. Chem., 113, p. 139 (1976).
(Te 4) T. Drakenberg, F. Fringuelli, S. Gronowitz, A.-B. Hörnfeldt, I. Johnson, and A. Taticchi, Chem. Scripta, 10, p. 139 (1976).

(Te 5) A. Willig, B. Sapoval, K. Leibler, and C. Vérié, J. Phys., C9, p. 1981 (1976).
(Te 6) G. J. Schrobilgen, R. C. Burns, and P. Granger, J. C. S. Chem. Commun., 957 (1978).
(Te 7) C. R. Lassigne and E. J. Wells, J. C. S. Chem. Commun., 956 (1978).
(Te 8) W. Koch, O. Lutz, and A. Nolle, Z. Physik A, 289, p. 17 (1978).
(Te 9) A. Nolle, Z. Physik, B34, p. 175 (1979).
(Te 10) G. V. Fazakerley and M. Celotti, J. Magn. Reson., 33, p. 219 (1979).
(Te 11) P. Granger and S. Chapelle, J. Magn. Res., 39, p. 329 (1980).
(Te 12) P. Granger, S. Chapelle, W. R. McWhinnie, and A. Al-Rubaie, J. Organomet. Chem., 220, p. 149 (1981).
(Te 13) S. Chapelle and P. Granger, Mol. Phys., 44, p. 459 (1981).
(Te 14) P. Granger, S. Chapelle, and C. Brevard, J. Magn. Reson., 42, p. 203 (1981).
(Te 15) W. Gombler, Z. Naturforsch., 36b, p. 535 (1981).
(Te 16) K. U. Buckler, J. Kronenbitter, O. Lutz, and A. Nolle, Z. Naturforsch, 32a, p. 1263 (1977).
(Te 17) R. Balz, M. Haller, W. E. Hertler, O. Lutz, A. Nolle, and R. Schafitel, J. Magn. Reson., 40, p. 9 (1980).
(Te 18) A. Nolle, Z. Phys., B34, p. 175 (1979).
(Te 19) P. Granger, Proc. 3rd Intern. Symp. Organic Selenium and Tellurium Compounds, Metz, July, 1979, p. 305.
(Te 20) R. J. Goodfellow, given in Ref. A 3.
(Te 21) A. Willig and B. Sapoval, J. Phys. Lett., 38, p. L57 (1977).
(Te 22) N. Zumbulyadis and H. J. Gysling, J. Organomet. Chem., 192, p. 183 (1980).
(Te 23) W. Lohner and K. Praefcke, J. Organomet. Chem., 208, p. 39 (1981).
(Te 24) H. J. Gysling, N. Zumbulyadis, and J. A. Robertson, J. Organomet. Chem., 209, p. C41 (1981).
(Te 25) W.-W. du Mont and H.-J. Kroth, Z. Naturforsch., 36b, p. 332 (1981).

CHAPTER 20

THE HALOGENS--CHLORINE, BROMINE, AND IODINE

Torbjörn Drakenberg and Sture Forsén

Physical Chemistry 2, Chemical Center
University of Lund, S-220 07 LUND 7 Sweden

ABSTRACT

The stable isotopes of chlorine (^{35}Cl and ^{37}Cl), bromine (^{79}Br and ^{81}Br), and iodine (^{127}I) all are magnetic nuclei with electric quadrupole moments. The study of quadrupolar nuclei can provide unique and valuable information on a diversity of physicochemical and biological systems. The relaxation of quadrupolar nuclei is generally much simpler to interpret than the relaxation of nonquadrupolar nuclei. The relaxation of the former is in most cases totally dominated by the quadrupole relaxation, which is normally induced by purely intramolecular interactions modulated by the molecular motion. Studies of quadrupole relaxation have therefore provided very valuable information about molecular reorientation and association in liquids. The chemical exchange of the quadrupolar nucleus between two environments characterized by markedly different electric field gradients or correlation times can give unique information on exchange rates and the occurrence of weak interactions in inorganic as well as biological systems. In addition another important parameter may be obtained from NMR studies of quadrupolar nuclei in anisotropic environments: the quadrupole splitting. This parameter may be of great value for the characterization of the ordering of the system at a molecular level.

INTRODUCTION

We have not tried to make a complete coverage of chlorine, bromine, and iodine NMR but selected some areas which we find most interesting. More complete coverage of the literature have been published (1-3).

The earliest experimental observation of chlorine, bromine, and iodine NMR signals were achieved in the late 1940's, first for the ions in aqueous solutions where the lines are relatively narrow (4-8). Early attempts to observe NMR signals from covalent compounds were unsuccessful. The failure to observe bromine NMR spectra of liquid Br_2 and $CHBr_3$ was correctly interpreted by Pound (7) as due to very rapid quadrupole relaxation, which broadened the signals beyond detection.

The first systematic chlorine NMR studies of covalent compounds were published in 1956 by Diehl (9) and Masuda (10). In 1958 Myers made the first application of halogen NMR in the study of chemical exchange phenomena (11). From the linewidth of the ^{127}I NMR signal in KI solutions containing varying amounts of I_2 he calculated the rate of the iodide-triiodide ion exchange. A few years later Connick and Poppel (12) and Hertz (13) initiated more detailed halogen NMR studies of ligand exchange in metal-halide complexes.

Quadrupole splittings of a covalent halogen compound in liquid crystals were first observed in ^{35}Cl NMR spectra of CH_2Cl_2 in polybenzyl-L-glutamate by Gill, Klein, and Kotowycs (14). Quadrupole splittings of halide ions in amphiphilic mesophases were first observed in 1971 by Lindblom, Winnerström, and Lindman (15).

The first biological application of NMR spectroscopy on a quadrupolar halogen nucleus (^{35}Cl) was made by Stengele and Baldeschwieler in 1966 (16). They monitored the ^{35}Cl linewidth in a NaCl solution containing oxyhemoglobin as a function of the amount of $HgCl_2$ added. The latter binds strongly to free SH groups and the appearance of excess unbound $HgCl_2$ is directly seen from the changes in the ^{35}Cl linewidth.

ATOMIC AND NUCLEAR PROPERTIES OF HALOGENS

In Table 1 some atomic and nuclear properties of the halogens are summarized. Data on fluorine, although this nucleus is not treated here, are included for completeness and comparison. A more extensive collection of data is given in reference 1.

Concerning the nuclear quadrupole moments, it may be noted that the ratios of the moments of the two isotopes of chlorine and bromine have been determined with considerably higher accuracy than the individual quadrupole moments.

Table I. Atomic and Nuclear Properties of the Halogens

Property	^{19}F	^{35}Cl	^{37}Cl	^{79}Br	^{81}Br	^{127}I
Atomic number	9	17	17	35	35	53
Natural abundance (%)	100	75.53	24.47	50.54	49.46	100
Atomic mass ($^{12}C = 12.0000$)	18.9984	34.96885	36.9658	78.9183	80.9163	126.9004
Electronic configuration	$[He]2s^22p^5$	$[Ne]3s^23p^5$		$[Ar]3d^{10}4s^24p^5$		$[Kr]4d^{10}5s^25p^5$
Electronegativity (Pauling)	4.0	3.0	3.0	2.8	2.8	2.5
Single-bond covalent radius (Å)	0.71	0.99	0.99	1.14	1.14	1.33
Radius of X^- ion in Å (Pauling)	1.36	1.81	1.81	1.95	1.95	2.16
Van der Waals radius (Å)	1.35	1.80	1.80	1.95	1.95	2.15
Nuclear spin quantum number	1/2	3/2	3/2	3/2	3/2	5/2
Nuclear magnetic moment in multiples of the nuclear magneton	2.62727	0.82091	0.6833	2.0990	2.2626	2.7937
NMR frequency in MHz at a magnetic field of 1 Tesla	40.055	4.1717	3.472	10.667	11.498	8.5183
NMR sensitivity of equal number of nuclei at constant field relative to $^1H = 1.000$	0.833	4.70×10^{-3}	2.71×10^{-3}	7.86×10^{-2}	9.85×10^{-2}	9.34×10^{-2}

Table I (continued)

Property	^{19}F	^{35}Cl	^{37}Cl	^{79}Br	^{81}Br	^{127}I
Nuclear electric quadrupole moment (10^{-24} cm^2)	–	−0.0802	−0.0632	+0.332	+0.282	−0.785
Isotopic ratio of nuclear electric quadrupole moments	–		1.26879		1.19707	–

EXPERIMENTAL ASPECTS

NMR signals from the quadrupolar halogens in covalent compounds are generally very broad due to the efficient quadrupole relaxation. Until a few years ago special types of NMR spectrometers, "wideline spectrometers," which use phase sensitive detection and B_0 field modulation with amplitudes up to 10^{-4} Tesla (=1 gauss) were used in the study of covalent halogens. With the modern FT NMR spectrometers utilizing a high power rf pulse even the broad lines can be conveniently studied. The chloride ions in aqueous solution have, in contrast to the covalent compounds, relatively narrow NMR signals. Approximate chlorine linewidths at 25°C in a 0.5 M NaCl solution are $\Delta\nu_{1/2}$ = 8 Hz for ^{35}Cl and 5 Hz for ^{37}Cl. For bromine and iodine the lines are quite broad even for the ions in aqueous solutions (1). At Lund University ^{35}Cl and ^{37}Cl NMR spectra of chloride ions in aqueous solutions have been obtained on a modified Varian XL-100-15 spectrometer. With a 12 mm sample tube a signal-to-noise ratio of about 4:1 is obtained for the ^{35}Cl signal from a 5 mM solution of NaCl in H_2O in ca. 30 min (10^4 pulses, ca. 0.2 s acquisition time). On a modern high field spectrometer much better performance can be obtained; for example we have in Lund on a Nicolet 360 WB system with a home-built horizontal probe accommodating 17 mm OD samples got a better signal-to-noise ratio than above in less than one minute.

It should also be pointed out that the decay of the longitudinal as well as the transverse magnetization of nuclei with spin I > 1 does not follow a simple exponential decay under nonextreme narrowing conditions (17). For nuclei with I = 3/2 (^{35}Cl, ^{37}Cl, ^{79}Br, and ^{81}Br) the decay will be biexponential and for I = 5/2 nuclei (^{127}I) the decay will be described by the sum of three exponentials. Under nonextreme narrowing conditions the NMR signals of chlorine and bromine should thus be the sum of two Lorentzians, whereas the iodine NMR signals should be the sum of three Lorentzians. Nonextreme narrowing conditions are most likely to be encountered in biological systems which are further discussed in a later section.

Relaxation data of the quadrupolar halogens are directly obtained from pulsed NMR studies, although most data on T_2 are from measurements of linewidth or from peak-to-peak distances of the first derivative of absorption curves. Using pulse methods the lower limit of T_1 and T_2 depends on the dead time of the spectrometer. Presently the practical limits appear to be ca. 10 μs. A number of the available relaxation data on Cl, Br, and I compounds have been determined through the effects of the halogen relaxation on the relaxation of directly bonded spin 1/2 nuclei through modulation of scalar interaction; this mechanism is commonly termed "scalar relaxation of the second kind." Very short halogen relaxation times are accessible with this method. Examples of the use of this technique listed according to the $I = 1/2$ nuclei studied include ^1H (18-22), ^{13}C (23-29), ^{19}F (30-34), ^{31}P (19,35-37), ^{119}Sn (17,38-40), and ^{207}Pb (40). A more detailed discussion of the use of this technique and other methods to determine halogen relaxation times is found in reference 1.

CHEMICAL SHIFTS OF COVALENT HALOGEN COMPOUNDS

Chemical shift data of the quadrupolar halogens in covalent compounds are limited and with a few exceptions restricted to chlorine compounds. The experimental difficulties in the shift measurements are considerable, due to the efficient relaxation. A selection of reported shift data for ^{35}Cl is given in Table 2. A more extensive collection is given in reference 1.

With the exception of ClO_4^- the errors in the ^{35}Cl chemical shifts are considerable. Due to the tetrahedral symmetry of the perchlorate ion, its ^{35}Cl linewidth is only slightly above 1 Hz at low concentrations. The shift data for almost all the covalent chlorine compounds refer to the neat liquids and not to dilute solutions in a common inert solvent, and this fact makes a meaningful discussion of the data somewhat premature. Medium effects on chlorine chemical shifts in $TiCl_4$ have been observed (41). In CH_2I_2 solutions the chlorine shifts are some 15 ppm downfield and in hexane some 10 ppm upfield from the shift in neat $TiCl_4$.

The chemical shifts are, however, seen to range over about 1000 ppm. This value is of the same order as the total diamagnetic contributions to chlorine chemical shifts calculated for free chlorine atoms by Dickinson (42) and Bonham and Shand (43). In terms of shielding, σ, this value is $\sigma^d = 11.5 \times 10^{-4}$. Theoretical calculations of the average value of the diamagnetic shielding of the chlorine in the moleucles ClF (44) and HCl (45,46) have yielded values very close to that for the free atom. The availability of seemingly accurate values of the total diamagnetic shielding in these molecules raises the question whether an "absolute" shielding scale could be established for ^{35}Cl (or ^{37}Cl). Presently reasonable "absolute" shielding scales have been obtained for ^1H, ^{13}C, ^{14}N and ^{15}N, ^{17}O, ^{19}F and ^{31}P. These scales have been obtained using the circumstance first pointed out by Ramsey (47), that the paramagnetic components of the nuclear shielding tensor are related to the components of the spin-rotation interaction tensor. The latter components may in turn be determined through microwave spectroscopic studies. (A discussion of the theoretical background of this idea has been given by Flygare (48).) Spin-rotation interaction constants are in fact available for both ClF (44,49,50) and HCl (51), as well as for a few other covalent halogen compounds (48). In order to make use of these data to establish an "absolute" shielding scale for chlorine it would be necessary to determine the chlorine chemical shift of the <u>gaseous compounds</u> relative to a common reference such as aqueous sodium chloride. With presently available instrumentation such measurements are by no means trivial.

Table II. ^{35}Cl Chemical Shifts in Some Covalent Chlorine Compounds

Compound	^{35}Cl Shift[a]	Compound	^{35}Cl Shift[b]
CH_3Cl	50	$(CH_3)_2SiCl_2$	110
CH_2Cl_2	228	CH_3SiCl_3	151
$CHCl_3$	420	$C_2H_5SiCl_3$	115
CCl_4	540	$C_6H_5SiCl_3$	70
C_2H_5Cl	141	CCl_4	500
$n-C_3H_7Cl$	98	$SeCl_4$	174
$iso-C_3H_7Cl$	251	$GeCl_4$	170
$n-C_4H_9Cl$	125	$SnCl_4$	120
$iso-C_4H_9Cl$	182	PCl_3	370
$sec-C_4H_9Cl$	280	$POCl_3$	430
$tert-C_4H_9Cl$	310	$PSCl_3$	530
H_3SiCl	35	S_2Cl_2	480
H_2SiCl_2	97	$SOCl_2$	660
$HSiCl_3$	143	SO_2Cl_2	760
$SiCl_4$	185	$AsCl_3$	150
$(CH_3)_2SiHCl$	36	$VOCl_3$	791
$(CH_3)_3SiCl$	77	ClO_4^-	946
CH_3SiHCl_2	77		

[a]From Barlos et al., Chem. Ber., 111, p. 1839 (1978) and 113, p. 3716 (1980).
[b]From Johnson et al., J. Chem. Phys., 51, p. 4493 (1969).

The tetrachlorides of the Group IV elements, CCl_4, $SiCl_4$, $GeCl_4$, and $SnCl_4$ make an interesting series for which ^{35}Cl shift data are available. If we make the plausible assumption that the diamagnetic shielding for the chlorines is nearly constant in the series we may conclude that the paramagnetic contribution is largest in CCl_4 but less and nearly equal for the three other tetrachlorides. The paramagnetic shielding contribution approximately parallels the difference in Pauling electronegativity, $\chi_{Cl} - \chi_m$. This result would appear reasonable since an increase in the electronic density of the chlorine decreases the average value of $1/r^3$ for the 3p electrons and decreases σ^p, the paramagnetic shielding term, according to approximate theories of NMR chemical shifts.

SCALAR SPIN-SPIN COUPLING

Experimental spin coupling constants involving chlorine, bromine, and iodine are summarized in Table 3. Spin couplings are reported both as conventional coupling constants, J, and as reduced spin couplings, $K_{AB} = (2/h\gamma_A\gamma_B)J_{AB}$.

In general spin-spin couplings between $I = 1/2$ nuclei have a well recognized influence on their NMR spectra and a determination of the spin coupling constants is a relatively straightforward procedure. In the case when one of the nuclei involved in the scalar spin coupling has a relaxation rate which greatly exceeds the spin coupling constant in frequency units, no effect of the coupling is discernible in the NMR spectrum. The Cl, Br, and I nuclei in covalent environments usually have very high relaxation rates, and a direct determination of spin coupling constants from spectral fine structure is rarely possible. A few exceptions will be discussed below. Most of the available spin coupling constants for the quadrupolar halogens have therefore been determined indirectly from the scalar spin relaxation of a nonquadrupolar nucleus.

The scalar contribution to the relaxation rate of an $I = 1/2$ nucleus (I) spin coupled to a quadrupolar halogen nucleus (S) is given by eqs. 1 and 2 (52), in which S is the nuclear spin of the

$$\left(\frac{1}{T_{1I}}\right)^{SC} = \frac{8\pi^2 J^2}{3} S(S+1) \left(\frac{T_{2S}}{1 + (\omega_I-\omega_S)^2 T_{2S}^2}\right) \qquad (1)$$

$$\left(\frac{1}{T_{2I}}\right)^{SC} = \frac{4\pi^2 J^2}{3} S(S+1) \left(T_{1S} + \frac{T_{2S}}{1 + (\omega_I-\omega_S)^2 T_{2S}^2}\right) \qquad (2)$$

quadrupolar halogen nucleus, ω_I and ω_S are the Larmor frequencies of spins I and S, respectively, T_{1S} and T_{2S} are the relaxation times of the quadrupolar nucleus, and J is the scalar spin coupling constant in Hz.

For most combinations of nuclei I and S the frequency difference $\omega_I - \omega_S$ is normally large at ordinary fields; an exception is the combination $I = {}^{13}C$ and $S = {}^{79}Br$. Appreciable scalar contributions to T_{1I} are observed only when T_{2S} is very short ($<10^{-5}$ s) and the scalar coupling is large. Usually the scalar contribution to T_{2I} is more pronounced, especially when (ω_I -

Table III. Cl, Br, and I Spin Coupling Constants

Compound	Spin-spin coupling J_{ij} (in Hz)	Reduced spin-spin coupling K_{ij} ($=10^{+20}$ cm^{-3}) 10^{+19} Na^{-2}m^{-3}	Method of determination	Reference
HCl	$J(^1H^{35}Cl) = 41 \pm 2$	35 ± 2	$T_{1\rho}$	18
	$J(^1H^{37}Cl) = 35 \pm 2$			
HBr	$J(^1H^{79}Br) = 57 \pm 3$	19 ± 1	$T_{1\rho}$	18
	$J(^1H^{81}Br) = 62 \pm 3$			
PCl$_3$	$J(^{31}P Cl) = 112$	(235)	T_1 and T_2	36
PCl$_3$	$J(^{31}P^{35}Cl) = 120$	252	$T_{1\rho}$	19, 37
	$J(^{31}P^{37}Cl) = 100$			
PBr$_3$	$J(^{31}PBr) = 330$	(259)	T_1 and T_2	36
PBr$_3$	$J(^{31}P^{79}Br) = 350 \pm 17$	288 ± 14	T_1 and $T_{1\rho}$	35
	$J(^{31}P^{81}Br) = 380 \pm 19$			
CH$_3$I	$J(^{13}C^{127}I) < 60$	<96	T_1 and Overhauser	26
CHCl$_3$ (cf. also below)	$J(^{13}C^{35}Cl) = 23.3 \pm 0.8$	78.5 ± 2	T_1, T_2, and Overhauser	30
CHCl$_3$	$J(^{13}C^{35}Cl) = 49 \pm 10$	165 ± 34	T_2 (from linewidth)	23
SnCl$_4$	$J(^{119}Sn^{35}Cl) = 375$	856	T_1 and T_2	38
SnBr$_4$	$J(^{119}Sn^{81}Br) = 920$	762	T_1 and T_2	39
SnI$_4$	$J(^{119}Sn^{127}I) = 940$	1051	T_1 and T_2	38
PbCl$_4$	$J(^{207}Pb^{35}Cl) = 705$	2870	T_1 and T_2	40
ClF$_3$	$^1J(^{19}F^{35}Cl) = 260$	235	T_1 and T_2	35
ClF$_5$	$^1J(^{19}F^{35}Cl) = 192$	174	T_1 and T_2	35
	$J(^{19}F^{35}Cl)$ (eq.) ⩽ 20	18	T_1 and T_2	35
ClOF$_3$	$^1J(^{19}F^{35}Cl) = 195$	176	T_1 and T_2	35

Table III (continued)

Compound	Spin-spin coupling J_{ij} (in Hz)		Reduced spin-spin coupling K_{ij} $10^{+19}\mathrm{Na}^{-2}\mathrm{m}^{-3}$ $(=10^{+20}\mathrm{cm}^{-3})$	Method of determination	Reference
$FClO_3$	$J(^{19}F^{35}Cl)$	$= 278 \pm 5$	252 ± 5	Temp dep. of ^{19}F lineshape	104
ClF_6^+	$^1J(^{19}F^{35}Cl)$	$= 337$	305	Direct observation	105
	$^1J(^{19}F^{37}Cl)$	$= 281$			
BrF_6^+	$J(^{19}F^{79}Br)$	$= 1575$	558	Direct observation	106
	$J(^{19}F^{81}Br)$	$= 1697$			
IF_6^+	$J(^{19}F^{127}I)$	$= 2730 \pm 15$	1210 ± 6	Direct observation	107
IF_7	$J(^{19}F^{127}I)$	~ 2100	930	Direct observation and lineshape analysis	108
$CHCl_3$ (cf. also above)	$^2J(^1HCl)$	$= 5.5$	(5.1)	T_1 and T_2	36
$CHCl_3$	$^2J(^1HCl)$	$= 6.4$	(5.9)	T_1 at low fields	20
$CHFCl_2$	$^2J(^{19}F^{35}Cl)$	$= 27.6 \pm 0.5$	25.0 ± 0.5	$T_{1\rho}$	33
	$^2J(^{19}F^{37}Cl)$	$= 23.0 \pm 0.5$			
CF_2Cl_2	$^2J(^{19}FCl)$	$= 28$	(27.7)	$(T_{1\rho})$	30
CF_2Cl_2	$^2J(^{19}FCl)$	$= 12 \pm 3$	(12 ± 3)	T_1 and T_2	31
$CFCl_3$	$J(^{19}F^{35}Cl)$	$= 13.7 \pm 0.5$	12.4 ± 0.5	$T_{1\rho}$	33
	$J(^{19}F^{37}Cl)$	$= 11.4 \pm 0.5$			
$CFCl_3$	$^2J(^{19}F^{35}Cl)$	$= 11.9 \pm 0.4$	10.8 ± 0.4	T_1 and $T_{1\rho}$	32
$SiHCl_3$	$^2J(^1H^{35}Cl)$	$= 9.3 \pm 0.3$	8.6 ± 0.3	$T_{1\rho}$ and double resonance	22

$\omega_S)T_{2S} \gg 1$. When the scalar contribution to both T_{1I} and T_{2I} can be identified (this necessitates the substraction of other contributions to the relaxation of the I nucleus), eqs. 1 and 2 allow a determination of both J and the relaxation time of the quadrupolar nucleus (for moderate sized molecules it is generally safe to assume extreme narrowing conditions, so that $T_{1S} = T_{2S}$). When only one of the scalar contributions, $1/T_{1I}^{SC}$ or $1/T_{2I}^{SC}$, to the relaxation of the I nucleus can be obtained, the value of J can be calculated if the relaxation rate of the quadrupolar nucleus is known. The latter rate may for example be obtained from the linewidth of the NMR spectrum of the S nucleus.

When the I nucleus is a proton, contributions other than scalar (usually dipolar) to its relaxation rate are commonly equal both for T_{1I} and T_{2I}. In this case a measurement of the difference between T_{1I} and T_{2I} should be sufficient to calculate J, provided the relaxation of the quadrupolar nucleus is known.

At very low magnetic fields, of the order of 10^{-2} to 10^{-4} Tesla, the term $(\omega_I - \omega_S)^2 T_{2S}^2$ in eqs. 1 and 2 will eventually become smaller than unity and the scalar contribution to $1/T_{1I}$ will reach its maximum. Studies of $1/T_{1I}$ at low magnetic fields have also in a few cases been employed to determine J (20,21).

A second possibility of obtaining scalar halogen spin coupling constants is through measurements of the relaxation rate of the I nucleus in the rotating frame, $T_{1\rho}$ (53). Examples of the use of this technique are found in references 54 and 55.

As briefly mentioned above, direct observations of halogen spin couplings in NMR spectra are possible in a few symmetric halogen compounds, e.g., $FClO_3$, ClF_6^+, BrF_6^+, IF_6^+, and IF_7. In $FClO_3$ the chlorine nucleus, although not in a truly symmetric environment, has a relaxation rate that is sufficiently slow to allow the $^{19}F-^{35}Cl$ and $^{19}F-^{37}Cl$ spin couplings to produce resolvable fine structure in the fluorine NMR spectra. The ^{19}F NMR spectrum of the exotic cation BrF_6^+ is shown in Figure 1. At room temperature the spectrum consists of two overlapping 1:1:1:1 quartets at very low field. One quartet is assigned to $^{79}BrF_6^+$ and the other to $^{81}BrF_6^+$.

As further discussed in reference 1, it appears from the limited data available that the one-bond as well as the two-bond reduced spin coupling constants to chlorine, and possibly also to bromine, have values that are close to analogous reduced coupling constants involving fluorine. This has been taken to indicate that the coupling mechanism in these halides is essentially the same.

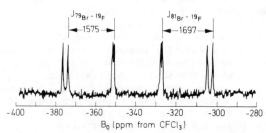

Figure 1. The ^{19}F NMR spectrum of the BrF_6^+ cation in HF solution. The two sets of quartets with almost the same intensities are assigned to ions $^{79}BrF_6^+$ and $^{81}BrF_6^+$, respectively. Reprinted with permission from Ref. 106. Copyright 1974, American Chemical Society.

CHEMICAL EXCHANGE OF HALIDE IONS

Let us consider the exchange given in eq. 3, in which M is

$$X^-_{(A)} + M^{n+} \longrightarrow MX^{(n-1)+}_{(B)} \tag{3}$$

a metal ion, X is a halide ion, (A) denotes the free halide ions, and (B) denotes the complexed ones. Let us further assume that there is a large excess of free ions, $p_A \gg p_B$ and therefore $p_A \sim 1$. For the quadrupolar halides we also have $1/T_{2A} \ll 1/T_{2B}$. Under these conditions only one NMR signal will be observed, whose linewidth is given by eq. 4 (56), in which τ_B is the life-

$$\frac{1}{T_2} = \frac{1}{T_{2A}} + \frac{p_B/\tau_B[1/T_{2B}(1/T_{2B} + 1/\tau_B) + \Delta\omega_{AB}^2]}{(1/T_{2B} + 1/\tau_B)^2 + \Delta\omega_{AB}^2} \tag{4}$$

time of the complexed ions and $\Delta\omega_{AB}$ is the chemical shift difference between free and complexed ions ($\Delta\omega_{AB} = \omega_A - \omega_B$). Following Lindman and Forsen (1) we can define the dimensionless parameters by eqs. 5 and 6 and rewrite eq. 4 as eq. 7, in which

$$\xi = \frac{1/\tau_B}{1/T_{2B}} = \frac{T_{2B}}{\tau_B} \tag{5}$$

$$\eta = \frac{\Delta\omega_{AB}}{1/T_{2B}} \tag{6}$$

$$\Delta\left(\frac{1}{T_2}\right) = \frac{1}{T_2} - \frac{1}{T_{2A}} = \frac{p_B}{T_{2B}} \left(\frac{\xi(1 + \xi + \eta^2)}{(1 + \xi)^2 + \eta^2}\right) \tag{7}$$

$\Delta(1/T_2)$ is the excess relaxation rate, which also defines the excess line broadening $\Delta\nu_e = \Delta(1/T_2)/\pi$. Let us now consider a few special cases.

(i) When $\xi \ll 1$, i.e., when the exchange rate is very much slower than the relaxation rate in site B, we see that the excess line broadening $\Delta\nu_e$ is independent of the value of η. In this region we simply have eq. 8. The excess broadening will thus be proportional to the chemical exchange rate $1/\tau_B$.

$$\Delta\nu_e = \frac{\xi}{\pi} \frac{P_B}{T_{2B}} = \frac{P_B}{\pi \cdot \tau_B} \tag{8}$$

(ii) When $\xi \gg 1$, i.e., fast chemical exchange, the excess line broadening is given by eq. 9.

$$\Delta\nu_e = \frac{P_B}{\pi \cdot T_{2B}} \tag{9}$$

(iii) For all values of ξ and for $\eta \ll 1$, i.e., small chemical shift difference, we have eq. 10. For other combinations of

$$\Delta\nu_e = \frac{P_B}{\pi \cdot (\tau_B + T_{2B})} \tag{10}$$

ξ and η, one has to use eq. 4 or 7. Figure 2 shows how the line broadening depends on the two dimensionless parameters ξ and η.

For both Cl and Br there are two quadrupolar isotopes, and information can be obtained regarding the ratio τ_B/T_{2B} when the line broadening for both isotopes is determined, since T_{2B} is given by eq. 11. From this expression is derived the ratios in

$$1/T_{2B} = \frac{3\pi^2}{10} \frac{2I+3}{I^2(2I-1)} \left(\frac{e^2qQ}{h}\right)^2 (1 + \eta^2/3)\tau_c \tag{11}$$

eqs. 12 and 13, in which $Q = (e^2qQ)/h$. In an exchanging system

$$\frac{1/T_{2B}^{35Cl}}{1/T_{2B}^{37Cl}} = \left(\frac{Q^{35Cl}}{Q^{37Cl}}\right)^2 = 1.61 \tag{12}$$

$$\frac{1/T_{2B}^{79Br}}{1/T_{2B}^{81Br}} = \left(\frac{Q^{79Br}}{Q^{81Br}}\right)^2 = 1.433 \tag{13}$$

we can, by observing the NMR spectra of the two different halogen isotopes at a given temperature, effectively change $1/T_{2B}$ without affecting the exchange rate (little or no isotope effects on the exchange).

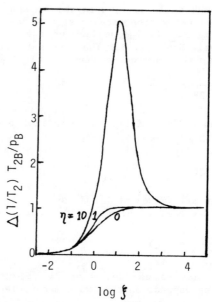

Figure 2. Graph showing the excess relaxation rate $(1/T_2)$ of the dominant NMR signal in a system where the observed nucleus is undergoing chemical exchange between two sites, A and B. The mole fraction of nuclei in site B, p_B, is much less than that of site A, p_A. The definitions of η and ξ are given by eqs. 5 and 6 (from Ref. 1). (with permission from Springer Verlag)

Let us define $(1/T_{2B}^{I})/(1/T_{2B}^{II}) = k$, in which I and II denote the two isotopes and $k = 1.61$ for chlorine and 1.433 for bromine. From eq. 7 we then obtain eq. 14 when $\eta \ll 1$, in

$$\frac{\Delta\nu_e^{I}}{\Delta\nu_e^{II}} = \frac{1 + k\xi^{I}}{1 + \xi^{I}} \tag{14}$$

which ξ^{I} is the dimensionless exchange parameter of eq. 5 referred to isotope I. When the inequality $\eta \ll 1$ is no longer fulfilled, the ratio $\Delta\nu_i^{I}/\Delta\nu_i^{II}$ is a function of η as well as of ξ. An equivalent of eq. 14 was first derived by Hertz (13b), who also pointed out the usefulness of studying the different halogen isotopes in studies of exchange rates. However, this technique has been very sparsely used during the last 20 years. The variation in the ratio $\Delta\nu_e^I/\Delta\nu_e^{II}$ is shown in Figure 3, where the variation has been calculated for $\eta = 0$ and $\eta = 1$. We may note that in the slow exchange limit the isotope ratio of the line broadening is unity, whereas in the fast exchange limit it approaches its maximum, k.

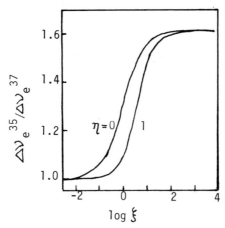

Figure 3. Graph showing the isotope ratio of the excess linewidth, $\Delta\nu_e$, for the ^{35}Cl and ^{37}Cl NMR signals in a two site system where the chloride nuclei are undergoing exchange between two sites, A and B ($p_B \ll p_A$) (from Ref. 1).

Another point regarding the isotope studies is that if the extreme narrowing limit may be safely assumed the isotope effect may equally well be studied at fixed frequency or fixed magnetic field. However, when nonextreme narrowing conditions apply the study has to be made at fixed frequency.

The rate constants which are accessible with this technique are mainly determined by two factors. Firstly, at slow exchange the accuracy in the determination of the line broadening is the determining factor. However, since the broadening of the observed signal is given by p_B/τ_B the accessible range can be changed by varying p_B. It must then be kept in mind that the above discussion is based on the assumption that $p_B \ll p_A$, or $p_A \sim 1$ and if this is no longer true more complex equations have to be used. Secondly, the fast exchange limit varies with T_{2B} and is again determined by the accuracy in the linewidth measurement.

The first more extensive use of halogen NMR to study ion exchange was published by Hertz (13). In these publications he pointed out how to use the different halogen isotopes in studies of exchange rates. Linewidth studies of ^{127}I, ^{79}Br, and ^{81}Br in solutions containing Cd^{2+}, Hg^{2+}, or Zn^{2+} ions were presented and the exchange rates evaluated by means of the equations valid for the two site exchange. Since in some of the systems studied a considerable number of species may be present and this is not considered in the two site model, it is not easy to judge how much this will influence the accuracy of the calculated rate constants. Some of the rate constants reported by Hertz are given in Table 4.

Table 4. Rate Constants for the Bromide Ion Exchange for the Cadmium-Bromide System in Water Solutions

$k_1 = 1.4 \times 10^9 \; M^{-1} s^{-1}$	$k_{-2} = 1 \times 10^6 \; s^{-1}$
$k_{-1} = 1 \times 10^7 \; s^{-1}$	$k_3 = 2.8 \times 10^7 \; M^{-1} s^{-1}$
$k_2 = 1 \times 10^8 \; M^{-1} s^{-1}$	$k_{-3} = 6 \times 10^6 \; s^{-1}$

In a more recent study Weingärtner et al. (56) have used ^{35}Cl, ^{37}Cl, ^{79}Br, and ^{81}Br NMR to study the interaction between halide ions and Ni^{2+} ions. From studies of the concentration and from the frequency and temperature dependence of both T_1 and T_2 for ^{35}Cl as well as ^{37}Cl (or ^{79}Br and ^{81}Br), they were able to show that the simple two site model was not capable of explaining all their data. However, using a three site model with eight parameters they were able to obtain good agreement between calculated and experimental data for all experiments. These parameters are given in Table 5. Starting from the dynamic parameters in Table 5 the authors tried to deduce structural information regarding the species with which the Cl^- ion is exchanging. After testing models containing chloride ions bound to $NiCl^+$ as well as $NiCl_2$ they finally concluded that there probably exists only one species with complexed chloride ions, $NiCl^+$. The two exchange processes observed are then given by eq. 15.

$$Ni^{2+} + Cl^- \rightleftharpoons NiCl^+ \tag{15a}$$

$$NiCl^+ + Cl^{*-} \rightleftharpoons NiCl^{*+} + Cl^- \tag{15b}$$

QUADRUPOLE SPLITTING IN LIQUID CRYSTALS

It has been emphasized above that relaxation of the quadrupolar halogens is dominated by the quadrupole relaxation due to the large quadrupole moment of these nuclei. It has so far been assumed that the molecular environment is isotropic, i.e., all orientations with respect to the applied magnetic field are equally probable. In this section we will deal with oriented

Table 5. Derived Parameters for the Exchange of Cl$^-$ Ions at 298 K in a 1 M NiCl$_2$ Solution

Parameter	Process 1	Process 2
$P_{Bi} = \tau_{Bi}/\tau_{Ai}$	0.030	0.049
$(1/T_{2B})$ s^{-1}	11,800	11,800
$(1/\tau_{Ai})$ s^{-1}	2.2×10^2	3.8×10^3
$(1/\tau_{Bi})$ s^{-1}	7.3×10^3	7.8×10^4
E_A kJ mol^{-1}	52.8	49.7
$(1/\tau_B°)$ s^{-1}	1.45×10^{13}	4.4×10^{13}
$(\Delta\omega_{AB})$ ppm	3.2×10^3	3.2×10^3

systems, which have some mobility. Let us, however, first consider the shape of the spectra for quadrupolar nuclei in solids. In a symmetric environment, e.g., tetrahedral or octahedral symmetry, the quadrupole interaction is averaged out and the energy levels of the spin states are evenly spaced. The NMR spectrum therefore consists of single lines. In less symmetric environments where the nucleus will experience an electric field gradient, the NMR spectrum will for a I = 3/2 nucleus consist of three lines. The central line, 40% of the total intensity, will to first order be unaffected by the quadrupole interaction. For stronger quadrupole interactions also second order effects will appear and result in a shift of the central line. This is, for I = 3/2, illustrated in Figure 4.

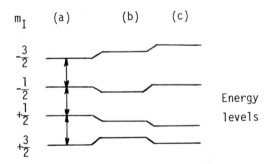

Figure 4. Energy level diagram for a spin I = 3/2 nucleus. (a) No quadrupolar effects, (b) first order quadrupolar effects, and (c) second order quadrupolar effects.

For an axially symmetric field gradient the frequency separation between the two peaks, Δ, in an NMR spectrum showing first order quadrupole effects is given by eq. 16, in which eq is the

$$\Delta = \Delta(\theta) = \left| \frac{3e^2qQ}{4hI(2I-1)} (3\cos^2\theta - 1) \right| \qquad (16)$$

largest component of the field gradient tensor, eQ is the quadrupole moment, and θ is the angle between the field gradient and the magnetic field. For a sample where all directions of the field gradient are equally probable a so-called powder pattern is obtained as shown schematically in Figure 5. The frequency separation between two adjacent peaks is given by eq. 17.

$$\Delta_p = \left| \frac{3e^2qQ}{4hI(2I-1)} \right| \qquad (17)$$

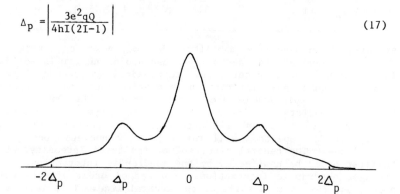

Figure 5. Schematic representation of a powder pattern for a spin I = 3/2 nucleus.

In a liquid crystalline system there is a rapid molecular motion which will to some extent average out the quadrupole interaction. The residual interaction gives rise to NMR spectra analoguous to the one shown in Figure 5. A theoretical treatment of the quadrupolar effects in micellar solutions and amphiphilic liquid crystals have been published by Wennerström et al. (57), showing that eqs. 16 and 17 will be very much the same for this case as for the rigid solids (eqs. 18 and 19).

$$\Delta(\theta) = \left| \frac{3e^2qQ \cdot S}{4hI(2I-1)} (\cos^2\theta - 1) \right| \qquad (18)$$

$$\Delta_p = \left| \frac{3e^2qQ \cdot S}{4hI(2I-1)} \right| \qquad (19)$$

Only a few investigations of quadrupole splitting of covalent halogens in liquid crystals have so far been reported in the literature (14,58). Some more examples of studies of halide ions, especially chloride, in amphiphilic mesophases have appeared (1). However, the first observation of quadrupolar splittings was in the ^{35}Cl NMR spectrum of CH_2CCl_2 in polybenzylglutamate (14). The first system for which quadrupole splittings of halide ions in liquid crystals were observed was the octylammonium chloride-decanol-water system (15). In this system the liquid crystal is of the lamellar type, built up of alternating amphiphilic bilayers and water layers.

Contrary to the covalent compounds, where the field gradient is of intramolecular origin, for the monoatomic ions the field gradient is of intermolecular origin. The quadrupole coupling constant may therefore vary strongly with medium and there is in practice no way of obtaining the quadrupole coupling constant. Only the product of this one and the order parameter are accessible. The quadrupole splitting can in spite of this provide valuable information regarding ion binding.

In most cases it is reasonable to assume that the ion exchange is rapid compared to the difference in quadrupole splitting between the various ion sites. In these cases the first order splitting is given by eq. 20, in which p_i is the population

$$\Delta_p = \left| \Sigma \frac{3}{4} \frac{p_i e^2 q_i QS_i}{hI(2I-1)} \right| \qquad (20)$$

of halide ions in site i, characterized by the field gradient eq_i and the order parameter S_i.

First order quadrupole splittings have been determined for ^{35}Cl in the hexagonal mesophase of the dodecyltrimethylammonium chloride-water system (59) and for ^{81}Br in the hexagonal mesophase of the hexadecyltrimethylammonium bromide-water system (60). Attempts have been made (57) to rationalize the obtained quadrupole splitting in terms of a simple electrostatic model, where the field gradients are assumed to be due to the charges of the amphiphilic ions and the water dipoles. In this way the quadrupole coupling constant was estimated to be 0.6 MHz for ^{35}Cl$^-$ bound to a NH_3^+ group, 0.32 MHz for ^{35}Cl bound to a $N(CH_3)_3^+$ group, and 1.8 MHz for ^{81}Br$^-$ bound to a $N(CH_3)_3^+$ group.

BIOLOGICAL APPLICATIONS OF Cl, Br, AND I NMR

Theoretical Aspects

Biological applications of NMR studies on the quadrupolar halogens are of fairly recent date but today constitute a rapidly growing area. The biological applications invariably have involved studies of the relaxation of the aqueous ions taking part in the chemical exchange with a protein, another type of macromolecule, or some other molecule of biological significance. The basic idea behind the biological applications is simple. Unliganded quadrupolar halide ions in aqueous solutions are in an effectively symmetric environment and have slow relaxation rates, whereas liganded nuclei are in a nonsymmetric environment and generally have very rapid relaxation rates. If the nuclei are rapidly exchanging between the two environments the occurrence of ligand binding is clearly observed in the NMR relaxation rates even when only a very small fraction of the halide ions are liganded.

Two general types of biological halide NMR studies may be discerned.

(I) Studies of halide ion binding to native metal-containing or metal-free proteins or other macromolecules.

(II) Studies of halide ion binding to proteins or other macromolecules where an artificial metal label is attached.

The earliest biological applications of halide NMR were of the second type but presently the first type of applications is dominating. A number of parameters may be obtained from the halide NMR studies:

(i) stoichiometry and binding constants of halide ions;
(ii) stoichiometry (and accessibility) of metal reactive groups on a protein;
(iii) rotational correlation time(s) of the macromolecule;
(iv) quadrupole coupling constants(s) of the halide ion(s) in different macromolecular binding site(s);
(v) rates and activation parameters of halide ion exchange; and
(vi) binding constants and stoichiometry of other ions or molecules through competition experiments.

Information about the nature of the binding sites may be obtained through the actual values of one or more of the above parameters or through their dependence on variations in the experimental conditions--for example the addition of a substrate analogue or an inhibitor to an enzyme.

Chemical Exchange of a Spin I = 3/2 Nucleus Under Conditions of Extreme Narrowing and Nonextreme Narrowing

Since quadrupolar halide ions liganded to large macromolecules in many cases are under conditions when the extreme narrowing approximation ($\omega^2 \tau_c^2 \ll 1$) is no longer valid, we will here briefly discuss what effects chemical exchange will have on the NMR spectra in such a situation. Bull (61) has given a treatment of the chemical exchange effects on I = 3/2 nuclei in a two site system where the relaxation in each is nonexponential. The general solutions of the time dependence of the magnetizations are complex but considerable simplification is possible when the relaxation times and exchange times in one site, let us call this the B site, are much less than in the other site. The latter conditions are of prime interest in all biological applications of halide NMR. The time dependence of the longitudinal magnetization may now be written as in eq. 21, in which $\langle I_{Az}(t) \rangle$ and

$$\langle I_z(t) \rangle - \langle I_z^0 \rangle = \langle I_{Az}(t) \rangle - \langle I_{Az}^0 \rangle =$$
$$(\langle I_{Az}(0) \rangle - \langle I_{Az}^0 \rangle) \cdot [\frac{4}{5} e^{-c_1 t} + \frac{1}{5} e^{-c_2 t}] \quad (21)$$

$\langle I_{Az}^0 \rangle$ are the average expectation values of the z component of the nuclear spin operator of site A, the abundant site, at time t and at equilibrium, respectively. The exponents are given by eq. 22, in which the a_{ji}'s are the exponents a_i in normal equations

$$c_1 = a_{A1} + \frac{p_B}{1/a_{B1} + \tau_B} \quad (22a)$$

$$c_2 = a_{A2} + \frac{p_B}{1/a_{B2} + \tau_B} \quad (22b)$$

for the longitudinal and transverse relaxation rates without exchange, p_B is the fraction of spin I = 3/2 nuclei in site B and is assumed to be much less than p_A, and τ_B is the average lifetime of the exchanging nuclei in site B.

An analogous expression, but now also involving the difference in Larmor frequency, $\delta\omega$, between sites A and B, was obtained for the transverse magnetization.

The precise evaluation of a biological halogen NMR experiment may thus involve the determination of two time constants for the decay of both the longitudinal and the transverse magnetization. In an actual experiment this usually proves very difficult unless the time constants differ considerably. Otherwise the decay of the magnetization appears completely linear. Bull therefore also derived linearized approximations to his rigorous expressions. The results for a two site case when the nuclei in site A are under conditions of extreme narrowing are given by eqs. 23 and 24, in which a_{Bi} and b_{Bi} are the exponents as defined above (61) and $1/T_1$ and $1/T_2$ are the apparent relaxation times.

$$\frac{1}{T_1} \sim \frac{1}{T_{1A}} + p_B \left(\frac{0.8}{1/a_{B1} + \tau_B} + \frac{0.2}{1/a_{B2} + \tau_B} \right) \tag{23}$$

$$\frac{1}{T_2} \sim \frac{1}{T_{2A}} + p_B \left(\frac{0.6}{\tau_B} \left(\frac{b_{B1}(b_{B1} + \tau_B^{-1}) + \Delta\omega^2}{(b_{B1} + \tau_B^{-1})^2 + \Delta\omega^2} \right) + \right.$$

$$\left. + \frac{0.4}{\tau_B} \left(\frac{b_{B2}(b_{B2} + \tau_B^{-1}) + \Delta\omega^2}{(b_{B2} + \tau_B^{-1})^2 + \Delta\omega^2} \right) \right) \tag{24}$$

Eqs. 22 and 23 may be further simplified (62) under the assumption that (i) the time dependence of the electric field gradient causing the relaxation in the B site can be described by a single correlation time τ_c; (ii) the value of $\Delta\omega$ can be neglected; and (iii) the symmetry parameter γ at the B site may be put equal to zero. Then the following expressions are obtained for the longitudinal relaxation (eq. 25) and for the

$$\frac{1}{T_1} \sim \frac{1}{T_{1A}} + p_B \left(\frac{0.8}{T_{1B}' + \tau_B} + \frac{0.2}{T_{1B}'' + \tau_B} \right) \tag{25a}$$

$$\frac{1}{T_{1B}'} = \frac{2\pi^2}{5} \left(\frac{e^2qQ}{h} \right)^2 \left(\frac{\tau_c}{1 + 4\omega^2\tau_c^2} \right) \tag{25b}$$

$$\frac{1}{T_{1B}''} = \frac{2\pi^2}{5} \left(\frac{e^2qQ}{h} \right)^2 \left(\frac{\tau_c}{1 + \omega^2\tau_c^2} \right) \tag{25c}$$

transverse relaxation (eq. 26). It may be deduced from eqs. 25

$$\frac{1}{T_2} \sim \frac{1}{T_{2A}} + p_B \left(\frac{0.6}{T_{2B}' + \tau_B} + \frac{0.4}{T_{2B}'' + \tau_B} \right) \tag{26a}$$

$$\frac{1}{T_{2B}'} = \frac{\pi^2}{5} \left(\frac{e^2qQ}{h} \right)^2 \left(\tau_c + \frac{\tau_c}{1 + \omega^2\tau_c^2} \right) \tag{26b}$$

$$\frac{1}{T_{2B}''} = \frac{2}{5} \left(\frac{e^2qQ}{h} \right)^2 \left(\frac{\tau_c}{1 + 4\omega^2\tau_c^2} + \frac{\tau_c}{1 + \omega^2\tau_c^2} \right) \tag{26c}$$

and 26 that the relaxation rates $(T_1)^{-1}$ and $(T_2)^{-1}$ of the quadrupolar halide ions undergoing rapid chemical exchange with macromolecular binding sites (B) may become unequal when the B sites are characterized by large values of τ_c. This fact, experimentally verified in a number of systems, may be used to evaluate τ_c.

It should be noted that chemical exchange of a quadrupolar nucleus to and from a macromolecular binding site also produces fluctuations in the electric field gradient sensed by the nucleus and thus provides a second mechanism for quadrupole relaxation. The effective correlation time will now be given by eq. 27

$$1/\tau_c = 1/\tau_R + 1/\tau_{ex} \tag{27}$$

(63-65), in which τ_R is the rotational correlation time of the macromolecule (assuming isotropic rotational diffusion) and $(\tau_{ex})^{-1}$ is the rate of exchange for a halide nucleus from site B, all types of exchange processes being considered, including those between identical types of binding sites.

Internal motion at site B will also influence the halogen NMR spectrum as further discussed in reference 1.

Applications

The present subsection will be devoted to a few selected applications of quadrupolar halogen NMR to biological systems in order to illustrate some of the advantages and limitations of the technique. An up-to-date survey of the many diverse systems and problems studied is found in reference 3.

Carbonic Anhydrase. One of the first proteins studied by halogen NMR, where an artificial metal label is not used, was carbonic anhydrase. This enzyme has a molecular weight of about 30,000 and contains one zinc atom at, or near, the active center. The enzyme has great physiological significance primarily since it catalyses the reversible hydration of carbon dioxide (eq. 28).

$$H_2O + CO_2 \rightleftharpoons H_2CO_3 \rightleftharpoons H^+ + HCO_3^- \tag{28}$$

The enzyme, which in humans and horses exists in high activity ("C") and low activity ("B") forms, is present mainly in the erythrocyte.

In spite of the fact that the X-ray structure of carbonic anhydrase is known to a resolution of about 2 Å the detailed mechanism of the enzyme is still not known (66). Ward (67) studied the ^{35}Cl NMR linewidth on a 0.5 M NaCl solution containing about 4×10^{-5} M bovine carbonic anhydrase. The observed excess line broadening was attributed to a coordination of Cl^- to the zinc atom in the active cleft. Titration with acetazolamide, which is a strong inhibitor of the enzyme, and CN^- showed that these ligands almost entirely eliminated the excess line broadening. The results from the acetazolamide titration are shown in Figure 6. An equivalence point is obtained at a 1:1 molar ratio of acetazolamide to enzyme. Excess linewidth studies at different pH's indicated that anion binding disappears at high pH. The pH profiles of the $^{35}Cl^-$ excess linewidths showed also interesting differences between the high affinity and low affinity forms of the enzyme.

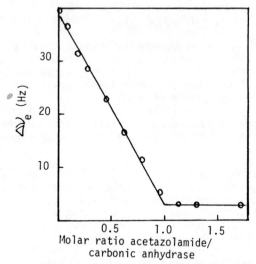

Figure 6. Excess ^{35}Cl linewidth of solutions containing 0.5 M NaCl and 4.05 x 10^{-5} M bovine carbonic anhydrase as a function of added acetazolamide. The pH was 8.05 and the temperature 32°C. Reprinted with permission from Ref. 67. Copyright 1969, American Chemical Society.

Ward assumed in his first study that the chloride exchange rate was faster than the relaxation rate in the enzyme binding site. Later Ward and co-workers studied the temperature dependence of the linewidth and also compared ^{35}Cl and ^{37}Cl linewidths (68,69). It was then found that the exchange rate actually was slower than the relaxation rate and that the excess linewidth thus is governed by the value of p_B/τ_B (eq. 8). It now appears that the changes in the Cl^- NMR linewidth with pH and the difference between the two isoenzymes B and C reflect variations in the exchange rate rather than in the field gradient or local mobility at the binding site. Ward and Cull (68) also studied the relative affinity for the binding of other anions to bovine carbonic anhydrase through competition experiments with $^{35}Cl^-$. The following affinity series was obtained: $CNO^- > N_3^- > SCN^- > I^- > Br^- > Cl^-$. Calculated binding constants were in good agreement with other data.

The conclusion of Ward that Cl⁻ binds directly to the Zn at the active cleft has been questioned by Koenig and Brown (70). From a comparison of water proton and Cl⁻ relaxation data they argued that competition between Cl⁻ and water for a common Zn site, as also implied in Ward's model, does not occur. They instead suggest that Cl⁻ may be bound to a cationic residue elsewhere on the protein. Norne et al. have reported that the ^{35}Cl excess linewidth observed in carbonic anhydrase solutions is markedly affected by the addition of the anion $Au(CN)_2^-$ (71). This anion binds very poorly to Zn^{2+} and other metal ions. This result also indicates that Cl⁻ is not directly coordinated to zinc.

Alcohol dehydrogenase. Alcohol dehydrogenase (ADH) is an enzyme found in vertebrates. It catalyses the reversible oxidation of alcohols by transfer of a hydrogen atom and an electron to a coenzyme, nicotinamide-adenine dinucleotide (NAD). The overall reaction may be represented by eq. 29. The alcohol

$$NAD^+ + R_1R_2CHOH \xrightleftharpoons{ADH} NADH + H^+ + R_1R_2CO \tag{29}$$

dehydrogenase from horse liver (LADH) has been studied in great detail and its 3D structure is known to a resolution of 2.4 Å (72). LADH has a molecular weight of about 80,000; it consists of two identical subunits, each with a catalytic site. Each subunit contains two Zn atoms. One of these atoms is found at the "bottom" of a crevice where coenzyme and substrate are bound and is considered to be involved in the catalytic function. The other Zn atom is assumed to have only a structure stabilizing function.

LADH was one of the first enzymes studied by halide NMR methods (73), and since 1969 a considerable number of studies have emanated (71,74-81). Through linewidth measurements of the isotope pairs ^{79}Br, ^{81}Br and ^{35}Cl, ^{37}Cl, both halide ion exchange rates have been shown to be faster than the relaxation rate of the binding sites (73,78,79). Comparison of T_1 and T_2 for chloride have also shown that the extreme narrowing approximation is not valid. Correspondingly it is possible to separate the relaxation effect of the field gradient and correlation time in the binding site.

Titrations of LADH with the reduced coenzyme NADH have been performed using both ^{35}Cl and ^{81}Br NMR relaxation rates as titration indicators. A strong competition between NADH and Cl⁻ or Br⁻ is evident from these studies and this in turn indicates that halide NMR studies may be helpful in the attempts to understand the function of the enzyme. An example of the ^{35}Cl T_1 study of the NADH–Cl⁻ competition is given in Figure 7. An equivalence point of two NADH units per enzyme molecule is obtained both in the absence and presence of the inhibitor isobutyramide. Competition has also been observed between Cl⁻ and the shortened NADH analogue adenosine diphosphate ribose, ADPR (= NADH minus the nicotinamide part) (81).

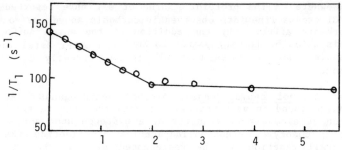

Figure 7. The ^{35}Cl longitudinal relaxation rate as a function of the molar ratio of reduced coenzyme (NADH) to horse liver alcohol dehydrogenase (LADH). The solution contained 0.13 mM LADH and 0.5 M KCl. The pH was 8.4 and temperatue 6°C. Reprinted with permission from Ref. 81. Copyright 1979, American Chemical Society.

The nature of the halide binding sites in LADH have been in dispute. Ward and Happe (74) assumed that Cl⁻ did coordinate directly to the zinc atom at the active site. On the other hand Lindman et al. (73,76,77) concluded that halide coordination to the zinc, if it occurs, has no influence on the halide ion relaxation rate. Their conclusions were based on results from halogen NMR studies in the presence of ligands believed or known to coordinate to the zinc in the active site. Thus neither oxyquinoline nor <u>ortho</u>-phenanthroline affect the chloride ion relaxation.

Norne et al. (78) have discussed the general problem of discriminating between metal-coordinative binding of a halide ion and other types of binding. Three types of experimental data were considered: (i) the exchange rate of the halide ion; (ii) the field gradient at the binding site(s); and (iii) competition studies with nonmetal coordinative anions like $Pt(CN)_4^{2-}$ and $Au(CN)_2^-$. In the case of metal-coordinative binding, the Cl⁻ exchange rate should in general be slower than in the case of noncoordinative binding. One would frequently expect the exchange rate for the metal-coordinative binding to be slower than the relaxation rate at the binding site. Some examples of slow halide ion exchange to nonmetallic binding sites on a protein have, however, been observed (81), and it is evident that care must be exercised before deductions are drawn from exchange rates. Field gradients may be assumed to be strongly dependent on the mode of halide-protein interaction. Studies by Norne et al. (71,78) on a couple of proteins indicate that the ^{35}Cl quadrupole coupling constants fall into two groups, the first comprising all nonmetalloproteins and two metalloenzymes, LADH and alkaline phosphatase, with quadrupole coupling constants of about 2 MHz, and the other comprising all other metalloenzymes with much higher quadrupole coupling constants (10 MHz). The implications of the competition studies referred to under (iii) above are simply that if competition is observed metal coordination of the halide ion is not likely. In the case of LADH, competition between Cl$^-$ and Br$^-$ and Pt(CN)$_4^{2-}$ or Au(CN)$_2^-$ has been observed (71,78,79). The collected evidence from the studies by Norne et al. therefore indicates that chloride binding to zinc is not of importance in LADH.

Ward and Cull have also reconsidered the site of chloride binding in LADH. They observed that elimination of the functional Zn^{2+} ion and its replacement with other divalent cations like Cd^{2+} or Co^{2+} did not affect the ^{35}Cl relaxation very much. These findings clearly point against metal coordination of the chloride ion (75).

So where on LADH do the halide ions bind? Here the competition studies with Pt(CN)$_4^{2-}$ and Au(CN)$_2^-$ have been most informative. These compounds have been used to prepare derivatives of LADH in the X-ray crystallographic studies and therefore are of value in correlation of the solution structure of the enzyme with the crystalline structure. Both Pt(CN)$_4^{2-}$ and Au(CN)$_2^-$ behave as competitive inhibitors with respect to the coenzyme NADH. In the presence of Pt(CN)$_4^{2-}$ it has furthermore been found that iodoacetate, an agent reactive towards SH groups, does not react with the SH group of Cys-46, an amino acid that is present on the "walls" of the NADH-binding crevice. In the absence of Pt(CN)$_4^{2-}$ the iodoacetate anion has been observed to bind reversibly to an anion binding site before it reacts with the SH group of Cys-46 (83). Since Pt(CN)$_4^{2-}$ prevents the iodoacetate reaction it was concluded that it interacts strongly with the positively charged side chain of Arg-47. This side chain is involved in the binding of the coenzyme and presumably also binds anions of fatty acids and, according to the NMR competition experiments, then also Cl$^-$. Later X-ray studies (84) show that Pt(CN)$_4^{2-}$ indeed is bound to Arg-47 in the coenzyme binding crevice of LADH and thus corroborate the above proposals. Titration with Pt(CN)$_4^{2-}$ does not completely eliminate the ^{35}Cl excess linewidth and there exists presumably also a second, weaker binding site for Cl$^-$ on LADH. In fact in a recent study by Andersson et al. (81) it is, based on various competition experiments, concluded that there are two Cl⁻ binding sites in the NADH binding region.

<u>Hemoglobin</u>. Hemoglobin and myoglobin are oxygen-carrying proteins in all vertebrates. Hemoglobin is a nearly spherical molecule with a molecular weight of about 65,000. It consists of four subunits, pairwise identical, each carrying a heme group capable of binding one O_2 molecule. Thanks to the pioneering X-ray studies of Perutz and co-workers, the 3D structure is known to high resolution. The amino acid sequence of the subunits differ somewhat in hemoglobin from different species. In human adults the major hemoglobin consists of two α chains with 141 amino acids and two β chains with 146 amino acids; the subunit structure is therefore usually designated $\alpha_2\beta_2$. A large number of mutant hemoglobins are also known, the most well known probably being the so-called sickle cell hemoglobin which differs from the normal in one amino acid residue in the β chains. In a number of vertebrates the blood has been observed to contain high percentages of several different hemoglobins. The physiological significance of these different hemoglobins is largely unknown at present.

The 3D structure of the deoxygenated and oxygenated forms of hemoglobin differ. The difference is most clearly observable in the region of contact between the subunits (85). The interaction between the subunits gives rise to the well known cooperative binding of oxygen to hemoglobin that plays such an important physiological role (85,86). The oxygen affinity of hemoglobin can be markedly affected by H^+ ions as well as by many anions like 2,3-diphosphoglyceric acid (DPG), adenosine triphosphate (ATP), and inositol hexaphosphate (IHP) (87).

Hemoglobin was the first molecule to be studied by the mercury "halide probe" technique introduced by Stengele and Baldeschwieler in 1966 (16). Human hemoglobin has six SH groups. In denatured hemoglobin all six SH groups may be reacted with SH binding reagents but in the native protein four of the SH groups become unreactive. Stengele and Baldeschwieler monitored the ^{35}Cl linewidth in a NaCl solution, containing about 4×10^{-5} M oxyhemoglobin as a function of the amount of $HgCl_2$ added. The $HgCl_2$ reacts with free R-SH groups to form R-S-HgCl. The chlorine on this "halide probe" exchanges fairly rapidly with Cl⁻ in solution, which affects the NMR relaxation. A change in slope at a ratio $HgCl_2$/hemoglobin of about two was observed. The break point was independent of the total NaCl concentration in the range 0.5 to 4.0 M. Since the hemoglobin $\alpha_2\beta_2$ tetramer is partly dissociated into αβ dimers at the highest NaCl concentrations, the result was taken as an indication that the four reactive SH groups are not involved in the interdimer binding.

In a theoretically interesting paper Collins et al. (88) have discussed the possiblity of determining not only the local correlation time characterizing the mercury halide probe bound to a macromolecle, but also the exchange rate of the halide ion. The requirement for such an evaluation is clearly that the mean lifetime of the halide nucleus at the mercury label, τ_B, should be of the same order of magnitude as the quadrupole relaxation time. Collins et al. titrated horse methemoglobin (i.e., hemoglobin with the heme iron as Fe^{3+}) and Hg^{2+} in a NaBr solution. From T_1 measurements of the ^{79}Br and ^{81}Br NMR signals at different pH, Collins et al. obtained the following rate parameters characterizing the protein binding site (with ν_Q = 320 MHz as the quadrupole coupling contstant of bromide in the halide probe and [Br$^-$] as the bromine concentration in M):

pH 7.0 $\tau_C = 1.5 \times 10^{-10}$ s
$\tau_B^{-1} = 3.3 \times 10^7$ [Br$^-$]s^{-1}
pH 10.0 $\tau_C = 0.8 \times 10^{-10}$ s
$\tau_B^{-1} = 1.2 \times 10^7$ [Br$^-$]s^{-1}

The Br$^-$ exchange to the Hg-Br label was assumed to be a second order process. The shorter τ_C values obtained at high pH is taken to indicate a conformational change which gives increased freedom of motion of the Cys-93 SH groups where the mercury labels are assumed to be attached (88).

A number of halogen NMR studies of hemoglobin have dealt with the direct interaction of anions with hemoglobin and the nature of the anion binding sites (62,82,89-92). Studies with a variety of physical methods have shown that small anions like perchlorate, phosphate, and chloride do not interact directly with the heme iron (86). Nevertheless, alkali salts of these and other small anions have a pronounced effect on the oxygen affinity of human hemoglobin and other hemoglobins (93,94). Thermodynamically the anion effect on the oxygen affinity implies that the anions interact differently with oxy- and deoxyhemoglobin.

In a series of papers Chiancone et al. (62,82,89-92) have described the use of ^{35}Cl and ^{37}Cl NMR to study the binding of chloride ions to hemoglobin. The leading theme in these studies has been the identification of the nature of the binding sites primarily through the pH profiles of the chloride linewidths in normal, mutant, and chemically modified hemoglobins. In the first paper the ^{35}Cl excess linewidth was studied as a function of sodium chloride concentration at constant hemoglobin concentration (about 1% by weight). The results are shown in Figure 8. Oxy- and deoxyhemoglobin (carbon monoxide hemoblogin is isomorphous with oxyhemoglobin and is frequently used in place of the latter for stability reasons) are seen to interact differently with Cl$^-$ ions. The experimental linewidth data were

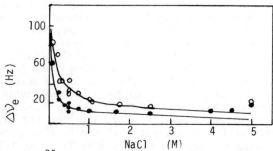

Figure 8. The ^{35}Cl excess linewidth as a function of NaCl concentration in solutions of human oxy- and deoxyhemoglobin at pH 7.45 to 7.50. The excess linewidths were proportional to the hemoglobin concentration and have been normalized to a protein concentration of 1.5%. Upper curve, oxyhemoglobin and lower curve deoxyhemoglobin. With permission from Ref. 89. Copyright: Academic Press Inc. (London) Ltd.

analysed in terms of two classes of independent binding sites, one with high and the other with low affinity for chloride. Variable temperature studies indicated that the chloride exchange was rapid compared to the quadrupole relaxation rate in the protein binding sites. The pH dependence of the ^{35}Cl excess linewidth in solutions of CO hemoglobin and deoxyhemoglobin was studied in the pH range 4.8 to 10.5. The results are shown in Figure 9. There is evidence for inflection points at pH 7.8 to 8.0 and possibly at pH 5.5. In this first study by Chiancone et al. these titration points could not be attributed to particular amino acid residues. The finding that the ^{35}Cl excess linewidth at neutral pH was markedly larger for carbon monoxide hemoglobin (HbCO) than for deoxyhemoglobin (Hb) was assumed to be connected to the X-ray finding that in the heme-liganded form of hemoglobin a number of intra- and interchain electrostatic "salt links" are broken up and the charged residues are therefore more available for the interaction with other ions.

Competition experiments with ATP performed at 0.2 M NaCl concentration showed that the ^{35}Cl linewidth markedly decreases with increasing ATP concentration at pH 6.0 to 6.3. This result indicates the ATP and Cl$^-$ compete for some, or most, of the high affinity sites. The binding constant for ATP to hemoglobin was estimated and found to be about an order of magnitude larger than for Cl$^-$.

In three subsequent papers (90-92) steps have been taken towards the identification of the chloride binding sites in

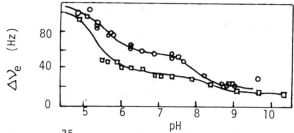

Figure 9. The ^{35}Cl excess linewidth as a function of pH in a solution of human oxy- and deoxyhemoglobin in 0.5 M NaCl. Circles: carbon monoxide hemoglobin. Squares: deoxyhemoglobin. With permission from Ref. 89. Copyright: Academic Press Inc. (London) Ltd.

hemoglobin. The strategy was to compare the pH profiles of normal hemoglobins and carboxypeptidase-digested hemoglobins. The A and B carboxypeptidases cut off specific amino acid residues at the COOH end of the α and β chains. The pH profiles of the ^{35}Cl excess line broadening in the presence of these modified hemoglobins were clearly different from that in the presence of normal hemoglobin. The removal of Arg-141 or His-146 was found to have very pronounced effects on the excess linewidth, and it was concluded that one chloride binding site is present at or near His-146 and a second at or near Val-1 and Arg-141, which are close to each other in space.

In a recent study Chiancone et al. employed ^{35}Cl NMR to study the anion binding to two main hemoglobin components in trout blood, Hb-Trout I and Hb-Trout IV. These two components have strikingly different functional properties. A pH dependent oxygen affinity is observed only for Hb-Trout IV. Organic phosphates also interact differently with the two forms--the oxygen affinity at Hb-Trout I is virtually unaffected by ATP while that of Hb-Trout IV increases in a similar way to human hemoglobin. Small anions like Cl$^-$ have a small but measurable effect on the oxygen affinity of both components. The pH dependence of the ^{35}Cl excess linewidth in the presence of the different components was found to be completely different. The linewidth in both the deoxy and carbon monoxide form of Hb-Trout I was independent of pH over the range studied (Figure 10). While the linewidth for Hb-Trout IV was found to decrease with increasing temperature, that for Hb-Trout I increased with temperature indicating that the ^{35}Cl linewidth in Hb-Trout I is determined by the chemical exchange rate rather than by the quadrupole relaxation rate at the binding site. This surprising finding was further checked by comparing the ^{35}Cl and ^{37}Cl excess linewidths. In the presence of Hb-Trout IV the ratio was 1.5 ± 0.1 while in Hb-Trout I the ratio was 1.0 ± 0.1. The latter result is expected if exchange dominates the relaxation.

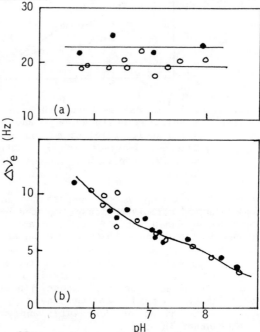

Figure 10. The ^{35}Cl excess linewidth as a function of pH in the presence of (a) Hb-Trout I, and (b) Hb-Trout IV. The solvent is 0.5 M NaCl. The protein concentration was 0.5% in (a) and 1% in (b). Open symbols for the carbon monoxide derivatives and filled symbols for the deoxy derivative (from Ref. 82).(with permission from Elsevier Biomedical Press Ltd.)

The interpretation of the slow exchange for Hb-Trout I in terms of structural features of the molecule is only tentative in view of the sparse structural data available. Chiancone et al. made the suggestion that chloride binds to positively charged residues in a cavity around the dyad axis between the β-type chains. Such a cavity has been found to be the binding site for organic phosphates in horse and human hemoglobin. One interpretation of the observation that the oxygen affinity of Hb-Trout I is unaffected by organic phosphates is that the entrance to the cavity, or the cavity as a whole, has become so small that it is no longer accessible to the phosphate groups. Still it could allow the entrance of a small anion like Cl$^-$, but hamper its exchange with chloride in the bulk. Further structural work will be needed to check this suggestion. Norne et al. (95) have studied the fractional change in the ^{35}Cl excess relaxation rate as a function of the fractional saturation of oxygen Y. This is shown in Figure 11. It is evident that the ionic interaction causing the ^{35}Cl line broadening is not linearly related to the fractional oxygen saturation. These results have been interpreted within the Monod, Wyman, and Changeux (96) allosteric two-state model of hemoglobin. It is characteristic of this allosteric model of hemoglobin, and also more refined models (97-99), that the transition from a predominant unreactive to a predominant reactive state takes place at a fractional oxygen saturation Y greater than 0.5. A simple interpretation of the ^{35}Cl results in Figure 11 is that the change in excess relaxation rate as we go from the deoxy to the oxy state of hemoglobin is directly proportional to the fraction of molecules in the reactive state. The dependence of the observed broadening on the fractional oxygen saturation agrees reasonably well with the allosteric two site model.

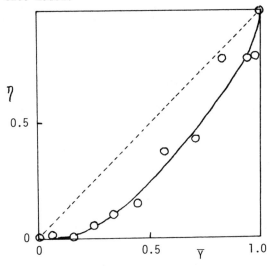

Figure 11. Fractional change in the ^{35}Cl excess linewidth, η, as a function of fractional oxygen saturation, Y, in human hemoglobin solutions (from Ref. 95).

Internal motion in the binding site. Proteins are not the rigid systems that, e.g., X-ray structures may make us believe, but are quite flexible with considerable degrees of internal motions or breathing motions. There can be no doubt that the dynamic phenomena have a biological significance. Many attempts, experimental as well as theoretical, have been made to deepen our understanding of the dynamics of protein molecules (100). The occurrence of internal motion of the anion binding sites in proteins was inferred at an early stage from the observation of correlation times for protein bound halide ions, which are considerably lower than those derived for the overall motion of the protein molecule from simple hydrodynamic theory. Also the observation that the correlation time was almost independent of temperature, and the observation of unusually small quadrupole coupling constants for protein bound halide ions points in the same direction. Bull et al. (101,102) have recently made an attempt to analyze the NMR relaxation data for $^{35}Cl^-$ in terms of a simplified model for the internal motion of the binding sites. Wallach (103) has derived expressions which can be used to describe how multiple internal motions will influence the correlation time and therefore also the NMR relaxation rate. These expressions however become difficult to handle when there are several internal motions. Furthermore a large number of adjustable parameters have to be introduced. Another complication is that the basic assumption made by Wallach of threefold or higher symmetry of the internal rotation may not be valid. Due to these complications, Bull (101) has developed a simplified model for internal motion. The internal motion is considered as the diffusion of a rod in a conical hole within the protein.

When the internal motion is combined with the overall correlation time, the correlation function is given by eq. 30, in

$$C(t) = (1 - A)\exp(-t(1/\tau_R + 1/\tau_i)) + A \exp(-t/\tau_R) \tag{30}$$

which τ_R is the the overall correlation time of the protein, τ_i is the correlation time characterizing the internal motion, and A is a function of the half angle Ψ of the cone in which the rod is free to move (eq. 31). Eq. 30 applies to a spherical protein

$$A = \cos^2\psi \, \sin^4\psi / 2(1-\cos\psi)^2 \tag{31}$$

molecule, and if the protein is instead a symmetric top the equation, also derived by Bull, becomes more complicated.

Near to the extreme narrowing or weakly nonextreme narrowing region, $\omega\tau_c \leqslant 1.5$, the relaxation time for the quadrupole relaxation is given by eqs. 32 and 33, in which χ is the quadru-

$$1/T_1 = \frac{2\pi^2}{5} \chi^2 \left(\frac{0.2 \, \tau_c}{1 + \omega^2\tau_c^2} + \frac{0.8 \, \tau_c}{1 + 4\omega^2\tau_c^2} \right) \tag{32}$$

$$1/T_2 = \frac{\pi^2}{5} \chi^2 \left(0.5 \, \tau_c + \frac{\tau_c}{1 + \omega^2\tau_c^2} + \frac{0.4 \, \tau_c}{1 + 4\omega^2\tau_c^2} \right) \tag{33}$$

pole coupling constant. In order to study the effect of the internal motion Bull substituted the expression within brackets in eqs. 32 and 33 with $0.2J(\omega) + 0.8J(2\omega)$ for T_1 and $0.6J(0) + J(\omega) + 0.4J(2\omega)$ for T_2. The spectral densities were then evaluated including internal motion as derived from the "rigid rod in a cone" model. Bull used these results to calculate an apparent correlation time, τ_{app}, and an apparent quadrupole coupling constant, χ_{app}. Such calculations showed that in the limit of very rapid internal motion the apparent correlation time becomes equal to the overall correlation time whereas the apparent quadrupole coupling constant is reduced--the more so the greater the half angle of the cone. When the correlation time for the internal motion and the overall motion are comparable in magnitude, both τ_{app} and χ_{app} are affected.

The above model has been used by Bull et al. (102) to analyse the chloride NMR relaxation data for several proteins. The overall correlation time(s) was estimated from the Debye-Stokes-Einstein equation or, when available, was taken from dielectric relaxation measurements. In order to perform the calculations, also the "true" value of the chloride quadrupole coupling constant in the site is needed. This is, however, not known and therefore Bull et al. estimated a value for Cl^- bound to a NH_3^- group of 3.6 MHz based on the electrostatic model of Cohen and Reif (103). In this way it was possible to calculate the values shown in Table 6. Obviously the model used is an oversimplification; however, it is noteworthy that all internal correlation times come out with a reasonable value of about 1 ns.

It is difficult to predict how useful this technique will be to study the motion of the halide ions in their binding sites. However, it is clear that more reliable values for the "true" quadrupole coupling constant for the bound ions are needed.

Table 6. Parameters Obtained in an Analysis of Chloride NMR Relaxation Data in Terms of the "Rigid Rod in a Cone" Model (Ref. 102)

Protein	$\tau_{c,app}$ (ns)	χ_{app} (MHz)	ψ (degrees)	Internal correlation time, τ_i (ns)
Hemoglobin	8	2.2	70	1.9
Human serum albumin	19	1.6	64	0.7
Alkaline phosphatase	8	1.4	79	0.7
Horse liver alcohol dehydrogenase	9	1.7	73	1.1
Pig heart lactate dehydrogenase	2	3.6	90	2.1
Pig heart LDH-NADH complex	17	3.3	53	6.7
Rabbit muscle lactate dehydrogenase	10	1.6	74	1.0
Rabbit muscle LDH-NADH complex	26	2.1	57	1.2
Aldolase	16	1.8	69	1.1

REFERENCES

(1) B. Lindman and S. Forsén, "Chlorine, Bromine and Iodine NMR. Physico-chemical and Biological Applications," Vol. 12 of the series "NMR Basic Properties and Progress," P. Diehl, E. Fluck, and R. Kosfeld, Eds., Springer Verlag, Berlin, 1976.
(2) B. Lindman and S. Forsén in "NMR and the Periodic Table," R. Harris and B. Mann, Eds., Academic Press, New York, 1978.

(3) S. Forsén and B. Lindman in "Methods of Biochemical Analysis," Vol. 27, D. Glick, Ed., John Wiley & Sons, Inc., 1981.
(4) F. Bitter, Phys. Rev., 75, p. 1326 (1949).
(5) W. H. Chambers and D. Williams, Phys. Rev., 76, p. 638 (1949).
(6) W. G. Proctor and F. C. Yu, Phys. Rev., 77, p. 716 (1950).
(7) R. V. Pound, Phys. Rev., 72, p. 1273 (1947).
(8) R. V. Pound, Phys. Rev., 73, p. 1112 (1948).
(9) P. Diehl, Helv. Phys. Acta, 29, p. 219 (1956).
(10) Y. Masuda, J. Phys. Soc. Japan, 11, p. 670 (1956).
(11) O. E. Myers, J. Chem. Phys., 28, p. 1027 (1958).
(12) R. E. Connick and C. P. Poppel, J. Am. Chem. Soc., 81, p. 6389 (1959).
(13) (a) H. G. Hertz, Z. Electrochem., 64, p. 53 (1960).
     (b) H. G. Hertz, Z. Electrochem., 65, p. 36 (1961).
(14) D. Gill, M. P. Klein, and G. Kotowycz, J. Am. Chem. Soc., 90, p. 6870 (1968).
(15) G. Lindblom, H. Wennerström, and B. Lindman, Chem. Phys. Lett., 8, p. 489 (1971).
(16) T. R. Stengele and J. D. Baldeschwieler, Proc. Natl. Acad. Sci. (USA), 55, p. 1020 (1966).
(17) R. T. Obermyer and E. P. Jones, J. Chem. Phys., 58, p. 1677 (1973).
(18) R. E. Morgan and J. H. Strange, Mol. Phys., 17, p. 397 (1969).
(19) J. H. Strange and R. E. Morgan, J. Phys. C., Solid State Phys., 3, p. 1999 (1970).
(20) H. Ottavi, Compt. Rend. Acad. Sci., Paris, 252, p. 1439 (1961).
(21) J. S. Blicharski and K. Krynicki, Acta Phys. Pol., 22, p. 409 (1962).
(22) A. Briguet, J.-C. Duplan, D. Graveron-Demilly, and J. Delman, Mol. Phys., 28, p. 177 (1974).
(23) R. Freeman, R. R. Ernst, and W. A. Anderson, J. Chem. Phys., 46, p. 1125 (1967).
(24) J. R. Lyerla, D. M. Grant, and R. D. Bertrand, J. Phys. Chem., 75, p. 3967 (1971).
(25) K. T. Gillen, M. Scwartz, and J. H. Noggle, Mol. Phys., 20, p. 889 (1971).
(26) T. C. Farrar, S. J. Druck, R. R. Shoup, and E. D. Becker, J. Am. Chem. Soc., 94, p. 699 (1972).
(27) G. C. Levy, Chem. Commun., p. 352 (1972).
(28) G. C. Levy, J. D. Cargioli, and F. A. L. Anet, J. Am. Chem. Soc., 95, p. 1527 (1973).
(29) R. R. Shoup and T. C. Farrar, J. Magn. Reson., 7, p. 48 (1972).
(30) R. E. J. Sears and E. L. Hahn, J. Chem. Phys., 45, p. 2753 (1966).
(31) J. S. Blicharski and B. Blicharska, Acta Phys. Pol., A 38, p. 289 (1970).

(32) K. T. Gillen, D. C. Douglass, M. S. Malmberg, and A. A. Maryott, J. Chem. Phys., 57, p. 5170 (1972).
(33) R. E. J. Sears, J. Chem. Phys., 56, p. 983 (1972).
(34) M. Alexandre and P. Rigny, Can. J. Chem., 52, p. 3676 (1974).
(35) M. Rhodes, D. W. Aksnes, and J. H. Strange, Mol. Phys., 15, p. 541 (1968).
(36) J. M. Winter, Compt. Rend. Acad. Sci., Paris, 249, p. 1346 (1959).
(37) J. H. Strange and R. E. Morgan, in "Magnetic Resonance and Radiofrequency Spectroscopy," Proc. XVth Colloque Ampere, Grenoble 1968, P. Averbuch, Ed., Amsterdam, 1969.
(38) R. R. Sharp, J. Chem. Phys., 57, p. 5321 (1972).
(39) R. R. Sharp, J. Chem. Phys., 60, p. 1149 (1974).
(40) R. M. Hawk and R. R Sharp, J. Chem. Phys., 60, p. 1009 (1974).
(41) S. Forsén, M. Gustavsson, B. Lindman, and N.-O. Persson, J. Magn. Reson., 23, p. 515 (1976).
(42) W. C. Dickinson, Phys. Rev., 80, p. 563 (1950).
(43) R. A. Bonham and T. G. Stand, J. Chem. Phys., 40, p. 344 (1964).
(44) J. McGurk, C. L. Norris, H. L. Tigelaar, and W. H. Flygare, J. Chem. Phys., 58, p. 3118 (1973).
(45) S. Rothenberg, R. H. Young, and H. F. Schaefer, III, J. Am. Chem. Soc., 92, p. 3243 (1970).
(46) T. D. Gierke and W. H. Flygare, J. Am. Chem. Soc., 94, p. 7277 (1972).
(47) N. F. Ramsey, Phys. Rev., 86, p. 243 (1952).
(48) W. H. Flygare, Chem. Rev., 74, p. 653 (1974).
(49) R. E. Davie and J. S. Muenter, J. Chem. Phys., 57, p. 2836 (1972).
(50) J. J. Ewing, H. L. Tigelaar, and W. H. Flygare, J. Chem. Phys., 56, p. 1957 (1972).
(51) E. W. Kaiser, J. Chem. Phys., 53, p. 1686 (1970).
(52) A. Abragam, "The Principles of Nuclear Magnetism," London, Oxford University Press, 1964.
(53) I. Solomon, Compt. Rend. hebd. Seance Acad. Sci., Paris, 249, p. 163k (1959).
(54) R. E. Morgan and J. H. Strange, Mol. Phys., 17, p. 397 (1969).
(55) J. H. Strange and R. E. Morgan, J. Phys. C, Solid State Phys., 3, p. 1999 (1970).
(56) H. Weingärtner, C. Müller, and H. G. Hertz, J. Chem. Soc., Faraday 1, 75, p. 2712 (1979).
(57) H. Wennerström, G. Lindblom, and B. Lindman, Chem. Scripta, 6, p. 97 (1974).
(58) B. M. Fung, M. J. Gerace, and L. S. Gerace, J. Phys. Chem., 74, p. 83 (1979).

(59) G. Lindblom, N.-O. Persson, and B. Lindman, in "Chemie, physikalische Chemie und Anwendungstechnik der grenzflachenaktiven Stoffe," Vol. II, Carl Hanser, München, 1972, p. 939.
(60) G. Lindblom, B. Lindman, and L. Mandell, J. Coll. Interface Sci., 42, p. 400 (1973).
(61) T. E. Bull, J. Magn. Reson., 7, p. 344 (1972).
(62) T. E. Bull, J. Andrasko, E. Chiancone, and S. Forsén, J. Mol. Biol., 73, p. 251 (1973).
(63) D. Beckert and H. Pfeifer, Ann. Phys., 7, p. 262 (1965).
(64) H. G. Hertz, Ber. Bunsenges. Phys. Chem., 71, p. 979 (1967).
(65) A. G. Marshall, J. Chem. Phys., 52, p. 2527 (1970).
(66) S. Lindskog, L. E. Henderson, K. K. Kannan, A. Liljas, P. O. Nyman, and B. Strandberg, in "The Enzymes," P. Boyer, Ed., 3d ed., Vol. V, Academic Press, New York, 1971, p. 587.
(67) R. L. Ward, Biochemistry, 8, p. 1879 (1969).
(68) R. L. Ward and M. D. Cull, Arch. Biochem. Biophys., 150, p. 436 (1972).
(69) R. L. Ward and P. L. Whitney, Biochem. Biophys. Res. Commun., 51, p. 343 (1973).
(70) S. H. Koenig and R. D. Brown, III, Proc. Natl. Acad. Sci. (USA), 69, p. 2422 (1972).
(71) J. E. Norne, H. Lilja, B. Lindman, R. Einarsson, and M. Zeppezauer, Eur. J. Biochem., 59, p. 463 (1975).
(72) H. Eklund, B. Nordström, E. Zeppezauer, G. Soderlund, I. Ohlsson, T. Boiwe, and C.-I. Brändén, FEBS Lett., 44, p. 200 (1974).
(73) M. Zeppezauer, B. Lindman, S. Forsén, and I. Lindqvist, Biochem. Biophys. Res. Commun., 37, p. 137 (1969).
(74) R. L. Ward and J. A. Happe, Biochem. Biophys. Res. Commun., 45, p. 1444 (1971).
(75) R. L. Ward and M. D. Cull, Biochim. Biophys. Acta, 365, p. 281 (1974).
(76) B. Lindman, M. Zeppezauer, and A. Åkeson, in "Structure and Function of Oxidation Reduction Enzymes," A. Akesson and A. Ehrenberg, Eds., Pergamon Press, Oxford, 1972.
(77) B. Lindman, M. Zeppezauer, and A. Akeson, Biochim. Biophys. Acta, 257, p. 173 (1972).
(78) J. E. Norne, T. E. Bull, R. Einarsson, B. Lindman, and M. Zeppezauer, Chem. Scripta, 3, p. 142 (1973).
(79) T. E. Bull, B. Lindman, R. Einarsson, and M. Zeppezauer, Biochim. Biophys. Acta, 377, p. 1 (1975).
(80) I. Andersson, D. Katzberg, B. Lindman, and M. Zeppezauer in "Energetics and Structure of Halophilic Microorganisms," S. R. Caplan and M. Ginzburg, Eds., Elsevier, Amsterdam, 1978.
(81) I. Andersson, M. Zeppezauer, T. E. Bull, R. Einarsson, J. E. Norne, and B. Lindman, Biochemistry, 18, p. 3407 (1979).
(82) E. Chiancone, J. E. Norne, S. Forsén, M. Brounori, and E. Antonini, Biophys. Chem., 3, p. 56 (1975).

(83) H. C. Reynolds and J. S. McKinley-McKee, Eur. J. Biochem., 14, p. 14 (1970).
(84) E. Zeppezauer, H. Jörnvall, and I. Ohlsson, Eur. J. Biochem., 58, p. 95 (1975).
(85) M. F. Perutz, Nature, 228, p. 726 (1970).
(86) E. Antonini and M. Brunori, "Hemoglobin and Myoglobin in Their Interaction with Ligands," North Holland, Amsterdam, 1971.
(87) R. Benesch and R. E. Benesch, Nature, 221, p. 618 (1969).
(88) T. R. Collins, Z. Starcuk, A. H. Burr, and E. J. Wells, J. Am. Chem. Soc., 95, p. 1649 (1973).
(89) E. Chiancone, J. E. Norne, S. Forsén, E. Antonini, and J. Wyman, J. Mol. Biol., 70, p. 657 (1972).
(90) E. Chiancone, J. E. Norne, J. Bonaventura, C. Bonaventura, and S. Forsén, Biochim. Biophys. Acta, 336, p. 403 (1974).
(91) E. Chiancone, J. E. Norne, S. Forsén, J. Bonaventura, M. Brunori, E. Antonini, and J. Wyman, Eur. J. Biochem., 55, p. 385 (1975).
(92) E. Chiancone, J. E. Norne, S. Forsén, A. Mansouri, and K. Winterhalter, FEBS Lett., 63, p. 309 (1976).
(93) J. Wyman, Adv. Protein Chem., 4, p. 407 (1948).
(94) J. Wyman, Adv. Protein Chem., 19, p. 223 (1964).
(95) J. E. Norne, E. Chiancone, S. Forsén, E. Antonini, and J. Wyman, FEBS Lett., 94, p. 410 (1978).
(96) J. Monod, J. Wyman, and J. P. Changeux, J. Mol. Biol., 12, p. 88 (1965).
(97) S. J. Edelstein, Nature, 230, p. 224 (1971).
(98) S. J. Edelstein, Biochemistry, 13, p. 4998 (1974).
(99) R. Ogata and H. M. McConnell, Cold Spring Harbor Symp. Quant. Biol., 36, p. 325 (1971).
(100) J. A. McCammon, B. R. Gelin, and M. Karplus, Nature, 267, p. 585 (1977).
(101) T. E. Bull, J. Magn. Reson., 31, p. 453 (1978).
(102) T. E. Bull, J. E. Norne, P. Reimarsson, and B. Lindman, J. Am. Chem. Soc., 100, p. 4643 (1978).
(103) M. H. Cohen and F. Reif, Solid State Phys., 5, p. 321 (1955).
(104) J. Bacon, R. J. Gillespie, and J. W. Quail, Can. J. Chem., 41, p. 3063 (1963).
(105) K. O. Christe, J. F. Hon, and D. Pilipovich, Inorg. Chem., 12, p. 84 (1973).
(106) R. J. Gillespie and G. J. Schrobilgen, J.C.S. Chem. Commun., p. 90 (1974).
(107) M. Brownstein and H. Selig, Inorg. Chem., 11, p. 656 (1972).
(108) R. J. Gillespie and J. W. Quail, Can. J. Chem., 42, p. 2671 (1964).

CHAPTER 21

TRANSITION METAL NMR SPECTROSCOPY

R. Garth Kidd

Faculty of Graduate Studies
The University of Western Ontario
London, Canada

ABSTRACT

Oxidation state is the variable exerting the largest influence on nuclear shielding of transition metal atoms. Sensitivity measured in $d\delta/d$(oxidation state) units is linear for a given atom and undergoes modest increases from left to right and from top to bottom within the transition metal rectangle. The sensitivity for ^{59}Co is anomalously high. The incidence of inverse halogen dependence is traced and attributed to the sign of the spin-orbit coupling constant. The $\sigma^p = 0$ standard state for absolute shielding of ^{59}Co is located using the extrapolation technique.

INTRODUCTION

On two occasions in the past several years (1,2) I have prepared a critical review on the NMR spectroscopy of the transition metals, and it is not a topic to which one can do full justice in a short chapter. Paul Ellis has very sensibly limited himself to the ^{113}Cd area at the end of the transition metal series where he has done most of the pioneering spectroscopic work. Rather than focusing on specific metals, I have selected three broadly based features of NMR spectroscopy and will choose from among the transition metals those specific examples that best illustrate the feature.

In Chapter 15 we saw a very pronounced correlation between Group III metal deshielding and metal atom oxidation state. The

Tl^{III} compounds span a range of 3000 ppm that is at higher δ values than the range for Tl^{I} compounds, with minimal overlap between the two ranges. Proceeding to the right in the Periodic Table, we saw this correlation gradually peter out until in ^{31}P spectroscopy no vestige of a δ correlation with oxidation state remains. The transition elements are, however, all metals, they all precede Group III in the Table, and the deshielding/oxidation state correlation for these elements, exhibiting such a rich variety of oxidation states, is alive and well. We will trace the incidence of this correlation among the transition metals and see that in ^{51}V, ^{95}Mo, ^{183}W, ^{55}Mn, ^{59}Co, and ^{195}Pt spectroscopies, it forms the basis upon which our understanding of their shielding rests.

The manner and extent to which metal shielding depends upon Cl$^-$, Br$^-$, and I$^-$ substitution is a central feature of transition metal spectroscopy. Not only was the phenomenon of inverse halogen dependence (IHD) first recognized among them (3), but in every case for which the halide complexes have been studied, they alone account for over half of the total shielding range exhibited by the metal. This is particularly evident for ^{47,49}Ti, ^{91}Zr, ^{93}Nb, ^{183}W, ^{67}Zn, and ^{113}Cd; for ^{195}Pt the shielding limit is marked by PtI$_6^{2-}$, the deshielded "limit" encompassing every other ^{195}Pt resonance save one is marked by PtCl$_6^{2-}$ some 6000 ppm to higher δ values, while the PtF$_6^{2-}$ resonance occurs an incredible 7000 ppm beyond. The basis for IHD and possible reasons for the extreme sensitivity to halogen substitution will be explored.

Cobalt-59 spectroscopy is noteworthy for two reasons. Its present shielding range of over 18,000 ppm (almost 2% !) is the widest to have been observed for any atom. It is also the first atom for which a relationship between deshielding and the existence of low-lying electronic excited states was noticed. In a classic paper by Freeman, Murray, and Richards from 1957 (4), the now familiar correlation between optical absorption spectrum and chemical shift for octahedral cobalt(III) complexes was established (Figure 1). This provided the springboard used by Leslie Orgel (5) to develop a theoretical model for chemical shifts in which electronic transition energy is used as an empirical substitute for the <u>average excitation energy</u> (AEE) in the Ramsey screening equation and to which σ^P is inversely proportional. The correlation was so successful that for the following 15 years, electronic excitation energy was the only variable with which spectroscopists attempted to correlate the chemical shifts of atoms heavier than hydrogen.

The standard state upon which an absolute shielding scale for ^{59}Co is founded is the free, gaseous Co^{+3} atom for which σ^P = 0. Although the resonance frequency for this state is not

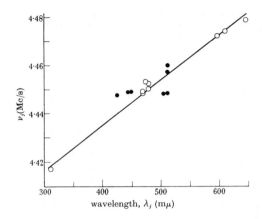

Figure 1. ^{59}Co resonance frequency plotted against lowest frequency optical absorption for octahedral Co^{III} complexes. Reproduced from Ref. 4 by permission of The Royal Society.

accessible by NMR techniques, and atomic beam or optical pumping studies have not been applied to this atom, extrapolation to zero wavelength of the chemical shift vs. optical wavelength plot provides a measure of the region in which the resonance frequency for the ^{59}Co standard state occurs (6,7). Because of its experimental inaccessibility and the large relative uncertainty in its resonance frequency, the standard state is of no value as a $\delta = 0$ reference state for chemical shifts. As a $\sigma^p = 0$ reference state from which to begin theoretical discussions of ^{59}Co shielding, however, it is invaluable.

OXIDATION STATE EFFECTS

For anyone whose initial contact with NMR is through ^{1}H spectroscopy, and this includes most of us, it is tempting to rationalize the chemical shift/oxidation state correlation through the σ^d term of the screening equation because (i) it works in the right direction and (ii) proton shifts do respond through σ^d, in both the direction and magnitude required, to relatively small changes in electron density. The Pauling electroneutrality principle (8) informs us that a one unit change in formal oxidation state represents an electron density change very much less than one electron unit, and operating through σ^d this is insufficient to cause the shifts of hundreds or thousands of parts per million that we observe with the transition metals. Only by operating through the σ^P term, and probably by changing the $\langle 1/r^3 \rangle$ factor in this term, can the relatively small oxidation state changes in electron density give rise to the extremely large variations in nuclear shielding that are observed.

Figure 2 provides a summary of the vanadium compounds that have been studied and of the 2450 ppm range covered by their ^{51}V chemical shifts measured against $VOCl_3$ as reference. The vanadium(V) compounds comprising the vanadium oxyhalides, the vanadyl esters, the tetrahedral oxyanions, and the iso- and heteropolyvanadates span 1225 ppm at the deshielded end of the range. The vanadium(I) compounds consisting of cyclopentadienylvanadium carbonyls with one or two P, As, or Sb donor heteroligands span 670 ppm at higher shieldings. The vanadium(-I) compounds of the type $V(CO)_{6-n}L_n^-$ absorb over 280 ppm at the most highly shielded end of the range.

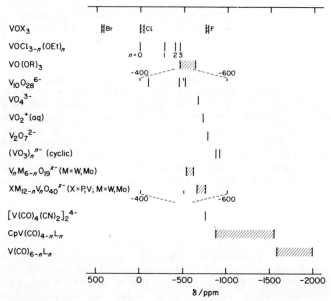

Figure 2. ^{51}V chemical shift ranges for V^V, V^I, and V^{-I} compounds. Note the x5 scale expansion for oxyanions. With permission from Ref. 3. Copyright: Academic Press Inc. (London) Ltd.

If we take the midpoint of each region as defining its chemical shift then V^V is at δ -1205, and V^{-I} is at δ -1810. It is instructive to note that the 2-electron oxidation from V^{-I} to V^I

deshields by 605 ppm and the 4-electron oxidation from V^I to V^V deshields by a further 1032 ppm, indicating a rough linearity between deshielding and oxidation state with a slope of about 280 ppm per oxidation state unit.

Figure 3 shows the chemical shifts for all three members of the chromium triad, each of which confirms the oxidation state correlation. In the ^{53}Cr and ^{95}Mo spectroscopies there is relatively little information available and we can thank Otto Lutz for most of the data that we do have. Fortunately the few ^{53}Cr and ^{95}Mo chemical shifts in Figure 3 include several oxidation states, and with uncanny prescience the 1800 ppm difference between Cr^{VI} and Cr^0 gives a slope of 300 ppm per oxidation state unit, almost identical with that of its neighbour ^{51}V.

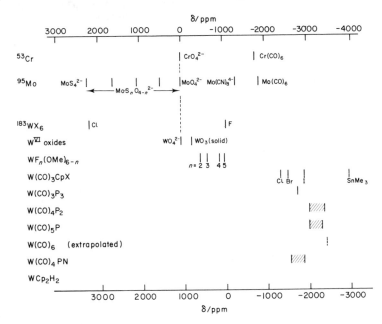

Figure 3. ^{53}Cr, ^{95}Mo, and ^{185}W chemical shift ranges all plotted to same scale. MO_4^{2-} shieldings arbitrarily equated. With permission from Ref. 2. Copyright: Academic Press Inc. (London) Ltd.

In ^{95}Mo spectroscopy, three oxidation states are covered with Mo^{VI} at δ 1100, Mo^{IV} at δ -1200, and Mo^0 at δ -1900. The full six unit difference in oxidation states spans a chemical shift range of 3000 ppm, for an average dependence of 500 ppm per oxidation state unit, 72% larger than was observed in the previous two cases as we would expect on moving from the first to the second transition series of elements.

In ^{183}W spectroscopy, the shieldings span 6850 ppm, with W^{VI} compounds occupying 3300 ppm at the deshielded end of the range. Tungsten(VI) and tungsten(0) compounds together occupy 2300 ppm at the top end and there is a gap of about 1250 ppm between the two. If we put W^{VI} at δ 1650 and W^0 at δ -3000, the average dependence is 775 ppm per oxidation state unit, 1.6 times that for ^{95}Mo. Previous attempts to establish relative shielding sensitivities (i.e., α indices) for ^{183}W and ^{95}Mo yielded a ratio of 1.8 based on $MO_4^{2-}-M(CO)_6$ differences, and one of 1.4 based on the WCl_6-WF_6, cf. $MoS_4^{2-}-MoO_4^{2-}$ comparison. With an index ratio of 1.6 based upon oxidation state dependence falling midway between the previous two, one has some confidence that ^{183}W is 1.6 more shiftable than ^{95}Mo.

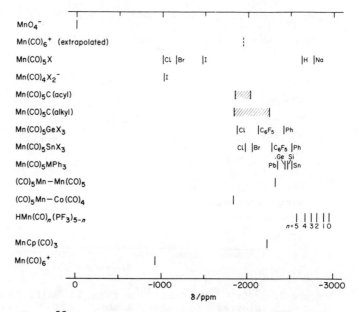

Figure 4. ^{55}Mn chemical shift ranges for Mn^{VII}, Mn^I, and Mn^{-I} compounds. With permission from Ref. 2. Copyright: Academic Press Inc. (London) Ltd.

Figure 4 shows the 3000 ppm shielding range spanned by ^{55}Mn resonances in Mn^{VII}, Mn^I, and Mn^{-I} compounds. The permanganate ion, the chemical shift reference that has been adopted by workers in the field, is the only Mn^{VII} compound to have been studied and its resonance is separated from that of all others by 1000 ppm. The Mn^I compounds span a range of 1800 ppm in the region of δ -2000, and Mn^{-I} compounds cover 750 ppm around δ -2625. The linearity of the correlation is again remarkable, giving a slope of 330 ppm per oxidation state unit and continuing the trend of gradually increasing shielding sensitivity as we proceed to the right along the first transition series.

On reaching ^{59}Co, we see in Figure 5 that while the correlation is maintained, it has expanded in scale beyond anything encountered anywhere else in the Periodic Table. If within the Co^{III} class we include $Co(CN)_6^{3-}$ and the phosphine complexes, both of which are atypical Co^{III} complexes and overlap the Co^I region, its range is an enormous 16,000 ppm centered on δ 7000. The Co^I compounds cover 1300 ppm around δ 1350 and Co^{-I} compounds span 4100 ppm at δ -2500. With three points available the linearity of the correlation is within ± 10% and the slope is 2600 ppm per oxidation state unit. To wonder at this extraordinary shielding sensitivity for a first transition series atom is not to explain it!

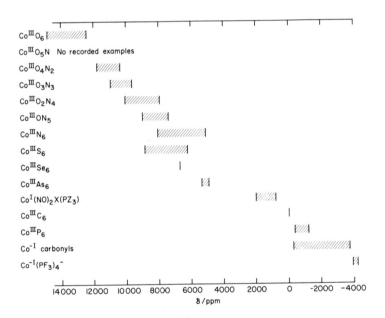

Figure 5. ^{59}Co chemical shift ranges for Co^{III}, Co^I, and Co^{-I} compounds referenced to $Co(CN)_6^{3-}$. With permission from Ref. 2. Copyright: Academic Press Inc. (London) Ltd.

Platinum-195 is the only other third transition series atom for which there are enough data to provide a comparison with ^{183}W, and here the situation is complicated by the enormous halogen dependence of both Pt^{II} and Pt^{IV} shifts. The only way to obtain the oxidation state dependence from the ^{195}Pt shifts in Figure 6 is to eliminate from the analysis all of the halide complexes and work with the remaining data. This leaves Pt^{II} compounds spanning 2000 ppm around δ 1700 and Pt^0 compounds covering 1500 ppm around δ -800. The slope is 1250 ppm per oxidation state unit, which represents a modest increase over the 775 ppm value for ^{183}W, not inconsistent with its position four triads to the right.

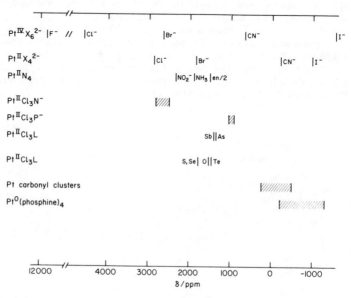

Figure 6. ^{195}Pt chemical shift ranges for Pt^{IV}, Pt^{II}, and Pt^0 compounds referenced to Ξ(^{195}Pt) = 21.4 MHz. With permission from Ref. 2. Copyright: Academic Press Inc. (London) Ltd.

NORMAL AND INVERSE HALOGEN DEPENDENCY

In the earliest days of ^{13}C spectroscopy when Paul Lauterbur and Jake Stothers were using rapid passage, dispersion mode scans to overcome the relaxation problem, the "anomalous" shielding property of iodine was already recognized through use of the precious ^{13}C-enriched methyl iodide sample as an external reference with a resonance at the extreme high field end of the ^{13}C range.

Bromine substituents by comparison were much less effective, and carbons bearing chlorine appeared at even lower fields, thereby establishing the shielding order Cl < Br << I as the normal halogen dependence (NHD). This shielding order is followed by the other atoms in Group IV, by all the atoms in Group III, and by many of the transition metals.

In 1971, the Russian group under Yuri Buslaev reported (3) a ^{93}Nb nucleus in NbCl$_6^-$ more highly shielded than that in NbBr$_6^-$, and shortly thereafter Garth Spinney and Ray Matthews working in my laboratory observed (9) a similar relationship for the ^{47,49}Ti resonances in TiCl$_4$ and TiBr$_4$. Since at that time all of the well documented cases of NHD were halides of main group elements, there was some speculation that IHD might be uniquely characteristic of transition metal halides with their d orbital participation in bonding, but this hypothesis was laid to rest when definitive studies showed that both ^{183}W halides and ^{195}Pt halides follow NHD.

Other instances of IHD are not so definitive because the evidence is not based upon kinetically stable and homogenously ligated complexes. The ^{51}V in VOCl$_3$ is more highly shielded than in VOBr$_3$, putting ^{51}V in the IHD category. Mixed ethylamine halide complexes of silver give ^{109}Ag shieldings in the order I<<Br<Cl, characteristic of IHD. Results from the solid cuprous halides are problematical but suggest that ^{63}Cu shows IHD. The counterion dependence of the concentration plots for M$^+$ alkali metal cations in solution, confirmed by the chemical shifts of the solid halide salts where available, indicate that ^{23}Na, ^{39}K, ^{87}Rb, and ^{133}Cs all show IHD. Evidence of a similar nature assigns ^{45}Sc and ^{89}Y to the IHD category.

This peculiar distribution of IHD atoms throughout the Periodic Table corresponds to the Groups whose valence shell electron configurations are s^1, d^3, d^4, d^5, and d^{10}s^1, all of which represent valence orbital sets less than or equal to half filled. These same electron configurations represent atoms whose spin-orbit coupling constants have negative signs. Although the spin-orbit coupling constant has not as yet been heavily implicated in the current crop of semi-empirical shielding models, it will undoubtedly be present in the first one that successfully explains why ^{13}C is shielded in CI$_4$ but ^{47}Ti is deshielded in TiI$_4$.

THE ABSOLUTE SHIELDING OF ^{59}Co

The formative years of NMR have been bedeviled by two problems that did not impede development of the traditional IR and optical spectroscopies. The first of these was the sign convention. Should a chemical shift that represents an increase in nuclear shielding be designated as positive or negative? This question was settled by an IUPAC ruling in 1972 (10), but the trilemma of whether a positive chemical shift should be described as a shift to a higher frequency, a shift to lower field, or a deshielding still remains. To the reader with an appropriate mental image of the relationship, the choice doesn't really matter, but to the author attempting to convey an image, the choice can be critical.

The other problem is the choice of a zero reference for chemical shifts. In the early ^{13}C literature we find extensive compilations of data referenced to C_6H_6, others to CS_2, and yet others to CH_3I before TMS was finally agreed upon. The arithmetical inconvenience of converting δ values from one reference to another is not, however, the major reference-based vicissitude.

When an IR spectroscopist reports a 3600 cm^{-1} band, the number has immediate structural significance in terms of a stretching force constant and the masses undergoing vibration. The two limits between which the number lies, familiar because they are the same for all IR spectra, are implicitly present in the mind of the observer when he analyzes the significance of his 3600 cm^{-1}. The zero limit upon which his analysis is founded (and one which, incidentally, provides the ultimate criterion for whether a chemical bond does or does not exist) represents a state of very low stretching force constant or very large reduced mass.

By contrast, the $\delta = 0$ limit upon which chemical shifts are based has no structural significance whatsoever; its choice is based purely on the ease and reproducibility with which the resonance from a particular compound is obtained and in the case of ^{59}Co $\delta = 0$ happens to occur at the shielded end of the range while for ^{55}Mn it constitutes not only the deshielded end of the range but is 1000 ppm beyond any other ^{55}Mn resonance. In spite of this the NMR literature contains many examples of chemical shift analysis, based upon $\delta = 0$ for the particular nucleus, decked out as shielding analysis. In these instances the vocabulary used and the impression created depend entirely upon whether $\delta = 0$ occurs at the top or the bottom of the range, whereas shielding is invariant to the $\delta = 0$ choice. (I have chosen the words "top" and "bottom" by design to emphasize that even this usage contains an ambiguity.) It is not surprising therefore that a coherent theoretical model applicable simultaneously to, say, ^{13}C and ^{93}Nb shieldings has heretofore failed to emerge.

The way out of this difficulty lies in the uniform definition and adoption of a standard state for nuclear shielding, the interpretation of which will be absolute rather than relative and will be common throughout the NMR spectroscopies of all atoms. There are two choices for this standard state, and since the bare nucleus with $\sigma^d = \sigma^p = 0$ is experimentally inaccessible for all but the lightest atoms, it is not the best choice. The resonance frequency for the free, gaseous atom with a spherically symmetrical electron distribution and therefore with $\sigma^p = 0$ can be measured in favourable cases using atomic beam or optical pumping methods, and for these atoms a precise δ value for the standard state can be obtained. In other cases it must be obtained by the extrapolation technique, and this technique is here illustrated as it is applied in ^{59}Co spectroscopy.

The paramagnetic term which we are extrapolating to zero is viewed by the spectroscopist (as distinct from the ab initio theoretician) as having two variables (eq 1).

$$\sigma^p = \text{constant} \cdot \frac{\langle 1/r^3 \rangle}{\Delta E} \qquad (1)$$

Cobalt-59 chemical shifts are due primarily to variations in the ΔE variable, and any variation in $\langle 1/r^3 \rangle_d$ from compound to compound is small by comparison. Figure 7 shows the correlation between ^{59}Co chemical shift and $1/\Delta E$ representing the wavelength of the lowest energy optical transition, extrapolated to $1/\Delta E = 0$ and giving the δ value for the $\sigma^p = 0$ standard state. The slope of the line measures the value of $\langle 1/r^3 \rangle_d$ regarded as constant and the appearance of two lines in Figure 7 confirms what we already know about $\langle 1/r^3 \rangle$ and the nephelauxetic effect, that is, that second row ligating atoms exert a greater nephelauxetic effect upon a metal than do first row ligating atoms, and the correlation line for O-donor and N-donor ligands will have greater $\langle 1/r^3 \rangle_d$ and higher slope than will the correlation line for P-donor and S-donor ligands.

If the $\langle 1/r^3 \rangle_d$ factor were completely invariant among all of the compounds contributing to a single regression line, then the two lines would have a common intercept. As it is, the spread in intercept values covering about 2000 ppm is attributable to the small variation in $\langle 1/r^3 \rangle_d$ from compound to compound, magnified by the length of the extrapolation. The mean value of δ = -9265 ± 1000, while completely out of consideration as a chemical shift reference because of the ± 10% uncertainty limits, is invaluable as a standard state for ^{59}Co shielding because it identifies the $\sigma^p = 0$ position.

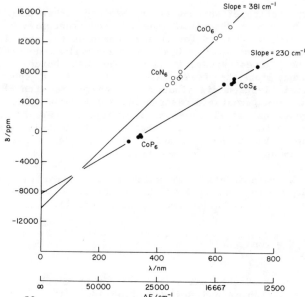

Figure 7. ^{59}Co chemical shifts referenced to Co(CN)$_6^{3-}$ correlated with $1/\Delta E$ from the electronic spectrum. With permission from Ref. 2. Copyright: Academic Press Inc. (London) Ltd.

REFERENCES

(1) R. G. Kidd in "NMR and the Periodic Table," R. K. Harris and B. E. Mann, Eds., Academic Press, London 1978, Chapter 8.
(2) R. G. Kidd, Ann. Rep. NMR Spectrosc., 10A, p. 1 (1980).
(3) Y. A. Buslaev, V. D. Kopanev, and V. P. Tarasov, Chem. Commun., p. 1175 (1971).
(4) R. Freeman, G. R. Murray, and R. E. Richards, Proc. Roy. Soc., A242, p. 455 (1957).
(5) J. S. Griffith and L. E. Orgel, Trans. Faraday Soc., 53, p. 601 (1957).
(6) R. G. Kidd, Ann. Rep. NMR Spectrosc., 10A, p. 33 (1980).
(7) N. Juranic, Inorg. Chem., 19, p. 1095 (1980); J. Chem. Phys., 74, p. 3690 (1981).
(8) L. Pauling, J. Chem. Soc., p. 1461 (1948).
(9) R. G. Kidd, R. W. Matthews, and H. G. Spinney, J. Am. Chem. Soc., 94, p. 6686 (1973).
(10) "Recommendations for the Presentation of NMR Data for Publication in Chemical Journals," Pure Appl. Chem., 29, p. 627 (1972); 45, p. 217 (1976).

CHAPTER 22

CADMIUM-113 NUCLEAR MAGNETIC RESONANCE SPECTROSCOPY IN BIOINORGANIC CHEMISTRY. A REPRESENTATIVE SPIN 1/2 METAL NUCLIDE

Paul D. Ellis

Department of Chemistry
University of South Carolina
Columbia, South Carolina 29208

ABSTRACT

   The main theme of this chapter involves the utilization of ^{113}Cd NMR spectroscopy to probe interesting structural and dynamic problems in inorganic and bioinorganic chemistry. Within the first portion of the chapter we have summarized most of the known ^{113}Cd chemical shifts and coupling constants. Further, we have also presented introductory comments with regard to relaxation mechanisms and the importance of chemical dynamics. Subsequently, we discuss in detail the consequences that chemical dynamics have upon the interpretation of relaxation parameters and chemical shifts. The discussion then shifts to solid state NMR methods. Our recent work on cadmium-substituted porphyrins and ^{113}Cd NMR of single crystals serve as examples. Finally, we bring all of these points into focus when we examine the utilization of cadmium as a surrogate probe for Ca^{+2} and Zn^{+2} in bioinorganic systems. Here, we have limited our discussion to our work on Concanavalin A and skeletal Troponin C.

INTRODUCTION

   In the past several years we have witnessed an enormous increase in the applicability of NMR methods to research problems in the areas of physics, chemistry, and biology. One of the many factors responsible for this explosion of interest is the routine acquisition of NMR data from almost any nucleus in the Periodic Table. It is the application of NMR methods to the study of spin 1/2 metal nuclides and the interpretation of these data that will be topics for this presentation. The examples that we shall employ will come from our efforts in utilizing ^{113}Cd NMR spectroscopy to investigate systems of bioinorganic interest. Even though the applications will use ^{113}Cd NMR, the principles are applicable to any spin 1/2 metal nuclide. The basic theme for this paper revolves around some of the potential pitfalls that can occur in extracting chemical information from multinuclear NMR data. These pitfalls generally arise when one ignores the basic chemistry and physics which differentiate between a metal in chemical systems and a lighter element such as ^{13}C.

Approximately ten years ago when we and others started to explore the applicability of multinuclear NMR spectroscopy to spin 1/2 metals, we were naive enough to base our understanding of chemical shifts and relaxation times upon our past experiences with ^{13}C NMR spectroscopy. For the most part this extrapolation served us well until we looked closely at the experimental data. If we chose to ignore the tried and true fields of chemistry and physics (as we are all prone to do at times) we would find ourselves "discovering" new relaxation pathways and novel mechanisms for producing chemical shielding that would force the physicists to find new Hamiltonians. This obvious confusion makes our initial efforts, and those of other groups in this area, look as though we have helped open an NMR spectroscopist's analogue to Pandora's box. However, if we recognize an old nemesis, chemical exchange, and borrow some understanding of the concept of spin-orbit couplings from our physics friends, we will see that some of our problems will disappear and as usual we have simply transformed the other problems to the proper rotating frame.

This presentation will be divided into four portions. Within the second section we will introduce ^{113}Cd NMR by summarizing the sensitivity of the method, the range of chemical shifts, indirect spin-spin couplings, relaxation times, and mechanisms. Next, we will turn our attention to the problem associated with interpretation of relaxation and chemical shielding data in the liquid state. We will then demonstrate how one can remove some of the ambiguity introduced by the solution dynamics by performing experiments in the solid state. These data will prove essential for proper interpretation of relaxation and shielding data. Finally, we will summarize some of our own efforts in utilizing ^{113}Cd NMR spectroscopy as a probe for Ca^{+2} in calcium-binding proteins. The proteins of interest will be Concanavalin A and Troponin C.

CADMIUM NMR SPECTRAL PARAMETERS

Sensitivity

The ^{113}Cd resonance of 0.1 M aq. $Cd(ClO_4)_2$ solution is shown in Figure 1a. This frequency spectrum is the Fourier transform of a single free induction decay obtained at 88.8 MHz in a 9.4 Tesla magnetic field. The large signal to noise ratio, 38:1, reflects the unusually high sensitivity of this heavy metal nucleus. Compared with the routinely observed natural abundance ^{13}C nucleus, the signal from ^{113}Cd is 7.6 times as intense. This comparison takes into account the responses of the carbon and cadmium isotopes to a given magnetic field strength and their relative natural abundances (1).

In certain cases, though, carbon and cadmium cannot be compared so straightforwardly. Carbon-13 is a spin 1/2 nucleus which is often strongly dipolar coupled to hydrogen nuclei. In such a situation, the nuclear Overhauser effect (2) can be used to enhance the carbon signal by a factor as large as 2.98. Cadmium-113 is a negative spin 1/2 nucleus and use of an Overhauser effect generally leads to a decrease in cadmium signal intensity. Other factors which could influence peak intensities, such as linewidths, are comparable. The conclusion, then, is that natural abundance cadmium peak intensities may be as much as 7.6 times greater than for carbon and in unfavorable comparisons are still more than 2.5 times as great.

One consequence of such favorable sensitivity is the use of cadmium as an NMR probe. In its divalent state, cadmium can replace ions such as calcium or zinc in molecules of biological origin. Figure 1b consists of ^{113}Cd spectra of the calcium binding subunit of whole Troponin, (TnC) (3), cadmium having replaced calcium in all structural and regulatory binding sites. The signal to noise ratio of 30 was achieved after 4.5 hours of accumulation time with 2.5 ml of 2.5 millimolar protein. This clearly demonstrates that detailed information concerning metalloproteins is to be had from cadmium NMR spectra using commercially available instruments and reasonable data acquisition time periods. The interpretation of cadmium resonance frequencies, lineshapes, and relaxation rates in such systems is deferred to a later section of this review.

In recent years, this high sensitivity has been taken advantage of in a number of studies involving the direct observation of cadmium in inorganic and organometallic compounds. Most such studies have employed the ^{113}Cd nucleus, although a few have utilized the spin 1/2 ^{111}Cd. This second isotope, having a natural abundance of 12.75 percent and a smaller sensitivity with respect to an applied field, generates a less intense NMR signal. The

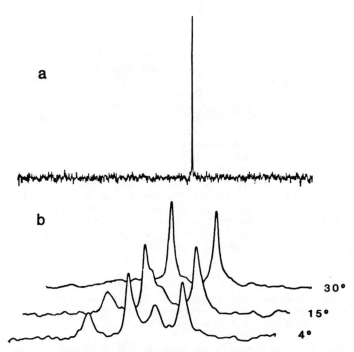

Figure 1. Natural abundance ^{113}Cd NMR spectra obtained at 9.4 Tesla (88 MHz). (a) This spectrum was obtained with a single 90° pulse from 0.1 M aqueous cadmium perchlorate. The linewidth is 0.4 Hz and signal-to-noise ratio 37.5:1. (b) These three spectra were obtained from Troponin C bound cadmium at the three temperatures indicated. A signal-to-noise ratio of ~30 was achieved with 4.5 hours of accumulation times (40,000 transients) using 2.5 ml of 2.5 millimolar protein. Five equivalents of cadmium were used for these spectra.

^{111}Cd spectrum analogous to Figure 1a occurs at 84.8 MHz and the peak is 9% less intense (signal to noise = 34:1) with a similar linewidth (0.4 Hz). Had scalar coupling to other nuclear spins been present, peak splittings would have been smaller for this nucleus by a factor of $\varkappa(^{111}Cd)/\varkappa(^{113}Cd) = 902.8/944.5 = 0.9558$. A comparison of the relevant properties of cadmium and other selected nuclei is given in Table 1.

Table 1. Properties of Some Selected Nuclear Isotopes[a]

Isotope	Spin	Gyromagnetic ratio (Hz/gauss)	Natural abundance (percent)	Relative sensitivity (constant field)	Relative NMR signal	NMR frequency in 9.4 T field (MHz)
^{113}Cd	-1/2	944.5	12.26[b]	0.0109	7.58	88.8
^{111}Cd	-1/2	902.8	12.75[b]	0.00954	6.90	84.8
^{13}C	+1/2	1070.54	1.108	0.0159	1.00	100.6
^{1}H	+1/2	4257.59	99.985	1.00	5675.42	400.1

[a]This table has been prepared with data taken from "CRC Handbook of Chemistry and Physics," 53rd edition, CRC Press, 1972, pp. B-247-B-324.
[b]Naturally occurring cadmium isotopes not listed have zero spin.

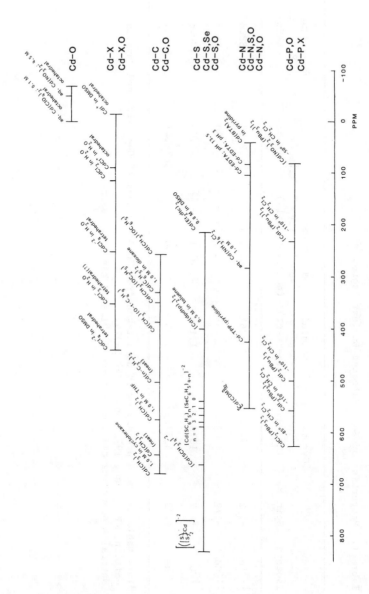

Figure 2. Chemical shift scales for various cadmium-containing compounds are presented here. Data were obtained from the footnote references of the following Chemical Shift Tables. The horizontal scale is cadmium chemical shift in ppm referred to 0.1 M aq.

Figure 2 (continued) Cd(ClO$_4$)$_2$. Positive values indicate deshielding of the cadmium nucleus. The vertical separation of scales emphasizes the chemical shift ranges expected for each ligand type, indicated at right. Abbreviations used include:

DMSO   dimethyl sulfoxide
Et$_2$dtc   N,N'-diethyldithiocarbamato
THF   tetrahydrofuran
EDTA   ethylenediaminetetraacetate
[Cd-CDM]$_2$ di-µ-chlorodichlorobis(6-mercaptopurine)diaquodicadmium(II)
TPP   tetraphenylporphyrin
PBu$_3$ tributylphosphine

For the purpose of this review, all scalar coupling constants will be quoted as if having been measured from ^{113}Cd spectra. The following discussion of chemical shifts will be in parts per million (ppm) of the applied magnetic field and require no such conversion.

Chemical Shifts

The resonance frequency of a nuclear spin in a NMR experiment is determined primarily by the nuclear magnetic moment of the observed nucleus and the applied magnetic field. In a molecular environment, this frequency is perturbed by contributions from electronic shielding (i.e., the chemical shift), by direct or indirect interactions by nearby nuclear magnetic dipoles (e.g., scalar coupling), and by the bulk magnetic susceptibility of the surrounding medium. The shift of cadmium resonance frequencies attributable to the differing bulk susceptibilities of solvents is on the order of a few Hertz. Other shift contributions are much larger and solvent bulk susceptibility corrections are therefore generally ignored in the literature, although a few careful investigations have included them, as noted in later tables.

The chemical shift range of solutions of cadmium inorganic and organometallic compounds is shown in Figure 2 to be several hundred ppm. This scalar chemical shift, σ, represents the isotropic part of the chemical shift tensor, $\underline{\sigma}$ (eq. 1). The

$$\sigma = \frac{1}{3} \text{Trace} (\underline{\sigma}) \tag{1}$$

anisotropic components of $\underline{\sigma}$ are averaged to zero by fast reorientational motions (3).

In liquids, the nonzero scalar chemical shift may be thought of as having a diamagnetic, σ^d, and a paramagnetic, σ^p, contribution (3-5). The diamagnetic part involves only the ground state

$$\sigma = \sigma^d + \sigma^p \tag{2}$$

electronic wavefunction of the molecule. This term corresponds to an induced circulation of electrons about the observed nucleus; the circulation of electric charge creates a magnetic field which opposes the static applied field.

The paramagnetic part is a second-order effect caused by the mixing of ground and excited electronic states by the electron oribtal angular momentum. Although mathematical expressions describing this contribution exist, they are generally difficult to evaluate quantitatively because of approximations made in the description of excited state electronic energy levels and wavefunctions. Semi-empirical evaluation, however, is capable of predicting, within experimental error, the chemical shifts of the lighter nuclei. These methods may be found in the literature (6-8).

For compounds of nuclei such as carbon, nitrogen, and fluorine, investigation has shown variations in the diamagnetic term to contribute significantly to the chemical shift (9-11). Nevertheless, the 200-900 ppm chemical shift ranges of these nuclei are dominated by the paramagnetic shielding of the p electrons. The chemical shift ranges of nuclei heavier than fluorine are expected also to be dominated by the effects of their p and d electrons.

Table 2 compares chemical shifts of cadmium and the other group IIB metals in several compounds. In each case, the chemical shift reference is an aqueous solution of the metal's perchlorate salt; the symmetric hydration of the solvated ions produces only a small paramagnetic contribution of the observed shielding. The relatively small negative shifts shown here result from increased shielding, as from diamagnetic effects, which lower the apparent field strength at the nucleus. The large positive shifts demonstrate the effectiveness of the zinc 3d, cadmium 4d, and mercury 5d electrons in deshielding nuclei, particularly for the dialkylmetals.

Straightforward qualitative interpretation of the chemical shifts of heavy nuclei is complicated by the presence of electron spin-orbit coupling. A discussion of the effects of this shielding contribution will appear later in this review, but it is appropriate to mention the effect here. Spin-orbit coupling is most widely appreciated for the case of the Group VIIIA halogens; iodine often causes anomalous behavior in carbon and hydrogen chemical shifts. Similar difficulties may be true for cadmium and any reader attempting to find simple trends in the following chemical shift tables should keep this in mind.

Figure 2 shows a schematic chemical shift scale for ^{113}Cd similar to the one previously prepared by Holm et al. (12). Although the recent interest in ^{113}Cd NMR has expanded the amount of data available, the comments of this early paper still accurately reflect what is known of cadmium chemical shifts.

The uppermost horizontal portion of Figure 2 depicts the chemical shift range of cadmium with oxygen ligands, originally determined by Maciel and Borzo (13) and soon thereafter extended by Kostelnik and Bothner-by (14) and Cardin et al. (15). The solution species have been determined by Ohtaki et al. (16a),

Table 2. Chemical Shifts of Group IIB Metal Compounds[a]

	$^{67}Zn(II)$	$^{113}Cd(II)$	$^{199}Hg(II)$	footnote
$(ClO_4)_2^{-2}$	0.00[b]	2.2 (H_2O)[b] -1.13 (D_2O)[b]	0.00[c]	1-4
SO_4^{-2}	0.00	-2.81	d	1,3
$(NO_3)_2^{-2}$	0.00	-49.41[e]	-176[f]	1,6,7
Cl_2^{-2}	~70	97.84	782.4[g]	1,3-5
Br_2^{-2}	~35	108.64		1,3,5
I_2^{-2}	~11	55.13		1,3,5
$-(CH_3)_2$		642.93[h]	2284.0[h]	3,4
$-(C_2H_5)_2$		543.20[h]	2004.[h] 1954[h]	3,4,8
$-(C_3H_7)_2$		504.28[h]	2044[h]	3,8
$-(C_4H_9)_2$		489.11[h]	1644[h]	3,8
$-(C_6H_5)_2$		328.80[i]	1542[j]	3,4

[a]Chemical shifts are in ppm with respect to dilute aqueous perchlorate salts, as indicated. Concentrations are 1.0 M in D_2O unless otherwise stated. A negative chemical shift indicates greater shielding of the observed nucleus.
[b]Chemical shift standard is the 0.1 M aq. perchlorate salt.
[c]3.78 g in 3.0 ml of 1.0 M $HClO_4$ (3.15 M Hg^{+2}).
[d]Hg(II) sulfate decomposes in H_2O.
[e]4.5 M in H_2O.
[f]Saturated aqueous solution.
[g]1 M in DMSO.
[h]Neat compound.
[i]1 M in 1,4-dioxane.
[j]1 M in methylene chloride.

1. B. W. Epperlein, H. Krüger, O. Lutz, and A. Schwenk, Z. Naturforsch., 29a, pp. 1553-1557 (1974).
2. R. A. Haberkorn, L. Que, Jr., W. O. Gullum, R. H. Holm, C. S. Liu, and R. C. Lor, Inorg. Chem., 15, p. 2408 (1976).
3. A. D. Cardin, P. D. Ellis, J. D. Odom, and J. W. Howard, Jr., J. Am. Chem. Soc., 97, p. 1672 (1975).
4. M. A. Sens, N. K. Wilson, P. D. Ellis, and J. D. Odom, J. Magn. Reson., 19, pp. 323-336 (1975).
5. B. W. Epperlein, H. Krüger, O. Lutz, and A. Schwenk, Z. Naturforsch., 29a, pp. 660-661 (1974).
6. G. E. Maciel and M. Borzo, J. C. S. Chem. Commun., p. 394 (1973).
7. W. G. Schneider and A. D. Buckingham, Disc. Faraday Soc., 34, p. 147 (1962).
8. R. E. Dessy, T. J. Flautt, H. H. Jaffe, and G. F. Reynolds, J. Chem. Phys., 30, p. 1422 (1959).

from X-ray diffraction, to be the octahedral hexaaquo complex, $Cd(H_2O)_6^{+2}$. Methanol, dimethyl formamide, and dimethyl sulfoxide solvents also yield six-coordinate complexes with "oxygen type" ligands whose cadmium chemical shifts fall in the range indicated here (16).

The linear concentration dependence (14) of the inorganic cadmium salts in aqueous solution could perhaps be explained by inner- or outer-sphere complexation with the given anion (12). At "infinite dilution," these cadmium resonances approach that for 0.1 molar aqueous cadmium perchlorate, the generally accepted standard for cadmium chemical shift measurements.

In Table 3 are given the resonance frequencies of several representative Cd-O compounds. While this compilation does not exhaust the literature, it does provide additional quantitative data. The references associated with this table, and references therein, will serve the reader in locating further studies.

The second topmost section of Figure 2 shows the chemical shift range of cadmium-halide complexes in solvents which provide oxygen ligands. The shift of CdX_2 is a distinctly nonlinear function of concentration (14). Addition of the alkali metal halides (17) or ammonium halides (18,19) to solutions of cadmium salts results in deshielding in the order CdX_3^-, CdX_4^{-2}, CdX_2, CdX^+ (cf., Figure 2 and Table 4). At this time it remains unclear whether this dependence is the result of ligand rearrangement (octahedral to tetrahedral) or ligand type (oxygen or halogen), although it appears that the most shielded resonances are those with the greatest number of oxygen ligands.

The most deshielded resonance for cadmium with carbon ligands is that of dimethylcadmium in the nonassociating solvent, cyclohexane. In solvents which provide some ligation through oxygen or nitrogen, the cadmium may be shielded by as much as 100 ppm (15) (see Table 5a). Alkyl groups other than methyl cause shielding with respect to dimethylcadmium (15); diphenylcadmium in dioxane solvent is the most extreme ($\Delta\sigma = -314$ ppm). From Table 5a approximately 50 ppm of this shielding could arise from the dioxane solvent. The oxygen of the methylcadmium alkoxides again appears to cause appreciable shielding of cadmium (22) (see Table 5b). These compounds can self-associate in solution to form dimers (e.g., $(CH_3CdO-t-C_4H_9)_2$: benzene), tetramers (e.g., $(CH_3CdOC_2H_5)_4$:benzene), and hexamers (e.g., $(CH_3CdS-i-C_3H_7)_6$) (21).

The cadmium-thiolate complexes (Table 6) have the greatest known shifts to lower shielding (24). Magnetic circular dichroism indicates near tetrahedral symmetry about cadmium, and it is suggested that the greatest deshielding is from total

Table 3. Cadmium-113 Chemical Shifts--Oxygen Ligands

Compound	Chemical shift[a]	Conditions	Footnote reference
$Cd(ClO_4)_2$	0.00	0.1 M in H_2O; $\Delta\nu_{1/2}$ = 3 Hz	1
$Cd(ClO_4)_2$	-1.13	1.0 M in D_2O	2
$Cd(ClO_4)_2$	-29.4	1.0 M in CH_3OH; 25°; $\Delta\nu_{1/2}$ = 5 Hz	3
$Cd(ClO_4)_2$	-26.0	0.5 M in DMSO; 25°; $\Delta\nu_{1/2}$ = 8 Hz	3
$Cd(NO_3)_2$	-3.72	0.1 M in H_2O; $\Delta\nu_{1/2}$ = 4 Hz	1
$Cd(NO_3)_2$	-49.41	4.5 M in H_2O; $\Delta\nu_{1/2}$ = 4 Hz	1
$Cd(SO_4)$	-0.0	0.1 M in H_2O	4
$Cd(SO_4)$	-5.0	3.0 M in H_2O	4
$Cd(NO_2)_2$	-2.3	0.1 M nitrite in 2M $CdSO_4$	4
$Cd(NO_2)_2$	-11.4	0.5 M NO_2^- in 2M $CdSO_4$	4
$Cd(CH_3COO)_2$	-1.0	0.1 M acetate in 2M $CdSO_4$	4
$Cd(HCOO)_2$	-0.7	0.1 M formate in 2M $CdSO_4$	4
Cd:glycine	+1.9	0.1 M glycine in 2M $CdSO_4$	4
Cd(II)glycine	-77	0.1 M $Cd(ClO_4)_2$; 0.05 M glycine pH 7; -50°C	5

[a]Chemical shifts are in ppm with respect to 0.1 M $Cd(ClO_4)_2$. A positive shift denotes lower shielding.

1. G. E. Maciel and M. Borzo, J. C. S. Chem. Commun., p. 394 (1973).
2. A. D. Cardin, P. D. Ellis, J. D. Odom, and J. W. Howard, Jr., J. Am. Chem. Soc., 97, pp. 1672-1679 (1975).
3. R. A. Haberkorn, L. Que, Jr., W. O. Gillum, R. H. Holm, C. S. Liu, and R. C. Lord, Inorg. Chem., 15, pp. 2408-2414 (1976).
4. R. J. Kostelnik and A. A. Bothner-by, J. Magn. Reson., 14, pp. 141-151 (1974). This reference contains extensive information on the concentration dependence of chemical shifts of ^{113}Cd containing compounds.
5. M. J. B. Ackerman and J. J. H. Ackerman, J. Phys. Chem., 84, 3151-3153 (1980).

Table 4. Cadmium-113 Chemical Shifts--Halogen Ligands

Compound	Chemical shift[a]	Conditions	Footnote reference
$CdCl_2$	97.84	1.0 M in D_2O	1
$CdCl_2$	130.8	5.0 M in H_2O	2
$CdBr_2$	108.6	1.0 M in D_2O	1
CdI_2	55.1	1.0 M in D_2O	1
$CdCl_2$	294.1	0.1 M in 12 M aq. HCl	3
$CdBr_2$	349.7	0.1 M in 9 M HBr	3
CdI_2	71.2	0.1 M in 5.5 M HI	3

Halide[b]	σCdX^- (ppm)	σCdX_2 (ppm)	σCdX_3^- (ppm)	σCdX_4^{-2} (ppm)
Chloride	89	114	292	495
Bromide	72	75	365	379
Iodide	47		140	71

[a]Chemical shifts are in ppm with respect to 0.1 M aq. $Cd(ClO_4)_2$. Positive shifts denote shielding.
[b]In aqueous solution. Estimated uncertainties in shifts are ±20 ppm. Information is taken from footnote reference 3 of this table.

1. A. D. Cardin, P. D. Ellis, J. D. Odom, and J. W. Howard, Jr., J. Am. Chem. Soc., 97, p. 1672 (1975).
2. R. J. Kostelnik and A. A. Bothner-by, J. Magn. Reson., 14, pp. 141-151 (1974).
3. J. J. H. Ackerman, T. V. Orr, V. J. Bartuska, and G. E. Maciel, J. Am. Chem. Soc., 101, p. 341 (1979).

Table 5a. Cadmium-113 Chemical Shifts of $Cd(CH_3)_2$ for Various Solvents[a]

Solvent	Chemical shift	Solvent	Chemical shift
Cyclohexane	34.67	Ethyl acetate	-33.25
Cyclopentane	33.08	Methyl formate	-36.29
Methylene chloride	3.13	Acetonitrile	-37.38
Toluene	2.03	Acetone	-40.94
Benzene	2.03	1,4-Dioxane	-50.77
(neat)	0.00	N,N-DMF	-55.18
Diethyl ether	-5.32	Diglycine	-65.93
Pyridine	-28.26	THF	-66.65

[a]Chemical shifts (ppm) are referenced to neat external $Cd(CH_3)_2$. σ(neat dimethylcadmium) = σ(0.1 M aq. cadmium perchlorate)$\mp$2 642.93 ppm. A negative shift indicates higher shielding. Data are taken from A. D. Cardin, P. D. Ellis, J. D. Odom, and J. W. Howard, Jr., J. Am. Chem. Soc., 97, p. 1672 (1975).

Table 5b. Cadmium-113 Chemical Shifts--Organometallic

Solvent	Chemical shift	Conditions	Footnote reference
$Cd(CH_3)_2$	642.9	Neat	1
$Cd(C_2H_5)_2$	543.2	Neat	1
$Cd(n-C_3H_7)_2$	504.3	Neat	1
$Cd(n-C_4H_9)_2$	489.1	Neat	2
$Cd(C_6H_5)_2$	328.8	1 M in 1,4-dioxane	1
$Cd(CH_3)(t-OC_4H_9)$	259.1	0.8 M in benzene; 24°	3
$Cd(CH_3)(n-OC_4H_9)$	299.0	0.8 M in benzene; 24°	3
$Cd(CH_3)(OC_2H_5)$	293	0.8 M in benzene; 24°	3
$Cd(CH_3)(OCH_3)$	323	0.8 M in benzene; 24°	3
$Cd(CH_3)(OC_6H_5)$	383	0.8 M in benzene; 24°	3

[a]Chemical shifts are in ppm with respect to 0.1 M aq. $Cd(ClO_4)_2$. Positive shifts denote deshielding.

1. A. D. Cardin , P. D. Ellis, J. D. Odom, and J. W. Howard, Jr., J. Am. Chem. Soc., 97, p. 1672 (1975).
2. G. E. Maciel and M. Borzo, J. C. S. Chem. Commun., 394 (1973).
3. J. D. Kennedy and W. McFarlane, J. Chem. Soc. Perkin 2, 1187 (1977).

Table 6. Cadmium-113 Chemical Shifts--Sulfur Ligands

Compound	Chemical shift[a]	Conditions	Footnote reference
$[Cd(SCH_2CH_2S)_2]^{-2}$	829	aq. solution; 308 K	1
$[Cd(S\text{-}C_6H_4\text{-}CH_3\text{-}S)_2]^{-2}$	796	aq. solution; 308 K	1
$[CH(SCH_3)_4]^{-2}$	663	aq. solution; 308 K	1
$[Cd(SC_6H_5)_n(SeC_6H_5)_{4-n}]^{-2}$			
n = 0	541	methanol solvent; 213 K	2
1	554		
2	566		
3	578		
4	590		
$[Cd(dpdtp)_2]_2$[b]	401	toluene solvent; 0.5 M $\Delta\nu_{1/2}$ = 5 Hz; 25°C	3
$Cd(Et_2dtc)_2$[c]	215	0.08 M in DMSO; 25°C $\Delta\nu_{1/2}$ = 45 Hz	3

[a] Chemical shifts are in ppm with respect to aq. 0.1 M $Cd(ClO_4)_2$. Positive shifts are to lower shielding.
[b] dpdtp is O,O'-diisopropyldithiophosphinato.
[c] Et_2dtc is N,N'-diethyldithiocarbamato.

1. G. K. Carson, P. A. W. Dean, M. J. Stillman, Inorg. Chim. Acta, 56, pp. 59-71 (1981).
2. G. K. Carson and P. A. W. Dean, Inorg. Chim. Acta, 66, pp. 37-39 (1982).
3. R. A. Haberkorn, L. Que, Jr., W. O. Gillum, R. H. Holm, C. S. Liu, and R. C. Lord, Inorg. Chem., 15, p. 2408 (1976).

sulfur ligation; the surrounding water is completely excluded. The thiophenol complex, $[Cd(SC_6H_5)_4]^{-2}$ observed by Carson et al. (24) produces the most shielded cadmium resonance (583 ppm) for sulfur ligation (i.e., $Cd-S_4$) and the bidentate $[Cd(SCH_2CH_2S)_2^{-2}]$ the most deshielded (829 ppm).

The remaining 60% of this chemical shift scale includes the mixed ligand complexes, Cd-(S,Se) and Cd-(S,O), the heteroligands causing higher shielding. The selenium ligands, studied by Carson and Dean (25) and Dean (27) for their ability to moderate the extreme toxicity of cadmium, shield cadmium slightly (~50 ppm) with respect to sulfur. Mixed (sulfur, oxygen) ligands are capable of several hundred ppm shielding relative to tetrahedral $Cd-S_4$ (12,26).

The best representative of pure cadmium-nitrogen ligation (Table 7) is the pyridine adduct of cadmium-tetraphenylporphyrin (Cd-TPP:py) in chloroform solvent (28), the cadmium resonance occurring at 436 ppm. Modification of the pyridine adduct induces small changes in the cadmium resonance frequency (29). Strong shielding is provided by the mixed nitrogen-oxygen ligands of aqueous cadmium-ethylenediaminetetraacetic acid (Cd-EDTA) (34) and cadmium-benzoyltrifluoroacetone (Cd-BTA$_2$) in pyridine (35). Weak shielding may be present in aq. $Cd(NH_3)_6Cl_2$ (15). Studies of polydentate amine ligands (33) reveal bridged compounds with octahedral symmetry about cadmium in the solid state. Solution ^1H NMR measurements in D_2O, CD_3OD, and DMSO-d_6 indicate geometries insensitive to temperature or solvent change, but explicit statements as to solution structure are not given (33). The mixed Cd-N, S, O complex, di-μ-chlorodichlorobis(6-mercaptopurine)diaquodicadmium(II) (Cd-CdM$_2$) was studied in DMSO-d_6 by ^{113}Cd NMR and in solid state by X-ray diffraction (36). Dimers are found in both solution and crystal with octahedral symmetry about cadmium. The 554 ppm cadmium shift in DMSO is deshielded from Cd-TPP:py, illustrating the deshielding ability of sulfur ligands relative to nitrogen and oxygen. An equilibrium mixture of isomers was inferred from the cadmium linewidth (ca. 20 Hz).

The final chemical shift scale in Figure 2 is of mixed-ligand cadmium-phosphine complexes (Table 8). The experiments were performed at low temperature in dichloromethane to slow the chemical exchange of phosphine and anions (37,38). Consistent with previous examples, adducts with oxygen ligands, e.g., NO_3^-, are most shielded (σ ~ 100 ppm). Shifts to higher shielding are also seen for the cadmium dimers, which typically form when only one phosphine per cadmium is present. These 1:1 adducts contain cadmium in distorted tetrahedral sites (37). The monomeric 1:2 adduct, $CdX_2(PR_3)_2$, is also tetrahedral about cadmium and represents the most deshielded resonance (38).

Table 7. Cadmium-113 Chemical Shifts--Nitrogen Ligands

Compound	Chemical shift[a]	Conditions	Footnote reference
Cd(CDM)$_2$[b]	554	aq. solution; $\Delta\nu_{1/2} \simeq 20$ Hz	1
Cd-TPP:pyridine[c]	426.8	chloroform solvent	2
Cd(NH$_3$)$_6$Cl$_2$	287.4	1.0 M aq. solution of NH$_3$	3
Cd-EDTA[d]	104	pH 13.5	4
	85	pH 3	4
Cd(BTA)$_2$[e]	41	0.5 M in pyridine: 297 K	5

[a]Chemical shift is in ppm with respect to 0.1 M aq. Cd(ClO$_4$)$_2$. Positive shift denotes deshielding.
[b]Cd(CDM)$_2$ is di-μ-chlorodichlorobis(6-mercaptopurine)diaquocadmium(II).
[c]TPP is tetraphenylporphyrin.
[d]EDTA is ethylenediaminetetraacetic acid.
[e]BTA is benzoyltrifluoroacetone. The pyridine supplies the nitrogen ligand.

1. E. A. H. Griffith and E. L. Amma, J.C.S. Chem. Commun., p. 1013 (1979).
2. J. H. Jakobsen, P. D. Ellis, R. R. Inners, and C. F. Jensen, J. Am. Chem. Soc., in press.
3. A. D. Cardin, P. D. Ellis, J. D. Odom, and J. W. Howard, Jr., J. Am. Chem. Soc., 97, p. 1672 (1975).
4. C. F. Jensen, S. Deshmukh, J. H. Jakobsen, R. R. Inners, and P. D. Ellis, J. Am. Chem. Soc., 103, p. 3659 (1981).
5. T. Maitani and K. T. Suzuki, Inorg. Nucl. Chem. Lett., 15, p. 213 (1979).

Table 8. Cadmium-113 Chemical Shifts--Phosphorus Ligands

Compound	Chemical shift[a]	Conditions	Footnote reference
$CdCl_2(PBu_3)_2$[c]	625	-85°C; CH_2Cl_2 solvent	1,2
$CdCl_2(PBu_3)_3$	530	-100°C; CH_2Cl_2 solvent	1
$[CdI_2(PBu_3)]_2$	230	-110°C; CH_2Cl_2 solvent	1
$CdI_2(PBu_3)_2$	560	-110°C; CH_2Cl_2 solvent	1
$CdI_2(PBu_3)_3$	495	-110°C; CH_2Cl_2 solvent	1
$[Cd(NO_3)_2(PBu_3)]_2$	80	-50°C; CH_2Cl_2 solvent	1
$[Cd(NO_3)(PBu_3)_3]^+$	375	-70°C; CH_2Cl_2 solvent	1

[a] Chemical shift is in ppm with respect to 0.1 M aq. $Cd(ClO_4)_2$. Positive shift denotes deshielding.
[b] These references contain extensive information on chemical shifts and coupling constants of adducts of other cadmium salts and phosphines. The data presented in this table are representative of chemical shifts of the adducts investigated.
[c] PBu_3 is tributylphosphine.

1. D. Dakternieks, Austral. J. Chem., 35, pp. 469-481 (1982).
2. R. Colton and D. Dakternieks, Austral. J. Chem., 33, pp. 1677-1684 (1980).

In summary, these investigations of cadmium inorganic and organometallic compounds indicate that at least three important factors may influence cadmium chemical shifts. Two have been emphasized in the previous discussion; ligand type may be responsible for shifts of several hundred ppm and ligand geometry about cadmium can be a source of somewhat smaller shifts (100-200 ppm). The third factor, that of chemical exchange, will be discussed later.

These conclusions have been employed in a number of studies of metalloproteins. In these studies, the large chemical shift range of cadmium has been a useful tool in obtaining information concerning calcium and zinc binding sites. Figure 3 and Table 9 summarize the known chemical shifts of these cadmium-protein systems.

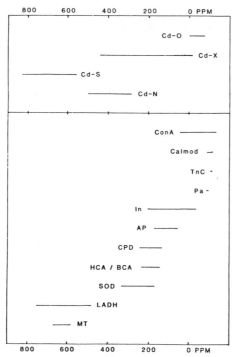

Figure 3. Chemical shift ranges of cadmium metalloproteins are compared here with chemical shifts determined for cadmium complexes. The shift ranges for the complexes are for pure ligand type, taken from Figure 2. Abbreviations used here include
Con A   Conconavalin A
Pa   Paravalbumin
TnC   Troponin C
Calmod   Calmodulin
In   Insulin
CPD   Carboxypeptidase A
HCA   Human Carbonic Anhydrase
BCA   Bovine Carbonic Anhydrase
AP   Alkaline Phosphatase
SOD   Superoxide Dimutase
LADH   Liver Alcohol Dehydrogenase
MT   Metallothionein

Table 9. Cadmium-113 Chemical Shifts—Metalloproteins

Metalloprotein	Approx. coordination	Ligands	pH sensitive	Counterion sensitive	^{113}Cd chemical shift	Footnote reference
Concanavalin A	S2:6-coordinate (?)	6 oxygen	no	no	−125 to −133	1,2
	S1:6-coordinate	5 oxygen; 1 nitrogen	no	no	43	1,2
Parvalbumin	6-coordinate (?)	6 oxygen			−90 to −100	3
Troponin C	6,7,8 coordinate (?)	oxygen	yes	no	−111.0	4
	6,7,8 coordinate (?)	oxygen	yes	no	−107.5	4
Calmodulin	6,7,8 coordinate (?)	oxygen		no	−115.0	5
	6,7,8 coordinate (?)	oxygen		no	−88.5	5
Insulin	6-coordinate	6 oxygen	no	no	−36	6
	6-coordinate	3 oxygen, 3 nitrogen	yes	yes	165−201[a]	6
Carboxy-peptidase A	4-coordinate	2 nitrogen, 1 oxygen, substrate donor atom (usually oxygen)	yes	yes	217,[b] 240[c]	7
					132[d]	8

Table 9 (continued)

Metalloprotein	Approx. coordination	Ligands	pH sensitive	Counterion sensitive	^{113}Cd chemical shift	Footnote reference
Human Carbonic Anhydrase B	4- or 5-coordinate[e]	3 nitrogens, substrate donor atom (usually oxygen)	yes	yes	228[f], 410[g], 355[h] 223-165[i] 145	9,10 11 8
Human Carbonic Anhydrase C	4- or 5-coordinate[e]	3 nitrogens, substrate donor atom (usually oxygen)	yes	yes	231.7-216.3[j] 225	11 8
Bovine Carbonic Anhydrase B	4- or 5-coordinate[e]	3 nitrogens, substrate donor atom (usually oxygen)	yes	yes	220-225 214	12 8
Alkaline Phosphatase	4-coordinate	3 nitrogens, substrate donor atom usually oxygen	yes	yes	55-170	8,9,13
Superoxide Dismutase	4-coordinate	3 nitrogens, 1 oxygen	no	no	310-330 170.2	14 8

Table 9 (continued)

Metalloprotein	Approx. coordination	Ligands	pH sensitive	Counterion sensitive	113Cd chemical shift	Footnote reference
Liver Alcohol Dehydrogenase	4-coordinate 4-coordinate	4 sulfur 2 sulfur, 1 nitrogen, 1 oxygen	no yes (pH > 10)		748-752 484	15 15
Metallothionein (MT)	possibly 4-coordinate	sulfur				
Calf liver MT1					612.3-668.2[k]; 590.4-606.2[l]	16
Calf liver MT2					613.1-668.0[k]; 603.4[l]	16
Rabbit liver MT1,2					611.2-670.3[k]; 643.5-665.1[l]	17,18
Rat liver					581-670	19
Crab MT1,2					631.0-660.6	20

[a]pH varies 8.0–10.4.
[b]In the presence of dℓ-benzylsuccinate inhibitor; pH 8.0; Cl⁻ buffer.

[c] In the presence of dℓ-benzylsuccinate inhibitor; pH = 8.0; ClO$_4^-$ buffer. [d] In the presence of β-phenylpropionate inhibitor; no resonance in absence of inhibitor; pH = 6.5; Tris-HCl buffer. [e] P. H. Haffner and J. E. Coleman, J. Biol. Chem., 248, p. 6626 (1973). [f] Without inhibitor; pH = 9.7. [g] With 1 equiv. KCN inhibitor. [h] Benzenesulfonamide inhibitor. [i] pH = 8.3; [HCO$_3^-$] = 0.0 to 0.8 M. [j] pH = 8.1 [HCO$_3^-$] = 0.0 to 0.4 M. [k] Cluster A. [l] Cluster B.

1. A. R. Palmer, D. B. Bailey, W. D. Behnke, A. D. Cardin, P. P. Yang, and P. D. Ellis, Biochem., 19, pp. 5063-5070 (1980).
2. D. B. Bailey, P. D. Ellis, A. D. Cardin, and W. D. Behnke, J. Am. Chem. Soc., 100, p. 5236 (1978).
3. T. Drakenberg, B. Lindman, A. Cavé, and J. Parello, FEBS Lett., 92, p. 346 (1978).
4. S. Forsén, E. Thulin, and H. Lilja, FEBS Lett., 104, pp. 123-126 (1979).
5. S. Forsén, E. Thulin, T. Drakenberg, J. Krebs, and K. Seamon, FEBS Lett., 117, pp. 189-194 (1980).
6. J. L. Sudmeier, S. J. Bell, M. C. Storm, and M. F. Dunn, Science, 212, pp. 560-562 (1981).
7. D. B. Bailey and P. D. Ellis, unpublished results.
8. I. M. Armitage, A. J. M. Schoot-Uiterkamp, J. F. Chlebowski, and J. E. Coleman, J. Magn. Reson., 29, pp. 375-392 (1978).
9. J. L. Sudmeier and S. J. Bell, J. Am. Chem. Soc., 99, pp. 4499-4500 (1977).
10. J. L. Evelhoch, D. F. Bocian, and J. L. Sudmeier, Biochem., 20, pp. 4951-4954 (1981).
11. N. B. H. Jonsson, L. E. A. Tibell, J. L. Evelhoch, S. J. Bell, and J. L. Sudmeier, Proc. Natl. Acad. Sci. (USA), 77, pp. 3269-3272 (1980).
12. A. J. M. Schoot-Uiterkamp, I. M. Armitage, and J. E. Coleman, J. Biol. Chem., 255, pp. 3911-3917 (1980).
13. I. M. Armitage, R. T. Pajer, A. J. M. Schoot-Uiterkamp, J. F. Chlebowski, and J. E. Coleman, J. Am. Chem. Soc., 98, pp. 5710-5711 (1976).
14. D. B. Bailey, P. D. Ellis, and J. A. Fees, Biochem., 19, pp. 591-596 (1980).
15. B. R. Bobsein and R. J. Myers, J. Biol. Chem., 256, pp. 5313-5316 (1981).
16. R. W. Briggs and I. M. Armitage, J. Biol. Chem., 257, pp. 1259-1262 (1982).
17. J. D. Otvos and I. M. Armitage, J. Am. Chem Soc., 101, pp. 7734-7736 (1979).
18. K. T. Suzuki and T. Maitani, Experientia, 34, pp. 1449-1450 (1978).
19. P. J. Sadler, A. Bakka, and P. J. Beynon, FEBS Lett., 94, pp. 315-318 (1978).
20. J. D. Otvos, R. W. Olafson, and I. M. Armitage, J. Biol. Chem., 257, pp. 2427-2431 (1982).

The chemical shifts in Figure 3 are qualitatively consistent with the general trends found in nonbiological systems. The binding sites containing sulfur ligands, e.g., metallothionein (47) and liver alcohol dehydrogenase (45f), are most deshielded (>600 ppm) while those sites with only oxygen ligands, concanavalin A (42a) and parvalbumin (42b, 43b), are most shielded (<-100 ppm). The greater shielding (~100 ppm) of metalloprotein "oxygen ligand" environments over that of inorganic "oxygen ligand" environments is quantitatively inconsistent, however. The greater rigidity of the metalloprotein binding sites may be responsible. The solid state cadmium NMR currently being used to explore this possibility is discussed in a later section of this review.

A second difficulty in interpreting the cadmium NMR data of metalloproteins is whether the cadmium environment being measured is similar to the same site when the biologically appropriate metal binds. The partial and sometimes enhanced activity of many cadmium-substituted proteins (40) argues strongly for high similarity in the environments of the metals.

Chemical Exchange

Lability of ligands in cadmium complexes is known to affect NMR spectral parameters. The exchange contributes to cadmium lineshapes and peak splittings in phosphine adducts (37-39) and thiolatocadmates (24-27), to changes in observed cadmium-carbon scalar coupling, carbon linewidths, and cadmium chemical shifts in Cd-EDTA complexes (34), to enzymatic activity of metalloproteins (44), and to shifts in or disappearance of cadmium resonances in metalloproteins, e.g., carbonic anhydrase (45).

The magnitude of scalar coupling constants provides valuable information concerning rates of exchange and the existence of inner- or outer-sphere coordination of ligands. For this reason, representative cadmium scalar coupling constants have been collected in Tables 10-14. The footnote references of these Tables contain additional values.

In Table 15, the rates of methyl group exchange in various systems are presented. Self-exchange lifetimes in various solvents are typically on the order of tenths of a second. Exchange between methyl derivatives of Group II and Group III metals is generally slightly faster (see Table 15 footnote references).

Relaxation Rates

Dynamic NMR parameters such as spin-lattice relaxation rates and the nuclear Overhauser enhancement (NOE) have also been used as sources of information. Cardin et al. (15) have measured

Table 10. Scalar Coupling Constants[a] ^{113}Cd-^{1}H

Compound	^2J(Hz)	^3J(Hz)	Solvent, temp.	Footnote
Cd(CH$_3$)$_2$	51.0 ± 0.5		0.8 M in benzene, 24°C	1
	52.0		neat	2
	52		neat	4
Cd(C$_2$H$_5$)$_2$	51.6		neat	2
Cd(n-C$_3$H$_7$)$_2$	53.5		neat	2
(CH$_3$)Cd(OCH$_3$)	86.3 ± 1	9 ± 1	saturated in benzene, 78°C	1
	85	9	pyridine solvent	3
(CH$_3$)Cd(OC$_2$H$_5$)	83.2 ± 0.5	7.8 ± 0.5	0.8 M in benzene, 24°C	1
	84	7	benzene, 0°C	3
(CH$_3$)Cd(O-n-C$_3$H$_7$)	83 ± 2	b	0.8 M in benzene, 24°C	1
(CH$_3$)Cd(O-i-C$_3$H$_7$)	82.4 ± 0.5	6.4 ± 0.3	0.8 M in benzene, 24°C	1
	82	6	benzene	3
(CH$_3$)Cd(O-n-C$_4$H$_9$)	84.0 ± 0.5	7.8 ± 0.5	0.8 M in benzene, 24°C	1
(CH$_3$)Cd(O-i-C$_4$H$_9$)	84.6 ± 0.5	7.8 ± 0.3	saturated in benzene, 24°C	1
(CH$_3$)Cd(O-s-C$_4$H$_9$)	80 ± 1	7 ± 2	0.8 M in benzene, 24°C	1
(CH$_3$)Cd(O-t-C$_4$H$_9$)	80.8 ± 0.5		0.8 M in benzene, 24°C	1
(CH$_3$)Cd(O-neo-C$_5$H$_{11}$)	84.9 ± 0.5	8.3 ± 0.3	0.8 M in benzene, 24°C	1
	94		benzene, -20°C	3
(CH$_3$)Cd[OCH$_2$(C$_6$H$_5$)]	86.0 ± 0.5	7.8 ± 0.3	0.8 M in benzene, 24°C	1
(CH$_3$)Cd[OCH(C$_6$H$_5$)$_2$]	88.8 ± 0.5		0.8 M in benzene, 24°C	1
(CH$_3$)Cd[OC(C$_6$H$_5$)$_3$]	90.5 ± 1		0.8 M in benzene, 24°C	1
[(CH$_3$)Cd(OCH$_2$)]$_2$CH$_2$	84 ± 4	b	0.8 M in benzene, 24°C	1
(CH$_3$)CdBr	80		neat	4
Cd-TPP(py)[c]	[^4J(^{113}Cd-Hβ) = -5.0 ± 0.2]		0.023 M in CdCl$_3$	5

Table 10 (continued)

aUnless explicitly stated by sign, only the magnitude of the coupling constant is given.
bNot well resolved.
cTPP(py) is the pyridine adduct of tetraphenylporphyrin.

1. J. D. Kennedy and W. McFarlane, J. Chem. Soc. Perkin 2, p. 1187 (1977).
2. A. D. Cardin, P. D. Ellis, J. D. Odom, and J. W. Howared, Jr., J. Am. Chem. Soc., 97, pp. 1672-1679 (1975).
3. E. A. Jeffery and T. Mole, Austral. J. Chem., 21, pp. 1187-1196 (1968).
4. W. Bremser, M. Winokur, and J. D. Roberts, J. Am. Chem. Soc., 92, p. 1080 (1970).
5. H. J. Jakobsen, P. D. Ellis, R. R. Inners, and C. F. Jensen, J. Am. Chem. Soc., in press.

Table 11. Scalar Coupling Constants ^{113}Cd–^{13}C

Compound	^1J(Hz)	^2J(Hz)	^3J(Hz)	Condition	Footnote
Cd(CH$_3$)$_2$	-537.3			neat	1
Cd(C$_2$H$_5$)$_2$	-498.0^a	19.0		neat	2
Cd(n-C$_3$H$_7$)$_2$	-509.2^a		44.8	neat	2
Cd-EDTA		13.0 ± 0.1		aq. Cd(NO$_3$)$_2$ with 0.4 M EDTA	3
[(t-C$_4$H$_9$)$_3$Si]$_2$Cd		31.0	11.4	50% in C$_6$D$_6$	4
(CH$_3$)$_2$P–CH$_2$\\N–P(CH$_3$)$_2$ / Cd / (CH$_3$)$_2$P–CH$_2$\\N–P(CH$_3$)$_2$	271.0			37°C benzene solvent	5
Cd-[^{14}N$_4$]-TPP:pyridine		$\|^2J(^{113}\text{Cd}-^{13}\text{C}\alpha)\| = 2.6$ Hz		30°C; 0.023 M Cd-TPP in CDCl$_3$ 10:1 ligand; porphyrin ratio	3
Cd-[^{15}N$_4$]-TPP:pyridine		$^2J(^{113}\text{Cd}-^{13}\text{C}\gamma) = -10.7$ Hz	$^3J(^{113}\text{Cd}-^{13}\text{C}\beta) = -12.2$ Hz		6

Table 11 (continued)

aThe sign of this coupling constant is assumed negative based on the sign determination in Ref. 1 of this Table.
bEDTA is ethylenediaminetetracetic acid.
cTPP is tetraphenylporphyrin.

1. H. Dreeskamp and K. Hildenbrand, Z. Naturforsch, Teil A., 23, p. 940 (1968).
2. A. D. Cardin, P. D. Ellis, J. D. Odom, and J. W. Howard, Jr., J. Am. Chem. Soc., 97, p. 1672 (1975).
3. R. Hagan, J. P. Warren, D. H. Hunter, and J. D. Roberts, J. Am. Chem. Soc., 95, pp. 5712-5716 (1973).
4. H. Müller and L. Rösch, J. Organometal. Chem., 133, pp. 1-6 (1977).
5. H. Schmidbaur, H. J. Füller, V. Bejenke, A. Franck, and G. Huttner, Chem. Ber., 110, pp. 3536-3543 (1977).
6. H. J. Jakobsen, P. D. Ellis, R. R. Inners, and C. F. Jensen, J. Am. Chem. Soc., in press.

Table 12. Scalar Coupling Constants ^{113}Cd–^{15}N

Complex	^1J(Hz)	Condition	Footnote
Cd–EDTA[a]	83 ± 1	aq. 0.4 M EDTA with Cd(NO$_3$)$_2$	1
^{113}Cd–[^{15}N]–TPP[b]: (4-cyanopyridine)	147.6	8–10 mg Cd-TPP/0.5 ml CDCl$_3$; >10:1 ligand: porphyrin ratio	2
^{113}Cd–[^{15}N]–TPP: (3-chloropyridine)	146.4	same	2
^{113}Cd–[^{15}N]–TPP: pyridine	142.5	same	2
^{113}Cd–[^{15}N]–TPP: (4-methylpyridine)	141.0	same	2
^{113}Cd–[^{15}N]–TPP: (4-aminopyridine)	137.4	same	2
^{113}Cd–[^{15}N]–TPP: pyridine	+150.1 ± 0.1	39°C; 0.023 M Cd-TPP in CDCl$_3$; 10:1 ligand: porphyrin ratio	3
^{113}Cd–HCAB[c]:BSA[d]	210 ± 3		4
^{113}Cd–HCAC[e]:BSA	190 ± 3		4
^{113}Cd–BCA[f]:BSA	190 ± 3		4
^{113}Cd–BSA: Neoprontosil[g]	190 ± 3	3.7 mM BCA; 1 equiv. ligand; pH 8.9	4

[a]EDTA is ethylenediaminetetraacetic acid.
[b]TPP is tetraphenylporphyrin.
[c]HCAB is human carbonic anhydrase B.
[d]BSA is benzenesulfonamide.
[e]HCAC is human carbonic anhydrase C.
[f]BCA is bovine carbonic anhydrase.

[g]Neoprontosil is

H_3CCNH ... OH ... $N=N$ ... SO_2NH_2
^-O_3S ... SO_3^-

1. R. Hagen, J. P. Warren, D. H. Hunter, and J. D. Roberts, J. Am. Chem. Soc., 95, pp. 5712-5716 (1973).
2. D. D. Dominguez, M. M. King, and J. H. C. Yeh, J. Magn. Reson., 32, pp. 161-165 (1978).
3. H. J. Jakobsen, P. D. Ellis, R. R. Inners, and C. F. Jensen, J. Am. Chem. Soc., unpublished.
4. J. L. Evelhoch, D. F. Bocian, and J. L. Sudmeier, Biochem., 20, pp. 4951-4954 (1981).

Table 13. Scalar Coupling Constants $^{113}Cd-^{31}P$

Compound	^1J (Hz)[a]	Conditions	Footnote reference[b]
$CdCl_2(PBu_3)$	1690	-30°C; CH_2Cl_2 solvent	1
$CdCl_2(PBu_3)_2$	1635	-85°C; CH_2Cl_2 solvent	1
$CdCl_2(PBu_3)_3$	1600	-100°C; CH_2Cl_2 solvent	1
$CdI_2(PBu_3)$	1770	-110°C; CH_2Cl_2 solvent	1
$CdI_2(PBu_3)_2$	1350	-110°C; CH_2Cl_2 solvent	1
$CdI_2(PBu_3)_3$	1525	-110°C; CH_2Cl_2 solvent	1
$[Cd(NO_3)_2(PBu_3)]_2$	2840	-50°C; CH_2Cl_2 solvent	1
$CdCl_2[P(C_6H_{11})_3]$	2300	30°C; CH_2Cl_2 solvent	1
$CdBr_2(n-BuPSe)_2$	^2J = 40 Hz	-72°C; CH_2Cl_2 solvent	2

[a] Three coupling constants were observed with both ^{113}Cd and ^{31}P NMR.
[b] The chemical shifts and coupling constants of Ref. 1 include those for various cadmium salts and phosphine groups. The values reported here are representative.

1. D. Dakternieks, Austral. J. Chem., 35, pp. 469-481 (1982).
2. S. O. Grim, E. D. Walton, and L. C. Satek, Can. J. Chem., 58, pp. 1476-1479 (1980).

Table 14. Scalar Coupling Constants to Other Nuclei

Compound	J (Hz)[a]	Conditions	Footnote reference[b]
$[Cd(SPh)_n(SePh)_{4-n}]^{-2}$	$^1J(^{113}Cd-^{77}Se)$		
n = 0	126 ± 3	213 K in 4:1 $CH_3OH:CD_3OD$	1
1	99 ± 3		1
2	72 ± 3		1
3	46 ± 4		1
4	–		1
$Cd(ClO_4)_2:D_2^{17}O$	$^1J(^{113}Cd-^{17}O)$		
	248[a]	25°C; 6.7 mol% $Cd(ClO_4)_2$ in $D_2^{17}O$	2

[a] Calculated from scalar coupling contribution to spin-lattice relaxation.

1. G. K. Carson and P. A. W. Dean, Inorg. Chim. Acta, 66, pp. 37-39 (1982).
2. M. Holz, R. B. Jordan, and M. D. Zeidler, J. Magn. Reson., 22, pp. 47-52 (1976).

relaxation rates of 0.16-0.012 sec^{-1} (T_1 = 6.5-84.4 sec) for aqueous solutions of simple inorganic salts in H_2O and D_2O. Cadmium-proton dipolar relaxation was determined to be 48% of that observed (R_1 = 0.032 sec^{-1}; T_1 = 30.7 sec) in 1.0 M aqueous perchlorate salt, the chemical species being $[Cd(H_2O)_6]^{+2}$. Dipolar relaxation contributes 20% to observed rates for 1.0 M aq. $CdCl_2$ and negligibly to aq. $CdBr_2$ and CdI_2 rates. Other contributing relaxation mechanisms are thought to be spin-rotation, scalar coupling, and chemical shielding anisotropy. Maciel and Borzo (13) have measured T_1 = 16 sec for saturated aqueous $CdCl_2$.

Holz et al. (57) have similarly determined dipolar relaxation to dominate in aqueous perchlorate solution over a 100°C temperature range. At 25°C, a rotational correlation time of

Table 15. Ligand Exchange for Dimethylcadmium

Part A. Concentration Dependence of Self-Exchange in Diethyl ether (+33.3°C)[a]

[Cd(CH$_3$)$_2$] Molar	Inverse lifetime 1/τ (Cd-CH$_3$) sec^{-1}	Rate constant k_1, liter mol^{-1} sec^{-1}
0.681	2.20	51.2
1.07	3.46	51.3
1.36	4.40	52.9
1.90	5.34	43.8
		average = 51 ± 2

Part B. Ligand Exchange of Cd(CH$_3$) in Various Chemical Systems

Reaction	Solvent	E_a, kcal mol^{-1}	ΔS, eu	k_1, liter mol^{-1} sec^{-1}
self-exchange	O(C$_2$H$_5$)$_2$[a]	6.9 ± 0.2		
self-exchange	N(C$_2$H$_5$)$_3$[b]	5.6 ± 0.2	-38	38.5
self-exchange	THF[b]	6.8 ± 0.1	-28.3 ± 0.4	245
self-exchange	pyridine[c]	13		
self-exchange	toluene[c]	16		
self-exchange	neat[b]	15.6 ± 0.1	-9.8	0.4
Zn(CH$_3$)$_2$-Cd(CH$_3$)$_2$	methylcyclohexane[d]	17.0 ± 1	-3	1.7
In(CH$_3$)$_3$-Cd(CH$_3$)$_2$	dichloromethane[d]			21.8 ± 3.8

[a] J. Soulati, K. L. Henold, and J. P. Oliver, J. Am. Chem. Soc., 93, pp. 5694-5698 (1971).
[b] W. Bremser, M. Winokur, and J. R. Roberts, J. Am. Chem. Soc., 92, p. 1080 (1970).
[c] E. A. Jeffery and T. Mole, Austral. J. Chem., 21, p. 1187 (1968).
[d] K. Henold, J. Soulati, and J. P. Oliver, J. Am. Chem. Soc., 91, 3171-3174 (1969).

3.9×10^{-11} sec was calculated for the spherically hydrated cadmium ion. In isotropically enriched water, $D_2^{17}O$, scalar coupling strongly dominates relaxation. A coupling constant of 248 Hz is calculated.

Spin-lattice relaxation rates for cadmium in organometallic compounds have been measured for neat dibutylcadmium (13) ($T_1 \simeq 0.2$ sec; $R_1 \simeq 5$ sec^{-1}; Temp = 228 K; B_0 = 2.1 Tesla) and dimethyl-, diethyl-, and dipropylcadmium (15) (T_1 = 3.5-0.6 sec; B_0 = 2.35 Tesla). From NOE measurement, neat dimethylcadmium experiences no dipolar relaxation (15). Mechanisms contributing to relaxation are thought to be chemical exchange and chemical shielding anisotropy; however, their relative importance has not been published for these compounds.

Cadmium relaxation in Cd-EDTA complexes have been measured at 2.3, 4.7, and 9.4 Tesla (34). The dipolar contribution dominates at low field strength and chemical shielding anisotropy at high fields. A third, frequency independent, contribution is also measured. The data are consistent with a 1.1×10^{-10} sec correlation time for the low pH form. Relaxation rates and NOE of Cd-cyclohexanediaminetetraacetate at 2.3 Tesla are also dominated by the dipolar mechanism (32).

Relaxation of the cadmium-tetraphenylporphyrin:pyridine adduct studied by Jakobsen et al. (28) is strongly dominated by the chemical shielding anisotropy mechanism. Even at the moderate field strength of 4.7 Tesla (^{113}Cd 44 MHz), this mechanism is responsible, within experimental error of 10%, for all observed relaxation (T_1 = 28.5 sec).

Recent improvements have increased the sensitivity of NMR instrumentation, making relaxation rate experiments of cadmium-metalloprotein systems feasible if not commonplace. Bailey et al. (41) have measured a 1.2 sec (±20%) relaxation time for cadmium occupying the zinc binding sites in bovine superoxide dismutase. Theoretical considerations indicate nearly equal contributions to the relaxation rate from proton-cadmium dipolar and from chemical shielding anisotropy mechanisms.

In summary, several relaxation mechanisms are found to contribute to cadmium spin-lattice relaxation. In aqueous solutions of cadmium salts, cadmium-proton dipolar relaxation, and possibly spin-rotation at high temperatures, will dominate. Chemical shielding anisotropy becomes an important mechanism when working with larger molecules undergoing slow, anisotropic tumbling, particularly at high magnetic field strengths. For these large molecules, such as porphyrins and metalloproteins, chemical shielding anisotropy and dipolar mechanisms are important and spin-rotation unimportant.

INTERPRETATION OF METAL NUCLIDE CHEMICAL SHIFTS IN THE PRESENCE OF CHEMICAL DYNAMICS--A STATEMENT OF THE PROBLEM AND SOME POTENTIAL SOLUTIONS

From the discussions within the previous section it should be clear that ^{113}Cd chemical shifts are sensitive to subtle structural changes. This sensitivity is not limited to cadmium, but is representative of most of the spin 1/2 metal nuclides. However, associated with this chemical shift sensitivity is the potential caveat associated with the time scale for ligand dynamics. Such a situation can bring about intermediate or rapid exchange between the various ligand environments, and the observed spectrum becomes the result of an average over all the ligand sites. In analyzing such a spectrum one is forced to apply only the simplest of models to attempt to explain the data, the validity of which often rests upon data from other experimental techniques. To this extent, one has lost the ability to obtain from the NMR experiment new and independent information about the nuclide that could complement other techniques.

One of the experimental facts of life that plagues the practitioner of spin 1/2 metal NMR spectroscopy is the disappearance of NMR signals. The origins of this larceny, as alluded to above, often arise from unforeseen dynamics within the experimental system or simply the appearance of an insoluble mass at the bottom of a NMR tube. Until recently, if the experimental system was a biological one, the solution to these experimental problems was simple but frustrating--find another system! The reasons for such a decision are simple: if the origin of the problem lies in dynamics, then the biological system resists low temperature experiments because of the necessity of the aqueous solution. One of the objectives of the present discussion is to point out that for the most part the preceding reasoning is no longer valid. The problem concerning unfortunate dynamic time scales has been solved in an elegant and, in a way, surprisingly simple procedure by the Ackermans (58)!

They employed the method of Rasmussen and MacKenzie (59), based on aqueous emulsions first suggested by Turnbull (50), for supercooling aqueous solutions. By dividing a macroscopic sample into micron size drops through emulsification, it is possible to obtain the NMR spectra of these aqueous emulsions at temperatures as low as $-50°C$. Employing this methodology, the Ackermans (58) obtained the ^{113}Cd NMR spectrum of 0.1 M $Cd(ClO_4)_2$ in the presence of various amounts of glycine. By varying the concentration of the glycine and the solution pH, they were able to observe separate resonances for what appeared to be $CdGly^+$, $CdGly_2$, $CdGly_3^-$, and $CdGly^{-2}$. The tentative assignment of these resonances was confirmed by Jakobsen and Ellis (61). Using the procedure as described by the Ackermans and with ^{15}N-enriched glycine, Jakobsen and Ellis observed the cadmium-nitrogen coupling pattern expected for the proposed complexes, Figure 4. This figure depicts the ^{113}Cd NMR spectrum of a mixture of 0.25 M glycine and 0.1 M Cd(ClO$_4$)$_2$ at -40°C. At ambient temperature this solution (or one of normal composition) would have yielded, at best, a broad featureless line. Yet, we see three relatively narrow resonances representing CdGly$_2$, CdGly^{+1}, and free Cd^{+2}. Clearly, this approach presents several exciting possibilities to the experimentalists. A summary of these data are presented in Table 16. The general applicability of these methods is unclear at the present time. However, the technique is certainly promising and deserves further investigation.

Table 16. ^{113}Cd Chemical Shifts[a] and ^{113}Cd-^{15}N Coupling Constants[b] for Cd(II) ^{15}N-Gly$_n$$^{(2-n)+}$ Complexes Supercooled to -40°C

	[Cd(II)][c] M	[^{15}N-Gly] M	pH	δ(^{113}Cd) ppm	^{1}J(^{113}Cd-^{15}N) Hz
Cd(II) "Free"	0.10	0.25	7.0	-35.9	
Cd(II)Gly$^+$	0.10	0.25	7.0	53.6	170
Cd(II)Gly$_2$	0.10	0.25	7.0	153.9	165
Cd(II)Gly$_3^-$	0.10	1.00	8.0	262.8	>140[d]

[a]^{113}Cd chemical shifts (±1 ppm) are referenced to 0.1 M Cd(ClO$_4$)$_2$.
[b]In Hz with an estimated error of ± 5 Hz.
[c]Total concentration of Cd(II) using Cd(ClO$_4$)$_2$.
[d]Minimum value of the coupling constant extracted from the two inner lines of the exchange-broadened quartet.

Recognizing the presence of ligand exchange processes in a particular system is not always intuitively obvious. For example, in either (CH$_3$)$_2$Cd or (C$_2$H$_5$)$_2$Cd their respective ^{113}Cd NMR spectra yield sharp resonances with well defined proton-cadmium and carbon-cadmium coupling constants (15). It is not until you mix these two compounds and obtain the ^{113}Cd NMR spectrum of the mixture that the ligand exchange processes become obvious: one observes three resonances, from (CH$_3$)$_2$Cd, (CH$_3$)Cd(C$_2$H$_5$), and (C$_2$H$_5$)$_2$Cd (15).

In the preceding organometallic system, the ligand exchange in this system became a novel example of how one could study such processes. However, in the Cd-EDTA system such ligand exchange processes represented a serious pitfall, particularly if they went unnoticed (34). In this case, we studied the pH dependence

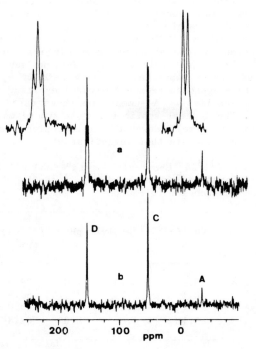

Figure 4. Natural abundance ^{113}Cd NMR spectra of -40°C supercooled aqueous (D$_2$O/H$_2$O, 1:1) solutions of 0.1 M Cd(ClO$_4$)$_2$ and 0.25 M glycine in 5 M NaNO$_3$ at pH 7.0. (a) Obtained using 95% ^{15}N-enriched glycine and 2300 transients; expansions of the doublet and triplet for the resonances C and D are inserted. (b) Obtained using isotopically normal glycine and 1000 transients. Gated ^{1}H decoupled and a line broadening of 20 Hz have been applied for both spectra.

of the ^{113}Cd T$_1$ in Cd-EDTA. At pH values in excess of 11, there was a precipitous decrease in the value of the ^{113}Cd T$_1$. After an extensive analysis of the field dependence of the ^{113}Cd T$_1$ it was concluded that this sudden drop in T$_1$ resulted from a rapid chemical exchange between Cd-EDTA and, in all probability, a hydroxylated form of Cd-EDTA. From the relaxation data it was concluded that this latter species had a much shorter T$_1$ and hence the measured T$_1$ was reflecting the exchange dynamics to the relaxation sink and not the "true" spin-lattice relaxation properties of the system of interest. Hence, one could invent some novel relaxation mechanisms to explain the data when the relaxation data were actually reflecting the chemical dynamics of the system.

In this same paper we developed some simple equations to demonstrate the significant contribution that exchange processes could make to the measured T_1. In the present discussion, we would like to illustrate these relations with a concrete example of the cadmium exchange that occurs in a biochemical system, specifically, cadmium-substituted Concanavlin-A (Con A) (42). Our interest in this system was stimulated, in part, by the need to clarify the importance that spin-lattice relaxation plays in determining the high field sensitivity of ^{113}Cd NMR spectroscopy in bioinorganic chemistry. During initial studies conducted at 2.3 T and 4.7 T, we noted that the resonance corresponding to free cadmium had an anomalously short value for T_1 (42a). We suspected that chemical exchange processes could be important in this system. We will demonstrate how one can separate exchange and relaxation processes in this special exmaple of a three-site exchange system. The method we have employed is a <u>double</u> saturation transfer experiment coupled with a measurement of T_1. Within the context of this system, we will examine the field dependence of the ^{113}Cd T_1 for each site in Con A and make some general comments about ^{113}Cd relaxation in cadmium-substituted proteins.

The direct effects of spin exchange resulting in magnetization transfer between exchange coupled sites have been analyzed in detail (2). The particular exchange network of interest to us here is depicted in the scheme below. This network is a special case of the generalized three site exchange system. In the general case, site C would be allowed to couple directly with site A. For concreteness, in this diagram sites A, B, and C denote the conventional transition metal binding site (S1), free cadmium, and the calcium-binding site (S2) in cadmium-substituted Con A, respectively. The symbols W_{1A}, W_{1B}, and W_{1C} are the total probabilities (by all NMR relaxation mechanisms) for spin transitions in sites A, B, and C, respectively. Further, W_{OAB} is the probability for spin transfer from site A to B. Finally, P_A^α is the ^{113}Cd population of Site A for the spin state α.

The exchange network depicted in the scheme is a special case, but it is one of general interest in chemistry. Most ligand exchange processes in inorganic, organometallic, and bioinorganic chemistry can be schematically represented by such a network. Hence, an experimental procedure that would allow one to extract the four kinetic parameters and three relaxation rates would be significant.

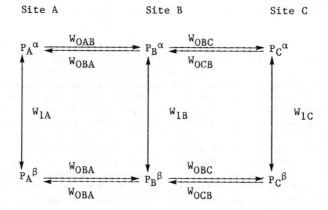

By invoking the master equations for populations (63), the preceding exchange network can be transformed into coupled expressions for the time evolution of the magnetization at sites A, B, and C (eqs. 1-3). Here, $\langle I_{AZ} \rangle$, $\langle I_{ZB} \rangle$, and $\langle I_{ZC} \rangle$ are the

$$\langle \dot{I}_{ZA} - I_{ZA}°\rangle = -(R_{1A} + W_{OAB})[\langle I_{ZA}\rangle - I_{ZA}°] + W_{OBA}[\langle I_{ZB}\rangle - I_{ZB}°] \tag{1}$$

$$\langle \dot{I}_{ZB} - I_{ZB}°\rangle = -(R_{1B} + W_{OBA} + W_{OBC})[\langle I_{ZB}\rangle - I_{ZB}°] + W_{OAB}[\langle I_{ZA}\rangle - I_{ZA}°] + W_{OCB}[\langle I_{ZC}\rangle - I_{ZC}°] \tag{2}$$

$$\langle \dot{I}_{ZC} - I_{ZC}°\rangle = -(R_{1C} + W_{OCB})[\langle I_{ZC}\rangle - I_{ZC}°] + W_{OBC}[\langle I_{ZB}\rangle - I_{ZB}°] \tag{3}$$

time dependent line intensities denoting Z magnetization in the three environments. The symbols $I_{ZA}°$, $I_{ZB}°$, and $I_{ZC}°$ indicate the equilibrium value of the magnetization in a given site. In arriving at these equations we have also used the identity that $2W_{1A}$, $2W_{1B}$, and $2W_{1C}$ are equal to R_{1A}, R_{1B}, and R_{1C}, that is, R_{1A} denotes the spin-lattice relaxation rate of A in the absence of chemical exchange.

Because of the special nature of the exchange network presented in the scheme, we have the following simple equilibrium boundary conditions (eqs. 4 and 5). However, even with this special exchange network these equations in general cannot be unambiguously decoupled.

$$\frac{I_{ZA}°}{I_{ZB}°} = \frac{W_{OBA}}{W_{OAB}} \tag{4}$$

$$\frac{I_{ZB}°}{I_{ZC}°} = \frac{W_{OCB}}{W_{OBC}} \tag{5}$$

Forsén and Hoffman (63) in a classic series of papers demonstrated that the analogous two site problem can be decoupled by employing a saturation transfer experiment. We can decouple the preceding set of coupled differential equations by employing a series of <u>double</u> saturation transfer experiments. If we perform this experiment in conjunction with a T_1 experiment, all of the relevant kinetic and relaxation parameters can be determined.

In the presence of saturating rf at two sites, the evolution of the magnetization at the remaining site can be described by eqs. 6-8, which are of the general form given by eq. 9, in which

A{B,C}:

$$\langle \dot{I}_{ZA} - I_{ZA}°\rangle = -(R_{1A} + W_{OAB})[\langle I_{ZA}\rangle - I_{ZA}°] - W_{OBA}I_{ZB}° \tag{6}$$

B{A,C}:

$$\langle \dot{I}_{ZB} - I_{ZB}°\rangle = -(R_{1B} + W_{OBA} + W_{OBC})[\langle I_{ZB}\rangle - I_{ZB}°] - W_{OAB}I_{ZA}° - W_{OCB}I_{ZC}° \tag{7}$$

C{A,B}:

$$\langle \dot{I}_{ZC} - I_{ZC}°\rangle = -(R_{1C} + W_{OCB})[\langle I_{ZC}\rangle - I_{ZC}°] - W_{OBC}I_{ZB}° \tag{8}$$

$$\langle \dot{I} - I°\rangle + E\langle I - I°\rangle = -F \tag{9}$$

E and F are constants which are simple linear combinations of exchange and relaxation parameters. The formal soluton of eq. 9 is given by eq. 10 (64), in which D is a constant which depends

$$\langle I - I°\rangle = -F/E + D\exp[-Et] \tag{10}$$

upon the initial condition of the experiment.

A weighted three parameter nonlinear least squares analysis of the data arising from either inversion recovery or saturation recovery T_1 sequences yield (F/E), D, and E. The constant D depends upon the choice of the experiment that has been selected to follow the time evolution of the magnetization. For an inversion recovery experiment, D is given as $-I^*-I°+F/E$ at $t = 0$.

Here I^* is the line intensity when the other two lines are saturated at $t = 0$. For the saturation recovery experiment, the constant D is equal to $-I° + F/E$ at $t = 0$.

If we further limit discussion to the saturation recovery experiment, eq. 10 for the three cases of interest can be explicitly written as eqs. 11-13. In addition to these coefficients,

A{B,C}

$$\langle I_{ZA} - I_{ZA}° \rangle = - \frac{W_{OBA} I_{ZB}°}{R_{1A} + W_{OAB}} + \left(-I_{ZA}° + \frac{W_{OBA} I_{ZB}°}{R_{1A} + W_{OAB}}\right) \exp[-(R_{1A} + W_{OAB})t] \tag{11}$$

B{A,C}

$$\langle I_{ZB} - I_{ZB}° \rangle = - \frac{W_{OCB} I_{ZC}° + W_{OAB} I_{ZA}°}{R_{1B} + W_{OBC} + W_{OBA}} + \left(-I_{ZB}° + \frac{W_{OCB} I_{ZC}° + W_{OAB} I_{ZA}°}{R_{1B} + W_{OBC} + W_{OBA}}\right) \times$$

$$\exp[-(R_{1B} + W_{OBC} + W_{OBA})t] \tag{12}$$

C{A,B}

$$\langle I_{ZC} - I_{ZC}° \rangle = - \frac{W_{OBC} I_{ZB}°}{R_{1C} + W_{OCB}} + \left(-I_{ZC}° + \frac{W_{OBC} I_{ZB}°}{R_{1C} + W_{OCB}}\right) \exp[-(R_{1C} + W_{OCB})t] \tag{13}$$

one must use the equilibrium boundary conditions, eqs. 4 and 5, to solve for <u>all</u> of the necessary parameters. Hence, from the analysis of the time evolution of A and a knowledge of $I_{ZB}°$ one can determine W_{OBA}, and with eq. 4, R_{1A} can be determined. A similar procedure can be applied to the time evolution of C to yield W_{OBA}, W_{OBC}, W_{OCB}, and R_{1C}. These can now be employed for the determination of R_{1B} from eq. 12. As a guide for the <u>overall</u> relative experimental error within the saturation recovery experiment, the boundary conditions require that $I° - (F/E) + D = 0$. Deviations from zero are related to the signal-to-noise ratio of the data and the difference, relative to D, is a reliable estimate of the figure of merit for the various values of the derived parameters.

Even though this procedure results in a determination of the various parameters embodied in eqs. 1-3, we have traded this mathematical complexity for a decrease in sensitivity or dynamic range in transforming eqs. 1-3 to eqs. 11-13. The loss in sensitivity or dynamic range can be regained by more accumulations.

Hence, without much difficulty, one can imagine circumstances where this procedure is simply impractical to apply. Clearly, biological systems represent potential candidates. In an attempt to learn the limitations of this approach and to solve an interesting kinetic and relaxation problem, we have applied this methodology to the ^{113}Cd exchange dynamics in cadmium-substituted Con A (42, 65).

The data summarized in Table 17 demonstrate the importance that chemical exchange processes can have upon the measured "time constants" obtained from spin-lattice relaxation experiments. Exchange processes of the type described in the above scheme will be encountered more often with inorganic nuclides than with organic systems. An obvious exception to this statement is the case where one is looking for differential relaxation rates for a substrate in the presence or absence of protein. For the system discussed here, exchange processes accounted for 14%, 75%, and 20% of the observed relaxation rates for the ^{113}Cd resonances corresponding to the S1 site, free cadmium, and the S2 site, respectively. Therefore, if one is to perform a detailed relaxation analysis with inorganic nuclides, one must perform experiments to determine the relative importance of the potential exchange dynamics in the system.

Table 17. Cadmium-113 Rate Constants (s^{-1}) and Time Constants(s)[a]

W_{OAB}	W_{OBA}	W_{OCB}	W_{OBC}	R_{1A}	R_{1B}	R_{1C}
0.018	0.012	0.035	0.024	0.109	0.012	0.093
τ_{OAB}	τ_{OBA}	τ_{OCB}	τ_{OBC}	τ_{1A}	τ_{1B}	τ_{1C}
56	80	28	42	9	80	11

[a]From P. D. Ellis, P. P. Yang, and A. R. Palmer, submitted to J. Magn. Reson.

The first step in such a procedure is to determine the "T_1's" by normal methods. With these data in hand, one can then perform a spin-counting experiment. For example, in the metal nuclide experiment, can all of the metal in the system be accounted for in terms of the observed resonances? In those cases where all of the spins can be accounted for, and where multiple resonances are observed, the existence of exchange dynamics can be probed by single saturation transfer experiments. In the case where only a single resonance is observed, one has to be more cautious because of the possibility of a minor species (less than 5%) playing the role of a relaxation sink. Examples of such a relaxation sink are paramagnetic ions in a diamagnetic sample and the Cd-EDTA system (34). The only way this situation can be probed is by examining the temperature and field dependence of all the relaxation parameters (T_1, η, and possibly T_2). If at all possible, three magnetic fields should be employed. This same potential problem can occur with the multiple line case.

One of the principal reasons for this study was to inquire if ^{113}Cd spin-lattice relaxation represented a serious problem for high field applications of ^{113}Cd NMR spectroscopy in bioinorganic chemistry. To this end, we have investigated the field dependence of the ^{113}Cd T_1 in cadmium substituted Con A. The exchange corrected ^{113}Cd spin-lattice relaxation times for Con A are summarized in Table 18. If Con A is a representative protein, then it is clear from the data in Table 18 that spin-lattice processes will not represent a serious problem in the acquisition of ^{113}Cd spectra at high (>9T) magnetic fields.

Table 18. Cadmium-113 Spin-Lattice Relaxation in Concanavalin A: The Effect of Varying Magnetic Field

Site	2.3 T (22.2 MHz)[a]	4.7 T (44.4 MHz)[c]	9.4 T (88.8 MHz)[c]
S_1	6.2[e]	10.3[e]	9.2
S_2	6.2[f]	14.2[f]	10.7

[a] Relaxation times are in seconds.
[b] The NOE (1 + η) for the S1 and S2 site are 0.5 and 0.6, respectively.
[c] The NOE (1 + η) for both sites was 0.6.
[d] The NOE (1 + η) was not determined at this field (see discussion).
[e] Corrected by the W_{OAB} exchange rate determined from the 88.8 MHz data.
[f] Corrected by the W_{OCB} exchange rate determined from the 88.8 MHz data.

It is also of interest to discuss the possible mechanisms which give rise to the relaxation times and whether these processes can be discussed in terms of simple motional arguments. Examination of the T_1's alone leads to arguments of the increasing importance of CSA processes as one increases the magnetic field. Given the symmetry at each site, such discussions are reasonable. However, inclusion of the NOE for each resonance leads to considerable uncertainty in such a simple picture. In going from 2.3 to 4.7 T, there appears to be a little change in the NOE for each site, an interesting observation for a protein with an apparent correlation time in the range of 10 to 100 nsec. This weak field dependence may arise because several internal motions contribute to the relaxation time and NOE for cadmium in a given site. However, these motions are not the traditional internal motions one associates with the twofold hopping motion in tyrosine or similar residues. Rather, these motions may be related to the detailed mechanism of how the metal dissociates from the protein.

It is assumed that only in rare circumstances does a metalloprotein expectorate the metal in some concerted process. But rather, the metal may leave when a solvent molecule or anion displaces a single ligand at the metal binding site. With the solvent molecule or anion trapped in the ligand sphere, the energetics of this process may lead to eructation of the new ligand or to sufficient distortion such that other anions or solvent molecules can now have access to the metal, eventually leading to its extrusion from the protein. The initial ligand displacement may result from a fortuitous collision between the exogenous ligand and the metal complex with one (or more) of its natural ligands at the extreme of its own librational motion. If the motion associated with the formation or destruction of these "abortive" complexes (those complexes which do not lead to the metal leaving the protein) has Fourier components in the range of the Larmor frequency, than all of the spin-lattice relaxation processes would be modulated by this complicated internal motion, in addition to the overall motion of the protein. It would be difficult, at best, to describe properly a mathematical representation for this "motion", but it certainly would lead to the generation of several correlation times.

Otvos and Armitage (46a) have observed T_1 values similar to those presented here in their study of alkaline phosphatase. They attempted to apply simple models to their T_1 data using rotational correlaton times for alkaline phosphatase. These models failed to explain the observed T_1 and NOE, and their conclusion was that the relaxation was probably dominated by dipolar processes mediated by internal motions within the protein. This conclusion may also be consistent with the data presented here. For metalloproteins where one can "readily" exchange the metal(s) out of the protein, the "internal motions" may be the formation of "abortive" complexes similar to those described above.

The preceding discussion has ignored the possible contribution that exchange processes could have upon the heteronuclear

NOE and whether such processes could lead to the observed field dependence in the NOE. From eqs. 11 and 13 we see that in the double saturation transfer experiment the equilibrium magnetization of A and C has decreased by the following factors (eqs. 17 and 18). Now let us consider the situation where cadmium at

$$\langle I_{ZA}\rangle = I_{ZA}^\circ - \frac{W_{OBC}I_{ZB}^\circ}{R_{1A}+W_{OAB}} \qquad (17)$$

$$\langle I_{ZC}\rangle = I_{ZC}^\circ - \frac{W_{OBC}I_{ZB}^\circ}{R_{1C}+W_{OCB}} \qquad (18)$$

sites A, B, and C is dipole-coupled to protons and its equilibrium magnetization in the presence of saturating rf applied to the protons is described by eq. 19. Here η_A, η_B, and

$$I_{ZA}^\circ\{^1H\} = I_{ZA}^\circ(1 + \eta_A) \qquad (19a)$$

$$I_{ZB}^\circ\{^1H\} = I_{ZB}^\circ(1 + \eta_B) \qquad (19b)$$

$$I_{ZC}^\circ\{^1H\} = I_{ZC}^\circ(1 + \eta_C) \qquad (19c)$$

η_C are the respective nuclear Overhauser enhancements for those sites in the absence of exchange. Placing eq. 19 into eqs. 17 and 18, we can write equations for the apparent NOE at sites A and C caused by exchange with site B, eqs. 20 and 21, which

$$\frac{\langle I_{ZA}\{^1H\}\rangle}{\langle I_{ZA}\rangle} = 1 + \frac{I_{ZA}^\circ \eta_A}{\langle I_{ZA}\rangle} - \frac{W_{OBA}I_{ZB}^\circ \eta_B}{(R_{1A}+W_{OAB})\langle I_{ZA}\rangle} \qquad (20)$$

$$\frac{\langle I_{ZC}\{^1H\}\rangle}{\langle I_{ZC}\rangle} = 1 + \frac{I_{ZC}^\circ \eta_C}{\langle I_{ZC}\rangle} - \frac{W_{OBC}I_{ZB}^\circ \eta_B}{(R_{1C}+W_{OCB})\langle I_{ZC}\rangle} \qquad (21)$$

clearly show how the NOE can be perturbed by the presence of chemical exchange. Hence, the NOE cannot be used in this case to partition the relative amount of dipole-dipole processes from other relaxation mechanisms. The fact that there is a weak field dependence observed for the NOE at the protein sites is clear from eqs. 21 and 22. Since B is a small ion, its relaxation parameters (T_1, T_2, and NOE) should be representative of a species under the conditions of extreme narrowing. Further, if the η's for the Cd at the S1 and S2 sites are small, there would be an apparent weak field dependence observed for both sites.

It is essential to note the importance of knowing whether chemical exchange is present within a given metal nuclide system, before one draws any firm conclusions about the relative importance of the various relaxation mechanisms or of the presence of internal motions. Although the present analysis and its conclusions are applicable to Con A, it is not clear whether they can be applied to any other cadmium-substituted protein, unless it is known that chemical exchange processes are operational and that the exchange rates are comparable to the relaxation rates. At this point, we can put aside the question of chemical dynamics and turn our attention to the significance of the observed chemical shifts.

From the discussions presented in the previous section, it should be recognized that the ^{113}Cd chemical shift is reflective of its ligand environment. For example, there would be little difficulty in distinguishing Cd^{+2} in an all oxygen environment as compared to Cd^{+2} where one or two of the oxygen atoms have been replaced by nitrogen atoms. Hence, to a first order approximation the ^{113}Cd chemical shift is a measure of the number and type of ligands within the cadmium's first coordination sphere. However, to be able to draw any specific conclusions about a particular system, one has to be sure that ligand exchange processes are absent. With these considerations in mind, the ^{113}Cd chemical shift can be employed as a "black box" by experimentalists to yield information related to the overall symmetry of the binding environment and the number and type of ligands at a particular site. Further, because of the sensitivity of the ^{113}Cd chemical shift to subtle changes, the chemical shift can be used to probe the existence of allosteric pathways in proteins. We will see examples of such utilizations in the next section.

Given this sensitivity of the ^{113}Cd chemical shift to structural parameters, we now turn our attention to the possibility of using this sensitivity to probe the details of the metal binding site. What are the detailed factors which give rise to the ^{113}Cd chemical shift? Are these factors easily translated into such terms as ligand donor or acceptor properties, bond distances, coordination number, and overall complex topology? To address these questions unambiguously, our experimental data must be free of the effects of chemical dynamics; hence we will employ chemical shifts determined by solid state methods. This approach has the further advantage that one can correlate these data with well defined structures obtainable by standard X-ray methods.

Before we proceed further, we must clarify two points concerning the additivity of substituent effects and the magnitude of these effects and whether we can translate our experiences from ^{13}C NMR to metal nuclide chemical shifts. Maciel summarized the basis of ^{13}C chemical shift additivity as follows: "A priori, one can consider that nearly any organic species, except perhaps some highly strained compounds, can be viewed as derivable by substitution upon a simple molecular framework" (66). The extrapolation of this reasoning to the analysis of metal nuclide chemical shifts in general inorganic systems is ill-advised because of ligand dynamics in the solution phase and the sensitivity of the metal nuclide chemical shift to subtle changes in metal geometry and symmetry. As we will see later, with some care one can prepare systems which should allow investigations of electron substituent effects devoid of structure and symmetry effects.

The magnitude and sign of substituent effects upon ^{13}C chemical shifts were found to be consistent with our "chemical common sense" of inductive and resonance effects. Therefore, for example, the substituent effects caused by halogens, e.g., $\delta(CH_3I) > \delta(CH_3Br) > \delta(CH_3Cl) > \delta(CH_3F)$, was considered normal. In Table 19 we summarize some solid state ^{113}Cd chemical shifts measured by Nolle (67) for CdI_2, CdF_2, $CdBr_2$, $CdCl_2$, CdO, and CdS. It is clear from these data that halogen substituent effects can neither by considered to be "normal" or "inverse." Clearly, other factors are influencing these chemical shifts. Is it reasonable to assume that metal nuclides such as cadmium should share similar patterns of chemical shieldings with nuclei such as carbon? The answer, simply put, is no. The fundamental reason for this lies within our understanding of the concept of spin-orbit couplings.

Table 19. Aberrant ^{113}Cd Chemical Shifts (Neither "Normal" nor "Inverse" Halogen Effects)

Compound	Chemical shift
CdI_2	0.0
CdF_2	449
$CdBr_2 \cdot 4H_2O$	713
$CdCl_2$	883
CdO	1091
CdS	1222

From A. Nolle, Z. Naturforsch., 33a, p. 666 (1978).

Atomic spectroscopists have known for a long time that selection rules based upon the quantum numbers L and S progressively break down as the atomic number increases beyond 35 (Br) (68). This difficulty arises because of spin-orbit effects, $\underline{L}\cdot\underline{S}$. An electron in a nonspherically symmetric state is equivalent to a current. This current produces a magnetic field that interacts with the spin of the electron and produces a change in energy in somewhat the same way as the interaction of a current with an applied magnetic field does. The inclusion of the spin-orbit interaction term in molecules invalidates the concept of the spin molecular orbital, since the presence of the operators S_x and S_y will lead to a mixing of both α and β spin character into a single molecular orbital. This brings about the notion of nonperfect pairing. The effect is small but it has important consequences upon chemical shifts and coupling constants. This effect occurs in the absence of an applied magnetic field and for a closed shell diamagnetic molecule has the effect of mixing triplet character into the ground state. In the presence of an applied magnetic field, the triplet character results in a spatial imbalance of α and β spin character of the electrons which results in an induced spin density throughout the molecule. This resulting spin density generates a local field at a given nucleus. These local fields arise via nuclear-spin-dipole, electron-spin-dipole, and Fermi contact pathways. Hence, the spin-orbit contributions to the local field at a given nucleus are transmitted throughout the molecule by mechanisms which are analogous to these for the indirect nuclear spin-spin coupling constant.

It is not our intent to present a detailed theoretical account of spin-orbit effects on chemical shifts (69), but rather to point out that such effcts are important for a proper discussion of metal nuclide chemical shifts and spin coupling constants. The existence of these processes should force experimentalists to be more careful in the analysis of their data. With these precautionary notes in mind, we will illustrate our approach to the possible separation of electronic and structural substituent effects on ^{113}Cd chemical shifts.

From a detailed analysis of the theory of chemical shifts and rotational properties of angular momentum operators, one can make some simple assertions concerning the shielding tensor. In planar aromatic systems or in aromatic systems where nonplanarity exists along an axis of threefold or higher rotational symmetry, the shielding tensor for an atom which lies along this axis of symmetry will be axially symmetric. Further, any diagonal element of the paramagnetic portion of the shielding tensor results from a mixing of orbitals which are orthogonal to that direction of the shielding tensor. Molecular examples of systems which should follow these basic ideas of orbital symmetry are metalloporphyrins. In the case of substituted porphyrins in the solid state one has to be careful about the maintenance of the threefold or higher axis of symmetry, since the addition of the fifth ligand could lower the symmetry at the metal, unless the ligand rotated rapidly about this symmetry axis. In this case, the metal will lie along a pseudofourfold axis. Serendipitously, 4-substituted pyridines are such a ligand (70, 28).

The cross polarization (71) ^{113}Cd powder spectrum of cadmium meso-tetraphenylporphyrin (CD-TPP) and its pyridine adduct Py-Cd-TPP are shown in Figure 5. The lineshape for both powder patterns confirms the expectation based on molecular symmetry that the ^{113}Cd shielding tensors are axially symmetric for both compounds (vide supra). The MAS spectra (72,73) (Figure 6) were obtained from high-speed sample spinning at a maximum rate of 4.0 kHz about an axis oriented at 54.7° with respect to the static magnetic field. Spinning sidebands, with a separation of ±4.0 kHz, are clearly observable in the CP MAS spectrum of Cd-TPP but are absent in the MAS spectrum for Py-Cd-TPP (not shown). The complexity of spinning sidebands can be avoided only if the spinning rate is comparable to or larger than the chemical shielding anisotropy (expressed in frequency units). Obviously, this is not possible for Cd-TPP with a spinning rate of 4.0 kHz at a field of 4.7 T. Generally, the linewidth obtained in a MAS experiment is determined by instrumental limitations such as a magnet inhomogeneity and offsets from the magic angle setting; for our system ^{113}Cd linewidths of ca. 1 ppm (40 Hz) are usually obtained. The much larger linewidths of approximately 400-500 Hz observed in MAS spectra for both Cd-TPP and Py-Cd-TPP can be ascribed to the scalar $^1J(^{113}Cd-^{15}N)$ interactions which are not averaged out by magic angle spinning. Actually, spectral expansions of the individual spinning sidebands for Cd-TPP reveal the lineshape of a broadened 1:4:6:4:1 quintet with $^1J(^{113}Cd-^{15}N) =$ 150.1 Hz observed for Py-Cd-TPP in the liquid state.

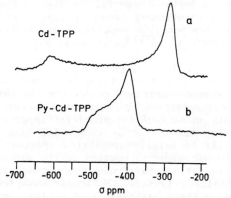

Figure 5. Cross polarization (CP) solid state ^{113}Cd NMR powder spectra of (a) "free" Cd-TPP and (b) the ^{15}N-pyridine adduct of Cd-TPP. The shielding constant (σ) scale is positive to larger shieldings and is referred to a sample of solid $Cd(ClO_4)_2 \cdot 6H_2O$.

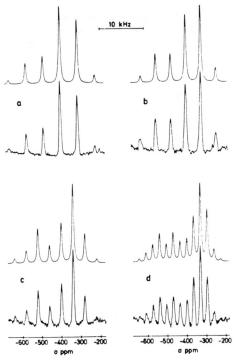

Figure 6. Experimental (lower) and simulated (upper) magic angle spinning (MAS) CP solid state ^{113}Cd NMR spectra of "free" Cd-TPP with spinning rates of (a) 3.936 kHz, (b) 3.417 kHz, (c) 2.685 kHz, and (d) 1.546 kHz. A value $\Delta\sigma = -341$ ppm has been used in the simulation of all theoretical MAS spectra.

Lineshapes of NMR powder patterns are well understood and have been the subject of several reviews (74). From the prominent features of these spectra, such as for those in Figure 5, it is often possible to derive the principal components of the shielding tensor (σ_{xx}, σ_{yy}, σ_{zz}) directly. However, in most cases, the values derived in this way are used as starting parameters for a theoretical simulation of the experimental powder spectrum (74) in order to obtain more reliable parameters. We used another approach in that these parameters may also be determined from theoretical computer simulations of spinning sideband intensities for the MAS spectra observed at different spinning rates (75). Figure 6 shows a series of such computer-simulated spinning spectra, generated from the theoretical treatment by

Marciq and Waught (76), along with the corresponding experimental MAS spectra for Cd-TPP. However, a more direct procedure, which appears to be the method of choice, for obtaining the principal elements of the shielding tensor has recently appeared by Hertzfeld and Berger (77). The ^{113}Cd chemical shielding tensor parameters extracted from the spectra of Cd-TPP and Py-Cd-TPP are tabulated in Table 20.

Table 20. ^{113}Cd Chemical Shielding Tensors[a] for Cd-TPP and Py-Cd-TPP

	$\bar{\sigma}$[b]	$\sigma_\parallel$	$\sigma_\perp$	$\Delta\sigma$[c]
Cd-TPP[d]	−399	−626	−285	−341
Py-Cd-TPP[e]	−432	−502	−397	−105

[a] All shifts are in ppm from external $Cd(ClO_4)_2 \cdot 6H_2O$; negative shfits to lower shielding; no bulk susceptiblity corrections.
[b] $\bar{\sigma} = (\sigma_\parallel + 2\sigma_\perp)/3$. [c] $\Delta\sigma = \sigma_\parallel - \sigma_\perp$.
[d] Estimated errors ± 3 ppm. [e] Estimated errors ±2 ppm.

The isotropic shift of 432 ppm observed for Py-Cd-TPP is similar to that observed from liquid state NMR using $CHCl_3$ as solvent. Furthermore, the solid state isotropic shift data in Table 20 show that addition of pyridine as a fifth ligand to Cd-TPP shifts the ^{113}Cd resonance to lower shielding by 33 ppm. The power of solid state NMR spectroscopy is, however, best appreciated when it is realized that this relatively small isotropic chemical shift difference is, in fact, the result of rather large changes in the individual shielding tensor elements moving in the opposite directions; that is, the unique tensor element ($\sigma_\parallel$) for Py-Cd-TPP is shifted to higher shielding by 124 ± 5 ppm while the in-plane element ($\sigma_\perp$) is shifted to lower shielding by 112 ± 5 ppm relative to Cd-TPP. From these changes in the shift tensor elements, the structural factors that give rise to the observed isotropic chemical shift of 33 ppm may be argued qualitatively. The argument would have been meaningless without the solid state NMR information. If the pyridine (fifth) ligand is simply coordinating the in-plane cadmium of Cd-TPP via its $5p_z$ orbital (or an appropriate combination of cadmium orbitals) and without geometrical distortions, then large changes in the in-plane components of the ^{113}Cd shielding tensor and only minor changes for the unique element are expected relative to Cd-TPP. However, if the fifth ligand pulls the cadmium out of the molecular plane, then gross changes are expected for all tensor elements. Obviously, from the data in Table 20 the latter situation prevails. This conclusion is consistent with the structural data available on five coordinate metalloporphyrins (78) and with a recent X-ray structure for the piperidine adduct of Cd-TPP (79). It would be formally incorrect to conclude that nitrogen ligand binding to the cadmium $5p_z$ orbital causes a shift of the unique tensor element to higher shielding, but rather, the observed shift for $\sigma_\parallel$ represents the combination of both ligand binding and the effect of the ligand moving the metal out of the ring. A similar rationale exists for the in-plane element, $\sigma_\perp$.

This "orthogonal" interaction is expected if the so-called "local paramagnetic" term is the dominant contributing mechanism to the observed shifts in the individual shielding tensor elements. This term is proportional to matrix elements of the form (80) $\sigma_{zz} \propto \langle\mu|(L_z/3)|\nu\rangle$, in which μ and ν are orbitals on the cadmium (for sake of the present discussion μ, ν are 5p orbitals) and L_z is the z component of angular momentum. From the symmetry of Py-Cd-TPP, the $5p_z$ orbital belongs to a different irreducible representation from that of the $5p_x$ and $5p_y$ orbitals. Hence, because of this symmetry constraint and the nature of the angular momentum operator, perturbations to the $5p_z$ orbital cannot directly contribute to σ_{zz}. Likewise, it is only perturbations to $5p_x$ and $5p_y$ that contribute to σ_{zz}. Again, it must be recalled that such symmetry arguments assume the dominance of the one-center term. Clearly, more experimental data are needed to check such an assertion.

Presently, our knowledge of Cd chemical shifts is based solely on experimental observations. One such observation is that a 100 ppm change in the ^{113}Cd chemical shift can occur by seemingly trivial changes of the molecular system; for example, in going from $(CH_3)_2Cd$ to $(C_2H_5)_2Cd$ (15) or by substituting an oxygen for a nitrogen ligand (42). Hence, one has to exercise great caution in trying to interpret small chemical shift changes (less than 50 ppm) in terms of orbital polarization arguments which are valid for lighter elements (e.g., ^{13}C). Until a general theoretical picture emerges for chemical shifts of heavier elements, one must rely upon a close empirical relationship between the benchmarks of solid state chemical shift data and well known structures. Here chemical shifts mean not only the isotropic chemical shift, but also the elements of the Cd chemical shift tensor.

Our investigation on Py-Cd-TPP represents a special case of ^{113}Cd solid state NMR studies in general. The ^{113}Cd chemical shift tensor elements are axially symmetric and can be easily assigned to a molecule fixed coordinate system because of the molecular symmetry. Hence, one can envision a study with different axial ligands which would allow the separation of structural effects (e.g., the degree of nonplanarity of the cadmium with respect to the porphyrin ring) versus electronic effects (e.g., ligand polarization of metal orbitals perpendicular or parallel to the ring plane) (81). In general, the ^{113}Cd chemical shift tensors are not symmetrical and therefore the only way that a ^{113}Cd chemical shift tensor element can be assigned relative to a molecule fixed coordinate system is via the analysis of Cd shift tensors for selected single crystals. Therefore, let us turn our attention to some single crystal experiments that are being carried out in our laboratories.

Because of our interest in utilizing ^{113}Cd NMR spectroscopy as a probe for calcium sites in biological systems (42,43,49,51), we have been interested in the isotropic ^{113}Cd chemical shifts in these compounds. In the systems studied (Concanavalin A (42), Parvalbumin (43), Troponin C (50), Calmodulin (49), and Insulin (51)), replacing calcium with cadmium places cadmium in an environment in which all of the atoms in the first coordination sphere are oxygens. The cadmium is assumed to be six coordinate. The isotropic chemical shifts for these compounds (82) fall into a characteristic range of chemical shifts from -85 to -130 ppm with respect to 0.1 M $Cd(ClO_4)_2$. To date, there are no model compounds that can be studied in aqueous solutions which have isotropic shifts within this range.

If, however, one performs the ^{113}Cd NMR experiments in the solid state (83), then one finds a bewildering array of cadmium-oxo compounds that could be employed to model these biological systems. The coordination numbers for cadmium in these compounds is 6, 7, or 8, while the isotropic chemical shifts span an awesome range from 150 to -100 ppm. There is no apparent relationship between Cd and coordination geometry or number and the isotropic chemical shift. Such a situation is reflective of the degree of sensitivity of the ^{113}Cd shielding tensor to distortions in the cadmium coordination symmetry and to subtle fluctuations in cadmium-oxygen distances. The only remaining approach that can be employed to understand this enigmatic situation is a detailed investigation of selected cadmium-oxo compounds via single crystal NMR methods (84,85).

CADMIUM-113 NMR SPECTROSCOPY OF SKELETAL TROPONIN C

One of the principal driving forces behind our efforts with ^{113}Cd NMR spectroscopy has been the desire to develop a clean spectroscopic probe for zinc and calcium in biological systems. The first observation of ^{113}Cd signals from a metalloprotein was made by the Yale group headed by Armitage and Coleman (45d). Other ^{113}Cd observations appeared quickly by Sudmeier (45c) at the University of California at Riverside, our group (42b), Drakenberg and co-workers (43b) at Lund, and Myers (45f) at the University of California at Berkeley. Rather than present an extensive review of all this work, we will focus, instead, on our recent research on the calcium-binding muscle protein, Troponin C (94). This work is being performed in collaboration with Professor James D. Potter at the University of Cincinnati College of Medicine.

Troponin C (95), TnC, is the calcium-binding subunit of the protein Troponin. Troponin contains three subunits: the tropomyosin binding subunit, TnT, the actomyosin ATPase inhibitory subunit, TnI, and TnC. Troponin forms a protein complex with tropomyosin to form native tropomyosin. Various studies have shown that in skeletal muscle thin filaments a protein aggregate forms that contains one mole of native tropomyosin to seven moles of actin. It is Ca^{+2} binding to skeletal TnC, STnC, within this protein aggregate which results in the activation of muscle contraction. As a result of this key biological function there has been an enormous effort directed toward a detailed understanding of the nature of the calcium binding sites in TnC and how calcium binding at these sites can regulate muscle contraction.

Based upon these numerous studies, a reasonably detailed picture of these Ca^{+2} binding sites has evolved. Skeletal TnC contains two classes of calcium binding sites (96). The first category contains two high affinity Ca^{+2} sites with $K_{Ca} = 2.1 \times 10^7$. Furthermore, it has been shown that these sites can also bind magnesium competitively with $K_{Mg} \simeq 200$. Hence, the sites are often referred to as the Ca/Mg sites. The second category of sites contains two low affinity Ca^{+2} sites with $K_{Ca} = 3.2 \times 10^5$. Under normal physiological concentrations of Mg^{+2}, these sites appear to bind only Ca^{+2}. Hence, they are commonly referred to as calcium specific sites. Based upon normal cellular concentrations of Ca^{+2} and Mg^{+2} in resting and Ca^{+2} pulsed cells, Potter and co-workers (95) have concluded that the high affinity Ca^{+2} sites are structural sites in STnC. The low affinity Ca^{+2} sites, on the other hand, are the regulatory sites. It is calcium binding to these sites that is responsible for the regulatory processes in muscle contraction. In an effort to avoid confusion with the cadmium binding studies that we will report here, we will employ a slightly more general nomenclature for the Ca^{+2} binding sites in STnC. The high affinity Ca^{+2} sites will be referred to as the structural sites and the low affinity Ca^{+2} sites will be discussed as regulatory sites.

During the past twenty years, much has been learned about the role calcium plays in these calcium-binding muscle proteins. However, in the absence of a high resolution X-ray structure of

STnC, little is known about the individual calcium binding sites. A priori, the best approach to learn about the individual Ca^{+2} sites would be to investigate these sites with a spectroscopic probe that would allow such site selective resolution, i.e., an NMR experiment. The first nucleus to come to mind might be ^{43}Ca. Such innovative experiments have been performed by Forsén and associates (97). Even though these experiments represent the forefront in NMR technology with respect to ^{43}Ca NMR spectroscopy, the biochemical conclusions have been disappointing, in the sense that they only confirm what was known from previous Ca^{+2} binding studies (96) and, further, did not manifest any individual site resolution. The lack of site differentiation was to be expected based upon the fact that ^{43}Ca is a quadrupolar nuclide with I = 7/2. The problem of site resolution can be overcome by the use of a surrogate Ca^{+2} probe, namely cadmium.

Cadmium-113 NMR spectroscopy has been demonstrated to be an excellent probe for metal binding sites in metalloproteins and calcium-binding proteins. Forsén and co-workers have been one of the leading groups in applying these methods to calcium-binding proteins, Parvalbumin (43), STnC (50), and Calmodulin (49). With Parvalbumin, they were able to observe resolvable resonances for each of the high affinity Ca^{+2} binding sites. However, in STnC they were only able to observe the resonances for the structural sites. They were able to assign the observed resonances to the structural sites by noting the nearly equal and high affinity of each resonance for added cadmium or added calcium to the cadmium-substituted STnC. Even with this disappointment, they clearly demonstrated that each structural site has nearly the same affinity for Ca^{+2}.

As important as these pioneering experiments were, they shed doubt on the utility of using ^{113}Cd NMR spectroscopy as a calcium probe in these systems. We wish to summarize here our own experiments employing ^{113}Cd NMR spectroscopy to study cadmium substituted STnC, the results of which are dramatically different than those reported by Forsén et al. (50). We observe resonances for all four calcium binding sites within STnC. By a series of Cd^{+2} and Ca^{+2} titrations, we have unambiguously assigned the resonances to the two classes of binding sites. Our assignment of the structural sites are consistent with those reported by Forsén and co-workers (50). Furthermore, we demonstrate that the Ca^{+2} binding sites are allosterically coupled to one another and we propose that the coordination number for Ca^{+2} in each of these sites is greater than six and probably eight.

<u>Temperature Dependence of the ^{113}Cd Spectra of STnC.</u> The ^{113}Cd NMR spectra of cadmium-substituted STnC (2.5 mM apo STnC in the presence of a <u>five</u> equivalents of ^{113}Cd) as a function of sample temperature are depicted in Figure 7 and the chemical

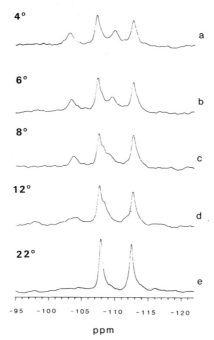

Figure 7. The temperature dependence of the ^{113}Cd NMR of STnC at 2.5 mM in the presence of five equivalents of ^{113}Cd (added as $CdCl_2$) is summarized in this figure. Each spectrum is a result of 20,000 accumulations and a negative chemical shift denotes resonances to higher shielding with respect to a 0.1 M $Cd(ClO_4)_2$ solution at 22°C.

shifts of the various resonances are summarized in Table 21. Figure 7e is the ^{113}Cd NMR spectrum of STnC at 22°C and is virtually identical to the spectrum reported by Forsén et al. (50). The resonances appearing at -107.8 and -112.7 ppm presumably arise from cadmium bound to the two high calcium sites within the protein. At the decreased sample temperature of 4°C, Figure 7a produces a profound difference in the ^{113}Cd spectrum of STnC. A total of four resonances are clearly observed at -103.1, -107.2, -109.8, and -112.7 ppm. The two new resonances at -103.1 and -109.8 ppm arise from cadmium bound at the two calcium regulatory sites. The temperature variation of the ^{113}Cd spectrum of STnC exhibits some exciting features. First, the chemical shift or the normalized integrated intensity of the resonances associated with the high affinity calcium sites appears to be a function of the occupancy of the regulatory sites or upon the sample temperature and hence the conformation of the protein. Secondly, the chemical shifts of the regulatory sites are dependent upon the sample temperature. Finally, there are dramatic differences in the exchange rates of cadmium for the two classes of binding sites, and, furthermore, the exchange processes have been slowed down for cadmium binding to the regulatory sites but not stopped with respect to the chemical shift time scale at 4°C.

Table 21. Cadmium-113 Chemical Shifts of Cadmium Substituted STnC as a Function of Sample Temperature[a]

Temperature (°C)	R_1[b]	ST_1[b]	R_2[b]	ST_2[b]
4	-103.09	-107.19	-109.85	-112.66
6	-103.22	-107.27	-109.66	-112.69
8	-103.31	-107.30	-109.44	-112.63
10	-103.53	-107.38	-109.22	-112.63
12	-103.61	-107.49	-109.00	-112.66
14	-104.02	-107.54	-108.8[c]	-112.66
16	-104.00	-107.63	-108.8[c]	-112.66
22	-	-107.82	-	-112.66

[a]All chemical shifts are in ppm with respect to 0.1 M $Cd(ClO_4)_2$ at 22°C. A negative sign denotes resonances to higher shielding. All temperatures are ±1°C.
[b]The nomenclature R_1, R_2, ST_1, ST_2 refers to the first and second regulatory and structural sites within STnC.
[c]Appears as a shoulder of the resonance corresponding to ST_1.

The results summarized in Figure 7 provide the key to clarify the uncertainties raised in the previous work on cadmium-substituted STnC. These data also provoke several questions which can be addressed unambiguously. The most important one is whether cadmium-substituted STnC represents an accurate model for calcium STnC. Calcium STnC possesses two high affinity binding sites and two low affinity sites. The former are the regulatory sites and the latter represent the structural sites. From Figure 7 it appears the cadmium-substituted STnC is in agreement with this qualitative picture. This conclusion can be critically tested by performing a titration of apo STnC with cadmium followed by a titration of this sample with calcium. Furthermore, these titrations should be performed at 4°C to address the question properly of whether the occupancy at the regulatory sites is modulating the chemical shifts of one of the structural sites or whether the observed changes in chemical shifts are the result of subtle structural changes that occur around the metal binding site as a result of temperature dependent conformation changes. Finally, such titrations may provide further insight into the origins of the decrease in the normalized cadmium intensity of the structural sites as a result of populating the regulatory sites.

Cadmium-113 Titration of Apo STnC. Figure 8 summarizes the titration of a 2.5 mM sample of apo STnC with $^{113}CdCl_2$ at 4°C. Within Figure 8, spectra 8a through 8e correspond to successive additions of one equivalent of cadmium, i.e., Figure 8a denotes one equivalent of cadmium present and Figure 8e represents five equivalents of cadmium. A close examination of relative intensities of the ^{113}Cd resonances in Figures 8a through 8e allows one to address questions concerning the nature of the metal binding sites in the presence of only a single structural site or a single regulatory site. However, we will defer an amplification of these points until we have measured the ^{113}Cd spin-lattice relaxation times of the various resonances in a subsequent paper.

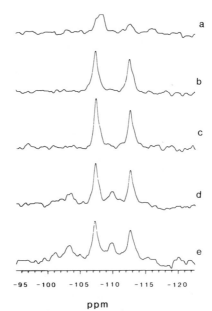

Figure 8. The titration of a 2.5 mM solution of apo STnC with ^{113}Cd is depicted in this figure. Spectra a-e represent successive additions of one equivalent of ^{113}Cd (added as $Cd(ClO_4)_2$.

It is evident from Figure 8 that the resonances at -107.8 and -112.7 ppm correspond to metal binding sites with a preferential affinity for cadmium. From the data in Figures 8a through 8c, it is clear that one can readily prepare cadmium-substituted STnC with the regulatory sites empty. Addition of more cadmium leads to a population of the regulatory sites. It is important to note that when the regulatory sites are being titrated at 4°C, the chemical shifts of the structural sites are independent of the population of the regulatory sites. These data demonstrate that the chemical shifts observed for the structural site (-107.8 ppm) for a fixed number of cadmiums as a function of temperature (Figure 7) arise from temperature dependent conformational changes within STnC. This conclusion is consistent with other spectroscopic data which attest to the conformational flexibility of STnC. The present data cannot distinguish between conformational changes occurring at or near both classes of metal-binding sites or a conformational change occurring at or near one site, the remaining sites being coupled to this change via an allosteric network.

It is apparent from Figure 8 that the resonance at -107.8 ppm can be a complicated composite of several diverging species (one cadmium at either structural site with the other sites empty or one cadmium at either regulatory site with the two structural sites occupied). For this reason, we have plotted, in Figure 9, the normalized area of the most shielded ^{113}Cd resonance, -112.7 ppm, as a function of added cadmium. These data indicate that the decrease in the peak height of this resonance (Figure 7) is due to a small increase in linewidth and not a significant decrease in peak area. The total area of this resonance does not change significantly after the addition of three cadmiums per mole of STnC. That this is not two cadmiums per mole is reflecting the overall lower affinity that cadmium has relative to calcium.

<u>Calcium Titration of Cadmium-Subsituted STnC</u>. Throughout the previous discussions we have made the reasonable assumption that the resonances which correspond to binding sites with a high affinity for cadmium are the same sites that have a high affinity for calcium, the so-called calcium/magnesium sites. Likewise, parallel assumptions have been made with calcium specific or regulatory sites. The validity of these assumptions can be tested unambiguously by titrating the cadmium protein with calcium. If the preceding assumptions are correct, the resonances corresponding to the sites with a high affinity for cadmium should be the first to be titrated by the calcium. Because of the nearly two orders of magnitude difference in the affinity for calcium that the structural sites have over the regulatory sites, cadmium in the regulatory sites should be unaffected by the calcium titration until the structural sites have been saturated by calcium.

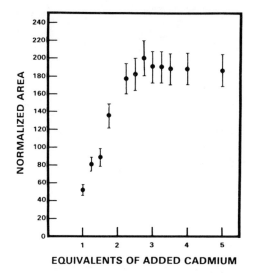

Figure 9. This is a plot of the normalized area of the resonances at -112.7 ppm as a function of added Cd^{+2}. The error bars represent approximately 10% of the normalized area.

The results of such a titration are summarized in Figure 10. The data in Figure 10 are in excellent agreement with the beforementioned assumptions. However, what is also clear from Figure 10 is the sensitivity of ^{113}Cd chemical shifts to the allosteric coupling that is present within STnC. Clearly, the ^{113}Cd spectra depicted in Figure 10 represent a composite of several independent proteins. If we use the notation $[Cd_2]_{St}[Cd_2]_R$STnC to denote two cadmiums in the structural sites, $[Cd_2]_{St}$, and two cadmiums in the regulatory sites, $[Cd_2]_R$, then depending upon how much calcium is present we could have the following species in solution: $[Cd_2]_{St}[Cd_2]_R$STnC, $[CaCd]_{St}[Cd_2]_R$STnC, $[CdCa]_{St}[Cd_2]_R$STnC, $[Ca_2]_{St}[Cd_2]_R$STnC, $[Ca_2]_{St}[CaCd]_R$STnC, $[Ca_2]_{St}[CdCa]_R$STnC, and $[Ca_2]_{St}[Ca_2]_R$STnC. To quantitate how much of each of these species is present in each of the spectra depicted in Figure 10 would be exceptionally difficult in the absence of the ^{113}Cd T_1's of each subsystem. However, from these data we can create a series of simulated spectra which can account for the overall appearance of Figure 10. These subspectra are summarized in Figure 11. In the analysis of Figure 10 we need only consider the presence of four species: $[Cd_2]_{St}[Cd_2]_R$STnC, $[CaCd]_{St}[Cd_2]_R$STnC, $[CdCa]_{St}[Cd_2]_R$STnC, and $[Ca_2]_{St}[Cd_2]_R$STnC. Figure 11a ($[Cd_2]_{St}[Cd_2]_R$STnC) is a schematic representation of Figure 7a, and Figure 11c ($[Ca_2]_{St}[Cd_2]_R$STnC) is an analogous reconstruction of Figure 7d. Figure 11b was

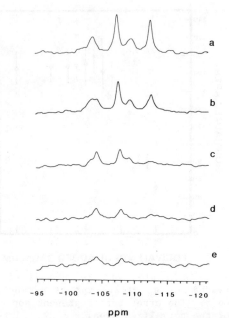

Figure 10. The titration of a 2.5 mM solution of apo STnC in the presence of five equivalents of Cd^{+2} as a function of added Ca^{+2} is summarized in this figure. Spectrum a denotes no added Cd^{+2}, whereas spectra b-e denote successive additions of one equivalent of Ca^{+2}.

constructed from a simple analysis of the linewidths observed in the spectra for cadmium-substituted (5 equivalents) STnC in the presence of 0.25, 0.50, and 0.75 equivalents of calcium (not shown). Figure 10 now can be reconstructed from a statistical combination of each of these subspectra. Hence, in the presence of one equivalent of calcium, one would expect to have zero percent of the $[Cd_2]_{St}[Cd_2]_R$STnC and equal amounts of the other three species. Upon the addition of another equivalent of calcium, the relative ratios of $[CaCd]_{St}[Cd_2]_R$STnC, $[CdCa]_{St}[Cd_2]_R$STnC, and $[Ca_2]_{St}[Cd_2]_R$STnC would be 1:1:4. Within the experimental error and our explicit assumption of equal calcium-binding affinity for all of the species, the data summarized in Figure 10 are in complete agreement with this approach. Further addition of calcium leads to a reduction of the resonances of corresponding to $[Ca_2]_{St}[Cd_2]_R$STnC. Therefore, our assignments of the ^{113}Cd NMR spectrum of cadmium-substituted STnC is in complete accord with the known calcium affinity for STnC.

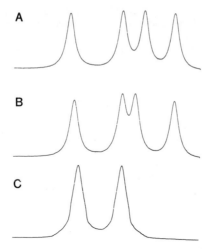

Figure 11. This figure summarizes the results of calcium substitution upon $[Cd_2]_{St}[Cd_2]_R$STnC. (A) We have simulated the ^{113}Cd spectrum of $[Cd_2]_{St}[Cd_2]_R$STnC in the absence of Ca^{+2}. (B) The simulated ^{113}Cd spectrum of $[CaCd]_{St}[Cd_2]_R$STnC plus an equal amount of $[CdCa]_{St}[Cd_2]_R$STnC. (C) The simulated ^{113}Cd spectrum $[Ca_2]_{St}[Cd_2]_R$StnC.

SUMMARY AND CONCLUSIONS

Within this chapter we have endeavored to demonstrate the utility of ^{113}Cd NMR spectroscopy in bioinorganic and inorganic chemistry. As with any spectroscpic technique, ^{113}Cd NMR has its pitfalls. We have shown that important dynamic problems may be addressed, providing unique data with regard to chemical rate constants and NMR relaxation times. We have illustrated, by the use of solid state NMR techniques, that one can study subtle changes in structure and bonding via the sensitivity of the shielding tensor. Furthermore, knowledge of these shielding tensors can then be used in liquid state NMR experiments to follow anisotropic motion in important bioinorganic systems. Finally, we have demonstrated that spin-lattice relaxation should not represent a serious problem for ^{113}Cd NMR investigations of metalloproteins at magnetic fields comparable to 9.4 T. We have highlighted this in our study of the chemical dynamics of Con A and in the demonstration of allosteric coupling in skeletal Troponin C.

Though we have limited our specific examples to ^{113}Cd, it is clear to us that there is nothing unique about ^{113}Cd NMR; our conclusions are generally applicable to other spin 1/2 nuclides. Within the context defined by their chemistry, all spin 1/2 metal nuclides should hold a rich and rewarding future as molecular structure probes for exciting research problems in bioinorganic and inorganic chemistry.

ACKNOWLEDGMENTS

The author wishes to thank Dr. Steven W. Sparks for his help in the preparation of this manuscript. Over the past several years several of my colleagues have made significant contributions to our resesarch progress in the area of ^{113}Cd NMR spectroscopy. I wish to acknowledge the important contributions made by Professors Jerry Odom, David Behnke, Hans Jakobsen, James Fee, and Jim Potter; Drs. Alan Cardin, David Bailey, Allen Palmer, Ping Yang, and Ruth Inners; and Misseurs Bob Honkonen and Paul Majors. A special thank you is needed for the untiring technical assistance from Helga Cohen. This manuscript was prepared with the patience of Terri Bolder. Finally, for the research reported here we also wish to acknowledge support from the NIH (GM 26295) and the NSF supported NMR facility at the University of South Carolina (CHE 78-18723).

REFERENCES

(1) E. D. Becker, "High Resolution NMR: Theory and Chemical Applications," Academic Press, 1969, Appendix A.
(2) J. H. Noggle and R. E. Shirmer, "The Nuclear Overhauser Effect," Academic Press, New York, 1971.
(3) A. Abragam, "The Principles of Nuclear Magnetism," from "The International Series of Monographs on Physics," W. C. Marshall and D. H. Wilkinson, editors, Clarendon Press, 1976, Chapter 5.
(4) N. F. Ramsey, Phys. Rev., 78, pp. 699-703 (1950).
(5) M. Mehring, "High Resolution NMR Spectroscopy in Solids," from "NMR Principles and Progress #11," P. Diehl, E. Fluck, and R. Kosfeld, Eds., Springer-Verlag, 1976, Chapter 5.
(6) A. Saika and C. P. Slichter, J. Chem. Phys., 22, p. 26 (1954).
(7) J. A. Pople, W. G. Schneider, and H. J. Bernstein, "High Resolution Nuclear Magnetic Resonance," McGraw-Hill, New York, 1959.
(8) M. Karplus and J. A. Pople, J. Chem. Phys., 38, p. 2803 (1963).
(9) R. Ditchfield, J. Chem. Phys., 56, p. 5688 (1972).
(10) R. Ditchfield, Chem. Phys. Lett., 15, p. 203 (1972).
(11) R. Ditchfield and P. D. Ellis, Chem. Phys. Lett., 17, p. 342 (1972).

(12) R. A. Haberkorn, L. Que, Jr., W. O. Gillum, R. H. Holm, C. S. Liu, and R. C. Lord, Inorg. Chem., 15, p. 2408 (1976).
(13) G. E. Maciel and M. Borzo, J. C. S. Chem. Commun., p. 394 (1973).
(14) R. J. Kostelnik and A. A. Bothner-by, J. Magn. Reson., 14, pp. 141-151 (1974).
(15) A. D. Cardin, P. D. Ellis, J. D. Odom, and J. W. Howard, Jr., J. Am. Chem. Soc., 97, p. 1672 (1975).
(16) (a) H. Ohtaki, M. Maeda, and S. Ito, Bull. Chem. Soc. Jpn., 47, p. 2217 (1974).
 (b) M. Tsurumi, M. Maeda, and H. Ohtaki, Denki Kagaku Oyobi Kogyo Butsuri Kagaku, 45, p. 367 (1977).
 (c) J. W. Macklin and R. A. Plane, Inorg. Chem., 9, p. 821 (1970).
(17) J. J. H. Ackerman, T. V. Orr, V. J. Bartuska, and G. E. Maciel, J. Am. Chem. Soc., 101, pp. 341-347 (1979).
(18) T. Drakenberg, N. O. Björk, and R. Portanova, J. Phys. Chem., 82, pp. 2423-2426 (1978).
(19) R. Colton and D. Dakternieks, Austral. J. Chem., 33, pp. 2405-2409 (1980).
(20) K. Hildenbrand and H. Dreeskamp, Z. Physik. Chem. N.F., 69, pp. 171-182 (1970).
(21) G. E. Coates and A. Lauder, J. Chem. Soc., A, p. 264 (1966).
(22) J. D. Kennedy and W. McFarlane, J. Chem. Soc. Perkin 2, 1187 (1977).
(23) C. J. Turner and R. F. M. White, J. Magn. Reson., 26, pp. 1-5 (1977).
(24) G. K. Carson, P. A. W. Dean, and M. J. Stillman, Inorg. Chim. Acta, 56, pp. 59-71 (1981).
(25) G. K. Carson and P. A. W. Dean, Inorg. Chim. Acta, 66, pp. 37-39 (1982).
(26) A. M. Bond, R. Colton, D. Dakternieks, M. L. Dillon, J. Hauenstein, and J. E. Moir, Austral. J. Chem., 34, 1393-1400 (1981).
(27) P. A. W. Dean, Can. J. Chem., 59, p. 3221 (1981).
(28) H. J. Jakobsen, P. D. Ellis, R. R. Inners, and C. F. Jensen, J. Am. Chem. Soc., in press.
(29) D. D. Dominguez, M. M. King, and H. J. C. Yeh, J. Magn. Reson., 32, pp. 161-165 (1978).
(30) R. Hagen, J. P. Warren, D. H. Hunter, and J. D. Roberts, J. Am. Chem. Soc., 95, p. 5712 (1973).
(31) E. H. Curzon, N. Herron, and P. Moore, J. Chem. Soc. Dalton Trans., pp. 721-725 (1980).
(32) C. C. Bryden and C. N. Reilley, J. Am. Chem. Soc., 104, pp. 2697-2699 (1982).
(33) M. Cannas, G. Marongiu, and G. Saba, J. Chem. Soc. Dalton, p1 2090 (1980).

(34) C. F. Jensen, S. Deshmukh, H. J. Jakobsen, R. R. Inners, and P. D. Ellis, J. Am. Chem. Soc., 103, pp. 3659-3666 (1981).
(35) T. Maitani and K. T. Suzuki, Inorg. Nucl. Chem. Lett., 15, pp. 213-217 (1979).
(36) E. A. H. Griffith and E. L. Amma, J. C. S. Chem. Commun., p. 1013 (1979).
(37) D. Dakternieks, Austral. J. Chem., 35, pp. 469-481 (1982).
(38) R. Colton and D. Dakternieks, Austral. J. Chem., 33, pp. 1677-1684 (1980).
(39) S. O. Grim, E. D. Walton, and L. C. Satek, Can. J. Chem., 58, pp. 1476-1479 (1980).
(40) B. L. Vallee and D. D. Ulmer, Ann. Rev. Biochem., pp. 91-128 (1972).
(41) D. B. Bailey, P. D. Ellis, and J. A. Fee, Biochem., 19, pp. 591-596 (1980).
(42) (a) A. R. Palmer, D. B. Bailey, W. D. Behnke, A. D. Cardin, P. P. Yang, and P. D. Ellis, Biochem., 19, pp. 5063-5070 (1980).
(b) D. B. Bailey, P. D. Ellis, A. D. Cardin, and W. D. Behnke, J. Am. Chem. Soc., 100, p. 5236 (1978).
(43) (a) A. Cavé, J. Parello, T. Krakenberg, E. Thulin, and B. Lindman, FEBS Lett., 100, p. 148 (1979).
(b) T. Drakenberg, B. Lindman, A. Cavé, and J. Parello, FEBS Lett., 92, p. 346 (1978).
(44) I. M. Armitage, A. J. M. S. Uiterkamp, J. F. Chlebowski, and J. E. Coleman, J. Magn. Reson., 29, pp. 375-392 (1978).
(45) (a) J. L. Evelhoch, D. F. Bocian, and J. L. Sudmeier, Biochem., 20, pp. 4951-4954 (1981).
(b) N. B. H. Jonsson, L. A. E. Tibell, J. L. Evelhoch, S. J. Bell, and J. L. Sudmeier, Proc. Natl. Acad. Sci. (USA), 77, pp. 3269-3272 (1980).
(c) J. L. Sudmeier and S. J. Bell, J. Am. Chem. Soc., 99, pp. 4499-4500 (1977).
(d) I. M. Armitage, R. T. Pajer, A. J. M. Schoot Uiterkamp, J. F. Chlebowski, and J. E. Coleman, J. Am. Chem. Soc., 98, p. 5710 (1976).
(e) A. J. M. S. Uiterkamp, I. M. Armitage, and J. E. Coleman, J. Biol. Chem., 255, pp. 3911-3917 (1980).
(f) B. R. Bobsein and R. J. Myers, J. Am. Chem. Soc., 102, pp. 2454-2455 (1980).
(46) (a) J. D. Otvos and I. M. Armitage, Biochem., 19, pp. 4031-4043 (1980).
(b) J. F. Chlebowski, I. M. Armitage, and J. E. Coleman, J. Biol. Chem., 252, pp. 7053-7061 (1977).
(47) (a) J. D. Otvos, R. W. Olafson, and I. M. Armitage, J. Biol. Chem., 257, pp. 2427-2431 (1982).
(b) R. W. Briggs and I. M. Armitage, J. Biol. Chem., 257, pp. 1259-1262 (1982).

(c) J. D. Otvos and I. M. Armitage, J. Am. Chem. Soc., 101, pp. 7734-7736 (1979).
(d) P. J. Sadler, A. Bakke, and P. J. Beynon, FEBS Lett., 94, pp. 315-318 (1978).
(e) K. T. Suzuki and T. Maitani, Experienta, 34, p. 1449 (1978).
(48) B. R. Bobsein and R. J. Meyers, J. Biol. Chem., 256, pp. 5313-5316 (1981).
(49) S. Forsén, E. Thulin, T. Drakenberg, J. Krebs, and K. Seamon, FEBS Lett., 117, pp. 189-194 (1980).
(50) S. Forsén, E. Thulin, and H. Lilja, FEBS Lett., 104, pp. 123-126 (1979).
(51) J. L. Sudmeier, S. J. Bell, M. C. Storm, and M. F. Dunn, Science, 212, pp. 560562 (1981).
(52) J. Soulati, K. L. Henold, and J. P. Oliver, J. Am. Chem. Soc., 93, pp. 5694-5698 (1971).
(53) W. Bremser, M. Winokur, and J. D. Roberts, J. Am. Chem. Soc., 1080 (1970).
(54) K. Henold, J. Soulati, and J. P. Oliver, J. Am. Chem. Soc., 91, pp. 3171-3174 (1969).
(55) C. R. McCoy and A. L. Allred, J. Am. Chem. Soc., 84, pp. 912-915 (1961).
(56) E. A. Jeffery and T. Mole, Austral. J. Chem., 21, pp. 1187-1196 (1968).
(57) M. Holz, R. B. Jordan, and M. D. Zeidler, J. Magn. Reson., 22, pp. 47-52 (1976).
(58) M. J. B. Ackerman and J. J. H. Ackerman, J. Phys. Chem., 84, pp. 3151-3153 (1980).
(59) D. H. Rasmussen and A. P. MacKenzie in "Water Structure at the Water-Polymer Interface," H. H. G. Jellinek, Ed., Plenum Press, New York, 1971, pp. 126-144.
(60) D. Turnbull, J. Appl. Phys., 21, pp. 1022-1028 (1950).
(61) H. J. Jakobsen and P. D. Ellis, J. Phys. Chem., 85, pp. 3367-3369 (1981).
(62) Reference 3, Chapter VIII, p. 274.
(63) S. H. Forsén and R. A. Hoffman, J. Chem. Phys., 39, p. 2892 (1963); S. H. Forsén and R. A. Hoffman, ibid., 40, p. 1189 (1964); S. H. Forsén and R. A. Hoffman, ibid., 45, p. 2049 (1966).
(64) E. Kreyszig, "Advanced Engineering Mathematics," John Wiley and Sons, Inc., New York, 1962, Chapter 1, p. 53.
(65) R. E. Brown, III, C. F. Brewer, and S. H. Koenig, Biochem., 16, p. 3883 (1977); S. H. Koenig, C. F. Brewer, and R. D. Brown, III, ibid., 17, (1978); C. F. Brewer and R. D. Brown, III, ibid., 18, p. 2555 (1979).
(66) G. E. Maciel, Top. Carbon 13 NMR Spectrosc., 1, p. 54 (1974).
(67) A. Nolle, Z. Naturforsch., 33a, p. 666 (1978).
(68) J. C. Slater, "Quantum Theory of Atomic Structure," McGraw-Hill Co., 1960, Chapter 25.

(69) Y. Nomura, Y. Takeuchi, and N. Nakagawa, Tetrahedron Lett., 8, p. 639 (1969); I. Morishima, K. Endo, and T. Yonezawa, J. Chem. Phys., 59, p. 3356 (1973); A. A. Cheremisin and P. V. Sehastnev, J. Magn. Reson., 40, pp. 459-468 (1980); P. D. Ellis, unpublished results.
(70) P. D. Ellis, R. R. Inners, and J. H. Jakobsen, J. Phys. Chem., 86, pp. 1506-1508 (1982).
(71) A. Pines, M. E. Gibby, and J. S. Waugh, J. Chem. Phys., 59, p. 569 (1973).
(72) J. Schaefer and E. O. Stejskal, J. Am. Chem. Soc., 98, p. 1031 (1976).
(73) F. D. Doty and P. D. Ellis, Rev. Sci. Instrum., 52, p. 1868-1875 (1981).
(74) (a) U. Haeberlen, "High Resolution NMR in Solids," Supplement 1 in "Advances in Magnetic Resonance," J. S. Waugh, Ed., Academic Press, New York, 1976.
(b) M. Mehring, "High Resolution NMR Spectroscopy in Solids," Springer-Verlag, New York, 1976.
(75) P. D. Ellis and H. Bildsøe, unpublished results.
(76) M. Marciq and J. S. Waugh, J. Chem. Phys., 70, pp. 3300-3316 (1979).
(77) J. Herzfeld and A. E. Berger, J. Chem. Phys., 73, p. 6021 (1980).
(78) K. M. Smith, Ed., "Porphyrins and Metalloporphyrins," Elsevier, Amsterdam, Chapter 8, 1975.
(79) P. Rodesiler and E. L. Amma, unpublished results.
(80) R. Ditchfield and P. D. Ellis, Top. Carbon-13 NMR Spectrosc., 1 (1974).
(81) P. D. Majors and P. D. Ellis, unpublished results.
(82) The single exception to this chemical shift range is the shift observed for the calcium site in bovine insulin (51): its chemical shift is -36 ppm. The X-ray data are definitive in this case for the Ca^{+2} to be six-coordinate, T. Blundell, G. Dodson, D. Hodgkin, and D. Mercola, Adv. Protein Chem., 26, p. 279 (1972).
(83) (a) A. Nolle, Z. Naturforsch, 33a, p. 666 (1978).
(b) J. J. Ackerman, T. V. Orr, V. J. Bartuska, and G. E. Maciel, J. Am. Chem. Soc., 101, p. 341 (1979).
(c) T. T. P. Cheung, L. E. Worthington, L. E. Murphy, P. DuPois Murphy, and B. C. Gernstein, J. Magn. Reson., 41, p. 158 (1980).
(d) P. DuPois Murphy and B. C. Gernstein, J. Am. Chem. Soc., 103, p. 3282 (1981).
(e) P. DuPois Murphy, W. C. Stevens, T. T. P. Cheung, S. Lacelle, B. C. Gernstein, and D. M. Kurtz, Jr., J. Am. Chem. Soc., 103, p. 4400 (1981).
(f) P. G. Mennit, M. P. Shatlock, V. J. Bartuska, and G. E. Maciel, J. Phys. Chem., 85, p. 2086 (1981).
(g) R. R. Inners, F. D. Doty, A. R. Garber, and P. D. Ellis, J. Magn. Reson., 45, p. 503 (1981).

(h) P. D. Ellis, R. R. Inners, and H. J. Jakobsen, J. Phys. Chem., 86, p. 1506 (1982).
(i) H. J. Jakobsen, P. D. Ellis, R. R. Inners, C. F. Jensen, J. Am. Chem. Soc., in press.
(84) R. S. Honkonen, F. D. Doty, and P. D. Ellis, submitted to J. Am. Chem. Soc.
(85) R. S. Honkonen and P. D. Ellis, submitted to J. Am. Chem. Soc.
(86) D. A. Langs, C. R. Hare, Chem. Commun., 890 (1967). The authors would like to thank C. R. Hare for kindly supplying the atomic coordination for this structure.
(87) B. Malkovic, B. Rebov, B. Zelenko, and S. W. Peterson, Acta Cryst., 21, p. 719 (1966).
(88) J. A. Weil, T. Buck, and J. E. Clapp, Advan. Magn. Reson., 7, p. 183 (1973).
(89) In this paper, distinguishable indicates tensors which have, at minimum, differing eigenvectors. Distinct tensors are defined to have different eigenvalues. Thus, tensors with identical eigenvalues may be distinguishable.
(90) This statement follows from the definition of the shielding tensor (80) and the nature of angular momentum operators.
(91) M. L. Post and J. Trotter, J. Chem. Soc., Dalton Trans., 674 (1972).
(92) A. Hempell, S. E. Hull, R. Ram, and M. P. Gupta, Acta Cryst., B35, p. 2215 (19798); D. E. Woessner, J. Chem. Phys., 36, p. 1 (1962); D. E. Woessner, ibid., 37, p. 647 (1962); H. Shimizu, ibid., 40, p. 754 (1964); D. Wallach, ibid., 47, p. 5258 (1967); T. T. Bopp, ibid., 47, p. 3621 (1967); D. E. Woessner, B. S. Snowden, Jr., and E. T. Strom, Mol. Phys., 14, p. 265 (1968); W. T. Huntress, Jr., J. Chem. Phys., 48, p. 3524 (1968); W. T. Huntress, Jr., J. Phys. Chem., 73, p. 103 (1969); D. Wallach and W. T. Huntress, Jr., J. Chem. Phys., 50, p. 1219 (1969); J. Jonas and T. M. DiGennaro, ibid., 50, p. 2392 (1969); W. T. Huntress, Adv. Magn. Reson., 4, p. 1 (1970).
(94) P. D. Ellis and J. D. Potter, submitted to J. Biol. Chem.
(95) There are several extensive reviews of Troponin C biochemistry: we will refer the reader to a recent review by J. D. Potter and J. D. Johnson in "Calcium and Cell Function", Vol. II, Academic Press, New York, 1982, Chapter 5, and references cited within.
(96) J. D. Potter and J. Gergely, J. Biol. Chem., 250, pp. 4628-4683 (1975).
(97) T. Anderson, T. Drakenberg, S. Forsén, E. Thulin, and M. Sward, J. Am. Chem. Soc., 104, pp. 576-580 (1982) and references cited within.

PARTICIPANTS

A. C. P. Alves, Chemistry Department, Departamento de Química, FCT-Universidade de Coimbra, 3000 Coimbra, Portugal

Kenneth K. Andersen, Parsons Hall, University of New Hampshire, Durham, New Hampshire 03824, U.S.A.

Steven Jon Bachofer, Box H, Brown University, Providence, Rhode Island 02912, U.S.A.

Craig E. Barnes, Department of Chemistry, Stanford University, Stanford, California 94305, U.S.A.

J. R. Barnes, Shell Research Ltd., Thornton Research Centre, P.O. Box 1, Chester, England

Lynne S. Batchelder, National Institutes of Health, Bldg. 30, Room 106, Bethesda, Maryland 20205, U.S.A.

Simona Biagini, Istituto Chimica Fisica, Via Ospedale 72, 09100 Cagliari, Italy

M. J. Buckingham, Queen Mary College, University of London, Department of Chemistry, Mile End Road, London E1 4NS, England

Eugene A. Cioffi, University of Connecticut, Department of Chemistry, Box U-60, Storrs, Connecticut 06268, U.S.A.

Deirdre Cleary, Institut de Chimie Minérale et Analytique, Place du Château 3, 1005 Lausanne, Switzerland

I. J. Colquhoun, City of London Polytechnic, 31 Jewry Street, London EC3N 2EY, England

Ray Colton, Chemistry Department, University of Bristol, Bristol BS6 1TS, England

Claude-Cécile Coupry, Centre National de la Recherche Scientifique, Groupe de Laboratoires de Vitry-Thiais, Laboratoire de Spectrochimie Infrarouge et Raman, 2, rue Henri-Dunant, B.P. 28, 94320 Thiais, France

Jane Cox, School of Chemical Sciences, University of East Anglia, Norwich NR4 7TJ, England

Esin Çurgunlu, Chem. Dept. of the Istanbul State Academy of Engineering and Architecture, I.D.M.M. Akademisi Temel Bilimler, Fakültesi Istanbul, Turkey

Alfred Delville, Université de Liège, Institut de Chimie,
Sart Tilman par 4000, Liège 1, Belgium

Peter Domaille, CR&DD, DuPont, E356/33 DuPont Exp. Sta.,
Wilmington, Delaware 19898, U.S.A.

John H. Enemark, University of Arizona, Department of Chemistry,
Tucson, Arizona 85721, U.S.A.

P. Filippone, Istituto di Scienze Chimiche,
Piazza Rinascimento 6, 61029 Urbino, Italy

Anne-Marie Frisque-Hesbain, Université de Louvain-la-Neuve,
1, Place Louis Pasteur, Lab. Orsy, 1348 Louvain-la-Neuve,
Belgium

Douglas J. Gale, C.S.I.R.O., Division of Protein Chemistry,
343 Royal Parade, Parkville, Victoria 3052, Australia

J. H. Gamelkoorn, Free University de Boelelaan 1083,
1081 hv Amsterdam, Holland

Carlos F. Geraldes, Chemistry Department, University of Coimbra,
3000 Coimbra, Portugal

Ioannis Gerothanassis, Institut de Chimie Organique, Université
de Lausanne, Rue de la Barre 2, 1005 Lausanne, Switzerland

Véronique Gibon, Facultés Universitaires, Notre Dame de la Paix,
Rue de Bruxelles 61, 5000 Namur, Belgium

Julius Glaser, Royal Institute of Technology, Inorganic
Chemistry, S-10044 Stockholm 70, Sweden

David S. B. Grace, Norwegian National NMR Laboratory,
Kjemisk Institutt, NLHT Rosenborg, Trondheim N7000, Norway

Pierre Granger, Institut Universitaire de Technologie de Rouen,
Boîte Postale No. 47, 76130 Mont-Saint-Aignan, France

Ruth A. Grieves, 4, Marine Drive, Goring-by-Sea, Worthing,
West Sussex BN124QN, England

M. C. Grossel, Bedford College, University of London,
Regent's Park, London NW1 4NS, England

Turgut Günduz, G.M.K. Bulvari, 116/3 Maltepe, Ankara, Turkey

Anna-Maija Häkkinen, Department of Medical Physics, University of
Helsinki, Siltavuorenpenger 10, 00170 Helsinki 17, Finland

PARTICIPANTS

Mahajit Kaur Hayer, Chemistry Department, University of Stirling, Stirling FK9 4LA, Scotland

Nikolaus Hebendanz, Technische Universität München, Lichtenbergstrasse 4, D-8046 Garching, West Germany

Lothar Helm, Inst. Chimie Minérale et Analytique, University Lausanne, Place du Château, CH 1005 Lausanne, Switzerland

Norbert Hertkorn, Technische Universität München, Lichtenbergstrasse 4, D-8046 Garching, West Germany

Alain Hugi, Institut de Chimie Minérale et Analytique, Place du Château 3, 1005 Lausanne, Switzerland

J. D. Kennedy, Department of Inorganic and Structural Chemistry, University of Leeds, Leeds LS2 9JT, England

Shakil A. Khan, Mobay Chemical Corporation, Penn-Lincoln Parkway West, Pittsburgh, Pennsylvania 15205, U.S.A.

Doris Köhnlein, Kleiststrasse 10, D-7400 Tübingen, West Germany

Günther Kössler, Sudetenstrasse 17, D 7400 Tübingen, West Germany

Águst Kvaran, Science Institute, University of Iceland, Dunhaga 3, IS-107 Reykjavík, Iceland

Kerstin Larsson, Dept. of Physical Chemistry, University of Stockholm, Arrhenius Laboratory, Fack S-10405 Stockholm, Sweden

Gerard A. Lawless, T.C.D. Chemistry Department, 72 Irishtown Road, Dublin 4, Ireland

Prem P. Mahendroo, Alcon Laboratories, Inc., 6201 S. Freeway, P.O. 1959, Fort Worth, Texas 76101, U.S.A.

H. Marsmann, Gesamthochschule Paderborn, Fachbereich Chemie, Warburger Str. 100, 479 Paderborn, West Germany

Joan Mason, Open University, 152 Castelnau, London SW13 9ET, England

W. McFarlane, Chemistry Department, City of London Polytechnic, Jewry Street, London EC3N 2EY, England

William P. McKenna, University of Utah, Box 80, Department of Chemistry, Salt Lake City, Utah 84112, U.S.A.

Martin Minelli, Department of Chemistry, University of Arizona, Tucson, Arizona 85721, U.S.A.

Maura Monduzzi, Istituto Chimica Fisica, Via Ospedale 72,
09100 Cagliari, Italy

Karen J. Mordecai, Bedford College, University of London,
Regent's Park, London NW1 4NS, England

R. P. Moulding, Inorganic Chemistry, University of Oxford,
South Parks Road, Oxford OX1 3QR, England

Kathryn M. Nicholls, Department of Inorganic and Organic
Chemistry, University Chemical Laboratory, Lensfield Road,
Cambridge CB2 1EW, England

Werner Offerman, Universität München, Institut für Organische
Chemie, Karlstrasse 23, D-8000 München 2, West Germany

Sigurjón N. Ólafsson, Science Institute, University of Iceland,
Dunhaga 3, IS-107 Reykjavík, Iceland

Saim Özkâr, Department of Chemistry, Middle East Technical
University, Ankara, Turkey

Mohindar S. Puar, Schering Corporation, 60 Orange Street,
Bloomfield, New Jersey 07003, U.S.A.

Paul Rademacher, Universität Essen - GHS, Fachbereich 8 - Chemie,
Postfach 6843, D-4300 Essen 1, West Germany

Roberto L. Rittner, Instituto de Química, UNICAMP,
Caixa Postal 6154, 13100 Campinas SP, Brazil

Pirkko Anneli Ruostesuo, University of Oulu, Department of
Chemistry, SF-90570 Oulu 57, Finland

David S. Rycroft, University of Glasgow, Department of
Chemistry, Glasgow G12 8QQ, Scotland

Helena Santos, Instituto Superior Técnico, Centro de Química
Estrutural, Complexo Interdisciplinar, 1096-Lisboa, Portugal

G. L. Schrobilgen, Department of Chemistry, McMaster University,
Hamilton, Ontario L8S 4M1, Canada

Regina Schuck, c/o Prof. Kessler, Universität Frankfurt,
Niederurseler Hang, D-6000 Frankfurt 50, West Germany

Herbert M. Schwartz, Department of Chemistry, Rensselaer
Polytechnic Institute, Troy, New York 12181, U.S.A.

Bernard L. Shapiro, Department of Chemistry, Texas A&M
University, College Station, Texas 77843, U.S.A.

PARTICIPANTS

J. Simpson, Chemistry Department, Royal Military College of
Science, Shrivenham Swindon, Wilts SN6 8LA, England

Tore Skjetne, University of Trondheim, Dept. of Chemistry,
Unit/NLHT, N. 7055 Dragvoll, Norway

Olle Söderman, Physical Chemistry II, University of Lund,
Chemical Center, P.O.B. 740, S-220 07 Lund, Sweden

Catriona M. Spencer, Chemistry Department, University of
Sheffield, Sheffield, Yorkshire S3 7H5, England

Ruth E. Stark, Amherst College, Department of Chemistry,
Amherst, Massachusetts 01002, U.S.A.

Daniel Stec III, Northwestern University, Department of
Chemistry, Evanston Illinois 60201, U.S.A.

Alan S. Tracey, Simon Fraser University, Department of
Chemistry, Burnaby, British Columbia V5A 1S6, Canada

David L. Turner, Department of Chemistry, The University,
Southampton S09 5NH, England

Luce Vander Elst, Université de Mons, Avenue Maistriau, 24,
7000 Mons, Belgium

Harold Walderhaug, Physical Chemistry 1, Chemical Center,
P.O.B. 740 S-22007 Lund 7, Sweden

Rodney L. Willer, Code 3853, Naval Weapons Center,
China Lake, California 93555, U.S.A.

Brian Wood, City of London Polytechnic, 31 Jewry Street,
London EC3N 2EY, England

Nicholas Zumbulyadis, Research Laboratories, Eastman Kodak Co.,
Bldg. 82, Rochester, New York 14650, U.S.A.

INDEX

Strict alphabetical order is used. Spaces within phrases are ignored. Punctuation, numerals, Greek letters, and other modifiers in names, e.g., *cis*, *meso*, are ignored for purposes of alphabetization, unless they are needed to distinguish entries. American spelling is used throughout, even when English spelling is used within the text.

Ab initio calculation of shielding, 32,33
Abundance, natural, 2
Accordion spectroscopy, 98-100
[^{3}H]Acetate, 195
N-Acetyl-D-glucosamine, 147
Acholeplasma laidlawii
 see A. laidlawii
Acoustic ringing, 22,310
Actinomycin D, 183-85
Adamantanone, 137-38
Adiabatic demagnetization in the rotating frame, 129-30
Adsorption, 374
A. laidlawii B, 162,166
Alcohol dehydrogenase, 429-32,440,478,480
Alkali metal anions, 272-74
Alkali metal NMR, 261-91
 also see individual nuclei, e.g., Lithium-7
Alkaline earth metal NMR, 297-306
Alkyllithium compounds, ^{7}Li spectra of, 286-88
Allosteric model, 437
Aluminosilicates, 375
Aluminum
 chemical shifts, 334-38
 magnetic properties, 333,337
 relaxation, 340
Aluminum-27 NMR, 40,330,332-35,337-38,340
Amber, 374
Amide rotation, 103
Amides, ^{15}N chemical shifts, 216-17
Amines, ^{15}N chemical shifts, 212-16
2-Amino-5-methylbenzenesulfonic acid, 366
Ammonium ions, ^{15}N chemical shifts, 212-16
Anilines, 215-16
Anisotropic indirect spin-spin coupling, 400
Anisotropic motion, 102,147
Anisotropy of shielding, 40,375,397,399-400,462,505-507
 also see Relaxation, chemical shielding anistropy
Anthraquinone derivative, 367-68
Antimony
 chemical shifts, 385-85
 magnetic properties, 380

Antimony-121,123 NMR, 12-13,380-82,384-86
Arsenic
 chemical shifts, 384
 magnetic properties, 380
Arsenic-75 NMR, 380-82,384
Aryl rotation, 93-95
ASIS, 287
Asymmetry factor (η), 85,143-44,153,277-78
Atomic beam, 455
Auto correlation function, 76-78
Auto-peak, 96-98
Average excitation energy, 38-39,446,455
Azo compounds, ^{15}N chemical shifts, 220-21

Bacteriophage coat protein, 156-57
Barium
 acetate, 366-67
 aqueous solutions of, 299-300
 magnetic properties, 298-99
 relaxation, 306
Barium-135,137 NMR, 298-300
Benzaldehyde-d$_3$, 147
2-Benzamidonorborneols, 141-42
Benz[a]anthracene, 192-93
Benzene, polycrystalline, 119
Benz[cd]imidazole, 232-33
Beryllium
 aqueous solutions of, 299-300
 chemical shifts, 302-304
 magnetic properties, 298-99
 relaxation, 306
Beryllium-9 NMR, 297-300,302-304
Bicyclobutane, 55
Bismuth, magnetic properties, 380
Bismuth-209 NMR, 380-82
Bloch equations, 72,87,311
Boron
 chemical shifts, 334-38
 magnetic properties, 333,337
Boron hydrides, 34
Boron-10,11 NMR, 34,40,99,330,332-35,337
Bovine insulin A, 238
Broad component, 87
Bromine
 chemical exchange of, 416-20
 coupling constants, 413-14
 isotope effects, 429
 magnetic properties, 406-408
Bromine-79,81 NMR, 405-20,429,433

INDEX

Bullvalene, 92-93
2-Butyl cation, 106,108

Cadmium
  chemical shifts, 462-80
  coupling constants, 480-87
  exchange reactions, 480,488,490-508
  halogen effect, 502
  magnetic properties, 459-61
  relaxation, 480,487,489
Cadmium-113--carbon-13 couplings, 483-84
Cadmium-113--nitrogen-15 couplings, 485-86,491,504
Cadmium-111,113 NMR, 400,446,457-518
Cadmium-113--phosphorus-31 couplings, 486
Cadmium-113--proton couplings, 481-82
Calcium
  aqueous solutions of, 299-301
  binding to phospholipase, A_2, 318-20
  binding to troponin C, 320-25
  biological applications, 309-26
  DNA interactions with, 325-26
  magnetic properties, 298-99
  protein complexes, 314-17
  ratio of Larmor frequencies, 300-302
  relaxation of, 306
Calcium-43 NMR, 298-302,305-306,309-26
Calmodulin, 476
Carbon-13--cadmium-113 couplings
  see Cadmium-113--carbon-13 couplings
Carbon-13--carbon-13 couplings, 55-56,356
Carbonic anhydrase, 427-29
Carbon-13--lithium-7 couplings
  see Lithium-7--carbon-13 couplings
Carbon-13--nitrogen-15 couplings
  see Nitrogen-15--carbon-13 couplings
Carbon-13 NMR, 16-17,19,33,40-42,44,46,92-93,95,99,127,129-30,
  144,334,345,362-74,410,459,461,502
Carbon-13--silicon-29 couplings
  see Silicon-29--carbon-13 couplings
Carbonyl groups, ^{17}O chemical shifts, 250-51,253
γ-Carboxyglutamic acid, 314-15,317-18
Carboxypeptidase A, 236-37,476
Cephaloridine, 371
Cesium
  magnetic properties, 261-64
  relaxation, 282,284
Cesium-133 NMR, 261-64,274-76,282,284,290
Charge transfer, 268
Chemical exchange
  see Dynamic processes

Chemical shielding anisotropy
  see Relaxation, chemical shielding anisotropy
Chemical shift anisotropy
  see Relaxation, chemical shielding anisotropy
Chemical shifts
  see individual cases, e.g., Nitrogen-15
Chlorine
  chemical exchange of, 416-20,424-39
  chemical shifts, 410-12
  coupling constants, 413-14
  isotope effects, 418-19,429
  magnetic properties, 406-408
Chlorine-35,37 NMR, 40,334,405-40
Cholesterol, 161-64
Chromium chemical shifts, 449
Chromium-53 NMR, 449
CNDO calculations, 37-40
Coalescence
  of relaxation times, 104-105
  temperature, 103,106
Cobalt
  absolute shielding, 453-56
  chemical shifts, 451
Cobalt-59 NMR, 3,8,446-47,451,453-56
Concanavalin A, 476,493-501
Conformational analysis, 224,233-34,369
Contact pulse, 124
Contact term, 51,53-54,222,225-26,356,503
Coordination effects, 330,355,381
Correlation functions, 76-78
Correlation time, effective (τ_c), 77-80,146-47,310-13,316,426-27,433
Correlation time, internal (τ_i), 438-40
Correlation time, rotational (τ_R), 427,487,489
Coupling, nuclear spin-spin
  anisotropic, 400
  calculations of, 49-60
  contact term, 51,53-54,222,225-26,356,503
  dipolar term
    see Dipole-dipole interaction
  orbital term, 50,53
  also see individual cases, e.g., Nitrogen-15--carbon-13
    couplings
Cross-correlation, 81,86
Cross-peak, 96-98
Cross polarization, 105-108,123-27,167,238-39,361,504-505
Cross-relaxation, 81,86,97,130
Cryptand, 272-73,290
Crystallographic nonequivalence, 364,366
Cycle time, 121-22

INDEX

Cycloartenol, biosynthesis of, 197-99
Cyclopropanes, protonated, 140-42

D
 see Diffusion constant
Dead-time, receiver, 154-55
trans-Decahydroquinoline, 215
cis-Decalin, 98-100,104
Decoupling
 heteronuclear dipolar, 122-23
 noise, 178
Dehydrogriseofulvin, 139-40
Density matrix, 74
2-Deoxyglucose, 191-92
De-Paking, 164-66
Detectability, 4,12
Deuterium
 magnetic properties, 134-36
 relaxation, 143-48
Deuterium–deuterium couplings, 136
Deuterium NMR, 11,102,133-49,151-67,201
Deuterium–proton couplings, 136
Deverell approach, 274
Diamagnetic shielding, 30,32,35,40-42,273,274,410,447,462,465
 local, 35,40-42
Diastereoisomerism, 369,371-72
Diastereotopic tritons, 190
1,1-Diazene, 232
Diffusion coefficients, 72
Diffusion constant (D), 101-102
Diffusion, discrete, 156-58
Dihydrosterculic acid, 166
Dilute spins, 344,361-62
p-Dimethoxybenzene, 105
Dimethylcadmium, 467,470,491,507
Dimethylformamide, 104-105
1,4-Dinitrosopiperazine, 93
Dipolar order, 128-30
Dipolar relaxation
 see Relaxation, dipole-dipole
Dipole-dipole interaction
 broadening by, 117-21
 coupling by, 51-52,112-15,143
[16-^{3}H]Diprenorphine, 181
DNA, 155-56,235-36,325-26
DNMR5, 92
Double resonance in solids, 123-25
Dynamic processes, 91-108,234-35,311,363-64,416-20,424-39,480,488,490-508
Dynamic shift, 87,283-84,311-12

η
  see Asymmetry factor
Edge interchange, 94-95
Effective correlation time
  see Correlation time
Electric field gradient, 115,143,145,152-54,159,208,277-80,306, 310-11
Electric quadrupole coupling, 115-16,143,152
Electrostatic model, 278-80
Electrostriction, 269-70
Emission, spontaneous, 70-71
β Enolization, 137-38
Ethers, ^{17}O chemical shifts, 250-52
[1',2'-^{3}H]Ethylbenzene, 179-80
Ethyl β-carboline-3-carboxylate, 182-83
Excess linewidth, 321,416-20,427-28,434-37
Exchange effects
  see Dynamic processes
Exchange matrix
  see Statistical matrix
Extreme narrowing condition, 78,276-77,312,316,365,424,438

Fermi contact term
  see Contact term
Field gradient, 85
Finite perturbaton theory, 31,54-55
Flip-flop transitions, 114
Fluorine-19--fluorine-19 couplings, 56,58
Fluorine-19--nitrogen-15 couplings
  see Nitrogen-15--fluorine-19 couplings
Fluorine-19 NMR, 33,40,46,94-95,100,334,410,416
Fluorine-19--phosphorus-31 couplings
  see Phosphorus-31--fluorine-19 couplings
Fluorine-19--tritium couplings
  see Tritium--fluorine-19 couplings
4-Fluoro[2-^{3}H]benzoic acid, 188
Formamide, 44
Fourier transform, 5,101
FPT
  see Finite perturbation theory

Gallium
  chemical shifts, 334-38
  magnetic properties, 333,337
  relaxation, 340
Gallium-71 NMR, 329,332-34,336-37,340
Gamma shielding effect, 355
Gauge invariant atomic orbitals, 32
Geochemistry, 374

INDEX

Germanium
 chemical shifts, 352
 magnetic properties, 344-46
 reference standard, 351
Germanium-73 NMR, 343-48,352,357,362
Glassy polymers, 130
L-Glutamic acid, 147-48
Glutathione, 97-98
Guest molecules, 162-64
Gutmann donor number, 275
Gyromagnetic ratio, 262-344
 also see individual nuclei, e.g., Lithium-6

Halogen NMR, 405-40
Hartmann-Hahn condition, 124,376
Heat capacity of spin systems, 125
Heavy atom effects, 337-38,351
Helicity reversal, 94-95
Hemoglobin, 432-37,440
Heptamethylbenzenium ion, 96-97
Heterogeneity, 373-74
Hill formalism, 288-89
Histidine, 239
Hydration numbers, 265-67,269,271-72,275
Hydrazones, ^{15}N chemical shifts, 220-21
Hydrogen-1
 see Proton
Hydrogen-2
 see Deuterium
Hydrogen-3
 see Tritium
3β-Hydroxy-[7-^{3}H]androst-5-en-17-one, 172

I, 1
Imines, ^{15}N chemical shifts, 220-21
INADEQUATE, 8
Independent electron model, 34-35
Indium
 chemical shifts, 334-38
 magnetic properties, 333,337
 relaxation, 339
Indium-115 NMR, 332-34,336-37,339
INDO calculations, 34,37,39-40,55-60
Indoles, ^{15}N chemical shifts, 218
INEPT, 8,16-21,24-25,347-48
Inhomogeneity, magnetic field, 71
Inverse halogen dependence, 338,382,446,452-53,502
Iodine
 coupling constants, 413-14
 magnetic properties, 406-408

Iodine-127 NMR, 405-19
Ion pairs, 287-88
IQ values, 263,277
Iron-57 NMR, 25
Isochromat, 71
Isoguvacine, 183-84
Isonitriles, ^{15}N chemical shifts, 217
Isopropylideneuridine, 253
Isotope, 1

Jeener-Broekaert pulse sequence, 128-29

Karplus equation, 57,224

β-Lactam, 138-39
Lactams, ^{15}N chemical shifts, 216
Lamellar lyotropic phase, 160
Lattice, 66
 also see Relaxation, spin-lattice
Laurate, potassium, 160
Lead
 chemical shifts, 351-52,354-55
 couplings, 355-56
 magnetic properties, 344-45
 relaxation, 350-51
 solid state spectra, 362
Lead-207 NMR, 23,343-48,350-52,354-57,362,410
Libration model, 280-81
Ligand substitution, 330-31
Linewidth factor, 346
Linoleic acid, 146
Liquid crystals, 119,159-64,420-23
Lithium
 affinity, 268
 chemical shift, 287-88
 magnetic properties, 261-64
 relaxation, 280-82
Lithium-7--carbon-13 couplings, 286-87
Lithium-6--lithium-7 couplings, 286
Lithium-7--lithium-6 couplings
 see Lithium-6--lithium-7 couplings
Lithium-6,7 NMR, 40,261-64,275,280-82,286-88
Lithium-7--proton couplings, 286
Local field
 static, 112-13
 time dependent, 114
Lone pair electrons, 55,223-25
Longitudinal relaxation
 see Relaxation, spin-lattice
Lysergic acid diethylamide, 181-82

INDEX

Magic angle rotation
  see Magic angle spinning
Magic angle spinning, 105-108,120-21,126-27,167,238-39,361, 504-506
Magnesium
  aqueous solutions of, 299-300
  ATP complex, 313
  biological applications, 309-26
  DNA interactions with, 325-26
  magnetic properties, 298-99
  relaxation, 304-306
Magnesium-25 NMR, 298-300,304-306,309-26
Magnet, 8,10
Magnetization vector, 67,72,103-104,494-96,500
Magnetogyric ratio
  see Gyromagnetic ratio
[^{3}H]Malonate, 197
Manganese chemical shifts, 450-51
Manganese-55 NMR, 446,450-51,454
Master equation for relaxation, 74
Matrix isolation, 363
Medium effects
  of alkali ions, 288
  on ^{35}Cl chemical shifts, 410
  on ^{15}N chemical shifts, 234
  on shielding, 30,43-46,364
  on spin-spin coupling, 58-60
Mercury-199 NMR, 23,100
Metal carbonyl complexes, ^{17}O chemical shifts, 254
Metalloproteins
  see Proteins
Methanol-d, 136-37
Methyl group, chiral, 197,199
[Methyl-^{3}H]methionine, 186-87
Methyl rotation, 102
Methyltriphenylphosphonium iodide, 187-88
MINDO calculations, 37,39-40
Molecular complexes, 369
Molecular mobility, 363-64
  also see Dynamic processes and Rotation
Molybdate ions, ^{17}O chemical shifts, 255-56
Molybdenum chemical shifts, 449-50
Molybdenum-95,97 NMR, 7,9,10,12-13,446,449-50
Moments, spectral, 114,161-62
Monte Carlo technique, 272,279
Motional averaging, 117,119-20
MREV, 122
Multiple contact, 126
Multiple pulse cycles, 121-22
Multiple quantum NMR, 285

Muscle contraction, 320,323
Myoglobin, 432
Myristic acid, 162

Nalidixic acid, 139-41
Narrow component, 87
Nephelauxetic effect, 330,338,455
Neutron diffraction, 271
Niobium-93 NMR, 446,454
Niobium-93--oxygen-17 couplings
  see Oxygen-17--niobium-93 couplings
Nitriles, ^{15}N chemical shifts, 217
Nitroalkanes, 44-46
Nitro compounds, ^{15}N chemical shifts, 219-20
Nitrogen
  chemical shift reference, 211,380
  chemical shifts, 210-21
  coupling to, 222-31
  experimental methods, 210
  magnetic properties, 208-10,380
  quadrupole moment of ^{14}N, 208
  relaxation, 208-10,229-31
Nitrogen-15--cadmium-113 couplings
  see Cadmium-113--nitrogen-15 couplings
Nitrogen-15--carbon-13 couplings, 55-57,59-60,225-27
Nitrogen-15--fluorine-19 couplings, 228
Nitrogen-15--nitrogen-15 couplings, 227-28
Nitrogen-14 NMR, 11,21,207-10,334,379-80,410
  also see Nitrogen-15 NMR
Nitrogen-15 NMR, 9,11,26,33,40,44-46,99,207-40,379-80,410
  also see Nitrogen-14 NMR
Nitrogen-15--phosphorus-31 couplings, 229
Nitrogen-15--proton couplings, 222-25,234
Nitroso compounds, ^{15}N chemical shifts, 219-20
N-Nitrosoproline, 141,143
Nocardicin A, 138-39
NOE
  see Nuclear Overhauser effect
Noise decoupling
  see Decoupling, noise
Nonadecane, 157-58
Nonbonding electrons
  see Lone pair electrons
Noncontact coupling
  see Dipole-dipole interaction and Orbital motion
Nonexponential relaxation
  cross-correlation, 81,86
  cross-relaxation, 81,86,97,130
  also see Relaxation functions, nonexponential
Nonlorentzian lineshape, 283,325

INDEX

Nonradiative processes, 70
2-Norbornyl cation, 107-108
Normal halogen dependence, 338,382,384,452-53,502
Nuclear Overhauser effect, 15,102,178-79,210,212,230-31,346,351,
  394,459,480,489,498-500
Null signal problem, 346

Optical pumping, 455
Orbital motion, coupling by, 50,53
Order parameter S, 159-64
Order-position profile, 160
Organometallic chemistry, 23-24
DL-[2,3-^{3}H]Ornithine, 190-91
Orsellinic acid, 195-96
Overhauser effect
  see Nuclear Overhauser effect
Oxidation state effects on shielding, 329-31,381,445-52
Oxomolybdenum(VI) compounds, ^{17}O chemical shifts, 255-57
Oxygen
  chemical shifts, 249-57
  hydrogen bonding effects, 252-53
  quadrupole moment, 248
  signal to noise, 249
  spectral resolution, 249
Oxygen-17--niobium-93 couplings, 246,248
Oxygen-17 NMR, 14,33,40,44,46,245-59,334,389,410

$\pi \to \pi^*$ transitions, 35
Pairwise additive model, 385-86
Paramagnetic shielding, 30,32,35-36,38-42,273,338,410,446,455,
  462,465
  local, 35-36,38-42,507
Penicillic acid, biosynthesis of, 195-96
Phase transition, 161-63
Phenanthridine, 193
Phenylmercuric triflate, 100
Phosphates, ^{17}O chemical shifts, 254-55
Phosphatidylethanolamine, 161,163
Phospholipase A$_2$, 318-20
Phospholipid, 162-64
Phosphorus
  chemical shifts, 382-83
  magnetic properties, 380
Phosphorus-31--cadmium-113 couplings
  see Cadmium-113--phosphorus-31 couplings
Phosphorus-31--fluorine-19 couplings, 57
Phosphorus-31--nitrogen-15 couplings
  see Nitrogen-15--phosphorus-31 couplings
Phosphorus-31 NMR, 9,25-26,30,99,155,334,380-83,410

Phosphorus-31--tritium couplings
  see Tritium--phosphorus-31 couplings
Platinum chemical shift, 452
Platinum-195 NMR, 3,8,101,446,452
Platinum-195--tin-119 couplings
  see Tin-119--platinum-195 couplings
Polarization transfer
  see Cross polarization and INEPT
Polyamides, ^{15}N chemical shift, 237-38
Polyelectrolyte, 325
Poly-L-glutamic acid, 147-48
Polymers, glassy, 365
Polymorphism, 371-72
Polypeptides, conformations of, 233-34
Polypropene, 369
Poly(vinylamine), 234-35
Potassium
  affinity, 268
  magnetic properties, 261-64
  relaxation, 281-82
Potassium-39,40,41 NMR, 14,261-64,275-76,281-82,290
Powder lineshape, 117-118,153-58,161-68,422
  also see Solid state spectra
Principal elements of the shielding tensor, 117
Probeheads, 8
Propane, substituted, 141-42
2-Propanol-d$_8$, 136
Proteins
  chloride ion binding to, 424,431
  complexes with calcium, 314-17
  complexes with cadmium, 474-79,493-501,508-17
  internal motion in, 437-39
Proton--cadmium-113 couplings
  see Cadmium-113--proton couplings
Proton--deuterium couplings
  see Deuterium--proton couplings
Proton--lithium-7 couplings
  see Lithium-7--proton couplings
Proton--nitrogen-15 couplings
  see Nitrogen-15--proton couplings
Proton NMR, 10,11,33,40,44,99,135,410,461
Proton--silicon-29 couplings
  see Silicon-29--proton couplings
Proton--tritium couplings
  see Tritium--proton couplings
Pulse width, 6
Pyrimidines, ^{15}N chemical shifts, 219
Pyridine, 40,43,135-36,147
Pyridines, ^{15}N chemical shifts, 217-19,235,239
Pyridinium ion, 40

Pyrrole, 40,43
Pyrroles, ^{15}N chemical shifts, 217-18,235,239

Q
 see Statistical matrix
Quadrupolar broadening, 365-66
Quadrupolar coupling constant (χ), 85-86,274,277-78,283,286,
 311-13,316,439
Quadrupolar relaxation
 see Relaxation, quadrupolar, and Electric quadrupole coupling
Quadrupole echo, 154-55
Quadrupole moment, 85,152,208,263,339,346
 also see Electric quadrupole coupling
Quadrupole shifts, 115
Quadrupole splitting, 153-55,162-63,420-23

Radial term, 332-34,338,448,455
Raleigh-Schrödinger theory, 31
Ramsey formulation, 30-31,338,410,446
Random process, 77
Rate processes
 see Dynamic processes
Receptivity, 4,263
Reduced coupling constant, 50,412
Relaxation
 anisotropy of spin-spin coupling tensor, 83
 chemical shielding anisotropy, 82-84,230-31,280,341,498
 dipole-dipole, 2,13-15,79-81,101-102,144,230-31,234-35,280-81,
  340,348-49,487,500
 electron-nuclear, 230-31
 in solids, 126-30
 in the rotating frame, 102-103,104,128-30,365,415
 quadrupolar, 2,70,83-87,102,143-48,209-10,229-30,246,249,276-84,
  306,310-11,338-40
 scalar coupling, 82,230-31,348,409,412,415
 spin-lattice, 2,22,63-87,101-104,128,143-48,208-209,229-31,
  234-35,304-306,311,346-47,349-50,365,393-95,412,420,426,480,
  492,497-99
 spin-rotation, 82,84,230-31,280-81,340,348,350
 spin-spin, 2,64,68,72,79-80,86,128,143-45,208-209,304-306,311,
  339,412,416-17,420,426
Relaxation functions, 72-73
 nonexponential, 72,81,86,104,283-84,409,425
 single exponential, 73,104
Relaxation mechanisms, 79-87
 also see Relaxation
Relaxation reagents, 81
Relaxation time
 apparent, 425
 coalescence of, 104-105

Relaxation time (continued)
  pH dependence of, 313-14
  temperature dependence of, 420
  also see Relaxation and individual nuclides, e.g., Nitrogen, relaxation
Rhodium complexes, ^{17}O chemical shifts, 257-58
Rhodium-103 NMR, 3,23,25
Ring reversal, 101,103-104
Rotating frame, relaxation in, 102-103,104,128-30,365,415
Rotation
  amide, 103
  aryl, 93-95
  methoxyl, 105
  methyl, 102
Rubidium
  magnetic properties, 261-64
  relaxation, 281-82,284
Rubidium-85,87 NMR, 261-64,274,281-82,284,290
Ruthenium-99,101 NMR, 5,8,10,12,25-26

σ^d
  see Diamagnetic shielding
σ^p
  see Paramagnetic shielding
Sagging pattern, 354
Satellite resonances, 348
Saturation, 67,70
Saturation transfer, 103-104, 493-500
Scalar coupling
  see Relaxation, scalar coupling
SCPT
  see Self consistent perturbation theory
Screening tensor, 82-83
  also see Relaxation, chemical shielding anisotropy, and Anisotropy of shielding
Second moment, 114
Selective population inversion, 21-22
Selenium
  aqueous solutions of, 394
  chemical shifts, 394
  magnetic properties, 389-90
  relaxation, 394-95
  solid state spectra, 396-99
Selenium-77 NMR, 389-90,394-99
Self consistent perturbation theory, 54-55
Semibullvalene, 106,108
Semi-empirical calculations of shielding, 32-38
Sensitivity, 3,24-25
  also see INEPT and individual nuclides
Serotonin, 194-95

INDEX

Shielding anisotropy
  see Relaxation, chemical shielding anisotropy, and Anisotropy of shielding
Shielding, nuclear, 29-46,115-17
  also see Diamagnetic shielding, Paramagnetic shielding, and Chemical shift
Shiftless relaxation reagents, 346
Silatranes, 353,355
Silica gel, 375-76
Silicates, 348,350
Silicon
  chemical shifts, 351-55
  couplings, 355-56
  magnetic properties, 344-45
  reference standard, 351
  relaxation, 346-51
  solid state spectra, 362,374-76
Silicon-29--carbon-13 couplings, 356
Silicones, 375
Silicon-29 NMR, 40,334,343-57,362,374-76
Silicon-29--proton couplings, 356
Silicon-29--silicon-29 couplings, 348,356
Siloxanes, 346-47
Silver-107,109 NMR, 9,10,23-24
Single crystal, 117
Single exponential
  see Relaxation functions
Sodium
  magnetic properties, 261-64
  relaxation, 281-84
  solid state spectra, 285
Sodium-23 NMR, 50,261-64,273,274-75,281-84,285,288-89
Solid echo, 121
Solid state spectra, 104-108,111-30,238-39,285,361-76,397-400, 504-508
  also see Powder lineshapes
Solvation, gas phase, 265-68
Solvaton model, 44-46,58-60
Solvent effects
  see Medium effects
SOS
  see Sum-over-states
Soybean lipoxygenase, 145-46
Spectral densities, 76-79
Spin diffusion, 114
Spin dilution
  see Dilute spins
Spin echo, 16
Spin-lattice relaxation
  see Relaxation, spin-lattice

Spin-lock, 102-103,124-26,128-30
Spin-orbit coupling, 351,453,465,502-503
Spin-rotation
 see Relaxation, spin-rotation
Spin-spin relaxation
 see Relaxation, spin-spin
Spin temperature, 73-74,124
Standard state for shielding, 446-47,454
Statistical matrix (Q), 92-93
Sternheimer antishielding factor, 86-87,264,277
Streptidine, 232-33
Streptomycine, 232-33
Strontium
 aqueous solutions of, 299-300
 magnetic properties, 298-99
 relaxation, 306
Strontium-87 NMR, 298-300
Structure-breaking ions, 270
Structure-making ions, 270
Strychnine, 194
Substituent parameters on amino ^{15}N chemical shifts, 213-14
Sulfones, ^{17}O spectra, 245-46
Sulfur
 chemical shifts, 390-92
 magnetic properties, 389-90
 relaxation, 393
 solid state spectra, 398
Sulfur-33 NMR, 40,334,389-94,398
Sum-over-states perturbation theory, 31-32,52-56

T_1
 see Relaxation, spin-lattice
$T_{1\rho}$
 see Relaxation, in the rotating frame
T_2
 see Relaxation, spin-spin
τ_c
 see Correlation time, effective
Tautomerism, 372-73
Tellurium
 aqueous solutions of, 394-95
 chemical shifts, 395-97
 magnetic properties, 389-90
 solid state spectra, 398-400
Tellurium-123,125 NMR, 23,389-90,394-400
Temperature of the sample, 23
Testosterone, 1,2-dehydrogenation of, 194
meso-Tetraphenylporphyrin, 504-508

INDEX

Thallium
 chemical shifts, 334-38
 magnetic properties, 333,337
 relaxation, 340-41
Thallium-205 NMR, 84,329-30,332-34,336-37,340-41
Tin
 chemical shifts, 351-52,354-55
 couplings, 355-56
 magnetic properties, 344-45
 relaxation, 350-51
 solid state spectra, 362
Tin-115,117,119 NMR, 3,343-48,351-52,354-57,362,410
Tin-119--platinum-195 couplings, 356
Tin-117--tin-119 couplings, 356
Tin-119--tin-117 couplings
 see Tin-117--tin-119 couplings
Tin-119--tin-119 couplings, 356
Titanium-47,49 NMR, 338,446
Transition metal complexes, ^{17}O chemical shifts, 255-58
Transition metal NMR, 445-56
Transition probability, 71
Transverse relaxation
 see Relaxaton, spin-spin
2',3',5'-Tri-$\underline{O}$-benzoyluridine, 235-36
Tritium
 chemical shifts, 174-77
 half-life, 170
 isotope effects, 175-77
 linewidth, 171
 magnetic properties, 170-71
 maximum specific radioactivity, 170
 noise decoupling of protons, 173
 relaxation, 171,173
 safety matters, 173-174
 sample preparation, 174-75
 sensitivity, 171,202-203
 test sample, 173
Tritium--fluorine-19 couplings, 178,186
Tritium NMR, 135,169-203
Tritium--phosphorus-31 couplings, 187-88
Tritium--proton couplings, 173,177-78,187
Tritium--tritium couplings, 183,187
Troponin C, 320-25,459-60,476,508-16
Tropylium azide, 93-95,100
Tungstate ions, ^{17}O chemical shifts, 255-58
Tungsten chemical shifts, 449-50
Tungsten-183 NMR, 9,25,446,449-50
Two dimensional spectra, 96-100,212

Urea, 103
Uridine triphosphate, 189-90

L-Valine, catabolism of, 199-200
Vanadate ions, ^{17}O chemical shift, 255-58
Vanadium chemical shifts, 448-49
Vanadium-51 NMR, 446, 448-49
Vincristine, 181-82

Wagner-Meerwein shift, 108
WAHUHA, 122
Water, ^{17}O spectrum of, 246-47
Wheat germ agglutinin, 145-47
Wilkinson's catalyst, 199-200
Woessner method, 101-102, 104

X-ray diffraction, 372, 427, 507, 509

Yttrium-89 NMR, 15

Zeolites, 375
Zinc-67 NMR, 446
Zirconium-91 NMR, 446